T0210899

# Lecture Notes in Physics

Founding Editors: W. Beiglböck, J. Ehlers, K. Hepp, H. Weidenmüller

Editorial Board

R. Beig, Vienna, Austria
W. Beiglböck, Heidelberg, Germany
W. Domcke, Garching, Germany
B.-G. Englert, Singapore
U. Frisch, Nice, France
F. Guinea, Madrid, Spain
P. Hänggi, Augsburg, Germany
W. Hillebrandt, Garching, Germany
R. L. Jaffe, Cambridge, MA, USA
W. Janke, Leipzig, Germany
H. v. Löhneysen, Karlsruhe, Germany
M. Mangano, Geneva, Switzerland
J.-M. Raimond, Paris, France
D. Sornette, Zurich, Switzerland
S. Theisen, Potsdam, Germany
D. Vollhardt, Augsburg, Germany
W. Weise, Garching, Germany
J. Zittartz, Köln, Germany

# The Lecture Notes in Physics

The series Lecture Notes in Physics (LNP), founded in 1969, reports new developments in physics research and teaching – quickly and informally, but with a high quality and the explicit aim to summarize and communicate current knowledge in an accessible way. Books published in this series are conceived as bridging material between advanced graduate textbooks and the forefront of research and to serve three purposes:

- to be a compact and modern up-to-date source of reference on a well-defined topic

- to serve as an accessible introduction to the field to postgraduate students and non-specialist researchers from related areas

- to be a source of advanced teaching material for specialized seminars, courses and schools

Both monographs and multi-author volumes will be considered for publication. Edited volumes should, however, consist of a very limited number of contributions only. Proceedings will not be considered for LNP.

Volumes published in LNP are disseminated both in print and in electronic formats, the electronic archive being available at springerlink.com. The series content is indexed, abstracted and referenced by many abstracting and information services, bibliographic networks, subscription agencies, library networks, and consortia.

Proposals should be sent to a member of the Editorial Board, or directly to the managing editor at Springer:

Christian Caron
Springer Heidelberg
Physics Editorial Department I
Tiergartenstrasse 17
69121 Heidelberg / Germany
christian.caron@springer.com

For other titles published in this series, go to
www.springer.com/series/5304

Anthony J. Guttman (Ed.)

# Polygons, Polyominoes and Polycubes

 Springer

Professor Anthony J. Guttman
Department of Mathematics and Statistics,
The University of Melbourne,
Victoria, 3010
Australia

ISSN 0075-8450
ISBN-13 978-94-017-7712-4
ISBN-13 978-1-4020-9927-4 (eBook)
DOI 10.1007/978-1-4020-9927-4

Published by Springer Science + Business Media B.V.
P.O. Box 17, 3300 AA Dordrecht, The Netherlands
In association with
Canopus Academic Publishing Limited,
15 Nelson Parade, Bristol BS3 4HY, UK

www.springer.com and www.canopusbooks.com

*To the memory of two distinguished scientists and wonderful colleagues, Pierre Leroux and Oded Schramm.*

# Preface

The problem of counting the number of self-avoiding polygons on a square grid, either by their perimeter or their enclosed area, is a problem that is so easy to state that, at first sight, it seems surprising that it hasn't been solved. It is however perhaps the simplest member of a large class of such problems that have resisted all attempts at their exact solution. These are all problems that are easy to state and look as if they should be solvable. They include percolation, in its various forms, the Ising model of ferromagnetism, polyomino enumeration, Potts models and many others. These models are of intrinsic interest to mathematicians and mathematical physicists, but can also be applied to many other areas, including economics, the social sciences, the biological sciences and even to traffic models. It is the widespread applicability of these models to interesting phenomena that makes them so deserving of our attention. Here however we restrict our attention to the mathematical aspects.

Here we are concerned with collecting together most of what is known about polygons, and the closely related problems of polyominoes. We describe what is known, taking care to distinguish between what has been proved, and what is certainly true, but has not been proved. The earlier chapters focus on what is known and on why the problems have not been solved, culminating in a proof of unsolvability, in a certain sense.

The next chapters describe a range of numerical and theoretical methods and tools for extracting as much information about the problem as possible, in some cases permitting exact conjectures to be made. Given that it is always easier to prove something if one is confident that it is true, this gives some direction as to what one should perhaps try to prove.

The subsequent chapters provide a range of numerical results, exact conjectures and rigorous results for properties of the generating function of polygons, counted by both area and perimeter, and a variant in which a number of polygons are placed on the grid, filling it completely. The final two chapters are devoted to powerful techniques of mathematical physics which can be used to tackle not just the properties of polygons and polyominoes, but also a broad range of related phenomena. Indeed, the earlier chapters on numerical and theoretical methods also have widespread applicability.

This is indeed a golden age for studying such problems. With powerful computers and new algorithms, unimaginable numerical precision in our estimates of properties of many of these models is now possible. On the mathematical side, we are developing tools for solving increasingly complex functional equations, while the theory of conformal invariance, and the developments around stochastic Löwner evolution have given us powerful tools to predict, and in some cases to prove, new results. The scientific community in this field is divided into those who think we will never solve the problem, of say the perimeter or area generating function of self-avoiding polygons in two dimensions, and those who think that we will. I am firmly in the latter camp, and hope that this book will trigger the imagination of some individual or team who will achieve just that.

I would like to thank all the contributors for their co-operation, patience, and above all the quality of their contributions. Many of my colleagues helped by providing diagrams, and I would particularly like to thank Jan de Gier, Iwan Jensen, Boris Peytchev, Jim Propp, Andrew Rechnitzer and Stuart Whittington for permission to use their diagrams, and I am indebted to Ian Enting and Andrew Rechnitzer for help with LaTeX, including making 29 packages co-exist in adequate harmony. I am also grateful to Yao-ban Chan, and Jason Doukas for reading some of the chapters. Finally, to Tom Spicer at Canopus books and the editorial team at Springer Verlag, my thanks for not giving up on me as one deadline after another slipped by.

Melbourne, November 2008                                                            *Anthony Guttmann*

# Contents

# List of Contributors

Mireille Bousquet-Mélou
CNRS, LaBRI, Université Bordeaux 1, 33405 Talence Cedex, France, e-mail: bousquet@labri.fr

Richard Brak
Department of Mathematics and Statistics, The University of Melbourne, Victoria, Australia, e-mail: r.brak@ms.unimelb.edu.au

Nathan Clisby
Department of Mathematics and Statistics, The University of Melbourne, Victoria, Australia, e-mail: n.clisby@ms.unimelb.edu.au

Jan de Gier
Department of Mathematics and Statistics, The University of Melbourne, Victoria, Australia, e-mail: degier@ms.unimelb.edu.au

Ian G. Enting
Department of Mathematics and Statistics, The University of Melbourne, Victoria, Australia, e-mail: ienting@unimelb.edu.au

Anthony J Guttmann
Department of Mathematics and Statistics, The University of Melbourne, Victoria, Australia, e-mail: tonyg@ms.unimelb.edu.au

Jesper Lykke Jacobsen
Laboratoire de Physique Théorique, École Normale Supérieure (LPTENS), 24 rue Lhomond, 75231 Paris Cedex 05, France, e-mail: jesper.jacobsen@ens.fr

E.J. Janse van Rensburg
Mathematics and Statistics, York University, Toronto, Ontario, Canada, e-mail: rensburg@yorku.ca

Iwan Jensen
Department of Mathematics and Statistics, The University of Melbourne, Victoria,
Australia, e-mail: i.jensen@ms.unimelb.edu.au

Wouter Kager
VU University Amsterdam, Department of Mathematics, De Boelelaan, 1081 HV
Amsterdam, The Netherlands, e-mail: wkager@few.vu.nl

Bernard Nienhuis
Institute for Theoretical Physics, University of Amsterdam, 1018 XE Amsterdam,
The Netherlands, e-mail: b.nienhuis@uva.nl

Aleks Owczarek
Department of Mathematics and Statistics, The University of Melbourne, Victoria,
Australia, e-mail: A.Owczarek@ms.unimelb.edu.au

Andrew Rechnitzer
Department of Mathematics, University of British Columbia, Vancouver, Canada,
e-mail: andrewr@math.ubc.ca

Christoph Richard
Fakultät für Mathematik, Universität Bielefeld, Postfach 10 01 31, 33501 Bielefeld,
Germany, e-mail: richard@math.uni-bielefeld.de

Gordon Slade
Department of Mathematics, University of British Columbia, Vancouver, Canada,
e-mail: slade@math.ubc.ca

Stuart G Whittington
Department of Chemistry, University of Toronto, Ontario, Canada, e-mail:
swhittin@chem.utoronto.ca

# Acronyms

| | |
|---|---|
| AF | anti-ferromagnetic |
| ASM | alternating-sign matrices |
| BCFT | boundary conformal field theory |
| BFACF | Berg-Foester-Aragao de Carvalho-Caracciolo-Fröhlich |
| CBL | conformal boundary loop model |
| CDA | constructible differentiably algebraic |
| CFT | conformal field theory |
| CG | Coulomb gas |
| D-finite | differentiably finite |
| FLM | finite-lattice method |
| FPL | fully-packed loop model |
| g.f. | generating function |
| HSFPL | horizontally symmetric fully-packed loop model |
| HTSFPL | half-turn symmetric fully-packed loop model |
| LERW | loop-erased random walk |
| MC | Monte Carlo |
| OPE | operator product expansion |
| PERM | pruned enriched Rosenbluth method |
| RG | renormalisation group |
| SAP | self-avoiding polygon(s) |
| SAW | self-avoiding walk(s) |
| SFL | semi-flexible loop model |
| SLE | stochastic Löwner evolution |
| UST | uniform spanning trees |

# Chapter 1
# History and Introduction to Polygon Models and Polyominoes

Anthony J Guttmann

## 1.1 Introduction

In this book we will primarily be concerned with the properties and applications of self-avoiding polygons (SAP). Two closely related problems are those of polyominoes, and the much broader one of tilings. We will describe and discuss polyominoes and, in the context of a discussion of SAP, will briefly mention relevant aspects of the subject of tilings. In passing we will also concern ourselves with a discussion of some properties of polycubes. It will also turn out to be appropriate to discuss self-avoiding walks (SAW), as SAP can be usefully and simply related to a proper subset of SAW.

In all cases we shall be considering paths on a regular lattice. This will most often be the two-dimensional square lattice $\mathbb{Z}^2$, or its three-dimensional counterpart, $\mathbb{Z}^3$, called the simple-cubic (sc) lattice, or its higher dimensional analogue $\mathbb{Z}^d$, called the ($d$-dimensional) hyper-cubic lattice. Other two-dimensional lattices, such as the triangular (t) and hexagonal (h) lattices, and other three-dimensional lattices, such as the body-centred cubic (bcc), face-centred cubic (fcc) and tetrahedral or diamond (d) lattices will also be mentioned.

### 1.1.1 Definitions and Notation

Let $\mathscr{L}$ be some regular $d$-dimensional lattice. Then an *n-step self-avoiding walk* (SAW) $\omega$ on $\mathscr{L}$ is a sequence of *distinct* points $\omega_0, \omega_1, ..., \omega_n$ in $\mathscr{L}$ such that each point is a nearest neighbour of its predecessor. We assume all walks to begin at

Anthony Guttmann
Department of Mathematics and Statistics, The University of Melbourne, Victoria, Australia, e-mail: tonyg@ms.unimelb.edu.au

the origin ($\omega_0 = 0$) unless stated otherwise. A typical SAW on the square lattice is shown below.

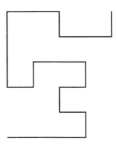

**Fig. 1.1** A self-avoiding walk of 21 steps.

If the end-point of a SAW, **x**, is adjacent to the origin $\omega_0$, an additional step joining the end-point to the origin will produce a *self-avoiding circuit*, which has been referred to in earlier literature as a *self-avoiding return*. An example is shown below:

**Fig. 1.2** A self-avoiding polygon of 14 steps.

The distinction between a self-avoiding circuit and a self-avoiding polygon is that the former is both rooted and directed. An $n$-step SAP can have any of its $n$ vertices as the root, or starting point, and it can be traversed in two directions. Thus if there are $p_n$ SAP of perimeter $n$, there are $2np_n$ self-avoiding circuits of perimeter $n$. More descriptively, a SAP is a closed, connected, non-intersecting, undirected path on a lattice. We consider two given SAW (or SAP) to be equivalent if they are translates of one another. Thus, on the square lattice, the first non-zero embedding of a SAP is the unit square, of perimeter 4 and area 1. If we denote by $p_m$ the number of SAP of perimeter $m$, by $a_n$ the number of SAP of area $n$, and by $p_{m,n}$ the number of SAP of perimeter $m$ and area $n$, we can define two single variable generating functions, for perimeter and area respectively, and a two-variable generating function, as follows:

$$P(x) = \sum_m p_m x^m, \tag{1.1}$$

$$A(q) = \sum_n a_n q^n \tag{1.2}$$

$$\mathscr{P}(x,q) = \sum_{m,n} p_{m,n} x^m q^n. \tag{1.3}$$

Turning briefly to SAW, let $c_n$ be the number of $n$-step SAW on a lattice $\mathscr{L}$ starting at the origin and ending anywhere. Then the SAW generating function is defined to be

$$C(x) = \sum_n c_n x^n.$$

If one joins an $n$-step and an $m$-step SAW end-to-end, one will either obtain an $n+m$ step SAW, or a non-self-avoiding walk. A moment's thought then yields the inequality $c_{n+m} \leq c_n c_m$. This is a sub-multiplicative inequality, apparently first discussed by Fekete [12], from which follows the existence of the *growth constant*[1] $\mu > 0$, given by

$$\mu = \lim_{n \to \infty} c_n^{1/n} = \inf_n c_n^{1/n}.$$

Kesten [30, 31] has proved the stronger result that $\mu^2 = \lim_{n \to \infty} c_{n+2}/c_n$, and O'Brien [41] has proved that $c_n > c_{n-1}$ for all $n$, yet there is still no proof that $\mu = \lim_{n \to \infty} c_{n+1}/c_n$, for $d = 2, 3, 4$. We note in passing that a number of authors have explicitly stated, or implicitly assumed, this to be true. While it probably is, it must be remembered that no proof exists.

Hammersley [20] similarly proved that $p_m$ grows exponentially with $m$; more precisely that

$$\mu = \lim_{m \to \infty} p_{2m}^{1/2m}.$$

While far from obvious, it *is* true that the growth constants $\mu$ that arise in the polygon case and the walk case are identical [20]. While unproven, a much stronger result is widely believed, notably that

$$p_m \sim \text{const} \times \mu^m m^{\alpha-3}, \tag{1.4}$$

where $\alpha$ is a *critical exponent*[2]. (Note that $p_{2m+1} = 0$ for SAP on $\mathbb{Z}^d$, as only polygons with even perimeter can exist on those lattices. For such lattices the above asymptotic form is of course only expected to hold for even values of $m$. For so-called *close-packed* lattices, such as the triangular or face-centred cubic lattices, polygons of all perimeters greater than two are embeddable, so eqn. (1.4) stands as stated.)

Another quantity of interest for SAW is $c_n(\mathbf{x})$, the number of $n$-step SAW on $\mathscr{L}$ starting at the origin and ending at $\mathbf{x}$. Then $c_n$ (defined above) and $c_n(\mathbf{x})$ are believed

---

[1] Some authors refer to this quantity as the connective constant, but the term *connective constant* originally referred to $\log \mu$. Contemporary usage seems to favour the former meaning.

[2] The notation $a_m \sim b_m$ means that $\lim_{n \to \infty} \frac{a_m}{b_m} = 1$.

to have the asymptotic behaviour as $n \to \infty$;

$$c_n \sim \text{const} \times \mu^n n^{\gamma-1} \tag{1.5}$$

$$c_n(\mathbf{x}) \sim \text{const} \times \mu^n n^{\alpha-2} \qquad (\mathbf{x} \text{ fixed} \neq 0) \tag{1.6}$$

where the growth constant $\mu$ is defined above, and $\gamma$ and $\alpha$ are *critical exponents* [3]. The growth constant depends on the lattice, so changes as one changes from, say, the square lattice to the triangular lattice. The critical exponents, by contrast, are expected to depend only on dimensionality, and so do not change from lattice to lattice (for lattices of the same dimensionality).

As shown above, the existence of the exponential growth term $\mu^n$ is known rigorously, while the existence of sub-dominant terms $n^{\gamma-1}$ and $n^{\alpha-2}$ is believed by all reasonable men (and women), but even the existence of the exponents remains unproved for lattices of dimensionality $d < 5$. Rigorous results concerning the asymptotic behaviour of the properties of SAW and SAP will be given in the next chapter by Whittington.

Indeed, in two dimensions, it is widely accepted, due to the Coulomb gas calculations (see Chapter 14) of Nienhuis [39], that $\gamma = 43/32$ and $\alpha = 1/2$ exactly. In three dimensions only numerical estimates are available[4], notably $\gamma \approx 1.156957$ and $\alpha \approx 0.23721$, and there is no reason to expect the exponents to be rational, while in four dimensions and more *mean-field* exponents are expected. That is to say, $\alpha = 0$ and $\gamma = 1$.

In exactly four dimensions there are also confluent logarithmic terms, which vanish in higher dimensions. In four dimensions, one expects $c_n \sim \text{const} \times \mu^n n^{\gamma-1} (\log n)^{1/4}$. Recently, for a hierarchical four-dimensional lattice, Brydges and Imbrie [3] proved the presence of the confluent logarithmic term, though no proof is currently known for SAW on a regular lattice. For $d > 4$, Hara and Slade [22, 23] have proved that $\gamma = 1$.

This can be understood heuristically, as with increasing lattice dimensionality, the self-avoiding constraint becomes less significant. If a direction is blocked, the walker can escape to another dimension. Above four dimensions, the self-avoiding constraint is sufficiently weakened that the walks behave like random walks, albeit an exponentially small subset. In four dimensions, which is the so-called *marginal dimensionality*, the self-avoiding constraint is just strong enough to modify the random walk behaviour by the addition of confluent logarithmic terms in the singular part of the generating function. These remarks will be quantified in Chapter 6.

To prove the existence and value of the critical exponents in two dimensions, the most promising route appears to be based on recent beautiful work by Cardy, Lawler, Schramm, Smirnov, Werner and others [35]. It will be necessary to prove the existence and conformal invariance of the scaling limit (these terms will be subsequently defined) for self-avoiding walks. If, as expected, the scaling limit is de-

---

[3] This is the same $\alpha$ as appears in equation (1.4)

[4] These estimates come from Monte Carlo estimates, series estimates and field theory estimates [36, 2, 34]. The three techniques give results that typically agree up to variations in the last quoted figure.

scribed by stochastic Löwner evolution, parameterised by 8/3, known as $SLE_{8/3}$, the existence and value of $\gamma$ for two-dimensional lattices would follow, as would a number of other results. These remarks will be explained in detail in Chapter 15.

The exact value of $\mu$ is not generally known, but for the special case of the two-dimensional hexagonal lattice, it is accepted [39] that $\mu = \sqrt{(2 + \sqrt{2})}$. For other two-dimensional lattices we have very accurate numerical estimates, notably $\mu = 2.63815853031..$ for the square lattice, and $\mu = 4.150797226..$ for the triangular lattice. Indeed, for the square lattice, note that the estimate above is indistinguishable from the root of the polynomial $581x^4 + 7x^2 - 13$, an observation first made by Conway et al. in 1993 [6]. For other lattices, in both two and three dimensions, see the Appendix.

Next we define several measures of the *size* of $n$-step SAW and SAP. For SAW:

- The *squared end-to-end distance* is

$$R_e^{2,(saw)} = \omega_n^2 . \tag{1.7}$$

- The *squared radius of gyration* is

$$R_g^{2,(saw)} = \frac{1}{2(n+1)^2} \sum_{i,j=0}^{n} (\omega_i - \omega_j)^2 . \tag{1.8}$$

- The *squared distance of a monomer from the endpoints* is

$$R_m^{2,(saw)} = \frac{1}{2(n+1)} \sum_{i=0}^{n} \left[ \omega_i^2 + (\omega_i - \omega_n)^2 \right] . \tag{1.9}$$

For SAP only the second of these three quantities is analogously defined:

- The *squared radius of gyration* of a SAP of perimeter $n$ is

$$R_g^2 = \frac{1}{2n^2} \sum_{i,j=0}^{n} (\omega_i - \omega_j)^2 . \tag{1.10}$$

The analogue of the end-to-end distance is usually taken to be the *caliper span* of the polygon, which is defined to be the maximum Euclidean distance between any two sites (monomers, in the language of polymers) of the polygon.

If one is considering *rooted* polygons (where one site is nominated as the origin), one could readily define the mean-square distance of a monomer from the origin, but this quantity has not been considered of much interest.

We then consider the mean values $\langle R_e^2 \rangle_n$, $\langle R_g^2 \rangle_n$ and $\langle R_m^2 \rangle_n$ in the probability distribution that gives equal weight to each $n$-step SAW. Very little has been proven rigorously about these mean values, but they are believed to have the leading asymptotic behaviour

$$\langle R_e^{2,(saw)} \rangle_n, \langle R_g^{2,(saw)} \rangle_n, \langle R_m^{2,(saw)} \rangle_n, \langle R_g^{2,(sap)} \rangle_n \sim \text{const} \times n^{2\nu} \tag{1.11}$$

as $n \to \infty$, where $\nu$ is another (universal) critical exponent. Hyperscaling [44] predicts that

$$dv = 2 - \alpha. \tag{1.12}$$

For SAW in two dimensions, Coulomb-gas arguments [39, 40] as well as arguments based on stochastic Löwner evolution (SLE) [35] predict that $v = 3/4$. Note that this result is not rigorous. In three dimensions we have only numerical estimates $v \approx 0.5876$,[5] a result which encompasses a variety of series, Monte Carlo and field theory estimates [36, 2, 34], while at the upper critical dimension of 4, it is believed (but not proved) that $v = 1/2$ with logarithmic corrections, notably $\langle R_e^{2,(saw)} \rangle_n \sim const. \times n(\log n)^{1/2}$ while for $d > 4$ the mean-field value $v = 1/2$ holds. This is a rigorous result (see the chapter by Clisby and Slade).

As noted above, because polygons are closed, it is almost as natural to ask for their enumeration by *area*, as by perimeter[6]. For specificity, consider polygons on the square lattice. Recall that $a_n$ denotes the number of polygons of area $n$. Then by concatenation arguments, similar to those given above for SAW, it is possible to prove that

$$\kappa = \lim_{n \to \infty} a_n^{1/n}$$

exists. Further, it is universally believed that

$$a_n \sim const \times \kappa^n n^{\tau}. \tag{1.13}$$

Only numerical estimates of $\kappa$ and $\tau$ are available. For the hexagonal, square and triangular latices respectively, the best current estimates[7] of $\kappa$ are $\kappa_{hexagonal} = 5.16193016(3)$, $\kappa_{square} = 3.970944(2)$, and $\kappa_{triangular} = 2.9446596(3)$, while $\tau$ is believed to be $-1$, so that the generating function

$$\mathscr{A}(x) = \sum a_n x^n \sim A \log(1 - \kappa x).$$

Consequently,

$$a_n \sim const. \times \kappa^n / n.$$

Of great interest is the two-variable generating function

$$\mathscr{P}(x, y) = \sum_m \sum_n p_{m,n} x^m q^n.$$

From this, we can define the free energy

$$\kappa(q) = \lim_{m \to \infty} \frac{1}{m} \log \left( \sum_n p_{m,n} q^n \right).$$

It has been proved [13] that the free energy exists, is finite, log-convex and continuous for $0 < q < 1$. For $q > 1$ it is infinite. The radius of convergence of $P(x, q)$,

---

[5] A newly developed Monte Carlo algorithm by Clisby, briefly described in Chapter 9, is dramatically more powerful, permitting him to estimate $v = 0.587597(7)$ and $\gamma = 1.156957(9)$.

[6] At least for polygons on two-dimensional lattices. The concept of area for higher dimensional lattices is not straightforward

[7] Kindly provided by Iwan Jensen, unpublished

which we denote $x_c(q)$, is related to the free energy by $x_c(q) = e^{-\kappa(q)}$. This is zero for fixed $q > 1$. A plot of $x_c(q)$ in the $x - q$ plane is shown (qualitatively) below. For $0 < q < 1$, the line $x_c(q)$ is believed to be a line of logarithmic singularities of the generating function $P(x, q)$. The line $q = 1$, for $0 < x < x_c(1)$, is believed to be a line of *finite* essential singularities [13]. At the point $(x_c, 1)$ we have more complicated behaviour, and this point is called a tricritical point. For the simpler polygon model of staircase polygons (see Chapter 3), the figure is qualitatively similar, though for $0 < q < 1$, the line $x_c(q)$ is known to be a line of simple pole singularities of the generating function $P(x, q)$, rather than logarithmic, as in the case of SAP. The line $q = 1$, for $0 < x < x_c(1)$, is also known [43] to be a line of *finite* essential singularities. As for SAP, at the point $(x_c, 1)$ we have more complicated tricritical behaviour, and this is discussed in more detail in Chapter 11.

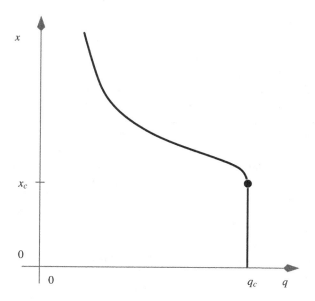

**Fig. 1.3** The phase diagram of self-avoiding polygons in the $x - -q$ plane.

Around the point $(x_c, 1)$ we expect tricritical scaling, so that

$$\mathscr{P}^{(sing)}(x, q) \sim (1 - q)^\theta F((x_c - x)(1 - q)^{-\phi}) \text{ as } (x, q) \to (x_c, 1^-).$$

Here the superscript *(sing)* means the singular part. There is an additional, additive part that is regular in the neighbourhood of $(x_c, 1)$, and so is not relevant to this discussion. For staircase polygons, Prellberg [43] has proved that

$$F(s) = \frac{1}{16} \frac{d}{ds} \log \text{Ai} \left( 2^{8/3} s \right),$$

where $s = (x_c - x)(1 - q)^{-\phi}$. For self-avoiding polygons, in a series of papers, Richard and co-authors [49, 50, 48] have provided abundant evidence for the surprisingly strong conjecture that

$$F(s) = -\frac{1}{2\pi} \log \mathrm{Ai}\left(\frac{\pi}{x_c}(4A_0)^{\frac{2}{3}} s\right) + C(q).$$

Here $C(q)$ is a function, independent of $x$, that arises as a constant of integration when moving from the rooted to the unrooted SAP scaling function [50]. For both models $\phi = 2/3$, while $\theta = 1/3$ for staircase polygons and $\theta = 1$ for SAP. Here $A_0 = 0.2811506(1)$ is the amplitude of the perimeter generating function, and $x_c$ is the perimeter generating function critical point. This result is described in much greater detail in Chapter 11 by Richard.

### 1.1.2 The Connection with Statistical Mechanics

Thus far we have discussed SAW and SAP as a combinatorial problem. However there is an alternative description as a statistical mechanical model. This both sheds light on some aspects of the behaviour of various quantities of interest, described above, such as critical points, critical exponents and critical amplitudes, and also gives us another way to tackle the problems that arise.

In statistical mechanics one starts with a Hamiltonian, which is the energy of a given configuration, taking into account the interactions, and then constructs the *partition function*. This assumes that configurations of a given energy and temperature are distributed according to the Boltzmann distribution. The Ising model and Heisenberg models of ferromagnetism have been known for decades, and were defined by Hamiltonians in which each spin was a classical unit vector of dimension 1 in the case of the Ising model, dimension 2 in the case of the classical planar Heisenberg model, and dimension 3 in the case of the classical Heisenberg model.

In 1968 Stanley [51] proposed the *n-vector model*, in which each spin was represented by a classical, $n$-dimensional vector. Clearly this reproduces the three cases mentioned above when $n = 1, 2, 3$ respectively. Stanley pointed out that the suitably defined limit as $n \to \infty$ reproduces the so-called spherical model, introduced by Berlin and Kac [4]. A more remarkable limit, $n \to 0$, was proposed by de Gennes [8], who showed that in this limit the model described was exactly self-avoiding walks and polygons. This is not at all intuitive. What after all do we mean by the interaction between two zero-component vectors? Nevertheless, proceeding with a curious mixture of formality and non-rigorous argument, a connection can be established. We will sketch the result. More details can be found in [7], and more recent, and perhaps more accessible accounts can be found in Madras and Slade [37] and Hughes [24].

One starts with a Hamiltonian

$$\mathscr{H} = - \sum_{<i,j>} \mathbf{s}_i \cdot \mathbf{s}_j - H \sum_i s_i(1), \tag{1.14}$$

where $\mathbf{s}$ is an $n$-component vector, which is located at the vertices of a regular lattice of arbitrary spatial dimension, and the subscript indexes a particular site on the lattice. The first sum is taken over all pairs of sites $(i,j)$ which are nearest neighbours on the lattice. The second sum is a scalar product of a magnetic field $H$, assumed w.l.o.g. to be in the '1' direction, and so couples to spin components in that direction. The vector is assumed to be of magnitude $\sqrt{n}$, so that $|\mathbf{s}_i|^2 = n$.

The partition function is obtained by the usual rules of statistical mechanics from

$$Z = \sum_{\text{all configurations}} \exp(-\mathscr{H}/kT). \tag{1.15}$$

In the above equation, $k$ is Boltzmann's constant and $T$ is the absolute temperature. The *Helmholtz free energy* is defined by

$$F(T,H) = -kT \log Z(T,H),$$

though we usually work in the *thermodynamic limit*,

$$f(T,H) = \lim_{N \to \infty} \frac{1}{N} F(T,H). \tag{1.16}$$

In the above equation, $N$ refers to the number of spins in the lattice, and should not be confused with the dimensionality of the spin vectors, denoted $n$, whereas the underlying dimensionality of space is denoted $d$. Next, we need to define the *magnetic susceptibility*,

$$\chi(T) = -\frac{\partial^2 f}{\partial H^2}|_{H=0} \tag{1.17}$$

The Ising model corresponds to the case $n = 1$. The susceptibility then behaves as

$$\chi(T) \sim const(1 - T_c/T)^{-7/4}$$

as $T \to T_c^+$. If one takes a formal series expansion of the susceptibility, after substituting equation (1.14) into equation (1.15) and then using equations (1.16) and (1.17) to calculate the susceptibility, it can be seen that each term in the expansion corresponds to sums over averages of dot products of the vector spins. These can be interpreted as graphs on the underlying lattice. In the limit as $n \to 0$, all graphs vanish except non-intersecting lattice paths joining site $i$ to site $k$. These are just self-avoiding walks from site $i$ to site $k$. Thus the susceptibility of the $n$ vector model in the $n \to 0$ limit is precisely the generating function for SAW. By the same token, the specific heat exponent for magnetic models can be related to the exponent characterising the generating function for SAP, so that if the specific heat is written

$$C(T) \sim const(1 - T_c/T)^{-\alpha}$$

as $T \to T_c^+$, we have, for the corresponding polygon generating function,

$$P(x) = \sum p_n x^n \sim const(1 - x/x_c)^{2-\alpha}.$$

The advantage of this formulation is that one can now bring to bear the full machinery of statistical mechanics to what would otherwise be a combinatorial problem. In statistical mechanics one has *scaling arguments* that link exponents. This is one justification for the hyperscaling relation (1.12) given above. Other tools from statistical mechanics are also available, such as the *renormalisation group* method, and other methods developed by mathematical physicists working in *field theory*. We won't discuss these techniques in any detail—each would justify a volume on its own—but will refer to some predictions made by these methods in later chapters.

### 1.1.3 Some History of the Problem

We conclude this section by a brief discussion of the chronological history of the SAW and SAP problem. The problem of enumerating self-avoiding walks (SAW) was initially proposed as a mathematical problem in a posthumously published paper by Orr in 1947 [42], and shortly thereafter was proposed as a model of long-chain polymers in dilute solution by Flory in 1949 [15]. Self-avoiding polygons (SAP), on the other hand, were first enumerated by Wakefield in 1951 [57]. Wakefield was not studying SAP in their own right, but rather enumerating them as part of a different project, notably the behaviour of the three-dimensional Ising model. In that study, SAP on the simple-cubic lattice contributed to the graphical expansion of the Ising model partition function.

Somewhat later, in 1954 and 1955, Wall and co-workers [54, 55, 56] calculated some properties of two- and three-dimensional SAP which arose as a by-product of their Monte Carlo study of SAW. Some of the SAW configurations they generated by Monte Carlo methods failed due to the coincidence of the end point of trial walks they generated with the origin. Such failures led them to study the probability of these occurrences, and hence they introduced the so-called *probability of initial ring closure*. This is defined as the probability that a SAW ends at a site adjacent to the origin, so that the addition of a single bond produces a self-avoiding circuit. This probability is just $2mp_m/((q-1)c_m)$, where $q$ is the *coordination number* of the lattice, or number of nearest-neighbour sites of a given site. For a $d$-dimensional hypercubic lattice, $q = 2d$.

In terms of the notation given above, and assuming the existence of the conjectured sub-dominant terms containing the critical exponents, we have $p_m \sim const \times \mu^m m^{\alpha-3}$, and $c_m \sim const \times \mu^m m^{\gamma-1}$. The probability of initial ring closure $p_m^0$ is $p_m^0 = 2mp_m/((q-1)c_m) \sim const \times m^{\alpha-\gamma-1}$. In two dimensions, from the above-quoted values of $\alpha$ and $\gamma$, we obtain $p_m^0 \sim const/m^{1.84375}$, while from the best numerical estimates for the exponents in the three-dimensional case, we find

$p_m^0 \sim \text{const}/m^{1.921}$. Wall and co-workers in 1955 estimated these exponents to be around 2 in both two and three dimensions, by Monte Carlo methods.

In 1959 Rushbrooke and Eve [47] studied this problem by direct enumeration, and ushered in the computer age of enumeration. They counted polygons up to perimeter 18 on the square lattice, and, more impressively, to order 14 on the simple cubic lattice. The coefficient $p_{14}$ on the sc lattice was counted on the Pegasus computer at Durham University, in a calculation that took 50 hours of CPU time. The exponent estimates obtained from an analysis of these enumerations were $2.4 \pm 0.2$ and 2.29 for the square and SC lattices respectively. At the same time, and quite independently, Fisher and Sykes [14] estimated these exponents to be $1.75 \pm 0.10$ and $1.810 \pm 0.007$ respectively. These estimates were improved by Hiley and Sykes in 1961 [21] based on further enumerations, and they found exponents $1.805 \pm 0.025$ and $1.92 \pm 0.08$ respectively. Remarkably, the estimate in the three-dimensional case is in precise agreement with current estimates to all quoted digits.

## 1.2 Polyominoes

The modern era of polyominoes began in 1953 when Golomb discussed them in a talk he gave to the Harvard Mathematics Club, which was subsequently published [17]. Martin Gardner added to the popularity of the subject when he discussed Golomb's article in his column in *Scientific American*. Polyominoes became a favourite topic of Gardner, and one he often returned to. As Golomb points out in the preface to the first edition of his book on the subject [16], there are many antecedents, either in the form of particular puzzles, or in discussions of the number of allowable patterns of a particular type in board games, such as *Go*. One notable antecedent appeared in 1907, when a puzzle involving 5-celled polyominoes, (of which there are twelve), was posed in the book *Canterbury Puzzles* [9]. In his book on polyomino puzzles, G. E. Martin [38] also points out that a variety of polyomino puzzle problems appeared in the British journal *Fairy Chess Review* in the 1930s. At that time, they were called *dissection problems*.

Formally, a polyomino (or, as they are sometimes known, a *square lattice site animal*[8]) of $n$ cells on the square lattice with origin 0 is a connected section graph[9] of the lattice containing the origin and having $n$ vertices. More intuitively, a polyomino is comprised of $n$ connected squares, which must be joined at an edge, and not just at a vertex. A *domino* is an example—indeed, the only example—of a 2-celled polyomino. Polyominoes may be similarly defined on other two-dimensional lattices, such as the triangular or honeycomb lattice. We will assume that we are referring to square lattice polyominoes in the following, unless stated otherwise.

---

[8] Other types of animals are discussed in the next chapter.

[9] A section graph $G^*$ of a graph $G$ is obtained from $G$ by deleting one or more of its vertices and any incident edges.

The difference between polyominoes and polygons is that a polyomino may have an internal hole. For example, see Fig. 1.4. So all polygons are polyominoes, but not vice-versa.

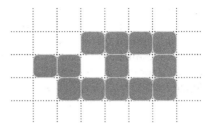

**Fig. 1.4** A polyomino. The presence of holes means that this is a polyomino, rather than a polygon.

Polyominoes are usually enumerated by *area*—or number of cells. While, like polygons, they can also be enumerated by perimeter, we will see below that this is less easily characterised, as the perimeter generating function has zero radius of convergence. Thus the more common asymptotic form (1.13) for lattice objects enumerated by perimeter does not hold when discussing the enumeration of polyominoes by perimeter.

The sort of results one can prove for polyominoes are not too different to those that can be proved for polygons. If $a_n$ denotes the number of polyominoes of area $n$ (or, if one prefers, of $n$ cells), then Klarner [32] proved that the growth constant $\alpha$, defined by

$$\alpha = \lim_{n \to \infty} a_n^{1/n} = \sup_n a_n^{1/n},$$

exists and is non-zero. The proof follows standard concatenation arguments, whereby polyominoes are uniquely concatenated to produce a subset of polyominoes of a size given by the sum of the sizes of the two individual polyominoes. Even before the existence of this limit was proved, it was widely assumed to exist, and indeed Eden, in 1961 [11], undertook a numerical analysis to estimate its value.

Such estimates rely on the enumeration of $a_n$, a computational problem of exponential complexity. As for polygons, the most effective method to date is the finite-lattice method (discussed in Chapter 7), which permits enumeration by finite-dimensional transfer matrices. The method was implemented for polyominoes by Conway and Guttmann in 1995 [5], giving rise to $a_n$ for $n \le 25$, which was dramatically improved by Jensen and Guttmann in 2000 [26], who obtained $a_n$ for $n \le 46$, then Knuth and Jensen played leap-frog for a few years, extending the series, with Jensen [28] currently holding the record with $a_{56} = 69,150,714,562,532,896,936,$ $574,425,480,218$.

Polyominoes are frequently referred to as *free* or *fixed*. Fixed polyominoes are considered distinct up to a translation. That is to say, fixed polyominoes means "an equivalence class of polyominoes under translation", whereas free polyominoes refers to "an equivalence class of polyominoes under translations, rotations and reflections". Asymptotically, the number of fixed polyominoes of $n$ cells, denoted $B_n$

on the square lattice is 8 times the corresponding number of free polyominoes of $n$ cells, denoted $b_n$. While the same distinction can be made for SAP, this is rarely done, and one usually considers fixed SAP, without explicitly saying so.

As for SAW and SAP, it is widely believed, but not proved, that $b_n \sim A\alpha^n n^\theta$ as $n \to \infty$, or, equivalently, that $B_n \sim 8A\alpha^n n^\theta$ as $n \to \infty$.

Like polygons, polyominoes on other two-dimensional lattices can be considered. Hexagonal and triangular lattice polyominoes were first introduced by Lunnon [33], and have also been exhaustively enumerated by Jensen [27], who has obtained the counts up to 75-celled animals on the triangular lattice and 46-celled animals on the hexagonal lattice.

As is the case for polygons, the growth constant $\alpha$ will change from lattice to lattice, while the critical exponent $\theta$ is expected to be the same for all two-dimensional lattices, and its value has been predicted by powerful, but non-rigorous, physical arguments [45] to be $\theta = -1$, corresponding to a logarithmic singularity in the generating function for polyominoes, while $\alpha \approx 4.062570$ [26].

## 1.3 Polycubes

A three-dimensional analogue of a polyomino on the square lattice is called a *polycube* on the simple-cubic lattice. Polycubes are composed of unit cubes joined at their faces. Many topologies are formed by this rule, including spheres, toroids and objects of higher genus. They are usually counted as *free* polycubes. That is to say, a particular polycube is counted only once irrespective of its orientation. Those that are mirror images of one another (so-called *chiral twins*) are usually (but not always) distinct.

Consideration of chiral twins is a slightly subtle distinction from the analogous two-dimensional case, where chirality is accommodated by reflection. But the reflection takes place by flipping the object through the third dimension. By convention, since one can't "see" the fourth dimension, the analogous reflection is forbidden, and that is why chiral twin polycubes are considered distinct.

As with polyominoes, it is customary to distinguish between *fixed* and *free* polycubes. To make this distinction clear, a tower of three cubes counts 1 as a free polycube, but counts 3 as a *fixed* polycube, as it may be oriented parallel to the $x$, $y$ or $z$ axis. Asymptotically, the number of fixed polycubes is 24 times the number of free polycubes.

In three dimensions, the longest series is due to Aleksandrowicz and Barequet [1], who give the first 18 series coefficients. We again have an exact prediction for the critical exponent from the powerful but non-rigorous physical arguments of Parisi and Sourlas [45], which is $\theta = -1.5$. We have analysed the available series coefficients, using the techniques described in Chapter 8, and find $\alpha \approx 8.3479$. (In 1978 Guttmann and Gaunt [19] predicted $\theta = -1.5$ and $\alpha = 8.34 \pm 0.02$, based on much less data.)

## 1.4 Tilings

The theory of tilings is a fascinating topic in its own right. Many books have been written on this subject, and we will have little to say about it, except to point out some of the more remarkable results, and indicate some of the open questions. An interesting, and still unanswered question is the necessary and sufficient conditions for a polyomino (or actually a polygon, as we don't want internal holes) to tile the plane. It is unknown if the tiling problem is decidable [46], and it is known that the problem of tiling a *finite* region is NP-complete [52].

To be more precise, by a *tiling of the plane*, we mean a covering of the Euclidean plane by a countable family of elementary tiles. The tiles may not overlap. If a single tile is used, we refer to the tiling as a *monohedral* tiling. The tile used is called a *prototile*. As a trivial example, it is clear that a $1 \times 1$ cell—indeed, a $k \times k$ cell—can tile $\mathbb{Z}^2$. For a discussion of tilings of the entire plane, the reader is referred to the discussion by Grünbaum and Shephard [18].

It is easy to see that any triangle or rectangle can tile the Euclidean plane, and with a bit more effort one can see that any quadrilateral can tile the plane. Conway [46] has a powerful method that provides a *sufficient*, but not necessary, condition for a prototile to tile the plane. Conway's method requires that if one can divide the boundary up into six segments, labelled sequentially as $A$, $B$, $C$, $D$, $E$ and $F$, such that:

(i) $A$ and $D$ are translates of one another,

(ii) the remaining four segments each possess a $180°$ rotational symmetry about their mid-point, and

(iii) while some segments may be empty, each pair $B$–$C$ and $E$–$F$ must be non-empty. Also, $A$ and $D$ can both be empty if at least three of the remaining four segments are non empty.

An example is shown in Fig. 1.5. Note that segments $A$ and $D$ are translates of one another, while the remaining segments each possess a $180°$ rotational symmetry about their respective mid-points. To tile the plane with this tile involves using both translates and rotated versions of the basic prototile, as can be seen in Fig. 1.6, illustrating the tiling.

If one restricts the movement of tiles to translations only, (that is to say, one forbids rotations and reflections), then there is a necessary and sufficient condition for a prototile to tile the plane. If the boundary can be divided into six consecutive segments, sequentially labeled $A$–$F$ as above, such that each element of the pairs $A$–$D$, $B$–$E$ and $C$–$F$ are translates of each other, the prototile tiles the plane. (One such pair may be empty, in which case the tiling forms a rectangular lattice, otherwise it forms a hexagonal lattice.) See [46] for a proof.

Rhoads [46] has made an interesting study of the effectiveness of Conway's criterion. He shows, among other results, that one needs to go to 9-cell polygons before one finds a polygon that does not satisfy the Conway criterion, but still tiles the plane. More precisely, apart from two 9-cell polygons, all polygons up to order 9 that tile the plane either satisfy the Conway criterion, or two copies form a patch that satisfies the Conway criterion.

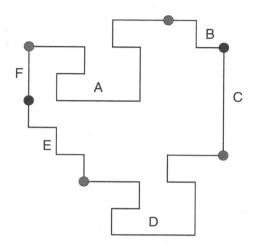

**Fig. 1.5** A tile that satisfies the Conway criterion for tiling the plane.

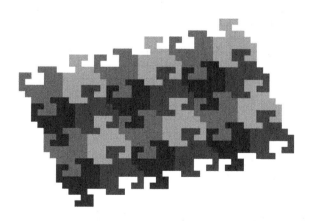

**Fig. 1.6** A tiling of the plane using the above tile. Notice that alternate tiles are rotated a half-turn.

A second question that is asked about tilings, once it has been established that a prototile can tile the plane is: *In how many ways can the plane be tiled?* One of the earliest non-trivial results in this direction is due to Kasteleyn [29], who used a Pfaffian to express the number of ways $2jk$ dominoes (or $2 \times 1$ tiles) can tile a $2j \times 2k$ rectangle. The same problem was treated by transfer matrix techniques, discussed in Chapter 7, by Fisher and Temperley [53]. The result is

$$4^{jk} \prod_{l=1}^{j} \prod_{m=1}^{k} \left( \cos^2 \frac{l\pi}{2j+1} + \cos^2 \frac{m\pi}{2k+1} \right).$$

It is far from obvious that this double product should even produce integers! The reader is invited to try it out for small values of $j$ and $k$ and confirm results by direct enumeration.

We will mention just one other remarkable exact result, the enumeration of *Aztec diamonds*. An Aztec diamond $AZ(k)$ is constructed by reflecting a pyramidal stack about its base. The stack base contains $2k$ cells, the next row $2k-2$ cells, the subsequent row $2k-4$ cells, and the top of the stack contains two cells. The Aztec diamond $AZ(10)$ is shown in the Fig. 1.7.

**Fig. 1.7** An Aztec diamond $AZ(10)$.

The number of domino tilings of $AZ(n)$ was found by Elkies, Kuperberg, Larsen and Propp [10], and is just

$$2^{(n^2+n)/2}.$$

This relatively simple result might suggest that there is a simple proof. Unfortunately that is not yet the case, despite the fact that the number of different proofs now runs to double figures.

Another typical question that is asked is: *What is the typical shape of an object?* Some tilings, or objects, display no particular structure, whereas others show quite remarkable properties. For example, the typical large Aztec diamond displays a surprising regularity at its corners, where, as can be seen from Fig. 1.9, it is tiled in a regular brickwork pattern, switching abruptly to apparent randomness in a circular region, corresponding to the largest possible circle that can be drawn inside the boundary. This geometry is sketched in Fig. 1.8.

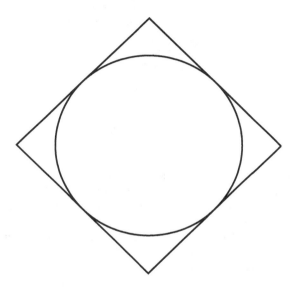

**Fig. 1.8** An Aztec diamond showing the so called Arctic circle.

Jokush, Propp and Shor [25] proved this result in probability. It is known as the *Arctic circle theorem,* The inscribed circle is the Arctic circle, the region at the poles, (and, perhaps disturbingly, at the equator), is "frozen". Hence the name. A typical such tiling is shown in Fig. 1.9.

## 1.5 The Rest of the Story

In the next chapter, Whittington describes the rigorous results that can be proved for SAP. While the SAP model remains unsolved, a number of simpler polygon models can be exactly solved, and methods for solving these are described in Chapter 3 by

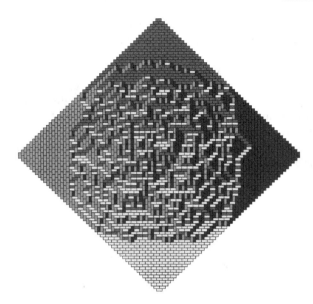

**Fig. 1.9** A typical tiling of the Aztec diamond, showing the occurrence of Arctic circle.

Bousquet-Mélou and Brak. In recent years we have gained a better understanding of just why the SAP model is so hard, and these ideas are described in Chapter 4. The key idea is that the solution is not holonomic, and in Chapter 5 Rechnitzer proves this, along the way developing tools that are useful for similar proofs. In Chapter 6 Clisby and Slade develop the lace expansion method that has proved such a powerful tool for both rigorous results and numerical results in higher dimensions ($d \geq 3$).

Much of our knowledge of SAP comes from numerical methods. In two dimensions the finite-lattice method (FLM) is, for most applications, the most powerful method, and it is described in Chapter 7 by Enting and Jensen. The result of the application of this method is a series expansion for the generating functions we have defined. Analysing this data to extract critical properties, such as critical exponents, critical points and critical amplitudes is the subject of Chapter 8 by Guttmann and Jensen, while the alternative numerical method, Monte Carlo, is described by Janse van Rensburg in Chapter 9. The advantage of Monte Carlo methods is that much larger systems can be sampled than is possible by direct enumeration. As an example, we show in Fig. 1.10 a SAP on the simple cubic lattice of size 92678 steps, produced by Clisby's new Monte Carlo algorithm, described in Chapter 9.

Interesting phenomena occur if SAW and SAP are restricted to confined geometries, such as polygons in strips, in slabs and in rectangles. These are described in Chapter 10 by Jensen and Guttmann. The underlying limit distributions of the area of SAP is the subject of Chapter 11 by Richard. Many physical, chemical and biological properties of polymers can be modelled by including monomer-monomer or

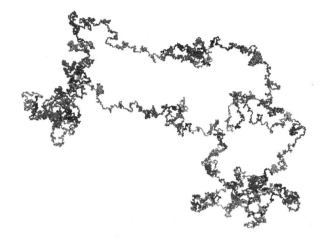

**Fig. 1.10** A typical self-avoiding polygon on the simple cubic lattice of 92678 steps.

monomer-surface interactions, and these are the subject of Chapter 12 by Owczarek and Whittington.

Recent developments that combine powerful physical concepts with advances in probability are the subject of the last three chapters. In chapter 13 de Gier describes fully packed loop models, in which the lattice is filled with polygons. These models impact on a variety of combinatorial and physical problems. In Chapter 14 Jacobsen describes the theory of conformal invariance, how it can be used to produce critical exponents, exact continuum limit partition functions, and many other powerful results, while in Chapter 15 Nienhuis and Kager describe the ideas behind stochastic Löwner evolution, which is believed to describe the scaling limit of a number of processes, including SAP and SAW.

# References

1. G. Aleksandrowicz and G. Barequet (2008) *Proc. 14th Ann. Int. Computing and Combinatorics Conf.,* Dalian, China Lecture Notes in Computer Science **5092** 100-109 June 2008.
2. Butera P and Comi M (1997) N-vector spin models on the simple cubic and body-centred cubic lattices Phys. Rev B **56** 8212–8240
3. D. Brydges and Y. Imbrie (2003) The Green's function for a hierarchical SAW in four dimensions Commun. Math. Phys **239** 549-584
4. T Berlin and M Kac (1952) The spherical model of a ferromagnet, Phys Rev. **86** 821-835
5. Conway A R and Guttmann A J (1995) On two-dimensional percolation J. Phys A:Math. Gen. **28** 891-904
6. Conway A R, Enting I G and Guttmann A J, Algebraic techniques for enumerating self-avoiding walks on the square lattice (1993), J. Phys. A:Math. Gen **26** 1519-34.

7. M Daoud, J P Cotton, B Farnoux, G Jannink, G Sarma, H Benoit R Dupplesix, C Picot and P-G de Gennes (1975) Solutions of flexible polymers. Neutron experiments and interpretation, Macromolecules **8** 804-818
8. P-G de Gennes, (1972) Exponents for the excluded volume problem as derived by the Wilson method, Phys. Letts A **38** 339-340.
9. H.E. Dudeney, *The Canterbury Puzzles and Other Curious Problems*, (Dover, New York 1958). (Reprint of 1907 edition)
10. N Elkies, G Kuperberg, M Larsen and J. Propp (1992) J. Alg. Comb, **1**, 111-132 and 219-234.
11. M. Eden: A two-dimensional growth process. In *Proc. Fourth Berkeley Symp. Math. Stat. Probab.* vol 4, ed by J. Neyman (Univ. Calif. Press 1961) pp 223-239
12. Fekete, M, Über die Verteilung der Wurzeln bei gewissenalgbraischen Gleichungen mit ganzzahligen Koeffizienten, (1923) Math. Z **17** 228-49
13. Fisher M E, Guttmann A J and Whittington S G Two-dimensional lattice vesicles and polygons (1991) J. Phys A:Math. Gen. **24** 3095-3106.
14. Fisher M E and Sykes M F, Excluded volume problem and the Ising model of ferromagnetism (1959) Phys. Rev. **114** 45-58
15. Flory P J The configuration of real polymer chains (1949) J. Chem. Phys. **17** 303-10.
16. S.W. Golomb: *Polyominoes*, (Charles Scribners' Sons, 1965)
17. S.W. Golomb (1954) Checkerboards and Polyominoes, American Math. Monthly, **61** No 10 675-82
18. B. Grünbaum and G C Shephard (1989) *Tilings and Patterns: An Introduction* W H Freeman and Co, New York
19. A.J. Guttmann and D.S. Gaunt (1978) On the asymptotic number of lattice animals in bond and site percolation, J. Phys A: Math. Gen, **11** 949-953
20. Hammersley, J M, (1961) On the number of polygons on a lattice, Proc. Cam. Phil. Soc, **57** 516-23.
21. Hiley B J and Sykes M F (1961) Probability of initial ring closure in the restricted random walk model of a macromolecule (1961) J. Chem. Phys. **34** 1531-37
22. T. Hara and G. Slade, (1992) Self-avoiding walk in five or more dimensions. I. The critical behaviour, Commun. Math. Phys., **147** 101–136, (1992)
23. T. Hara and G. Slade, (1992) The lace expansion for self -avoiding walks in five or more dimensions Reviews Math. Phys. **4**, 235-327
24. B D Hughes (1995) *Random walks and random environments, Volume 1: Random walks* (Clarendon Press, Oxford)
25. W Jockusch J. Propp and P Schor Random domino tilings and the arctic circle theorem. Preprint. Available at arxiv.org/abs/math.CO/9801068.
26. Jensen I and Guttmann A J Statistics of lattice animals (polyominoes) and polygons (2000) J. Phys A:Math. Gen. **33** L257-63.
27. Jensen I, see www.ms.unimelb.edu.au/~iwan.
28. I. Jensen: Counting polyominoes: A parallel implementation for cluster computing. Lecture Notes in Computer Science, **2659** 203-212 (2003)
29. P. Kasteleyn (1961) The statistics of dimers on a lattice I, The number of dimer arrangements on a quadratic lattice Physica **27** 1209-1225
30. Kesten H, On the number of self-avoiding walks, (1963) J. Math. Phys, **4**, 960-9
31. Kesten H, On the number of self-avoiding walks II, (1964) J. Math. Phys, **5**, 1128-37
32. Klarner D A Cell growth problems (1967) Cand. J. Math. **19** 851-63.
33. W. F. Lunnon Counting hexagonal and triangular polyominoes (1972) Graph Theory and Computing, ed. R. C Read, Academic Press pp. 87-100.
34. Li B, Madras N and Sokal A D Critical Exponents, hyperscaling and universal amplitude ratios for two-and three-dimensional self-avoiding walks, (1995) J. Stat. Phys. **80** 661-754
35. G.F. Lawler, O. Schramm and W. Werner, in *Fractal Geometry and Applications: A Jubilee of Benoît Mandelbrot*, Part 2, Proceedings of Symposia in Pure Mathematics #72 (American Mathematical Society, Providence RI, 2004), pp. 339–364, math.PR/0204277.
36. MacDonald D, Joseph S, Hunter D L, Moseley L L, Jan N and Guttmann A J (2000) Self-avoiding walks on the simple cubic lattice, (2000) J. Phys. A **33** 5973-83.

37. N Madras and G Slade *The Self-Avoiding Walk*, (Boston:Birkhäuser 1993)
38. G.E. Martin *Polyominoes: A guide to puzzles and problems in tiling*, (The Mathematical Association of America, 1991)
39. B. Nienhuis, (1982) Exact critical point and critical exponents of O(n) models in two dimensions, Phys. Rev. Lett. **49**, 1062-1065
40. B. Nienhuis, (1984) Critical behaviour of two-dimensional spin models and carge asymmetry in the Coulomb gas J. Stat. Phys. **34**, 731-761
41. O'Brien G L Monotonicity of the number of self-avoiding walks (1990) J. Stat. Phys, **59** 969-79
42. Orr, W J C Statistical treatment of polymer solutions at infinite dilution (1947) Trans. of the Faraday soc., **43**, 12-27.
43. Prellberg T Uniform $q$-series asymptotics for staircase polygons, (1995) J. Phys A:Math. Gen. **28** 1289-1304.
44. A. Pelissetto and E. Vicari, Critical phenomena and renormalization-group theory Phys. Reports **368**, 549-727 (2002),
45. G. Parisiand N. Sourlas (1981) Critical behaviour of branched polymers and the Lee-Yang edge singularity Phys. Rev. Letts, **46** 871-874
46. G C Rhoads (2005) Planar tilings by polyominoes, polyhexes, and polyiamonds, J. Comp and Appl. Math. **174** 329-353
47. Rushbrooke and Eve, (1959) On non-crossing lattice polygons J. Chem. Phys. **31**, 1333-4
48. C Richard (2002) Scaling behaviour of two-dimensional polygon models. J. Stat. Phys., **108** 459493
49. C Richard, A J Guttmann and I Jensen (2001) Scaling function and universal amplitude combinations for self-avoiding polygons. J. Phys. A: Math. Gen., 34:L495L501
50. C Richard, I Jensen and A J Guttmann (2004) Scaling function for self-avoiding polygons revisited. J. Stat. Mech.: Th. Exp., page P08007
51. H E Stanley (1968) Dependence of critical properties on dimensionality of spins, Phys Rev Letts, **20** 589-592.
52. D. Schattschneider (1990) *Visions of Symmetry*, W H Freeman and Co, New York
53. H.N.V. Temperley and M.E. Fisher (1961) Dimer problem in statistical mechanics—an exact result Phil Mag **6** 1061-1063
54. Wall F T, Hiller A L and Wheeler D J, Statistical Computation of Mean Dimensions of Macromolecules I, (1954) J. Chem. Phys, **22** 1036-41.
55. Wall F T, Hiller A L and Atchison W F, Statistical Computation of Mean Dimensions of Macromolecules II, (1955) J. Chem. Phys, **23** 913-921.
56. Wall F T, Hiller A L and Atchison W F, Statistical Computation of Mean Dimensions of Macromolecules III, (1955) J. Chem. Phys, **23** 2314-20.
57. Wakefield A J (1951) Statistics of the simple cubic lattice Proc. Camb. Phil. Soc., **47** 419-435

# Chapter 2
# Lattice Polygons and Related Objects

Stuart G Whittington

## 2.1 Introduction

Self-avoiding lattice polygons, i.e. embeddings of simple closed curves in a lattice, have been studied for more than fifty years. They are interesting as models of ring polymers in solution in good solvents and they appear in graphical expansions in, for instance, the Ising problem. They have been counted exactly (see e.g. [17]) and studied by Monte Carlo methods (see e.g. [22]). We understand many of their properties but rigorous results are scarce. In 1961 Hammersley [11] showed that they grow exponentially at the same rate as self-avoiding walks—a result which was by no means obvious at the time—but we still know very little about the sub-dominant asymptotic behaviour, except in dimensions higher than four where lace expansion techniques are useful.

This chapter will review some of the results which have been established rigorously. Apart from results about the numbers of polygons we also discuss counting polygons by both perimeter and area, which gives an interesting model of vesicles and how they respond to an osmotic force. This is closely related to the problem of self-avoiding surfaces. In three dimensions polygons can be knotted, and we investigate what is known rigorously about knot probabilities. There are many open questions in this area and we mention some of these. We then turn to polygons with a geometrical constraint and consider polygons confined to a wedge or to a slit or slab. The main interest has focussed on whether the constraint changes the exponential growth rate of the number of polygons. Finally we give a brief account of some results on lattice trees and lattice animals.

Stuart G Whittington
Department of Chemistry, University of Toronto, Ontario, Canada, e-mail: swhittin@chem.utoronto.ca

## 2.2 Counting Polygons

There are several different ways to think of and to define lattice polygons. To illustrate this we first consider a particular lattice, say the square lattice $\mathbb{Z}^2$. This is the lattice whose vertices are the integer points in Euclidean 2-space and whose edges join pairs of vertices which are unit distance apart. A lattice polygon is a connected subgraph of the lattice with all vertices of degree two. Alternatively we can think of a lattice polygon as an embedding of the circle graph. (We shall often drop the adjective *lattice* when the meaning is clear by the context.) If we stratify lattice polygons by the number ($n$) of edges in the embedding then the smallest polygon has four edges and there are no lattice polygons with an odd number of edges. We count polygons up to translation, i.e. we consider two polygons as identical if one can be superimposed on the other by a translation. This means that there is just one polygon with four edges, the unit square. If we write $p_n$ for the number of polygons with $n$ edges we have $p_4 = 1$. There are two polygons with six edges and seven with eight edges, so $p_6 = 2$ and $p_8 = 7$. See Fig. 2.1 for a sketch of the polygons with eight edges. Notice that these polygons are neither directed nor rooted. An alterna-

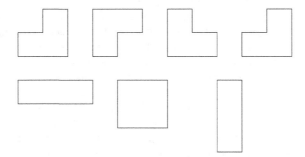

**Fig. 2.1** Polygons with eight edges on the square lattice.

tive approach is to think of a walk starting at the origin which is self-avoiding (i.e. it never revisits a vertex) except that it returns to the origin at its $n$th step. We call these objects *self-avoiding circuits* to distinguish them from the lattice polygons introduced above. Self-avoiding circuits are both directed and rooted (in this case they are rooted at the origin). We write $u_n$ for the number of self-avoiding circuits with $n$ edges. When $n = 4$ the first step can be taken in four ways and the second in two ways (since we are forming a unit square) and the third and fourth steps are then determined, so $u_4 = 8$. In fact $u_n$ and $p_n$ are closely related. Each $n$-edge polygon can be rooted in $n$ ways (i.e. at any vertex) and directed in two ways so

$$u_n = 2np_n. \tag{2.1}$$

Almost everything works in the same way for other lattices. An obvious extension is to the $d$-dimensional hypercubic lattice $\mathbb{Z}^d$, whose vertices are the integer points in $d$-dimensional Euclidean space. When $d = 3$ this is the simple cubic lattice. Again there are no polygons with an odd number of edges and the smallest polygon has four edges (since the girth of the lattice is 4). When $d = 3$ $p_4 = 3$ since the unit square can be in any one of the three coordinate planes. Similarly it is not difficult to see that $p_6 = 3 \times 2 + 3 \times 4 + 4 = 22$. The first term counts polygons which lie in a plane, the second counts *bent* polygons (pairs of unit squares in different planes which share a common edge) and the third counts *skew* polygons (six edges of a unit cube such that an opposite pair of vertices of the cube are not vertices of the polygon).

## 2.3 Asymptotic Behaviour

The values of $p_n$ are known exactly up to remarkably large values of $n$ in two [17] and three dimensions [3]. In two dimensions the results are obtained by methods described in Chapter 7, and in three dimensions the results are obtained from lace expansion ideas, as described in Chapter 6. Unfortunately knowing $p_n$ for a particular value of $n$ tells us nothing in principle about the value for larger values of $n$. Consequently we need some information about how $p_n$ grows for $n$ large. The main result is due to John Hammersley and dates back to 1961. Hammersley [11] used a concatenation argument to show that $p_n = \exp[\kappa n + o(n)]$ where $\kappa$ is a constant which depends on the lattice and where $\kappa$ is neither zero nor infinity. $\kappa$ is the *connective constant* of the lattice.

**Theorem 1.** *For the d-dimensional hypercubic lattice the limit*

$$\lim_{n \to \infty} n^{-1} \log p_n = \kappa$$

*exists and* $0 < \kappa < \infty$.

*Proof:* The proof relies on two observations. Since the coordination number of $\mathbb{Z}^d$ is $2d$ it is clear that $p_n \leq (2d)^n$ so

$$\limsup_{n \to \infty} n^{-1} \log p_n \leq \log(2d) < \infty. \tag{2.2}$$

Attach a coordinate system $(x, y, \ldots, z)$ to $\mathbb{Z}^d$. Consider a polygon whose vertices have coordinates $(x_i, y_i, \ldots, z_i), i = 1 \ldots n$. Using these coordinates we can construct a lexicographic order for the vertices. Call the vertex which is lexicographically first the *bottom vertex* of the polygon and the vertex which is lexicographically last the *top vertex* of the polygon. (In particular no vertex has smaller $x$-coordinate than the bottom vertex or larger $x$-coordinate than the top vertex.) Let the $x$-coordinate of the bottom vertex be $x^b$ and the $x$-coordinate of the top vertex be $x^t$. We call the plane $x = x^b$ the *bottom plane* of the polygon and the plane $x = x^t$ the *top plane* of the polygon. Each polygon has either one or two edges in the bottom plane, incident on the bottom vertex, and either one of two edges in the top plane, incident on the top vertex. If there is only one such edge we call it the *bottom edge* or *top edge*. If there are two we choose which one to call the bottom or top edge by the lexicographic order of the other vertex incident on the edge. Consider two polygons, $\pi_1$ with $n_1$ edges and $\pi_2$ with $n_2$ edges. $\pi_1$ has the top plane $x = x^t(\pi_1)$. If necessary rotate $\pi_2$ so that the bottom edge of $\pi_2$ is parallel to the top edge of $\pi_1$. With $\pi_1$ fixed, translate $\pi_2$ so that its bottom plane is $x = x^t(\pi_1) + 1$ and so that the bottom vertex of $\pi_2$ has all other coordinates equal to the corresponding coordinates of the top vertex of $\pi_1$. By deleting the top edge of $\pi_1$ and the bottom edge of $\pi_2$ and adding two edges to connect the two walks to produce a polygon, we obtain a polygon with $n_1 + n_2$ edges. See Fig. 2.2 for a sketch of the construction in two dimensions. This implies the inequality

$$p_{n_1} \frac{p_{n_2}}{d-1} \leq p_{n_1+n_2} \tag{2.3}$$

where the factor of $d-1$ comes from the possible rotation of $\pi_2$. Dividing both sides by $d-1$ gives the supermultiplicative inequality

$$\left(\frac{p_{n_1}}{d-1}\right)\left(\frac{p_{n_2}}{d-1}\right) \leq \frac{p_{n_1+n_2}}{d-1}, \tag{2.4}$$

so we see that $\log(p_n/(d-1))$ is a superadditive function. Together with (2.2) this implies that the limit $\lim_{n \to \infty} n^{-1} \log(p_n/(d-1))$ exists and that

$$\lim_{n \to \infty} n^{-1} \log(p_n/(d-1)) = \sup_{n>0} n^{-1} \log(p_n/(d-1)). \tag{2.5}$$

This proves the theorem and, in addition, gives the bound

$$\kappa \geq n^{-1} \log(p_n/(d-1)) \tag{2.6}$$

for every $n$.

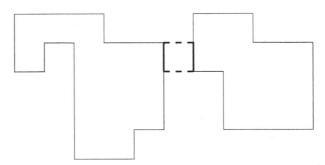

**Fig. 2.2** Concatenation of two polygons on the square lattice. The two thick edges (the top edge of one polygon and the bottom edge of the other) are deleted and the two dashed edges are added to connect and form a polygon.

## 2.4 Connection to Self-Avoiding Walks

In this section we consider self-avoiding walks, which are closely connected to lattice polygons. A *self-avoiding walk* is a walk on a lattice which never returns to a vertex already visited. If the walk has $n$ edges it is a sequence of $n + 1$ vertices, $i = 0, 1, \ldots, n$, such that

1. the vertices are distinct, and
2. the $(i - 1)$th and $i$th vertices are unit distance apart, $i = 1, 2, \ldots n$.

It is convenient to think of walks starting at the origin (although this is equivalent to counting walks up to translation). If we write $c_n$ for the number of $n$-edge self-avoiding walks then, for the square lattice, it is easy to see that $c_1 = 4$, $c_2 = 12$ and $c_3 = 36$. The first interesting event is at $n = 4$. Since the walk must be self-avoiding it cannot form a square so we have

$$c_4 = 3c_3 - u_4 = 3 \times 36 - 8 = 100.$$

As for polygons, self-avoiding walks can be counted exactly for modest values of $n$ and again there is a natural question about how $c_n$ grows for large $n$. If we concatenate a self-avoiding walk with $m$ edges and a self-avoiding walk with $n$ edges (by translating so that the zero'th vertex of the second walk coincides with the $m$th vertex of the first walk) we produce a set of objects which includes all the self-avoiding walks with $m + n$ edges. Hence

$$c_m c_n \geq c_{m+n} \tag{2.7}$$

and we see that $\log c_n$ is a sub-additive function. Since there is at least one $n$-edge self-avoiding walk (a straight line) $\liminf_{n \to \infty} n^{-1} \log c_n \geq 0$ and these two observations imply that the limit

$$\lim_{n \to \infty} n^{-1} \log c_n \equiv \hat{\kappa} \tag{2.8}$$

exists. Moreover $\hat{\kappa} \leq n^{-1} \log c_n$ for every $n > 0$.

We can obtain upper and lower bounds on $\hat{\kappa}$ by counting sets of walks which include the self-avoiding walks, and by counting subsets of self-avoiding walks. If we consider walks on $\mathbb{Z}^d$ which have no immediate reverse steps this includes all walks which are self-avoiding so $c_n \leq 2d(2d - 1)^{n-1}$. This immediately shows that $\hat{\kappa} \leq \log(2d - 1)$. If we count the subset of walks which only have steps in the positive coordinate directions we see that these are necessarily self-avoiding so $c_n \geq d^n$ and $\hat{\kappa} \geq \log d$. For a brief review of methods for deriving upper and lower bounds on $\hat{\kappa}$ see [23].

A natural question is how $\kappa$ and $\hat{\kappa}$ are related. If we delete the final edge of an $n$-edge self-avoiding circuit we obtain a self-avoiding walk with $n - 1$ edges so $c_{n-1} \geq u_n$. From (2.1) we see that $\lim_{n \to \infty} n^{-1} \log u_n = \kappa$ so it follows immediately that $\hat{\kappa} \geq \kappa$.

To get an inequality in the other direction requires rather more work. We introduce a new class of walks [12] which we call *x-unfolded walks*. These are $n$-

edge self-avoiding walks which satisfy the additional constraint that $x_0 < x_i \leq x_n$, $i = 1, 2, \ldots n - 1$. We write $c_n^\dagger$ for the number of $n$-edge $x$-unfolded walks. By inclusion $c_n^\dagger \leq c_n$. $x$-unfolded walks can be obtained from self-avoiding walks by a processive reflection of subwalks in the top and bottom planes and from this procedure one can show that [12]

$$c_n \leq c_n^\dagger e^{O(\sqrt{n})}. \tag{2.9}$$

Hence $\lim_{n \to \infty} n^{-1} \log c_n^\dagger = \hat{\kappa}$ and $x$-unfolded walks grow at the same exponential rate as self-avoiding walks.

The point of introducing unfolded walks is that they can be used to construct a subset of polygons. We shall explain the idea in two dimensions though the extension to higher dimensions isn't difficult. Consider $x$-unfolded walks in two dimensions. These walks can be unfolded again in the $y$-direction to produce *doubly unfolded walks* which satisfy the conditions that

1. $x_0 < x_i \leq x_n$ and
2. $y_0 \leq y_i < y_n$, $1 \leq i \leq n - 1$.

If the number of $n$-edge doubly unfolded walks is $c_n^\ddagger$ then $c_n^\ddagger \leq c_n \leq c_n^\ddagger e^{O(\sqrt{n})}$ so that $\lim_{n \to \infty} n^{-1} \log c_n^\ddagger = \hat{\kappa}$.

We define a *loop* with $n$ edges to be an $n$-edge self-avoiding walk with the additional constraints that

1. $x_0 = x_n < x_i$, $1 \leq i \leq n - 1$, and
2. $y_0 \leq y_i \leq y_n$, $1 \leq i \leq n - 1$.

We can construct a subset of loops from doubly unfolded walks as follows. Consider doubly unfolded walks with $n$ edges. Stratify these according to their span in the $x$-direction, and write $c_n^\ddagger(h)$ for the number with $x$-span equal to $h$. Suppose that $h_0$ is the smallest value of $h$ such that

$$c_n^\ddagger(h_0) = \max_h [c_n^\ddagger(h)]. \tag{2.10}$$

Then $c_n^\ddagger(h_0) \geq c_n^\ddagger/n$. If we concatenate a doubly unfolded walk with $x$-span equal to $h_0$ with another such walk reflected in the line $y = y_0$ we get a loop with $2n$ edges and hence

$$l_{2n} \geq c_n^\ddagger(h_0)^2 \geq [c_n^\ddagger/n]^2 \tag{2.11}$$

and since $l_n \leq c_n$ we see that

$$\lim_{n \to \infty} n^{-1} \log l_n = \hat{\kappa}. \tag{2.12}$$

Finally we construct a subset of polygons by concatenating two loops. Stratify loops according to their span in the $y$-direction and write $l_n(h)$ for the number of $n$-loops with $y$-span equal to $h$. Define $h_0$ to be the smallest value of $h$ such that

$$l_n(h_0) = \max_h [l_n(h)] \tag{2.13}$$

and concatenate two loops (more precisely a loop and a loop reflected in the line $x = x_0$) each with $n$ edges and with $y$-span equal to $h_0$. This yields the inequality

$$p_{2n} \geq l_n(h_0)^2 \geq [l_n/n]^2 \tag{2.14}$$

from which we can see that $\kappa = \hat{\kappa}$. Self-avoiding walks and polygons grow at the same exponential rate.

Although the value of $\kappa$ is not known rigorously for any lattice it is possible to obtain quite good upper and lower bounds. A lower bound can be obtained by enumerating polygons since we know that

$$\sup_{n>0} n^{-1} \log[p_n/(d-1)] = \kappa. \tag{2.15}$$

That is, if we know the number of polygons with $N$ edges then

$$\kappa \geq N^{-1} \log[p_N/(d-1)]. \tag{2.16}$$

Similarly, since $c_n$ is a sub-multiplicative function,

$$\kappa \leq N^{-1} \log c_N \tag{2.17}$$

for any $N \geq 1$.

An alternative scheme is to count exactly a subset of self-avoiding walks or polygons. For instance, on the square lattice, we can get a lower bound on $\kappa$ by enumerating partially directed walks. These are self-avoiding walks with no steps in the negative $x$-direction. Suppose that the number of such walks with $n$ edges is $b_n$. Clearly $b_n \leq c_n$ so

$$\lim_{n \to \infty} n^{-1} \log b_n \leq \kappa. \tag{2.18}$$

It is easy to see that

$$b_n = b_{n-1} + 2[b_{n-2} + b_{n-3} + \ldots] \tag{2.19}$$

from which we can derive the difference equation

$$b_{n+1} - 2b_n - b_{n-1} = 0 \tag{2.20}$$

from which it immediately follows that

$$\lim_{n \to \infty} n^{-1} \log b_n = \log(1 + \sqrt{2}). \tag{2.21}$$

## 2.5 Subdominant Asymptotic Behaviour

The results of Section 2.3 say that $p_n = \exp[\kappa n + o(n)]$ but say nothing about the $o(n)$ term. There is good reason to believe that

$$p_n \sim An^{\alpha-3}\mu^n \tag{2.22}$$

where $\mu = e^\kappa$. $\alpha$ is a *critical exponent* and critical exponents will play a major role in this book. The value of $\alpha$ is expected to be dependent on $d$ but not on the particular lattice of dimension $d$. That is, for instance, one expects $\alpha$ to be the same for the square, triangular and hexagonal lattices in two dimensions, whereas $\mu$ depends on the lattice. The fact that $p_n$ satisfies (2.4) implies that

$$p_n \leq (d-1)\mu^n \tag{2.23}$$

which is a useful upper bound on $p_n$ for general $d$. For $\mathbb{Z}^2$ this has been improved by Madras [19] who used an ingenious concatenation argument with the Loomis-Whitney inequality to show that

$$p_n \leq An^{-1/2}\mu^n \tag{2.24}$$

for some positive constant $A$. Results about the sub-dominant asymptotic behaviour for $d > 4$ are given in Chapter 6.

## 2.6 Counting Polygons by Perimeter and by Area

If we confine our attention to polygons in two dimensions we can stratify by both perimeter (the number of edges in the polygon) and by the enclosed area. Let $p_m(n)$ be the number of polygons with $m$ edges enclosing area $n$. Define the three generating functions [1, 7]

$$P_m(y) = \sum_n p_m(n)y^n, \tag{2.25}$$

$$A_n(x) = \sum_m p_m(n)x^m \tag{2.26}$$

and

$$P(x,y) = \sum_{m,n} p_m(n)x^m y^n = \sum_m P_m(y)x^m = \sum_n A_n(x)y^n. \tag{2.27}$$

One can equally well work in the constant perimeter or the constant area ensemble but we shall focus on the constant perimeter case [7]. This has a direct application as a two-dimensional model of a vesicle [1, 7] where $y > 1$ corresponds to the situation where large area is favoured (i.e. fluid flows from the surroundings into the vesicle under an osmotic pressure gradient and the vesicle swells) and where $y < 1$ corresponds to the situation where small area is favoured (i.e. fluid flows to the surroundings from the vesicle and the vesicle collapses).

For $y > 1$ the behaviour is dominated by polygons with maximum area. If $m$ is divisible by 4 the maximum area is $m^2/16$ and if $m$ is not divisible by 4 it is $(m^2 - 4)/16$. By taking the maximum area term in the partition function as a lower bound we have

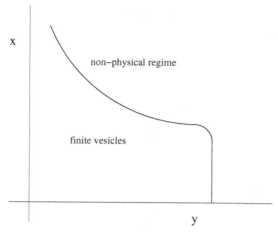

**Fig. 2.3** The phase diagram for square lattice polygons with perimeter fugacity $x$ and area fugacity $y$. The phase boundary is the boundary of convergence of $P(x,y)$. Below this phase boundary we have finite vesicles while on the phase boundary the vesicles are infinite.

$$P_m(y) \geq y^{(m^2-4)/16} \tag{2.28}$$

and it is clear that

$$P_m(y) \leq p_m y^{m^2/16}. \tag{2.29}$$

This implies that, for $y > 1$,

$$\lim_{m \to \infty} m^{-2} \log P_m(y) = \frac{\log y}{16} \tag{2.30}$$

and that the average area is $m^2/16 + o(m^2)$, so the vesicle is expanded. It also implies that, for $y > 1$, the partition function $P(x,y)$ converges only when $x = 0$. See Fig. 2.3 for a sketch of the boundary of convergence and the resulting phase diagram. The result in (2.30) is equivalent to

$$P_m(y) = y^{m^2/16 + o(m^2)}. \tag{2.31}$$

This has been refined [26] to show that

$$P_m(y) = A(y) y^{m^2/16} (1 + O(\rho^m)) \tag{2.32}$$

for some $0 < \rho < 1$, where $A(y)$ is $A_o(y)$ or $A_e(y)$ depending whether $n \to \infty$ through the odd or even integers. Explicit expressions are available for $A_o$ and $A_e$, both of which have an essential singularity at $y = 1$.

For $y < 1$ standard concatenation arguments can be used to prove the existence of the limit

$$\lim_{m \to \infty} m^{-1} \log P_m(y) \equiv \kappa(y) \tag{2.33}$$

and it can be shown (using Cauchy's inequality) that $\kappa(y)$ is a convex function of $\log y$. At fixed $m$ the minimum area is $(m-2)/2$ so, for $y < 1$,

$$P_m(y) \leq p_m y^{(m-2)/2} \tag{2.34}$$

and

$$\kappa(y) \leq \kappa + \frac{1}{2} \log y. \tag{2.35}$$

As a lower bound one can pick out the polygons with minimum area and write

$$P_m(y) \geq p_m^{\min} y^{(m-2)/2} \tag{2.36}$$

and one can show that the number of polygons with minimum area (which is the number of a class of site trees, see Section 2.10, on the dual lattice with $(m-2)/2$ vertices) satisfies $p_m^{\min} = \exp[\kappa_{\min} n + o(n)]$ so

$$\kappa(y) \geq \kappa_{\min} + \frac{1}{2} \log y. \tag{2.37}$$

The two inequalities (2.35) and (2.37) give bounds on the boundary of convergence of $P(x, y)$ for $y < 1$. One can show that the boundary of convergence is continuous for $0 < y \leq 1$ and the behaviour is consistent with the sketch in Fig. 2.3.

For $y < 1$ it can be shown that the average area $\langle n \rangle \sim m$ so the vesicles are highly ramified. This is sometimes called the *branched polymer phase* [6, 27]. The shape of the phase boundary as $y \to 1-$ is interesting since the behaviour is expected to be tricritical. This has been investigated by Richard and coworkers [27, 28]. Some simpler models such as staircase polygons can be solved exactly and this led to a conjecture [28] for the scaling form for self-avoiding polygons. See also [2]. This conjectured scaling form is in excellent agreement with numerical results [29].

## 2.7 Self-Avoiding Surfaces

A closely related problem is counting self-avoiding surfaces [9, 10, 15, 16]. First consider the two-dimensional case and the square lattice, $\mathbb{Z}^2$. Polygons embedded in the lattice enclose an area and it is only a small switch in our point of view if we stratify by area. The polygon and the area it encloses is an embedding of a disc in $\mathbb{Z}^2$, and the polygon is the boundary of the disc. We write $s_n(1)$ for the number of embeddings of discs with area $n$, where the argument 1 is to remind us that we have one boundary curve. It's easy to see that $s_1(1) = 1$, $s_2(1) = 2$ and $s_3(1) = 6$. Concatenation arguments can be used to show the existence of the limit

$$\lim_{n \to \infty} n^{-1} \log s_n(1) \equiv \log \beta_1. \tag{2.38}$$

The idea can be extended to embeddings of $k$-punctured discs with area $n$. These have $k+1$ boundary curves and we write $s_n(k+1)$ for the number of embeddings. It can be shown that

$$\lim_{n \to \infty} n^{-1} \log s_n(k+1) = \log \beta_1 \qquad (2.39)$$

for any fixed $k \geq 1$ so that the numbers of embeddings of punctured discs with any fixed number of punctures grow at the same exponential rate. If we consider discs with *any* number of punctures $s_n = \sum_{k \geq 0} s_n(k+1)$ then

$$\lim_{n \to \infty} n^{-1} \log s_n \equiv \log \beta \qquad (2.40)$$

and $\beta > \beta_1$.

The sub-dominant asymptotic behaviour is more difficult but Janse van Rensburg [15] proved bounds which, under the assumption that

$$s_n(k) \sim A n^{-\phi_k} \beta_1^n, \qquad (2.41)$$

imply that $\phi_{k+1} = \phi_1 - k$. Roughly speaking this means that an additional puncture can be added to a $k$-punctured disc in of order $n$ ways.

Similar but weaker results are available in three (and higher) dimensions [15]. When we consider punctured surfaces embedded in $\mathbb{Z}^3$ we need to keep track of the genus of the surface as well as the number of punctures. We write $s_n(k,g)$ for the number of embeddings, with area $n$, of a $k$-punctured surface of genus $g$, $k \geq 1$. The limit

$$\lim_{n \to \infty} n^{-1} \log s_n(k,g) \equiv \log \beta_k(g) \qquad (2.42)$$

exists and $\beta_k(g)$ is independent of $k$ and $g$ for all $k \geq 1$ and $g \geq 0$. Rather less is known about the sub-dominant behaviour than in two dimensions [15].

Guttmann et al. [9, 10] have considered embeddings of punctured discs in $\mathbb{Z}^2$ where the embeddings are stratified by the total number of edges in the boundary curves, i.e. by the total perimeter. If we write $\hat{p}_n^{(k)}$ for the number of embeddings (in $\mathbb{Z}^2$) of $k$-punctured discs with a total perimeter of $n$ we can write

$$\hat{p}_n = \sum_k \hat{p}_n^{(k)} \qquad (2.43)$$

for the number of embeddings of discs with an arbitrary number of punctures and total perimeter $n$. Guttmann et al. [10] show that

$$\lim_{n \to \infty} \frac{\log \hat{p}_n}{n \log n} = 1/4. \qquad (2.44)$$

This means that $\hat{p}_n$ grows faster than exponentially in $n$.

## 2.8 Knotted Polygons in Three Dimensions

In three dimensions there is the possibility that the polygon can be knotted. For the simple cubic lattice it is known that all polygons with less than 24 edges are unknotted and that polygons with 24 edges are either unknots or trefoils [5]. However, it is clear that other knots can occur on longer polygons and the possibility exists that polygons can be badly knotted (i.e. have high knot complexity according to some appropriate measure) when the polygons are large. Frisch and Wasserman [8] and, independently, Delbruck [4], conjectured that most sufficiently long polygons would be knotted, i.e. that unknots are rare in long polygons. This is known as the Frisch-Wasserman-Delbruck conjecture. The validity of this conjecture was established for lattice polygons in 1988, more than twenty years after the conjecture first appeared.

We confine our attention to the simple cubic lattice though similar results can be established for other three-dimensional lattices. Let $p_n^0$ be the number of unknotted polygons with $n$ edges. Diao [5] showed that $p_n^0 = p_n$ for $n < 24$. The Frisch-Wasserman-Delbruck conjecture says that $p_n^0/p_n \to 0$ as $n \to \infty$. The following, somewhat stronger result has been proved [25, 34].

**Theorem 2.** *The limit*

$$\lim_{n\to\infty} n^{-1} \log p_n^0 \equiv \kappa_0 \tag{2.45}$$

*exists and*

$$\kappa_0 < \kappa. \tag{2.46}$$

This means that the number of unknotted polygons, $p_n^0$, can be written as $p_n^0 = \exp[\kappa_0 n + o(n)]$ and that the probability that a polygon is unknotted, $P_n(\emptyset)$ is given by

$$P_n(\emptyset) = \frac{p_n^0}{p_n} = \exp[(\kappa_0 - \kappa)n + o(n)] \tag{2.47}$$

so, since $\kappa_0 < \kappa$, the probability of an unknot goes to zero exponentially rapidly as $n$ goes to infinity.

We shall not give a formal proof of Theorem 2 but we explain the three ideas which go into the proof. See [25, 30, 34] for details.

The first is that there is no *antiknot*. That is, if a simple closed curve is knotted it cannot be unknotted by adding an additional knot to cancel the first knot. Technically, if $k_1$ is a non-trivial knot type then there does not exist a knot type $k_2$ such that the connect sum of $k_1$ and $k_2$ is the unknot. This result follows directly from the additivity of genus.

The second component of the proof is the idea of a *knotted ball pair*. Consider a particular self-avoiding walk, $\omega$, with $m$ edges and vertices labelled $i = 0, 1, 2, \ldots m$. Associated with the $i$th vertex is a dual unit 3-cube, centred at the vertex, which we call $C_i$. Suppose that the union of these dual 3-cubes

$$C = \cup_{i=0}^{m} C_i \tag{2.48}$$

is a 3-ball. In that case, for the zero'th and $m$th vertices, extend the walk $\omega$ by adding a half-edge to form a walk $\hat{\omega}$. Then the walk $\hat{\omega}$ and $C$ form a ball-pair in which the 1-ball is properly embedded in the 3-ball. This ball pair can be *knotted*, i.e. not ambient isotopic the standard ball pair (the unit 3-ball with the 1-ball as a diameter). See Fig. 2.4 for an example. If the ball pair is knotted we say that $\omega$ is a *knotted arc*. The 3-ball is determined by the subwalk and no other part of the walk can enter the 3-ball. This means that any polygon which contains a knotted arc as a subwalk is necessarily knotted.

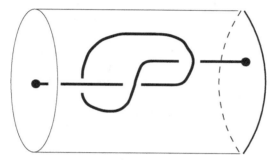

**Fig. 2.4** A knotted ball pair whose knotted arc is a trefoil.

It remains to show that all except exponentially few sufficiently long polygons contain at least one knotted arc. The key here is to use a pattern theorem. In 1963 Kesten considered the frequency of occurrence of certain patterns on self-avoiding walks. We say that a subwalk $\omega$ is a *Kesten pattern* if there exists a self-avoiding walk on which $\omega$ occurs three times. Kesten [18] proved the following theorem:

**Theorem 3.** *Suppose that $\omega$ is a Kesten pattern. Let $c_n(\bar{\omega})$ be the number of self-avoiding walks with n edges on which $\omega$ does not occur as a subwalk. Then*

$$\lim_{n \to \infty} n^{-1} \log c_n(\bar{\omega}) \equiv \kappa(\bar{\omega}) < \kappa. \tag{2.49}$$

From this it is easy to prove the corresponding theorem for polygons [34]:

**Theorem 4.** *Suppose that $\omega$ is a Kesten pattern. Let $p_n(\bar{\omega})$ be the number of self-avoiding walks with n edges on which $\omega$ does not occur as an (undirected) subwalk. Then*

$$\lim_{n \to \infty} n^{-1} \log p_n(\bar{\omega}) < \kappa. \tag{2.50}$$

Now we are ready to put the three components together. Suppose that $\omega$ is a knotted arc which is also a Kesten pattern. It is straightforward to construct a knotted arc whose knot type is a trefoil which is also a Kesten pattern, and Soteros et al. [30] show how this can be accomplished for any given knot type. Then

$$p_n^0 \leq p_n(\bar{\omega}) \tag{2.51}$$

and hence

$$\kappa_0 \leq \kappa(\bar{\omega}) < \kappa \tag{2.52}$$

which completes the proof.

The argument that we just gave says that unknotted polygons are exponentially rare but it can be extended to say rather more.

1. It is easy to construct a Kesten pattern which is also a knotted arc whose knot type is the connect sum of two trefoils. The argument given above says that this pattern appears on all except exponentially few sufficiently long polygons and this knot is composite. Hence prime knots are exponentially rare.
2. Since one can construct a Kesten pattern which is a knotted arc for any given (tame) knot type, every fixed knot type occurs on all except exponentially few sufficiently long polygons [30].
3. Kesten's theorem can be extended to show that Kesten patterns occur with positive density on all except exponentially few sufficiently long polygons and this implies that typical polygons are very badly knotted. For instance the crossing number (or span of the Alexander or Jones polynomial, etc.) all increase at least linearly with the length of the polygon [30].

There are some interesting open questions in this area. Although we know that most polygons are badly knotted we know very little about polygons with a fixed knot type. Suppose that $p_n(3_1)$ is the number of $n$-edge polygons whose knot type is the trefoil. (That is, the polygon contains exactly one trefoil and no other knots.) Then we know that

$$\kappa_0 \leq \liminf_{n \to \infty} n^{-1} \log p_n(3_1) \leq \limsup_{n \to \infty} n^{-1} \log p_n(3_1) < \kappa. \tag{2.53}$$

Does the appropriate limit exist for a fixed knot type? Is the connective constant of the (fixed) knot type equal to $\kappa_0$? These questions are open.

Polygons in three dimensions can be linked and one can ask similar questions about the numbers of embeddings in $\mathbb{Z}^3$ of links of various types. Suppose that $L_k$ is an unsplittable link of $k$ circles in which each circle is unknotted. Let $p_n(L_k)$ be the number of embeddings of this link in $\mathbb{Z}^3$ where each of the $k$ polygons has $n$ edges. Then Soteros et al. [31] showed that

$$\lim_{n \to \infty} (kn)^{-1} \log p_n(L_k) = \kappa_0. \tag{2.54}$$

This detailed information on embeddings of particular links contrasts with the open question discussed above for knots.

## 2.9 Polygons in Confined Geometries

In this section we examine the problem of polygons in confined geometries. We look at a relatively mild geometrical constraint, where the polygon is confined to a wedge, and a more severe constraint where the polygon is confined to a slab (in three or more dimensions), to a slit (in two dimensions) or to prism.

### 2.9.1 Polygons in Wedges

Consider the square lattice $\mathbb{Z}^2$ and define a *wedge* $W(\alpha)$ to be the set of integer points $\{(x,y)|x \geq 0, 0 \leq y \leq \alpha x + 1\}$ where $\alpha > 0$. Let $p_n(W)$ be the number of $n$-edge polygons with one vertex at $(0,0)$ and with all $n$ vertices in $W(\alpha)$. See Fig. 2.5.

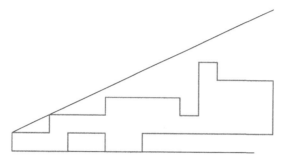

**Fig. 2.5** A polygon confined to a wedge on the square lattice.

We can ask if the limit $\lim_{n \to \infty} n^{-1} \log p_n(W)$ exists and how it depends on $\alpha$. The result is at first surprising. It turns out that

$$\lim_{n \to \infty} n^{-1} \log p_n(W) \equiv \kappa(W) \qquad (2.55)$$

exists and $\kappa(W) = \kappa$, for any positive $\alpha$ [13]. In fact the definition of a wedge can be generalized to

$$W(f) = \{(x,y)|x \geq 0, 0 \leq y \leq f(x)\} \qquad (2.56)$$

where $f(x)$ is a function of $x$ such that $f(0) = 1$ and $f(x) \geq 1$ for all $x > 0$. We call this an $f$-*wedge*. If $p_n(W)$ is the number of $n$-edge polygons with a vertex at $(0,0)$ and all other vertices in $W$ we have the following theorem:

**Theorem 5.** *The limit* $\lim_{n \to \infty} n^{-1} \log p_n(W) \equiv \kappa(W)$ *exists for all functions $f$ satisfying* $\lim_{x \to \infty} f(x) = \infty$, *and* $\kappa(W) = \kappa$, *independent of $f$.*

The existence of the limit is important in that one can construct functions which are unbounded but for which the exponential growth rate is strictly less than $\kappa$ [13].

This theorem also works for $\mathbb{Z}^d$ with suitably defined wedges. It has been extended to a more general definition of wedge by Soteros et al. [31].

### 2.9.2 Polygons in Slabs, Slits and Prisms

We first consider the $d$-dimensional hypercubic lattice $\mathbb{Z}^d$, $d \geq 3$, with coordinate system $(x,y,\ldots,z)$. Lattice vertices have integer coordinates. Consider polygons confined between the two parallel hyperplanes $z = 0$ and $z = L$, which we call a *slab*. Polygons are counted up to translation in all directions except the $z$-direction. Let $p_n(L)$ be the number of $n$-edge polygons constrained in this way. The limit defining the connective constant exists [13],

$$\lim_{n\to\infty} n^{-1} \log p_n(L) \equiv \kappa(L) \tag{2.57}$$

and

1. $\kappa(L) < \kappa(L+1)$ and
2. $\lim_{L\to\infty} \kappa(L) = \kappa$.

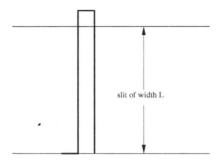

slit of width L

**Fig. 2.6** A pattern of width $L+1$ does not fit in a slit of width $L$.

The strict monotonicity comes from a pattern theorem argument, by considering the pattern shown in Fig. 2.6 which can occur on polygons in a slab of width $L+1$ but not on polygons in a slab of width $L$.

In two dimensions where we have the slit $0 \leq y \leq L$ the situation is somewhat different [32, 33]. For polygons in a slit of width $L$ there is a connective constant i.e. the appropriate limit exists) but the connective constant is strictly less than the corresponding connective constant for self-avoiding walks in the same slit. This also comes from a pattern theorem argument. One considers a pattern which "fills" the slit, which can occur on a walk but not on a polygon. The same result is true for polygons and walks in a prism in $\mathbb{Z}^3$.

## 2.10 Lattice Animals

In this section we give a brief account of some rigorous results about the numbers of lattice trees and lattice animals. The language used in the literature is rather cumbersome. A *bond animal* is a connected subgraph of the lattice. Bond animals are also called weakly embedded animals. In general bond animals can have cycles and the subset with no cycles are *bond trees*. *Site animals* are connected section graphs of the lattice and *site trees* are site animals with no cycles. Site animals and site trees are often called *strong embeddings*. The distinction between bond animals and site animals is that if two vertices of the lattice which are neighbours (i.e. are incident on a common edge of the lattice) are members of the vertex set of an animal then the edge is a member of the edge set for a site animal but this is not necessarily so for a bond animal. To add to the confusion, bond animals are sometimes counted by vertices and sometimes by edges, while site animals are essentially always counted by vertices. We write $a_n$ for the number (up to translation) of bond animals with $n$ vertices, $A_n$ for the number of site animals with $n$ vertices, $t_n$ for the number of bond trees with $n$ vertices and $T_n$ for the number of site trees with $n$ vertices. Clearly $T_n \leq t_n$, $t_n \leq a_n$, $T_n \leq A_n$ and $A_n \leq a_n$.

Concatenation arguments can be used to prove that the following limits exist:

$$\lim_{n \to \infty} n^{-1} \log a_n \equiv \log \lambda, \tag{2.58}$$

$$\lim_{n \to \infty} n^{-1} \log t_n \equiv \log \lambda_0, \tag{2.59}$$

$$\lim_{n \to \infty} n^{-1} \log A_n \equiv \log \Lambda, \tag{2.60}$$

and

$$\lim_{n \to \infty} n^{-1} \log T_n \equiv \log \Lambda_0. \tag{2.61}$$

The obvious bounds given above show that $\Lambda_0 \leq \lambda_0$, $\lambda_0 \leq \lambda$, $\Lambda_0 \leq \Lambda$ and $\Lambda \leq \lambda$. By using a pattern theorem due to Madras [20], these inequalities can be shown to be strict. In fact we know that

$$\Lambda_0 < \Lambda < \lambda_0 < \lambda. \tag{2.62}$$

There is a quite different proof that $\lambda_0 < \lambda$ due to Madras et al. [24].

One can also define bond animals with a prescribed cyclomatic index and ask how the number of animals depends on the number of cycles. Let $b_n(c)$ be the number of bond animals with $n$ edges and cyclomatic index $c$. Clearly $b_n(0) = t_n$. It is known that

$$\lim_{n \to \infty} n^{-1} \log b_n(c) = \log \lambda_0 \tag{2.63}$$

for every $c$, so bond animals with any fixed cyclomatic index are exponentially rare in the set of all bond animals.

Concatenation arguments also give upper bounds of the form

$$t_n \leq \lambda_0^n \tag{2.64}$$

and these bounds have been improved by Madras [19, 21], using the Loomis-Whitney inequality, to give

$$t_n \leq An^{-(d-1)/d}\lambda_0^n \tag{2.65}$$

for a positive constant $A$, for any $d \geq 2$, with similar bounds for bond and site animals. Janse van Rensburg [14] gave a lower bound for $t_n$ showing that

$$t_n \geq A\exp[-\delta(\log n)^2]\lambda_0^n \tag{2.66}$$

for positive constants $A$ and $\delta$.

# References

1. J. R. Banavar, A. Maritan and A. Stella: *Critical behaviour of 2-dimensional vesicles in the deflated regime* Phys. Rev. A **43**, 5752-5754 (1991)
2. J. L. Cardy: *Exact scaling functions for self-avoiding walks and branched polymers* J. Phys. A: Math. Gen. **34**, L665-L672 (2001)
3. N. Clisby, R. Liang and G. Slade: *Self-avoiding walk enumeration via the lace expansion* J. Phys. A: Math. Theor. **40**, 10973-11017 (2007)
4. M. Delbruck: Knotting problems in biology, in Mathematical Problems in the Biological Sciences, edited by R. E. Bellman, American Mathematical Society, Providence (1962)
5. Y. Diao: *Minimal knotted polygons on the cubic lattice* J. Knot Theory Ramifications **2**, 413-425 (1993)
6. M. E. Fisher: *Fractal and nonfractal shapes in two-dimensional vesicles* Physica D **38**, 112-118 (1989)
7. M. E. Fisher, A. J. Guttmann and S. G. Whittington: *Two-dimensional lattice vesicles and polygons* J. Phys. A: Math. Gen. **24**, 3095-3106 (1991)
8. H. L. Frisch and E. Wasserman: *Chemical topology* J. Am. Chem. Soc. **83**, 3789-3795 (1961)
9. A. J. Guttmann, I. Jensen, L. H. Wong and I. Enting: *Punctured polygons and polyominoes on the square lattice* J. Phys. A: Math. Gen. **33**, 1735-1764 (2000)
10. A. J. Guttmann, I. Jensen and A. L. Owczarek: *Polygonal polyominoes on the square lattice* J. Phys. A: Math. Gen. **34**, 3721o-3733 (2001)
11. J. M. Hammersley: *The number of polygons on a lattice* Proc. Camb. Phil. Soc. **57**, 516-523 (1961)
12. J. M. Hammersley and D. J. A. Welsh: *Further results on the rate of convergence to the connective constant of the hypercubic lattice* Quart. J. Math. Oxford **18**, 108-110 (1962)
13. J. M. Hammersley and S. G. Whittington: *Self-avoiding walks in wedges* J. Phys. A: Math. Gen. **18**, 101-111 (1985)
14. E. J. Janse van Rensburg: *On the number of trees in $Z^d$* J. Phys. A: Math. Gen. **25**, 3523-3528 (1992)
15. E. J. Janse van Rensburg: *Surfaces in the hypercubic lattice* J. Phys. A: Math. Gen. **25**, 3529-3547 (1992)
16. E. J. Janse van Rensburg and S. G. Whittington: *Self-avoiding surfaces with knotted boundaries* J. Phys. A: Math. Gen. **23**, 2495-2505 (1990)
17. I. Jensen: *A parallel algorithm for the enumeration of self-avoiding polygons on the square lattice* J. Phys. A: Math. Gen. **36**, 5731-5745 (2003)
18. H. Kesten: *On the number of self-avoiding walks* J. Math. Phys. **4**, 960-969 (1963)
19. N. Madras: *A rigorous bound on the critical exponent for the number of lattice trees, animals and polygons* J. Stat. Physics **78** 681-699 (1995)

20. N. Madras: *A pattern theorem for lattice clusters* Ann. Comb. **3**, 357-384 (1999)
21. N. Madras: *Enumeration bounds via an isoperimetric-type inequality* J. Physics: Conf. Series **42**, 213-220 (2006)
22. N. Madras, A. Orlitsky and L. A. Shepp: *Monte Carlo generation of self-avoiding walks with fixed endpoints and fixed length* J. Stat. Phys. **58**, 159-183 (1990)
23. N. Madras and G. Slade: The Self-Avoiding Walk, Birkhäuser, Boston (1993)
24. N. Madras, C. E. Soteros and S. G. Whittington: *Statistics of lattice animals* J. Phys. A: Math. Gen. **21**, 4617-4635 (1988)
25. N. Pippenger: *Knots in random walks* Discrete Appl. Math. **25**, 273-278 (1989)
26. T. Prellberg and A. L. Owczarek: *On the asymptotics of the finite perimeter partition function of two-dimensional lattice vesicles* Commun. Math. Phys. **201**, 493-505 (1999)
27. C. Richard: *Scaling behaviour of two-dimensional polygon models* J. Stat. Phys. **108**, 459-493 (2002)
28. C. Richard, A. J. Guttmann and I. Jensen: *Scaling functions and universal amplitude combinations for self-avoiding polygons* J. Phys. A: Math. Gen. **34**, L495-L501 (2001)
29. C. Richard, I. Jensen and A. J. Guttmann: *Scaling prediction for self-avoiding polygons revisited* J. Stat. Mech.: Theor. Exp. P08007 (2004)
30. C. E. Soteros, D. W. Sumners and S. G. Whittington: *Entanglement complexity of graphs in $Z^3$* Math. Proc. Camb. Phil. Soc. **111**, 75-91 (1992)
31. C. E. Soteros, D. W. Sumners and S. G. Whittington: *Linking of random p–spheres in $Z^d$* J. Knot Theory Ramifications **8**, 49-70 (1999)
32. C. E. Soteros and S. G. Whittington: *Polygons and stars in a slit geometry* J. Phys. A: Math. Gen. **21**, L857-L861 (1988)
33. C. E. Soteros and S. G. Whittington: *Lattice models of branched polymers: Effects of geometrical constraints* J. Phys. A: Math. Gen. **22**, 5259-5270 (1989)
34. D. W. Sumners and S. G. Whittington: *Knots in self-avoiding walks* J. Phys. A: Math. Gen. **21**, 1689-1694 (1988)

# Chapter 3
# Exactly Solved Models

Mireille Bousquet-Mélou and Richard Brak

## 3.1 Introduction

### 3.1.1 Subclasses of Polygons and Polyominoes

This chapter deals with the *exact enumeration* of certain classes of (self-avoiding) polygons and polyominoes. We restrict our attention to the square lattice. As the interior of a polygon is a polyomino, we often consider polygons as special polyominoes. The usual enumeration parameters are the *area* (the number of cells) and the *perimeter* (the length of the border). The perimeter is always even, and often refined into the horizontal and vertical perimeters (number of horizontal/vertical steps in the border). Given a class $\mathscr{C}$ of polyominoes, the objective is to determine the following *complete generating function* of $\mathscr{C}$:

$$C(x,y,q) = \sum_{P \in \mathscr{C}} x^{hp(P)/2} y^{vp(P)/2} q^{a(P)},$$

where $hp(P)$, $vp(P)$ and $a(P)$ respectively denote the horizontal perimeter, the vertical perimeter and the area of $P$. This means that the coefficient $c(m,n,k)$ of $x^m y^n q^k$ in the series $C(x,y,q)$ is the number of polyominoes in the class $\mathscr{C}$ having horizontal perimeter $2m$, vertical perimeter $2n$ and area $k$. Several specializations of $C(x,y,q)$ may be of interest, such as the *perimeter generating function* $C(t,t,1)$, its *anisotropic* version $C(x,y,1)$, or the *area generating function* $C(1,1,q)$. From such exact results, one can usually derive many of the asymptotic properties of the polyominoes of $\mathscr{C}$: for instance the asymptotic number of polyominoes of perimeter $n$,

Mireille Bousquet-Mélou
CNRS, LaBRI, Université Bordeaux 1, 33405 Talence Cedex, France, e-mail: bousquet@labri.fr

Richard Brak
Department of Mathematics and Statistics, The University of Melbourne, Victoria 3010, Australia, e-mail: r.brak@ms.unimelb.edu.au

or the (asymptotic) average area of these polyominoes, or even the limiting distribution of this area, as $n$ tends to infinity (see Chapter 11). The techniques that are used to derive asymptotic results from exact ones are often based on complex analysis. A remarkable survey of these techniques is provided by Flajolet and Sedgewick's book [33].

The study of sub-classes of polyominoes is natural, given the immense difficulty of the full problem (enumerate all polygons or all polyominoes). The objective is to develop new techniques, and to push the border between solved and unsolved models further and further. However, several classes have an independent interest, other than being an approximation of the full problem. For instance, the enumeration of *partitions* (Fig. 3.2(e)) is relevant in number theory and in the study of the representations of the symmetric group. The first enumerative results on partitions date back, at least, to Euler. A full book is devoted to them, and is completely independent of the enumeration of general polyominoes [2]. Another example is provided by *directed* polyominoes, which are relevant for directed percolation, but also occur in theoretical computer science as binary search networks [54].

All these classes will be systematically defined in Section 3.1.3. For the moment, let us just say that most of them are obtained by combining conditions of *convexity* and *directedness*.

From the perspective of subclasses as an approximation to the full problem, it is natural to ask how good this approximation is expected to be. The answer is quite crude: these approximations are terrible. For a start, all the classes that have been counted so far are *exponentially small* in the class of all polygons (or polyominoes). Hence we cannot expect their properties to reflect faithfully those of general polygons/polyominoes. Why would the properties of a staircase polygon (Fig. 3.2(b)) be similar to those of a general self-avoiding polygon? Indeed, the number of staircase polygons of perimeter $2n$ grows like $2^{2n}n^{-3/2}$ (up to a multiplicative constant), while the number of general polygons is believed to be asymptotically $\mu^{2n}n^{-5/2}$, with $\mu = 2.638\ldots$ [36]. The average width of a staircase polygon is clearly linear in $n$, while the width of general polygons is conjectured to grow like $n^{3/4}$ (see [42]). And so on! In this context, it may be a pure coincidence that the average area of polygons of perimeter $2n$ is conjectured to scale as $n^{3/2}$ (see [27]), *just as it does for staircase polygons*. But it is also conjectured that the limit distribution of the area of $2n$-step polygons (normalized by its average value) coincides with the corresponding distribution for staircase polygons, and for other exactly solved classes. The universality of this distribution may not be a coincidence (see Chapter 11 for more references and details).

## 3.1.2 Three General Approaches

In this chapter, we present three robust approaches that can be applied to count many classes $\mathscr{C}$ of polyominoes. The common principle of all of them is to translate a recursive description of the polyominoes of $\mathscr{C}$ into a functional equation satisfied by the generating function $C(x,y,q)$. Some readers may prefer seeing a translation in terms of the *coefficients* of $C(x,y,q)$, namely the numbers $c(m,n,k)$. This translation is possible, but it is usually easier to work with a functional equation than with a recurrence relation. The applicability of each of these three approaches depends on whether the polyominoes of $\mathscr{C}$ have, or don't have, a certain type of recursive structure.

The most versatile approach is probably the third one, as it virtually applies to any class of polyominoes having a convexity property. It was already used by Temperley in 1956 [51] and is often called, in the physics literature, *the Temperley approach*. However, it often produces functional equations that are non-trivial to solve, even when the solution finally turns out to be a simple rational or algebraic series (these terms will be defined in Section 3.1.3 below). From a combinatorics point of view, it is important to get a better understanding of the simplicity of these series, and this is what the first two approaches provide: the first one applies to classes $\mathscr{C}$ having a *linear* structure, and gives rise to rational generating functions. The second applies to classes having an *algebraic* structure, and gives rise to algebraic generating functions.

We have chosen to present these three approaches because, in our opinion, they are the most robust ones, and we want to provide effective tools to the reader. To our knowledge, almost all the classes that have been solved exactly can be solved using one (or several) of these approaches. Still, certain results have been given a beautiful combinatorial explanation via more specific techniques. Let us mention two tools that are often involved in those alternative approaches. The first tool is specific to the enumeration of polygons, and consists in studying classes of possibly self-intersecting polygons, and then using an inclusion-exclusion principle to eliminate the ones with self-intersections. This idea appears in an old paper of Pólya [43] dealing with staircase polygons, and was further exploited to count more general polygons [30, 31], including those in dimensions larger than two [35, 13]. The second tool is the use of *bijections* and is of course not specific to polyomino enumeration. The idea is to describe a one-to-one correspondence between the objects of $\mathscr{C}$ and those of another class $\mathscr{D}$, having a simpler recursive structure. In this chapter, even though we often use encodings of polyominoes by words, these encodings are usually very simple and do not use the full force of bijective methods, which is clearly at work in papers like [16] or [20].

The structure of the chapter is simple: the three approaches we discuss are presented, and illustrated by examples, in Sections 3.2, 3.3 and 3.4 respectively. A few open problems which we consider worth investigating are discussed in Section 3.5.

We conclude this introduction with definitions of various families of polyominoes and formal power series.

### 3.1.3 A Visit to the Zoo

All the classes studied in this chapter are obtained by combining several conditions of *convexity* and *directedness*. Let us first recall that a polyomino $P$ is a finite set of square cells of the square lattice whose interior is connected. The set of centres of the cells form an *animal A* (Fig. 3.1). The connectivity condition means that any two points of $A$ can be joined by a path made up of unit vertical and horizontal steps, in such a way that every vertex of the path lies in $A$. The animal $A$ is *North-East directed* (or *directed*, for short) if it contains a point $v_0$, called the source, such that every other point of $A$ can be reached from $v_0$ by a path made of North and East unit steps, having all its vertices in $A$. In this case, the polyomino corresponding to $A$ is also said to be NE-directed. One defines NW, SW and SE directed animals and polyominoes similarly.

A polyomino $P$ is *column-convex* if its intersection with every vertical line is connected. This means that the intersection of every vertical line with the corresponding animal $A$ is formed by consecutive points. The border of $P$ is then a polygon. Row-convexity is defined similarly. Finally, $P$ is $d_+$-convex if the intersection of $A$ with every line of slope 1 is formed by consecutive points. One defines $d_-$-convex polyominoes similarly.

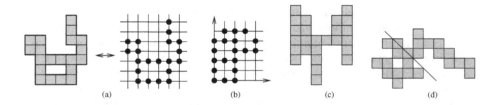

**Fig. 3.1** From left to right: (a) a polyomino and the corresponding animal, (b) a NE-directed animal, (c) a column-convex polygon, (d) a $d_-$-convex polyomino.

As discussed in [8], the combination of the four direction conditions and the four connectivity conditions gives rise to 31 distinct (non-symmetric) classes of polyominoes having at least one convexity property. To these 31 classes we must add the 4 different classes satisfying at least one directional property. Some prominent members of this zoo, which will occur in the forthcoming sections, are shown in Fig. 3.2:

- *convex* polyominoes (or polygons): polyominoes that are both column- and row-convex,
- *staircase* polyominoes (or polygons): convex polygons that are NE- and SW-directed,
- *bargraphs*: column-convex polygons that are NE- and NW-directed,
- *stacks*: row-convex bargraphs,

- *partitions,* a.k.a. *Ferrers diagrams,*: convex polygons that are NE-, NW- and SE-directed.

Finally, a formal power series $C(x) \equiv C(x_1,\ldots,x_k)$ with real coefficients is *rational* if it can be written as a ratio of polynomials in the $x_i$'s. It is *algebraic* if it satisfies a non-trivial polynomial equation

$$P(C(x),x_1,\ldots,x_k) = 0.$$

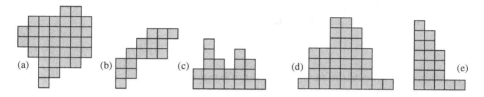

**Fig. 3.2** A photo taken at the zoo: (a) a convex polygon, (b) a staircase polygon, (c) a bargraph, (d) a stack, (e) a Ferrers diagram.

## 3.2 Linear Models and Rational Series

### 3.2.1 A Basic Example: Bargraphs Counted by Area

Let $b_n$ denote the number of bargraphs of area $n$. As there is a unique bargraph of area 1, $b_1 = 1$. For $n \geq 2$, there are two types of bargraphs:

1. those in which the last (i.e., rightmost) column has height 1,
2. those in which the last column has height 2 or more.

Bargraphs of the first type are obtained by adding a column of height 1 to the right of any bargraph of area $n - 1$. Bargraphs of the second type are obtained by adding one square cell to the top of the last column of a bargraph of area $n - 1$. Since a bargraph cannot be simultaneously of type 1 and 2, this gives

$$b_1 = 1 \quad \text{and for } n \geq 2, \quad b_n = 2b_{n-1},$$

which implies $b_n = 2^{n-1}$. The area generating function of bargraphs is thus a rational series:

$$B(q) := \sum_{n \geq 1} b_n q^n = \frac{q}{1 - 2q}.$$

### 3.2.2 Linear Objects

The above enumeration of bargraphs is based on a very simple recursive description of bargraphs. This description only involves the following two constructions:

1. taking disjoint unions of sets,
2. concatenating a new cell with an already constructed object.

In terms of generating functions (g.f.s), taking the disjoint union of sets means summing their g.f.s, while concatenating a new cell (of size 1) to all elements of a set means multiplying its g.f. by $q$. Hence the above description of bargraphs translates directly into a linear equation for the g.f. $B(q)$:

$$B(q) = q + qB(q) + qB(q).$$

This equation reflects the fact that the set of bargraphs is the union of three disjoint subsets (the unique bargraph of area 1, bargraphs of type 1, bargraphs of type 2), and that the second and third subsets are both obtained by adding a cell to any bargraph.

More generally, we will say that a class of objects, equipped with a size, is *linear* if these objects can be obtained from a finite set of initial objects using disjoint union and concatenation of one cell, or *atom*. It is assumed that the concatenation of an atom increases the size by 1. The construction must be *non-ambiguous*, meaning that each object of the class is obtained only once. The construction may involve several classes of objects simultaneously. For instance, the class $\tilde{\mathscr{B}}$ of bargraphs whose last column has height 1 is linear: the objects of $\tilde{\mathscr{B}}$, other than the one-cell bargraph, are obtained by adding one cell to the right of any bargraph. The associated series $\tilde{B}(q)$ is defined by the linear system:

$$\tilde{B}(q) = q + qB(q),$$
$$B(q) = q + qB(q) + qB(q).$$

In general, the generating function of a linear class of objects is the first component of the solution of a system of $k$ linear equations of the form

$$B_i(q) = P_i(q) + q \sum_{j=1}^{k} a_{i,j} B_j(q) \quad 1 \leq i \leq k, \tag{3.1}$$

where $a_{i,j} \in \mathbb{N}$ and each $P_i(q)$ is a polynomial in $q$ with coefficients in $\mathbb{N}$. The polynomial $P_i(q)$ counts the initial objects of type $i$, and there are $a_{i,j}$ ways to aggregate an atom to an object of type $j$ to form an object of type $i$. The system (3.1) uniquely defines each series $B_i(q)$, which is rational. The series obtained in this way are called $\mathbb{N}$-*rational*. Their study is closely related to the theory of *regular languages* [50].

### 3.2.3 *More Linear Models*

In this section we present three typical problems that can be solved via a linear recursive description. The first one is the perimeter enumeration of Ferrers diagrams (and stacks). The second one generalizes the study of bargraphs performed in Section 3.2.1 to all column-convex polygons (and to the subclass of directed column-convex polygons) counted by area. The third one illustrates the role of linear models in the approximation of hard problems, and deals with the enumeration of self-avoiding polygons confined to a narrow strip. In passing, we illustrate the two following facts:

1. it may be useful to begin by describing a size-preserving bijection between polyominoes and other objects (having a linear structure),
2. linear constructions are conveniently described by a directed graph when they become a bit involved.

#### 3.2.3.1 Ferrers Diagrams by Perimeter

The set of Ferrers diagrams can be partitioned into three disjoint subsets: first, the unique diagram of (half-)perimeter 2; then, diagrams of width at least 2 whose rightmost column has height 1; finally, diagrams with no column of height 1. The latter diagrams can be obtained by duplicating the bottom row of another diagram (Fig. 3.3).

**Fig. 3.3** Recursive description of Ferrers diagrams.

From this description, it follows that the set of words that describe the North-East boundary of Ferrers diagrams, from the NW corner to the SE one, admits a linear construction. This boundary is formed by East and South steps, and will be encoded by a word over the alphabet $\{e, s\}$. Any word over this alphabet that starts with an $e$ and ends with an $s$ corresponds to a unique Ferrers diagram. Let $\mathscr{F}$ be this class of words, and let $\mathscr{L}$ be the set of all non-empty prefixes of words of $\mathscr{F}$. Then $\mathscr{F}$ and $\mathscr{L}$ admit the following linear description:

$$\mathscr{F} = \mathscr{L}s \quad \text{and} \quad \mathscr{L} = \{e\} \cup \mathscr{L}e \cup \mathscr{L}s.$$

In these equations, the notation $\mathscr{L}s$ means $\{us, \, u \in \mathscr{L}\}$, and the unions are disjoint. The series that count the words of these sets by their length (number of letters) are thus given by the linear system

$$F(t) = tL(t) \quad \text{and} \quad L(t) = t + 2tL(t).$$

Since the length of a coding word is the half-perimeter of the associated diagram, this provides the length g.f.:

$$F(t) = \frac{t^2}{1 - 2t} = \sum_{n \geq 1} 2^{n-2} t^n.$$

By separately counting East and South steps, we obtain the equations

$$F(x,y) = yL(x,y) \quad \text{and} \quad L(x,y) = x + xL(x,y) + yL(x,y), \tag{3.2}$$

and hence the anisotropic perimeter g.f. of these diagrams:

$$F(x,y) = \frac{xy}{1-x-y} = \sum_{m,n \geq 1} \binom{m+n-2}{m-1} x^m y^n.$$

A similar treatment can be used to determine the perimeter g.f. of stack polygons: the construction schematized in Fig. 3.4 gives:

$$S(x,y) = xy + xS(x,y) + S_+(x,y), \quad S_+(x,y) = yS(x,y) + xS_+(x,y)$$

which yields

$$S(x,y) = \frac{xy(1-x)}{(1-x)^2 - y}.$$

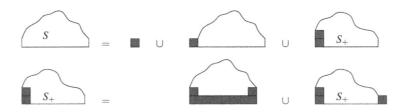

**Fig. 3.4** Recursive description of stack polygons.

### 3.2.3.2 Column-Convex Polygons by Area

Consider a column-convex polygon $P$ having $n$ cells. Let us number these cells from 1 to $n$ as illustrated in Fig. 3.5. The columns are visited from left to right. In the first column, cells are numbered from bottom to top. In each of the other columns, the lowest cell that has a left neighbour gets the smallest number; then the cells lying below it are numbered from top to bottom, and finally the cells lying above it are

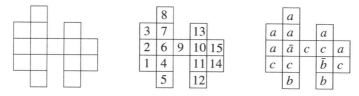

**Fig. 3.5** A column-convex polygon, with the numbering and encoding of the cells.

numbered from bottom to top. Note that for all $i$, the cells labelled $1, 2, \ldots, i$ form a column-convex polygon. This labelling describes the order in which we are going to aggregate the cells.

Associate with $P$ the word $u = u_1 \cdots u_n$ over the alphabet $\{a, b, c\}$ defined by
 $- u_i = c$ (like Column) if the $i^{\text{th}}$ cell is the first to be visited in its column,
 $- u_i = b$ (like Below) if the $i^{\text{th}}$ cell lies below the first visited cell of its column,
 $- u_i = a$ (like Above) if the $i^{\text{th}}$ cell lies above the first visited cell of its column.
Then, add a bar on the letter $u_i$ if the $i^{\text{th}}$ cell of $P$ has a South neighbour, an East neighbour, but no South-East neighbour. (In other words, the barred letters indicate where to start a new column, when the bottommost cell in this new column lies above the bottommost cell of the previous column.) This gives a word $v$ over the alphabet $\{a, b, c, \bar{a}, \bar{b}, \bar{c}\}$, and $P$ can be uniquely reconstructed from $v$.

We now focus on the enumeration of these coding words. Let $\mathscr{L}$ be the set of all prefixes of these words, including the empty prefix $\varepsilon$. By considering which letter can be added to the right of which prefix, we are led to partition $\mathscr{L}$ into five disjoint subsets $\mathscr{L}_1, \ldots, \mathscr{L}_5$, subject to the following linear recursive description:

$$
\begin{aligned}
&\mathscr{L}_1 = \{\varepsilon\}, \\
&\mathscr{L}_2 = \mathscr{L}_1 c \cup \mathscr{L}_2 a \cup \mathscr{L}_3 a \cup \mathscr{L}_4 c, \quad \mathscr{L}_4 = \mathscr{L}_2 \bar{a} \cup \mathscr{L}_3 \bar{a} \cup \mathscr{L}_4 a \cup \mathscr{L}_5 b, \\
&\mathscr{L}_3 = \mathscr{L}_2 c \cup \mathscr{L}_3 b \cup \mathscr{L}_3 c, \quad\quad\ \mathscr{L}_5 = \mathscr{L}_2 \bar{c} \cup \mathscr{L}_3 \bar{b} \cup \mathscr{L}_3 \bar{c} \cup \mathscr{L}_5 b.
\end{aligned} \tag{3.3}
$$

The words of $\mathscr{L}_4$ and $\mathscr{L}_5$ are those in which a barred letter (the rightmost one) still waits to be "matched" by a letter $c$ creating a new column. The words of $\mathscr{L}_2 \cup \mathscr{L}_3$ are those that encode column-convex polygons. This construction is illustrated by a directed graph in Fig. 3.6: every path starting from 1 and ending at $i$ corresponds to a word of $\mathscr{L}_i$, obtained by reading edge labels. The series counting the words of $\mathscr{L}_i$ by their length satisfy:

$$
\begin{aligned}
&L_1 = 1, \\
&L_2 = q(L_1 + L_2 + L_3 + L_4), \quad L_4 = q(L_2 + L_3 + L_4 + L_5), \\
&L_3 = q(L_2 + L_3 + L_3), \quad\quad\quad\ L_5 = q(L_2 + L_3 + L_3 + L_5).
\end{aligned}
$$

The area g.f. of column-convex polygons is $C(q) = L_2(q) + L_3(q)$. Solving the above system gives:

$$
C(q) = \frac{q(1-q)^3}{1 - 5q + 7q^2 - 4q^3}.
$$

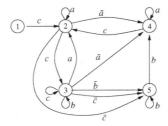

**Fig. 3.6** Linear construction of the words of $\mathscr{L}$. The words of $\mathscr{L}_i$ encode the paths starting from 1 and ending at $i$.

We believe that this result was first published by Temperley [51].

A column-convex polygon is directed if and only if its coding word does not use the letter $b$. We obtain a linear description of the prefixes of these words by deleting all terms of the form $\mathscr{L}_i b$ in the description (3.3). The class $\mathscr{L}_5$ becomes irrelevant. Solving the associated system of linear equations gives the area g.f. of directed column-convex polygons:

$$DC(q) = \frac{q(1-q)}{1-3q+q^2}.$$

As far as we know, this result was first published by Klarner [38].

### 3.2.3.3 Polygons Confined to a Strip

Constraining polyominoes or polygons to lie in a strip of fixed height endows them with a linear structure. This observation gives a handle to attack difficult problems, like the enumeration of general self-avoiding polygons (SAP), self-avoiding walks, or polyominoes [1, 5, 48, 55, 56]. As the size of the strip increases, the approximation of the confined problem to the general one becomes better and better. This widely applied principle gives, for instance, lower bounds on growth constants that are difficult to determine. We illustrate it here with the perimeter enumeration of SAP confined to a strip.

Before we describe this calculation, let us mention a closely related idea, which consists of considering anisotropic models (for instance, SAP counted by vertical and horizontal perimeters), and fixing the number of atoms lying in one direction, for instance the number of horizontal edges. Again, this endows the constrained objects with a linear structure. The denominators of the rational generating functions that count them often factor in terms $(1 - y^i)$. The number of exponents $i$ that occur can be seen as a measure of the complexity of the class. This is often observed only at an experimental level, and is further discussed in Chapter 4. However, this observation has been pushed in some cases to a proof that the corresponding generating function is not *D-finite* and in particular not algebraic (see for instance [49], and Chapter 5).

But let us return to SAP in a strip of height $k$ (a *k-strip*). A first observation is that a polygon is completely determined by the position of its horizontal edges. Consider the intersection of the polygon with a vertical line lying at a half-integer abscissa (a *cut*): the strip constraint implies that only finitely many configurations (or *states*) can occur. The number of such states is the number of even subsets of $\{0, 1, \ldots, k\}$. This implies that SAP in a strip can be encoded by a word over a finite alphabet. For instance, the polygon of Fig. 3.7 is encoded by the word $\tilde{b}\tilde{b}baabaaba$.

**Fig. 3.7** A self-avoiding polygon in a strip of height 2, encoded over a 3-letter alphabet.

It is not hard to see that for all $k$, the set of words encoding SAP confined to a strip of height $k$ has a linear structure. To make this structure clearer, we refine our encoding: for every vertical cut, we not only keep track of its intersection with the polygon, but also of the way the horizontal edges that meet the cut are connected to the left of the cut. This does not change the size of the alphabet for $k = 2$, as there is a unique way of coupling two edges. However, if $k = 3$, the configuration where 4 edges are met by the cut gives rise to 2 states, depending on how these 4 edges are connected (Fig. 3.8). The number of states is now the number of non-crossing couplings on $\{0, 1, \ldots, k\}$. This is also the size of our encoding alphabet $A$.

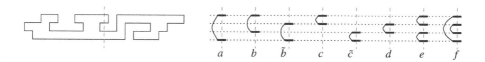

**Fig. 3.8** A self-avoiding polygon in a strip of height 3, encoded by the word $dba f \tilde{c}\tilde{c}eaa f \tilde{c}aceeabcc$.

Fix $k$, and let $\mathscr{S}$ be the set of words encoding SAP confined to a $k$-strip. The set $\mathscr{L}$ of *prefixes* of words of $\mathscr{S}$ describes incomplete SAP, and has a simple linear structure: for every such prefix $w$, the set of letters $a$ such that $wa$ lies in $\mathscr{L}$ only depends on the last letter of $w$. In other words, these prefixes are Markovian with memory 1. For every letter $a$ in the encoding alphabet, we denote by $\mathscr{L}_a$ the set of prefixes ending with the letter $a$. The linear structure can be encoded by a graph, from which the equations defining the sets $\mathscr{L}_a$ can automatically be written. This graph is shown in Fig. 3.9 (left) for $k = 2$. Every path in this graph starting from the initial vertex 0 corresponds to a word of $\mathscr{L}$, obtained by reading vertex labels. The linear structure of prefixes reads:

$$\mathscr{L}_a = (\varepsilon + \mathscr{L}_a + \mathscr{L}_b + \mathscr{L}_{\tilde{b}})a, \quad \mathscr{L}_b = (\varepsilon + \mathscr{L}_a + \mathscr{L}_b)b, \quad \mathscr{L}_{\tilde{b}} = (\varepsilon + \mathscr{L}_a + \mathscr{L}_{\tilde{b}})\tilde{b}.$$

From this we derive linear equations for incomplete SAP, where every horizontal edge is counted by $\sqrt{x}$, and every vertical edge by $\sqrt{y} = z$:

$$L_a = (z^2 + L_a + zL_b + zL_{\tilde{b}})x, \quad L_b = (z + zL_a + L_b)x, \quad L_{\tilde{b}} = (z + zL_a + L_{\tilde{b}})x.$$

These equations keep track of how many edges are added when a new letter is appended to a word of $\mathscr{L}$. They can be schematized by a weighted graph (Fig. 3.9, middle). Now the (multiplicative) weight of a path starting at 0 is the weight of the corresponding incomplete polygon. Finally, the completed polygons are obtained by adding vertical edges to the right of incomplete polygons. This gives the generating function of SAP in a strip of height 2 as:

$$S_2(x,y) = z^2 L_a + zL_b + zL_{\tilde{b}}.$$

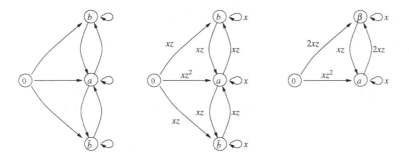

**Fig. 3.9** The linear structure of SAP in a 2-strip.

Clearly, we should exploit the horizontal symmetry of the model to obtain a smaller set of equations. The letters $b$ and $\tilde{b}$ playing symmetric roles, we replace them in the graph of Fig. 3.9 by a unique vertex $\beta$, such that the generating function of paths ending at $\beta$ is the sum of the g.f.s of paths ending at $b$ and $\tilde{b}$ in the first version of the graph (Fig. 3.9, right). Introducing the series $L_\beta = L_b + L_{\tilde{b}}$, we have thus replaced the previous system of four equations by

$$L_a = x(z^2 + L_a + zL_\beta), \quad L_\beta = x(2z + 2zL_a + L_\beta), \quad S_2(x,y) = z^2 L_a + zL_\beta,$$

from which we obtain

$$S_2(x,y) = \frac{xy(2-2x+y+3xy)}{(1-x)^2 - 2x^2y}.$$

Note that this series counts polygons of height 1 twice, so that we should subtract $S_1(x,y) = xy/(1-x)$ to obtain the g.f. of SAP of height at most 2, defined up to translation.

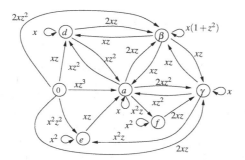

**Fig. 3.10** The linear structure of SAP in a 3-strip.

For $k = 3$, the original alphabet, shown in Fig. 3.8, has 8 letters, but two pairs of them play symmetric roles. After merging the vertices $b$ and $\tilde{b}$ on the one hand, $c$ and $\tilde{c}$ on the other, the condensed graph, with its $x, z$ weights, is shown in Fig. 3.10. The corresponding equations read

$$L_a = x\left(z^3 + L_a + zL_\beta + z^2L_\gamma + z^2L_d + zL_e\right),$$
$$L_\beta = x\left(2z^2 + 2zL_a + (1+z^2)L_\beta + zL_\gamma + 2zL_d\right),$$
$$L_\gamma = x\left(2z + 2z^2L_a + zL_\beta + L_\gamma + 2zL_f\right),$$
$$L_d = x\left(z + z^2L_a + zL_\beta + L_d\right),$$
$$L_e = x^2\left(z^2 + zL_\gamma + L_e\right),$$
$$L_f = x^2\left(zL_a + L_f\right),$$

and the generating function of completed polygons is

$$S_3(x,y) = z^3L_a + z^2L_\beta + zL_\gamma + zL_d + z^2L_f = \frac{xyN(x,y)}{D(x,y)}$$

where

$$N(x,y) = 3(x+1)^2(1-x)^5 + (5x+2)(2x-1)(x+1)^2(x-1)^3y$$
$$- (x-1)\left(6x^6 + 4x^5 - 18x^4 - 6x^3 + 11x^2 + 8x + 1\right)y^2$$
$$- x(x+1)\left(2x^4 + 6x^3 - 8x^2 + 4x + 1\right)y^3$$

and

$$D(x,y) = (x+1)^2 (x-1)^6 - x(1+4x)(x+1)^2 (x-1)^4 y$$
$$+ x^2 \left(3x^4 + 4x^3 - 6x^2 - 8x - 3\right)(x-1)^2 y^2$$
$$+ x^3 (x+1)\left(x^3 + 3x^2 - 5x + 3\right)y^3.$$

By setting $x = y = t$, we obtain the half-perimeter generating function of SAP in a 3-strip,

$$S_3(t) = \frac{t^2 \left(-8t^9 + 4t^8 + 10t^7 - 20t^6 - t^5 - t^4 + 7t^3 + 3t^2 - 7t + 3\right)}{4t^{10} - 2t^9 - 5t^8 + 8t^7 - t^6 + 2t^5 - 4t^4 + 2t^3 + 3t^2 - 4t + 1}$$

and, by looking at the smallest pole of this series, we also obtain the (very weak) lower bound $1.68\ldots$ on the growth constant of square lattice self-avoiding polygons.

The above method has been automated by Zeilberger [55]. It is not hard to see that the number of states required to count polygons in a $k$-strip grows like $3^k$, up to a power of $k$. This prevents one from applying this method for large values of $k$. Better bounds for growth constants may be obtained via the *finite lattice method* described in Chapter 7, and implemented in Chapter 10. A further improvement is obtained by looking at a cylinder rather than a strip [5].

### 3.2.4 q-Analogues

By looking at the height of the rightmost column of Ferrers diagrams, we have described a linear construction of these polygons that proves the rationality of their perimeter g.f. (Fig. 3.3). Let us examine what happens when we try to keep track of the area in this construction.

They key point is that the area *increases by the width of the polygon* when we duplicate the bottom row. (In contrast, the half-perimeter simply increases by 1 during this operation.) This observation gives the following functional equation for the complete g.f. of Ferrers diagrams:

$$F(x,y,q) = xyq + xqF(x,y,q) + yF(xq,y,q).$$

This is a *q-analogue* of the equation defining $F(x,y,1)$, derived from (3.2). This equation is no longer linear, but it can be solved easily by iteration:

$$F(x,y,q) = \frac{xyq}{1-xq} + \frac{y}{1-xq}F(xq,y,q)$$
$$= \frac{xyq}{1-xq} + \frac{y}{1-xq}\frac{xyq^2}{1-xq^2} + \frac{y}{1-xq}\frac{y}{1-xq^2}F(xq^2,y,q) \qquad (3.4)$$
$$= \sum_{n\geq 1}\frac{xy^n q^n}{(xq)_n}$$

with

$$(xq)_0 = 1 \quad \text{and} \quad (xq)_n = (1-xq)(1-xq^2)\cdots(1-xq^n).$$

Similarly, for the stack polygons of Fig. 3.4, one obtains:

$$\begin{aligned}
S(x,y,q) &= xyq + xqS(x,y,q) + S_+(x,y,q),\\
S_+(x,y,q) &= yS(xq,y,q) + xqS_+(x,y,q).
\end{aligned}$$

Eliminating the series $S_+$ gives

$$\begin{aligned}
S(x,y,q) &= \frac{xyq}{1-xq} + \frac{y}{(1-xq)^2}S(xq,y,q)\\
&= \sum_{n\geq 1}\frac{xy^n q^n}{(xq)_{n-1}(xq)_n}.
\end{aligned}$$

In Section 3.4 we present a systematic approach for counting classes of column-convex polygons by perimeter and area.


## 3.3 Algebraic Models and Algebraic Series

### 3.3.1 A Basic Example: Bargraphs Counted by Perimeter

Let us return to bargraphs. The linear description used in Section 3.2.1 to count them by area cannot be directly recycled to count them by perimeter: indeed, when we add a cell at the top of the last column, how do we know if we increase the perimeter, or not? Instead, we are going to scan the polygon from left to right, and *factor* it into two smaller bargraphs as soon as we meet a column of height 1 (if any). If there is no such column, deleting the bottom row of the polygon leaves another bargraph. This description is schematized in Fig 3.11.

**Fig. 3.11** A second recursive construction of bargraphs.

Let $\mathscr{B}$ be the set of words over the alphabet $\{n,s,e\}$ that naturally encode the top boundary of bargraphs, from the SW to the SE corner. Fig. 3.11 translates into the following recursive description, where the unions are disjoint:

$$\mathscr{B} = n\mathscr{L}s \quad \text{with} \quad \mathscr{L} = n\mathscr{L}s \cup \{e\} \cup e\mathscr{L} \cup n\mathscr{L}se \cup n\mathscr{L}se\mathscr{L}. \tag{3.5}$$

This implies that the anisotropic perimeter g.f. of bargraphs satisfies

$$\begin{cases} B(x,y) = yL(x,y), \\ L(x,y) = yL(x,y) + x + xL(x,y) + xyL(x,y) + xyL(x,y)^2. \end{cases}$$

These equations are readily solved and yield:

$$B(x,y) = \frac{1 - x - y - xy - \sqrt{(1-y)((1-x)^2 - y(1+x)^2)}}{2x}. \tag{3.6}$$

Thus the perimeter g.f. of bargraphs is algebraic, and its algebraicity is explained combinatorially by the recursive description of Fig. 3.11.

Note that one can directly translate this description into an algebraic equation satisfied by $B(x,y)$, without using the language $\mathscr{B}$. This language is largely a convenient tool to highlight the *algebraic structure* of bargraphs. The translation of Fig. 3.11 into an equation proceeds as follows: there are two types of bargraphs, those that have at least one column of height 1, and the others, which we call *thick* bargraphs. Thick bargraphs are obtained by duplicating the bottom row of a general bargraph, and are thus counted by $yB(x,y)$. Among bargraphs having a column of height 1, we find the single cell bargraph (g.f. $xy$), and then those of width at least 2. The latter class can be split into 3 disjoint classes:

- the first column has height 1. These bargraphs are obtained by adding a cell to the left of any general bargraph, and are thus counted by $xB(x,y)$,
- the last column is the only column of height 1. These bargraphs are obtained by adding a cell to the right of a thick bargraph, and are thus counted by $xyB(x,y)$,
- the first column of height 1 is neither the first column, nor the last column. Such bargraphs are obtained by concatenating a thick bargraph, a cell, and a general bargraph; they are counted by $xB(x,y)^2$.

This discussion directly results in the equation

$$B(x,y) = yB(x,y) + xy + xB(x,y) + xyB(x,y) + xB(x,y)^2. \tag{3.7}$$

### 3.3.2 Algebraic Objects

The above description of bargraphs involved two constructions:

1. taking disjoint unions of sets,
2. taking Cartesian products of sets.

For two classes $\mathscr{A}_1$ and $\mathscr{A}_2$, the element $(a_1, a_2)$ of the product $\mathscr{A}_1 \times \mathscr{A}_2$ is seen as the concatenation of the objects $a_1$ and $a_2$. We will say that a class of objects is *algebraic* if it admits a non-ambiguous recursive description based on disjoint unions and Cartesian products. It is assumed that the size of the objects is *additive* for the concatenation. For instance, (3.5) gives an algebraic description of the words of $\mathscr{L}$ and $\mathscr{B}$.

In the case of linear constructions, the only concatenations that were allowed were between one object and a single atom. As we can now concatenate two objects, algebraic constructions generalize linear constructions. In terms of g.f.s, concatenating objects of two classes means taking the product of the corresponding g.f.s. Hence the g.f. of an algebraic class will always be the first component of the solution of a polynomial system of the form:

$$A_i = P_i(t, A_1, \ldots, A_k) \quad \text{for} \quad 1 \le i \le k,$$

where $P_i$ is a polynomial with coefficients in $\mathbb{N}$. Such series are called $\mathbb{N}$-*algebraic*, and are closely related to the theory of *context-free languages*. We refer to [50] for details on these languages, and to [12] for a discussion of $\mathbb{N}$-algebraic series in enumeration.

### 3.3.3 More Algebraic Models

In this section we present three problems that can be solved via an algebraic decomposition: staircase polygons, then column-convex polygons counted by perimeter (and the subclass of directed column-convex polygons), and finally directed polyominoes counted by area.

#### 3.3.3.1 Staircase Polygons by Perimeter

In Section 3.1.3 we defined staircase polygons through their directed and convexity properties. See Fig. 3.2(b) for an example. We describe here a recursive construction of these polygons, illustrated in Fig. 3.12. It is analogous to the construction of bargraphs described at the end of Section 3.3.1 and illustrated in Fig. 3.11. Denote by $S(x, y)$ the anisotropic perimeter generating function of staircase polygons.

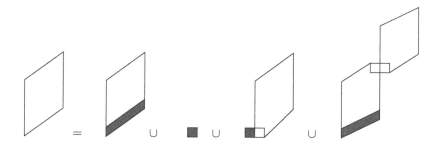

**Fig. 3.12** A recursive construction of staircase polygons.

We say that a staircase polygon is *thick* if deleting the bottom cell of each column gives a staircase polygon of the same width. These thick polygons are obtained by duplicating the bottom cell in each column of a staircase polygon, so that their generating function is $yS(x,y)$.

Among non-thick staircase polygons, we find the single cell polygon (g.f. $xy$), and then those of width at least 2. Let $P$ be in the latter class, and denote its columns $C_1,\ldots,C_k$, from left to right. The fact that $P$ is not thick means that there exist two consecutive columns, $C_i$ and $C_{i+1}$, that overlap by one edge only. Let $i$ be minimal for this property. Two cases occur:

- the first column has height 1. In particular, $i = 1$. These polygons are obtained by adding a cell to the bottom left of any general staircase polygon, and are thus counted by $xS(x,y)$.
- otherwise, the columns $C_1,\ldots,C_i$ form a thick staircase polygon, and $C_{i+1},\ldots,C_k$ form a general staircase polygon. Concatenating these two polygons in such a way that they share only one edge gives the original polygon $P$. Hence the g.f. for this case is $S(x,y)^2$.

This discussion gives the equation

$$S(x,y) = yS(x,y) + xy + xS(x,y) + S(x,y)^2$$

so that

$$S(x,y) = \frac{1}{2}\left(1 - x - y - \sqrt{1 - 2x - 2y - 2xy + x^2 + y^2}\right)$$

$$= \sum_{p,q \geq 1} \frac{1}{p+q-1}\binom{p+q-1}{p}\binom{p+q-1}{q}x^p y^q.$$

This expansion can be obtained using the Lagrange inversion formula [9]. The isotropic semi-perimeter g.f. is obtained by setting $t = x = y$:

$$S(t,t) = \frac{1}{2}\left(1 - 2t - \sqrt{1 - 4t}\right) = \sum_{n \geq 1} C_n t^{n+1}$$

where $C_n = \binom{2n}{n}/(n+1)$ is the $n^{\text{th}}$ Catalan number. The same approach can be applied to more general classes of convex polygons, like directed convex polygons and general convex polygons. See for instance [9, 23].

### 3.3.3.2 Column-Convex Polygons by Perimeter

We now apply a similar treatment to the perimeter enumeration of column-convex polygons (cc-polygons for short). Their area g.f. was found in Section 3.2.3. Let $\mathscr{C}$ denote the set of these polygons, and $C(x,y)$ their anisotropic perimeter generating function. Our recursive construction requires us to introduce two additional classes of polygons. The first one, $\mathscr{C}_1$, is the set of cc-polygons in which one cell of the last column is marked. The corresponding g.f. is denoted $C_1(x,y)$. Note that, by symmetry, this series also counts cc-polygons where one cell of the *first* column is marked. Then, $\mathscr{C}_2$ denotes the set of cc-polygons in which one cell of the first column is marked (say, with a dot), and one cell of the last column is marked as well (say, with a cross). The corresponding g.f. is denoted $C_2(x,y)$. Our recursive construction of the polygons of $\mathscr{C}$ is illustrated in Fig. 3.13.

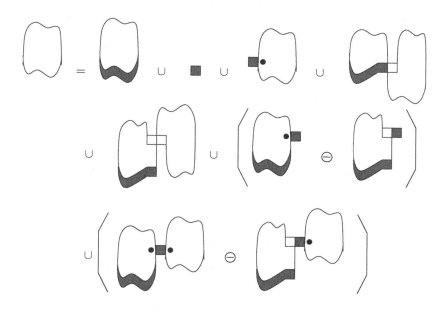

**Fig. 3.13** A recursive construction of column-convex polygons.

We say that a cc-polygon is *thick* if deleting the bottom cell of each column gives a cc-polygon of the same width. These thick polygons are obtained by duplicating the bottom cell in each column of a cc-polygon, so that their generating function is $yC(x,y)$.

Among non-thick cc-polygons, we find the single cell polygon (g.f. $xy$), and then those of width at least 2. Let $P$ be in the latter class, and denote its columns $C_1,\ldots,C_k$, from left to right. The fact that $P$ is not thick means that there exist two consecutive columns $C_i$ and $C_{i+1}$ that overlap by one edge only. Let $i$ be minimal for this property. Two cases occur:

– the first column has height 1. In particular, $i = 1$. These polygons are obtained by adding a cell to the left of any cc-polygon having a marked cell in its first column, next to the marked cell. They are thus counted by $xC_1(x,y)$.
– otherwise, the columns $C_1,\ldots,C_i$ form a thick cc-polygon $P_1$, and the columns $C_{i+1},\ldots,C_k$ form a general cc-polygon $P_2$. There are several ways of concatenating these two polygons in such a way they share only one edge:

- either the shared edge is at the bottom of $C_i$ and at the top of $C_{i+1}$. Such polygons are counted by $C(x,y)^2$.
- or the shared edge is at the top of $C_i$ and at the bottom of $C_{i+1}$. Such polygons are also counted by $C(x,y)^2$.
- if $C_{i+1}$ has height at least 2, there are no other possibilities. However, if $C_{i+1}$ consists of one cell only, this cell may be adjacent to *any* cell of $C_i$, not only to the bottom or top ones. The case where $C_{i+1}$ is the last column of $P$ is counted by $xy(C_1(x,y) - C(x,y))$. The case where $i+1 < k$ is counted by $x(C_1(x,y)^2 - C(x,y)C_1(x,y))$.

Let us drop the variables $x$ and $y$ in the series $C$, $C_1$ and $C_2$. The above discussion gives the equation:

$$C = yC + xy + xC_1 + 2C^2 + xy(C_1 - C) + x(C_1^2 - CC_1).$$

The construction of Fig. 3.13 can now be recycled to obtain an equation for the series $C_1$, counting cc-polygons with a marked cell in the last column. Note that the first case of the figure (thick polygons) gives rise to two terms, depending on whether the marked cell is one of the duplicated cells, or not:

$$C_1 = y(C + C_1) + xy + xC_2 + 2CC_1 + xy(C_1 - C) + x(C_1C_2 - CC_2).$$

We need a third equation, as three series (namely $C$, $C_1$ and $C_2$) are now involved. There are two ways to obtain a third equation:

- either we interpret $C_1$ as the g.f. of cc-polygons where one cell is marked in the *first* column. The construction of Fig. 3.13 gives:

$$C_1 = y(C + C_1) + xy + xC_1 + 2(C + C_1)C + xy((C_1 - C) + (C_2 - C_1))$$
$$+ x((C_1^2 - CC_1) + (C_2C_1 - C_1^2)).$$

  Note that now many cases give rise to two terms in the equation.
- or we work out an equation for $C_2$ using the decomposition of Fig. 3.13. Again, many of the cases schematized in this figure give rise to several terms. In particular, the first case (thick polygons) gives rise to 4 terms:

$$C_2 = y(C + 2C_1 + C_2) + xy + xC_2 + 2(C + C_1)C_1 + xy((C_1 - C) + (C_2 - C_1))$$
$$+ x((C_1C_2 - CC_2) + (C_2^2 - C_1C_2)).$$

Both strategies of course give the same equation for $C \equiv C(x,y)$, after the elimination of $C_1$ and $C_2$:

$$\left(-5xy - 18 + 2xy^2 - 18y^2 + 36y + 2x\right)C^4$$
$$+ (y-1)\left(5xy^2 - 21y^2 + 42y - 14xy + 5x - 21\right)C^3$$
$$+ 2(y-1)^2\left(-4y^2 + 2xy^2 + 8y - 7xy - 4 + 2x\right)C^2$$
$$+ (y-1)^3\left(xy^2 - y^2 + 2y - 6xy + x - 1\right)C - xy(y-1)^4 = 0.$$

This quartic has 4 roots, among which the g.f. of cc-polygons can be identified by checking the first few coefficients. This series turns out to be unexpectedly simple:

$$C(x,y) = (1-y)\left(1 - \frac{2\sqrt{2}}{3\sqrt{2} - \sqrt{1 + x + \sqrt{(1-x)^2 - 16\frac{xy}{(1-y)^2}}}}\right).$$

Feretić has provided direct combinatorial explanations for this formula [28, 29]. The algebraic equation satisfied by $C(t,t)$ was first[1] obtained (via a context-free language) in [22]. The method we have used is detailed in [26].

### 3.3.3.3 Directed Column-Convex Polygons by Perimeter

It is not hard to restrict the construction of Fig. 3.13 to *directed* cc-polygons. This is illustrated in Fig. 3.14. Note that the case where the columns $C_i$ and $C_{i+1}$ share the bottom edge of $C_i$ (the fourth case in Fig. 3.13) is only possible if $C_{i+1}$ has height 1. Moreover, only one additional series is needed, namely that of directed cc-polygons marked in the last column $(D_1)$.

One obtains the following equations:

$$D = yD + xy + xD + xD^2 + xyD + D^2 + xy(D_1 - D) + x(D_1 - D)D,$$
$$D_1 = y(D + D_1) + xy + xD_1 + xDD_1 + xyD + DD_1 + xy(D_1 - D) + x(D_1 - D)D_1.$$

Eliminating $D_1$ gives a cubic equation for the series $D \equiv D(x,y)$:

$$D^3 + 2(y-1)D^2 + (y-1)(x+y-1)D + xy(y-1) = 0.$$

This equation was first obtained in [21]. The first few terms of the semi-perimeter generating function are

$$D(t,t) = t^2 + 2t^3 + 6t^4 + 20t^5 + 71t^6 + 263t^7 + 1005t^8 + 3933t^9 + \cdots$$

---

[1] Eq. (32) in [22] has an error: the coefficient of $t^5 c^3$ in $p_2$ should be $-40$ instead of $+40$.

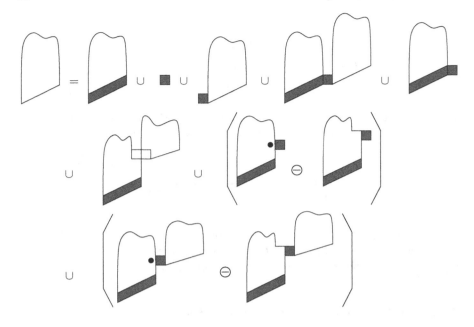

**Fig. 3.14** A recursive construction of directed column-convex polygons.

### 3.3.3.4 Directed Polyominoes by Area

Let us move to a class that admits a neat, but non-obvious, algebraic structure: directed polyominoes counted by area. This structure was discovered when Viennot developed the theory of *heaps* [52]. Intuitively, a heap is obtained by dropping vertically some solid pieces, one after the other. Thus, a piece lies either on the "floor" (when it is said to be *minimal*), or at least partially covers another piece.

Directed polyominoes *are*, in essence, heaps. To see this, replace every cell of the polyomino by a *dimer*, after a 45 degree rotation (Fig. 3.15). This gives a heap with a unique minimal piece. Such heaps are called *pyramids*. If the columns to the left of the minimal piece contain no dimer, we say we have a *half-pyramid* (Fig. 3.15, right).

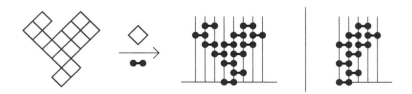

**Fig. 3.15** *Left*: A directed polyomino and the associated pyramid. *Right*: a half-pyramid.

The interest in heaps lies in the existence of a *product* of heaps: The product of two heaps is obtained by putting one heap above the other and dropping its pieces. Conversely, one can factor a heap by pushing upwards one or several pieces. See an example in Fig. 3.16. This product is the key in our algebraic description of directed polyominoes, or, equivalently, of pyramids of dimers, as we now explain.

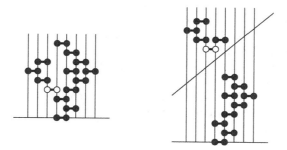

**Fig. 3.16** A factorization of a pyramid into a pyramid and a half-pyramid. Observe that the highest dimer of the pyramid moves up as we lift the white dimer.

A pyramid is either a half-pyramid, or the product of a half-pyramid and a pyramid (Fig. 3.17, top). Let $D(q)$ denote the g.f. of pyramids counted by the number of dimers, and $H(q)$ denote the g.f. of half-pyramids. Then $D(q) = H(q)(1 + D(q))$.

Now, a half-pyramid can be a single dimer. If it has several dimers, it is the product of a single dimer and of one or two half-pyramids (Fig. 3.17, bottom), which implies $H(q) = q + qH(q) + qH^2(q)$. Note that $D(q)$ is also the area g.f. of directed polyominoes. A straightforward computation gives:

$$D(q) = \frac{1}{2}\left(\sqrt{\frac{1+q}{1-3q}} - 1\right) \tag{3.8}$$

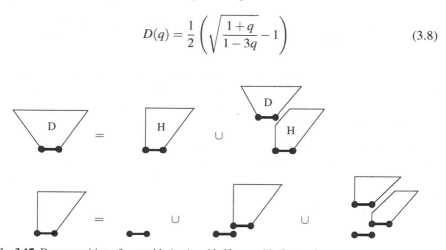

**Fig. 3.17** Decomposition of pyramids (top) and half-pyramids (bottom).

This was first proved by Dhar [24]. The above proof is adapted from [7].

### 3.3.4 q-Analogues

By looking for the first column of height 1 in a bargraph, we have described an algebraic construction of these polygons (Fig. 3.11) that proves that their perimeter g.f. is algebraic (Section 3.3.1). Let us now examine what happens when we try to keep track of the area of these polygons.

As in Section 3.2.4, the key observation is that the area behaves additively when one concatenates two bargraphs, but *increases by the width of the polygon* when we duplicate the bottom row. (In contrast, the half-perimeter simply increases by 1 during this operation.) This observation gives rise to the following functional equation for the complete g.f. of bargraphs:

$$B(x,y,q) = yB(xq,y,q) + xyq + xqB(x,y,q) + xyqB(xq,y,q)$$
$$+ xqB(xq,y,q)B(x,y,q). \quad (3.9)$$

This is a $q$-analogue of Equation (3.7) defining $B(x,y,1)$. This equation is no longer algebraic, and it is not clear how to solve it. It has been shown in [44] that it can be linearized and solved using a certain Ansatz. We will show in Section 3.4.1 a more systematic way to obtain $B(x,y,q)$, which does not require any Ansatz.

## 3.4 Adding a New Layer: a Versatile Approach

In this section we describe a systematic construction that can be used to find the complete g.f. of many classes of polygons having a convexity property [10]. The cost of this higher generalization is twofold:

- it is not always clear how to solve the functional equations obtained in this way,
- in contrast with the constructions developed in Sections 3.2 and 3.3, this approach does not provide combinatorial explanations for the rationality/algebraicity of the corresponding g.f.s.

This type of construction is sometimes called *Temperley's approach* since Temperley used it to write functional equations for the generating function of column-convex polygons counted by perimeter [51]. But it also occurs, in a more complicated form, in other "old" papers [6, 40]. We would prefer to see a more precise terminology, like *layered approach*.

### 3.4.1 A Basic Example: Bargraphs by Perimeter and Area

We return to our favourite example of bargraphs, and we now aim to find the complete g.f. $B(x,y,q)$ of this class of polygons. We have just seen that the algebraic description of Fig. 3.11 leads to the $q$-algebraic equation (3.9), which is not obvious to solve. The linear description of Section 3.2.1 cannot be directly exploited either: in order to decide whether the addition of a cell at the top of the last column increases the perimeter or not, we need to know which of the last two columns is higher.

We present here a variation of this linear construction that allows us to count bargraphs by area and perimeter, *provided we also take into account the right height* by a new variable $s$. By *right height*, we mean the height of the rightmost column. The g.f. we are interested in is now

$$B(x,y,q,s) = \sum_{h \geq 1} B_h(x,y,q)s^h,$$

where $B_h(x,y,q)$ is the complete g.f. of bargraphs of right height $h$.

**Fig. 3.18** A third recursive construction of bargraphs.

Our new construction is illustrated in Fig. 3.18. The class $\mathscr{B}$ of bargraphs is split into three disjoint subsets:

1. bargraphs of width 1 (columns). The g.f. of this class is $xysq/(1 - ysq)$,
2. bargraphs in which the last column is at least as high as the next-to-last column. These bargraphs are obtained by duplicating the last column of a bargraph (which boils down to replacing $s$ by $sq$ in the series $B(x,y,q,s)$), and adding a (possibly empty) column at the top of the newly created column. The corresponding g.f. is thus

$$\frac{x}{1 - ysq}B(x,y,q,sq).$$

3. bargraphs in which the last column is lower than the next-to-last column. To obtain these, we start from a bargraph, say of right height $h$, and add a new column of height $\ell < h$ to the right. The g.f. of this third class is:

$$x\sum_{h \geq 1}\left(B_h(x,y,q)\sum_{\ell=1}^{h-1}(sq)^\ell\right) = x\sum_{h \geq 1}\left(B_h(x,y,q)\frac{sq - (sq)^h}{1 - sq}\right)$$

$$= x\frac{sqB(x,y,q,1) - B(x,y,q,sq)}{1 - sq}. \qquad (3.10)$$

Writing $B(s) \equiv B(x, y, q, s)$, and putting together the three cases, we obtain:

$$B(s) = \frac{xysq}{1 - ysq} + \frac{xsq}{1 - sq}B(1) + \frac{xsq(y - 1)}{(1 - sq)(1 - ysq)}B(sq). \qquad (3.11)$$

This equation is solved in two steps: first, an iteration, similar to what we did for Ferrers diagrams in (3.4) (Section 3.2.4), provides an expression for $B(s)$ in terms of $B(1)$:

$$B(s) = \sum_{n \geq 1} \frac{(xs(y - 1))^{n-1}q^{\binom{n}{2}}}{(sq)_{n-1}(ysq)_{n-1}} \left( \frac{xysq^n}{1 - ysq^n} + \frac{xsq^n}{1 - sq^n}B(1) \right).$$

Then, one sets $s = 1$ to obtain the complete g.f. $B(1) \equiv B(x, y, q, 1)$ of bargraphs:

$$B(x, y, q, 1) = \frac{I_+}{1 - I_-} \qquad (3.12)$$

with

$$I_+ = \sum_{n \geq 1} \frac{x^n(y - 1)^{n-1}q^{\binom{n+1}{2}}}{(q)_{n-1}(yq)_n} \quad \text{and} \quad I_- = \sum_{n \geq 1} \frac{x^n(y - 1)^{n-1}q^{\binom{n+1}{2}}}{(q)_n(yq)_{n-1}}.$$

### 3.4.2 More Examples

In this section, we describe how to apply the layered approach to two other classes of polygons: staircase and column-convex polygons, counted by perimeters and area simultaneously. In passing we show how the *difference* of g.f.s resulting from a geometric summation like (3.10) can be explained combinatorially by an inclusion-exclusion argument.

#### 3.4.2.1 Staircase polygons

As with the bargraph example above, we define an extended generating function which tracks the height of the rightmost column of the staircase polygon,

$$S(x, y, q, s) = \sum_{h \geq 1} S_h(x, y, q)s^h,$$

where $S_h(x, y, q)$ is the generating function of staircase polygons with right height $h$. The set of all staircase polygons can be partitioned into two parts (Fig. 3.19):

1. those which have only one column. The g.f. for this class is $xyqs/(1 - yqs)$.
2. those which have more than one column. Their g.f. is obtained as the difference of the g.f. of two sets as follows.

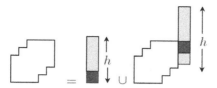

**Fig. 3.19** The two types of staircase polygons.

Staircase polygons of width $\ell \geq 2$ can be split into two objects: a staircase polygon formed of the $\ell - 1$ first columns, and the rightmost column. The left part has generating function $S(x,y,q,1)$ (ignoring the rightmost height), to which we then attach a column of cells. The attached column is constrained in that it must not extend below the bottom of the rightmost column of the left part. It is generated (see Fig. 3.19) by gluing a descending column (with g.f. $1/(1-qs)$) and an ascending column (with g.f. $1/(1-yqs)$) to a single square (with g.f. $xyqs$). The single square is required to ensure that the column is not empty and is glued to the immediate right of the topmost square of the left part. An important observation is that only the ascending column contributes in the increase of the vertical perimeter. This gives the generating function

$$S(x,y,q,1) \cdot xyqs \cdot \frac{1}{1-qs} \cdot \frac{1}{1-yqs}.$$

This construction however results in configurations which might have the rightmost column extending below the rightmost column of the left part. We must thus subtract the contribution of these "bad" configurations from the above g.f. We claim that they are generated by

$$S(x,y,q,sq) \cdot xyqs \cdot \frac{1}{1-qs} \cdot \frac{1}{1-yqs}.$$

The replacement of $s$ with $sq$ in $S(x,y,q,sq)$ is interpreted as adding a copy of the last column of the left part, as illustrated in Fig. 3.20. The $xyqs$ factor is interpreted as attaching a new cell to the bottom of the duplicated column (thus ensuring the rightmost column is strictly below the rightmost column of the left part). Finally, we add a descending and an ascending column. Again, the height of the latter must not be taken into account in the vertical perimeter.

Thus the final equation for the generating function is

$$S(x,y,q,s) = \frac{xyqs}{1-yqs} + \left(S(x,y,q,1) - S(x,y,q,sq)\right)\frac{xyqs}{(1-qs)(1-yqs)}.$$

It can also be obtained via geometric sums, as was done for (3.10). The equation is solved with the same two step process as for bargraphs. First we iterate it to obtain $S(x,y,q,s)$ in terms of $S(x,y,q,1)$, and then we set $s$ to 1, obtaining

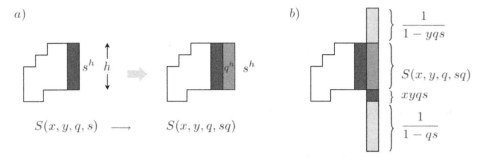

**Fig. 3.20** a) Replacing $s$ by $sq$ in $S(x,y,q,s)$ duplicates the last column of the polygon. b) Generating function of "bad" configurations.

$$S(x,y,q,1) = y\frac{J_1}{J_0},$$

where $J_0$ and $J_1$ are two $q$-Bessel functions [34]:

$$J_1(x,y,q) = \sum_{k\geq 1}(-1)^{k+1}\frac{x^k q^{\binom{k+1}{2}}}{(q)_{k-1}(yq)_k}$$

and

$$J_0(x,y,q) = \sum_{k\geq 0}(-1)^k\frac{x^k q^{\binom{k+1}{2}}}{(q)_k(yq)_k}.$$

Note, the appropriate limit as $q \to 1$ leads to standard Bessel functions which are related to the generating function for semi-continuous staircase polygons—see [18] for details.

### 3.4.2.2 Column-Convex Polygons

The case of column-convex polygons is more complex and we will not give all the details but discuss only the primary additional complication. We refer to [10] for a complete solution. As in the case of staircase polygons, a functional equation for column-convex polygons can be obtained by considering the rightmost (last) column. The position of the last column compared with the second last column must be carefully considered. Again there are several cases depending on whether the top (resp. bottom) of the last column is strictly above, at the same level or below the top (resp. bottom) of the second-last column. The case that leads to a type of term that does not appear in the equation for staircase polygons is the case where the top (resp. bottom) of the last column cannot be above (resp. below) the top (resp. bottom) of the second-last column. Thus we will only explain this case which we will refer to as the *contained* case, as the last column is somehow contained in the previous one.

If the generating function for the column-convex polygons is $C(s) = C(x,y,q,s)$ then we claim that the polygons falling into the contained case are counted by the generating function

$$\frac{xsq}{1-sq}\frac{\partial C}{\partial s}(1) - \frac{xs^2q^2}{(1-sq)^2}\big(C(1)-C(sq)\big). \tag{3.13}$$

Thus we see we now need a derivative of the generating function. As a polygon of right height $h$ contributes $h$ times to the series $\partial C/\partial s(1)$, this series counts polygons with a marked cell in the rightmost column.

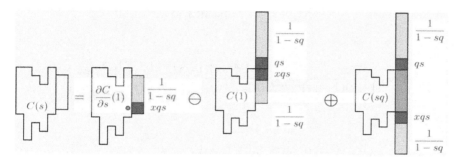

**Fig. 3.21** A schematic representation of the equation for the case where the rightmost column does not extend above or below the second-last column.

Let us now explain this expression, which is illustrated in Fig. 3.21. We consider a polygon as the concatenation of a left part with a new (rightmost) column $C$. In the left part, we mark the cell of the rightmost column that is at the same level as the bottom cell of $C$. So, starting from a marked polygon, we first add a single square to the right of the marked cell—this gives a factor $xsq$. Above this square we then add an ascending column which is generated by $1/(1-sq)$. However, as with the staircase polygons, the resulting series counts "bad" configurations, where the last column ends strictly higher than the second last column. We subtract the contribution of these bad configurations by generating them as shown on the second picture of Fig. 3.21. This results in subtracting the term $xq^2s^2C(1)/(1-sq)^2$. However, we have now subtracted too much! Indeed, some configurations counted by the latter series have a rightmost column that ends below the second last column. We correct this by adding the contribution of these configurations, which is $xq^2s^2C(sq)/(1-sq)^2$ (Fig. 3.21, right). This establishes (3.13) for the g.f. of the contained case.

The other cases are simpler, and in the same vein as what was needed for staircase polygons. Considering all cases gives

$$C(s) = \frac{xsyq}{1-syq} + \frac{xsq}{1-sq}\frac{\partial C}{\partial s}(1) + \frac{xs^2q^2(2y-syq-1)}{(1-sq)^2(1-syq)}C(1)$$

$$+ \frac{xs^2q^2(1-y)^2}{(1-sq)^2(1-syq)^2}C(sq). \quad (3.14)$$

In order to solve this equation, we first iterate it to obtain $C(s)$ in terms of $C(1)$ and $C'(1) = \partial C/\partial s(1)$. Setting $s = 1$ gives a linear equation between $C(1)$ and $C'(1)$. Setting $s = 1$ *after having differentiated with respect to s* gives a second linear equation between $C(1)$ and $C'(1)$. We end up solving a linear system of size 2, and obtain $C(1)$ as a ratio of two $2 \times 2$ determinants. The products of series that appear in these determinants can be simplified, and the final expression reads

$$C(x,y,q,1) = y\frac{(1-y)X}{1+W+yX}$$

where

$$X = \frac{xq}{(1-y)(1-yq)} + \sum_{n\geq 2} \frac{(-1)^{n+1}x^n(1-y)^{2n-4}q^{\binom{n+1}{2}}(y^2q)_{2n-2}}{(q)_{n-1}\,(yq)_{n-2}\,(yq)_{n-1}^2\,(yq)_n\,(y^2q)_{n-1}}$$

and

$$W = \sum_{n\geq 1} \frac{(-1)^n x^n(1-y)^{2n-3}q^{\binom{n+1}{2}}(y^2q)_{2n-1}}{(q)_n\,(yq)_{n-1}^3\,(yq)_n\,(y^2q)_{n-1}}.$$

The first solution, involving a more complicated expression, was given in [17]. The one above appears as Theorem 4.8 in [10].

### 3.4.3 The Kernel Method

In Sections 3.2 and 3.3, we have explained combinatorially why the area g.f. of bargraphs, $B(1,1,q)$, and the perimeter g.f. of bargraphs, $B(x,y,1)$, are respectively rational and algebraic. It is natural to examine whether these properties can be recovered from the construction of Fig. 3.18 and the functional equation (3.11).

As soon as we set $y = 1$ in this equation, the main difficulty, that is, the term $B(sq)$, disappears. We can then substitute 1 for $s$ and solve for $B(x,1,q,1)$, the width and area g.f. of bargraphs. This series is found to be

$$B(x,1,q,1) = \frac{xq}{1-q-xq}.$$

From this, one also obtains a rational expression for the series $B(x,1,q,s)$. The rationality of $B(x,1,q,1)$ also follows directly from the expression (3.12): setting $y = 1$ shrinks the series $I_+$ and $I_-$ to simple rational functions.

How the *perimeter* g.f. of bargraphs can be derived from the functional equation (3.11) is a more challenging question. Setting $q = 1$ gives

$$B(s) = \frac{xys}{1-ys} + \frac{xs}{1-s}B(1) + \frac{xs(y-1)}{(1-s)(1-ys)}B(s). \tag{3.15}$$

This equation cannot be simply solved by setting $s = 1$. Instead, the solution uses the so-called *kernel method*, which has proved useful in a rather large variety of enumerative problems in the past 10 years [3, 4, 14, 19, 32, 47]. This method solves, in a systematic way, equations of the form:

$$K(s,x)A(s,x) = P(x,s,A_1(x),\ldots,A_k(x))$$

where $K(s,x)$ is a polynomial in $s$ and the other indeterminates $x = (x_1,\ldots,x_n)$, $P$ is a polynomial, $A(s,x)$ is an unknown series in $s$ and the $x_i$'s, while the series $A_i(x)$ only depend on the $x_i$'s. (It is assumed that the equation uniquely defines all these unknown series.) We refer to [14] for a general presentation, and simply illustrate the method on (3.15). We group the terms involving $B(s)$, and multiply the equation by $(1-s)$ to obtain:

$$\left(1 - s - \frac{xs(y-1)}{1-ys}\right)B(s) = \frac{xys(1-s)}{1-ys} + xsB(1). \tag{3.16}$$

Let $S \equiv S(x,y)$ be the only formal power series in $x$ and $y$ that satisfies

$$S = 1 - \frac{xS(y-1)}{1-yS}.$$

That is,

$$S = \frac{1-x+y+xy-\sqrt{(1-y)((1-x)^2-y(1+x)^2)}}{2y}.$$

Replacing $s$ by $S$ in (3.16) gives an identity between series in $x$ and $y$. By construction, the left-hand side of this identity vanishes. This gives

$$B(1) \equiv B(x,y,1,1) = \frac{y(S-1)}{1-yS}$$

$$= \frac{1-x-y-xy-\sqrt{(1-y)((1-x)^2-y(1+x)^2)}}{2x},$$

and we have recovered the algebraic expression (3.6) of the perimeter g.f. of bargraphs.

## 3.5 Some Open Questions

We conclude this chapter with a list of open questions. As mentioned in the introduction, the combination of convexity and direction conditions gives rise to 35 classes of polyominoes, not all solved. But all these classes are certainly not equally interesting. The few problems we present below have two important qualities: they do not seem completely out of reach (we do not ask about the enumeration of all polyominoes) and they have some special interest: they deal either with large classes of polyominoes, or with mysterious classes (that have been solved in a non-combinatorial fashion), or they seem to lie just at the border of what the available techniques can achieve at the moment.

### 3.5.1 The Quasi-Largest Class of Quasi-Solved Polyominoes

Let us recall that the growth constant of $n$-cell polyominoes is conjectured to be a bit more than 4. More precisely, it is believed that $p_n$, the number of such polyominoes, is equivalent to $\mu^n n^{-1}$, up to a multiplicative constant, with $\mu = 4.06\ldots$ [37]. The techniques that provide lower bounds on $\mu$ involve looking at *bounded* polyominoes (for instance polyominoes lying in a strip of fixed height $k$) and a concatenation argument. See [5] for a recent survey and the best published lower bound, $3.98\ldots$. It is not hard to see that for $k$ fixed, these bounded polyominoes have a linear structure, and a rational generating function. This series is obtained either by adding recursively a whole "layer" to the polyomino (as we did for self-avoiding polygons in Section 3.2.3.3), or by adding one cell at a time. The latter approach is usually more efficient (Chapter 6).

What about solved classes of polyominoes that do not depend on a parameter $k$, and often have a more subtle structure? We have seen in Section 3.3.3 that the g.f. of directed polyominoes is algebraic, with growth constant 3. This is "beaten" by the growth constant $3.20\ldots$ derived from the rational g.f. of column-convex polyominoes (Section 3.2.3). A generalization of directed polyominoes (called multidirected polyominoes) was introduced in [15] and proved to have a fairly complicated g.f., with growth constant about 3.58. To our knowledge, this is the largest growth constant reached from exact enumeration (again, apart from the rational classes obtained by bounding column heights).

However, in 1967, Klarner introduced a "large" class of polyominoes that seems interesting and would warrant a better understanding [39]. His definition is a bit unclear, and his solution is only partial, but the estimate he obtains of the growth constant is definitely appealing: about 3.72. Let us mention that the triangular lattice version of this mysterious class is solved in [15]. The growth constant is found to be about 4.58 (the growth constant of triangular lattice animals is estimated to be about 5.18, see [53]).

### 3.5.2 Partially Directed Polyominoes

This is another generalization of directed polyominoes, with a very natural definition: the corresponding animal $A$ contains a source point $v_0$ from which every other point can be reached by a path formed of North, East and West steps, only visiting points of $A$ (Fig. 3.22(a)). This model has a slight flavour of *heaps of pieces*, a notion that has already proved useful in the solution of several polyomino models (see Section 3.3.3 and [7, 16, 15]). The growth constant is estimated to be around 3.6, and, if proved, would thus improve that of multi-directed animals [45].

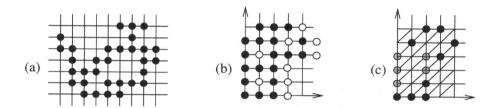

**Fig. 3.22** (a) A partially directed animal. The source can be any point on the bottom row. (b) A directed animal on the square lattice, with the right neighbours indicated in white. (c) A directed animal $A$ on the triangular lattice. The distinguished points are those having (only) their South neighbour in $A$.

### 3.5.3 The Right Site-Perimeter of Directed Animals

We wrote in the introduction that *almost all* solved classes of polyominoes can be solved by one of the three main approaches we present in this chapter. Here is one simple-looking result that we do not know how to prove via these approaches (and not via any combinatorial approach, to be honest).

Take a directed animal $A$, and call a *neighbour* of $A$ any point that does not lie in $A$, but could be added to $A$ to form a new directed animal. The number of neighbours is the *site-perimeter* of $A$. The *right site-perimeter* of $A$ is the number of neighbours that lie one step to the right of a point of $A$. It was proved in [11] that the g.f. of directed animals, counted by area and right site-perimeter, is a very simple extension of (3.8):

$$D(q,x) = \frac{x}{2}\left(\sqrt{\frac{(1+q)(1+q-qx)}{1-q(2+x)+q^2(1-x)}} - 1\right).$$

The proof is based on an equivalence with a one-dimensional gas model, inspired by [25]. It is easy to see that the right site-perimeter is also the number of vertices $v$

of $A$ whose West neighbour is not in $A$. (By the West neighbour, we mean the point at coordinates $(i-1,j)$ if $v = (i,j)$).

Described in these terms, this result has a remarkable counterpart for triangular lattice animals (Fig. 3.22). Let us say that a point $(i,j)$ of the animal has a West (resp. South, South-West) neighbour in $A$ if the point $(i-1,j)$ (resp. $(i,j-1)$, $(i-1,j-1)$) is also in $A$. Then the g.f. that counts these animals by the area and the number of points having a SW-neighbour (but no W- or S-neighbour) is easy to obtain using heaps of dimers and the ideas presented in Section 3.3.3:

$$\tilde{D}(q,x) = \frac{1}{2} \left( \sqrt{\frac{1+q-qx}{1-3q-qx}} \right).$$

What is less easy, and is so far only proved via a correspondence with a gas model, is that $\tilde{D}(q,x)$ also counts directed animals (on the triangular lattice) by the area and the number of points having a South neighbour (but no SW- or W-neighbour). Any combinatorial proof of this result would give a better understanding of these objects. One possible starting point may be found in the recent paper [41], which sheds some combinatorial light on the gas models involved in the proof of the above identities.

### 3.5.4 Diagonally-Convex Polyominoes

Let us conclude with a problem that seems to lie at the border of the applicability of the third approach presented here (the layered approach). In the enumeration of, say, column-convex polyominoes (Section 3.4.2), we have used the fact that deleting the last column of such a polyomino gives another column-convex polyomino. This is no longer true of a $d_-$-convex polyomino from which we would delete the last diagonal (Fig. 3.1(d)). Still, it seems that this class is sufficiently well structured to be exactly enumerable. Note that this difficulty vanishes when studying the restricted class of *directed* diagonally-convex polyominoes [8, 46], which behave approximately like column-convex polyominoes.

## References

1. S. E. Alm and S. Janson. Random self-avoiding walks on one-dimensional lattices. *Comm. Statist. Stochastic Models*, 6(2):169–212, 1990.
2. G. E. Andrews. *The theory of partitions*. Addison-Wesley Publishing Co., Reading, Mass.-London-Amsterdam, 1976. Encyclopedia of Mathematics and its Applications, Vol. 2.
3. C. Banderier, M. Bousquet-Mélou, A. Denise, P. Flajolet, D. Gardy, and D. Gouyou-Beauchamps. Generating functions for generating trees. *Discrete Math.*, 246(1-3):29–55, 2002.
4. C. Banderier and P. Flajolet. Basic analytic combinatorics of directed lattice paths. *Theoret. Comput. Sci.*, 281(1-2):37–80, 2002.

5. G. Barequet, M. Moffie, A. Ribó, and G. Rote. Counting polyominoes on twisted cylinders. *Integers*, 6:A22, 37 pp. (electronic), 2006.
6. E. A. Bender. Convex *n*-ominoes. *Discrete Math.*, 8:219–226, 1974.
7. J. Bétréma and J.-G. Penaud. Modèles avec particules dures, animaux dirigés et séries en variables partiellement commutatives. ArXiv:math.CO/0106210.
8. M. Bousquet-Mélou. Rapport scientifique d'habilitation. Report 1154-96, LaBRI, Université Bordeaux 1, http://www.labri.fr/perso/lepine/Rapports_internes.
9. M. Bousquet-Mélou. Codage des polyominos convexes et équations pour l'énumération suivant l'aire. *Discrete Appl. Math.*, 48(1):21–43, 1994.
10. M. Bousquet-Mélou. A method for the enumeration of various classes of column-convex polygons. *Discrete Math.*, 154(1-3):1–25, 1996.
11. M. Bousquet-Mélou. New enumerative results on two-dimensional directed animals. *Discrete Math.*, 180(1-3):73–106, 1998.
12. M. Bousquet-Mélou. Rational and algebraic series in combinatorial enumeration. In *Proceedings of the International Congress of Mathematicians*, pages 789–826, Madrid, 2006. European Mathematical Society Publishing House.
13. M. Bousquet-Mélou and A. J. Guttmann. Enumeration of three-dimensional convex polygons. *Ann. Comb.*, 1(1):27–53, 1997.
14. M. Bousquet-Mélou and M. Petkovšek. Linear recurrences with constant coefficients: the multivariate case. *Discrete Math.*, 225(1-3):51–75, 2000.
15. M. Bousquet-Mélou and A. Rechnitzer. Lattice animals and heaps of dimers. *Discrete Math.*, 258(1-3):235–274, 2002.
16. M. Bousquet-Mélou and X. G. Viennot. Empilements de segments et *q*-énumération de polyominos convexes dirigés. *J. Combin. Theory Ser. A*, 60(2):196–224, 1992.
17. R. Brak and A. J. Guttmann. Exact solution of the staircase and row-convex polygon perimeter and area generating function. *J. Phys A: Math. Gen.*, 23(20):4581–4588, 1990.
18. R. Brak, A. L. Owczarek, and T. Prellberg. Exact scaling behavior of partially convex vesicles. *J. Stat. Phys.*, 76(5/6):1101–1128, 1994.
19. A. de Mier and M. Noy. A solution to the tennis ball problem. *Theoret. Comput. Sci.*, 346(2-3):254–264, 2005.
20. A. Del Lungo, M. Mirolli, R. Pinzani, and S. Rinaldi. A bijection for directed-convex polyominoes. In *Discrete models: Combinatorics, Computation, and Geometry (Paris, 2001)*, Discrete Math. Theor. Comput. Sci. Proc., pages 133–144 (electronic). Maison Inform. Math. Discr., Paris, 2001.
21. M. Delest and S. Dulucq. Enumeration of directed column-convex animals with given perimeter and area. *Croatica Chemica Acta*, 66(1):59–80, 1993.
22. M.-P. Delest. Generating functions for column-convex polyominoes. *J. Combin. Theory Ser. A*, 48(1):12–31, 1988.
23. M.-P. Delest and G. Viennot. Algebraic languages and polyominoes enumeration. *Theoret. Comput. Sci.*, 34(1-2):169–206, 1984.
24. D. Dhar. Equivalence of the two-dimensional directed-site animal problem to Baxter's hard square lattice gas model. *Phys. Rev. Lett.*, 49:959-962, 1982.
25. D. Dhar. Exact solution of a directed-site animals-enumeration problem in three dimensions. *Phys. Rev. Lett.*, 51(10):853–856, 1983.
26. E. Duchi and S. Rinaldi. An object grammar for column-convex polyominoes. *Ann. Comb.*, 8(1):27–36, 2004.
27. I. G. Enting and A. J. Guttmann. On the area of square lattice polygons. *J. Statist. Phys.*, 58(3-4):475–484, 1990.
28. S. Feretić. The column-convex polyominoes perimeter generating function for everybody. *Croatica Chemica Acta*, 69(3):741–756, 1996.
29. S. Feretić. A new way of counting the column-convex polyominoes by perimeter. *Discrete Math.*, 180(1-3):173–184, 1998.
30. S. Feretić. An alternative method for *q*-counting directed column-convex polyominoes. *Discrete Math.*, 210(1-3):55–70, 2000.

31. S. Feretić. A $q$-enumeration of convex polyominoes by the festoon approach. *Theoret. Comput. Sci.*, 319(1-3):333–356, 2004.

32. S. Feretić and D. Svrtan. On the number of column-convex polyominoes with given perimeter and number of columns. In Barlotti, Delest, and Pinzani, editors, *Proceedings of the 5th Conference on Formal Power Series and Algebraic Combinatorics (Florence, Italy)*, pages 201–214, 1993.

33. P. Flajolet and R. Sedgewick. *Analytic Combinatorics*. Preliminary version available at http://pauillac.inria.fr/algo/flajolet/Publications/books.html.

34. G. Gasper and M. Rahman. *Basic hypergeometric series*, volume 35 of *Encyclopedia of Mathematics and its Applications*. Cambridge University Press, Cambridge, 1990.

35. A. J. Guttmann and T. Prellberg. Staircase polygons, elliptic integrals, Heun functions and lattice Green functions. *Phys. Rev. E*, 47:R2233–R2236, 1993.

36. I. Jensen and A. J. Guttmann. Self-avoiding polygons on the square lattice. *J. Phys. A*, 32(26):4867–4876, 1999.

37. I. Jensen and A. J. Guttmann. Statistics of lattice animals (polyominoes) and polygons. *J. Phys. A*, 33(29):L257–L263, 2000.

38. D. A. Klarner. Some results concerning polyominoes. *Fibonacci Quart.*, 3:9–20, 1965.

39. D. A. Klarner. Cell growth problems. *Canad. J. Math.*, 19:851–863, 1967.

40. D. A. Klarner and R. L. Rivest. Asymptotic bounds for the number of convex $n$-ominoes. *Discrete Math.*, 8:31–40, 1974.

41. Y. Le Borgne and J.-F. Marckert. Directed animals and gas models revisited. *Electron. J. Combin.*, R71, 2007.

42. N. Madras and G. Slade. *The self-avoiding walk*. Probability and its Applications. Birkhäuser Boston Inc., Boston, MA, 1993.

43. G. Pólya. On the number of certain lattice polygons. *J. Combinatorial Theory*, 6:102–105, 1969.

44. T. Prellberg and R. Brak. Critical exponents from nonlinear functional equations for partially directed cluster models. *J. Stat. Phys.*, 78(3/4):701–730, 1995.

45. V. Privman and M. Barma. Radii of gyration of fully and partially directed animals. *Z. Phys. B: Cond. Mat.*, 57:59–63, 1984.

46. V. Privman and N. M. Švrakić. Exact generating function for fully directed compact lattice animals. *Phys. Rev. Lett.*, 60(12):1107–1109, 1988.

47. H. Prodinger. The kernel method: a collection of examples. *Sém. Lothar. Combin.*, 50:Art. B50f, 19 pp. (electronic), 2003/04.

48. R. C. Read. Contributions to the cell growth problem. *Canad. J. Math.*, 14:1–20, 1962.

49. A. Rechnitzer. Haruspicy 2: the anisotropic generating function of self-avoiding polygons is not D-finite. *J. Combin. Theory Ser. A*, 113(3):520–546, 2006.

50. A. Salomaa and M. Soittola. *Automata-theoretic aspects of formal power series*. Springer-Verlag, New York, 1978. Texts and Monographs in Computer Science.

51. H. N. V. Temperley. Combinatorial problems suggested by the statistical mechanics of domains and of rubber-like molecules. *Phys. Rev. (2)*, 103:1–16, 1956.

52. G. X. Viennot. Heaps of pieces. I. Basic definitions and combinatorial lemmas. In *Combinatoire énumérative (Montréal, 1985)*, volume 1234 of *Lecture Notes in Math.*, pages 321–350. Springer, Berlin, 1986.

53. M. Vöge and A. J. Guttmann. On the number of hexagonal polyominoes. *Theoret. Comput. Sci*, 307(2):433–453, 2003.

54. T. Yuba and M. Hoshi. Binary search networks: a new method for key searching. *Inform. Process. Lett.*, 24:59–65, 1987.

55. D. Zeilberger. Symbol-crunching with the transfer-matrix method in order to count skinny physical creatures. *Integers*, pages A9, 34pp. (electronic), 2000.

56. D. Zeilberger. The umbral transfer-matrix method. III. Counting animals. *New York J. Math.*, 7:223–231 (electronic), 2001.

# Chapter 4
# Why Are So Many Problems Unsolved?

Anthony J Guttmann

## 4.1 Introduction

The problems discussed in this book, particularly that of counting the number of polygons and polyominoes in two dimensions, either by perimeter or area, seems so simple to state that it seems surprising that they haven't been exactly solved. The counting problem is so simple in concept that it can be fully explained to any schoolchild, yet it seems impossible to solve. In this chapter we develop what is essentially a numerical method that provides, at worst, strong evidence that a problem has no solution within a large class of functions, including algebraic, differentiably finite (D-finite) [27, 26] and at least a sub-class [7] of differentiably algebraic functions, called *constructible differentiably algebraic (CDA)* functions. Since many of the special functions of mathematical physics—in terms of which most known solutions are given—are differentiably finite, this exclusion renders the problem unsolvable within this class. Throughout this chapter the term *D-unsolvable* means that the problem has no solution within the class of D-finite functions as well as the sub-class of differentiably algebraic functions described above. In the next chapter, Rechnitzer shows how these ideas may be refined into a proof, in the case of polygons in two dimensions.

In fact, the exclusion is wider than D-finite functions, as we show that the solutions possess a natural boundary on the unit circle in an appropriately defined complex plane. This excludes not only D-finite functions, but a number of others as well—though we have no simple way to describe this excluded class.

We'll first give some definitions. Let $\mathbb{K}$ be a field with characteristic zero. A series $f(z) \in \mathbb{K}[[z]]$ is said to be *differentiably finite* if there exists an integer $k$ and polynomials $P_0(z), \cdots, P_k(z)$ with coefficients in $\mathbb{K}$ such that $P_k(z)$ is not the null polynomial and

Anthony J Guttmann
Department of Mathematics and Statistics, The University of Melbourne, Victoria, Australia, e-mail: tonyg@ms.unimelb.edu.au

$$P_0(z)f(z) + P_1(z)f'(z) + \cdots + P_k(z)f^{(k)}(z) = 0.$$

A series $f(z) \in \mathbb{K}[[z]]$ is said to be *differentiably algebraic* if there exists an integer $k$ and a polynomial $P$ in $k + 2$ variables with coefficients in $\mathbb{K}$, such that

$$P(z, f(z), f'(z), \cdots, f^{(k)}(z)) = 0.$$

A series $f(z) \in \mathbb{K}[[z]]$ is said to be *constructible differentiably algebraic* if there exists both series $f_1(z), f_2(z), \cdots, f_k(z)$ with $f = f_1$, and polynomials $P_1, P_2, \cdots, P_k$ in $k$ variables, with coefficients in $\mathbb{K}$, such that

$$
\begin{aligned}
f_1' &= P_1(f_1, f_2, \cdots, f_k), \\
f_2' &= P_2(f_1, f_2, \cdots, f_k), \\
&\cdots \\
f_k' &= P_k(f_1, f_2, \cdots, f_k).
\end{aligned}
\tag{4.1}
$$

A simpler, but non-constructive definition is that a function is *CDA* if it belongs to some finitely generated ring which is closed under differentiation [7]. Differentiably finite functions in several variables are discussed in [26].

A consequence of these definitions is that if a series in $z$, $f = \sum_n a_n(x)z^n$ with coefficients in the field $\mathbb{K} = \mathbb{C}(x)$ is algebraic, D-finite or *CDA*, then the poles of $a_n(x)$ lying on the unit circle cannot become dense on this circle as $n$ increases. This is because the poles must lie in a finite set, independent of $n$, which in turn is a consequence of the recurrence relations on $a_n(x)$ that follow from the above definitions. We make extensive use of this observation in the remainder of the chapter.

Note that algebraic, D-finite and *CDA* functions are all subsets of differentiably algebraic functions, and of course algebraic functions are both D-finite and *CDA*. However, D-finite functions are not necessarily *CDA*. For example the function $(e^t - 1)/t$ is not *CDA* as it fails to satisfy the Eisenstein criterion [7] though it is D-finite. Other functions, such as $1/\cos t$ are *CDA* but not D-finite.

The method which we shall describe and which can, in favourable circumstances, be sharpened into a formal proof, as in the next chapter, has been applied to a wide variety of problems in both statistical mechanics and combinatorics. Typically, the solution of the problem will require the calculation of the graph generating function in terms of some parameter, such as perimeter, area, number of bonds or sites. A key first step is to *anisotropise* the generating function. For example, if counting graphs, such as polygons, by the number of bonds on, say, an underlying square lattice, one distinguishes between horizontal and vertical bonds. In this way, one can construct a two-variable generating function, $G(x, y) = \sum_{m,n} g_{m,n} x^m y^n$ where $g_{m,n}$ denotes the number of graphs with $m$ horizontal and $n$ vertical bonds. Summing over one of the variables, we may write

$$G(x, y) = \sum_{m,n} g_{m,n} x^m y^n = \sum_n H_n(x) y^n \tag{4.2}$$

where $H_n(x)$ is the generating function for the relevant graphs with $n$ vertical bonds. It can be proved [28] that these generating functions are rational, with denominators given as products of cyclotomic polynomials (see the next Chapter).

In some cases one finds only a small finite number (typically one or two) of denominator zeros on the unit circle. Loosely speaking, this is the hallmark of a solvable problem. This remark is very loose. It does not mean that the solution is necessarily D-finite, though that is often the case, but rather that such problems seem to be solvable (we know of no counter-example, even though some of the solutions involve $q$-generalisations of standard functions). If, as is often observed, the denominator zeros become dense on the unit circle as $n$ increases, so that in the limit a natural boundary is formed, then this is the hallmark of a D-unsolvable problem.

The significance of this observation is substantial. It is observed in these cases that, as $n$ increases, the denominators of the rational functions $H_n(x)$ contain zeros given by steadily higher roots of unity. Hence the structure of the functions $H_n(x)$ is that of a rational function whose poles all lie on the unit circle in the complex $x$-plane, such that the poles become dense on the unit circle as $n$ gets large. This behaviour of the functions $H_n(x)$ implies that $G(x,y)$ (a) has a natural boundary (b) as a formal power series in $y$ with coefficients in the field $\mathbb{K} = \mathbb{C}(x)$ is neither algebraic nor D-finite, nor *CDA*. Further, provided that $G(x,c)$ is well-defined for a given complex value $c$, then, in the absence of miraculous cancellations, it follows that $G(x,c)$ also is neither D-finite nor *CDA*.

Of course, we are primarily interested in the solution of the *isotropic* case, when $x = y$, and it is clear that the anisotropic case can behave quite differently from the isotropic case. This is most easily seen by construction. Consider the function

$$f(x,y) = f_1(x,y) + (x-y)f_2(x,y), \tag{4.3}$$

where $f_1(x,y)$ is D-finite and $f_2(x,y)$ is not. Clearly, the function $f(x,y)$ is not D-finite, while $f(x,x)$ is D-finite. However, in all the cases we have studied where the solutions are known, the effect of anisotropisation *does not* change the analytic structure of the solution. Rather, it simply moves singularities around in the complex plane, at most causing the bifurcation of a real singularity into a complex pair. This can readily be seen from equation (4.4), given below, for the magnetisation of the square-lattice anisotropic Ising model [9].

$$M(x,y) = \left[ 1 - \frac{16xy}{(1-x)^2(1-y)^2} \right]^{1/8}. \tag{4.4}$$

Replacing $y$ by $\lambda x$ and varying $\lambda$ merely causes the singularities to move smoothly, and indeed initially linearly, with $\lambda$ in the complex plane. Further, for unsolved problems, numerical procedures indicate that similar behaviour prevails. Nevertheless, this remains an observation, rather than an established fact, and needs to be explicitly established for each new problem.

That being said, not all problems with a small number of denominator zeros have been solved, while some D-unsolvable problems have been solved. In the former case however we believe that it is only a matter of time before a solution is found for these problems, while in the latter case the solutions have usually been expressed in terms of modular functions or $q$-generalisations of the standard functions, which are of course not D-finite. As examples consider first the hard hexagon model [3]. Baxter's original solution was expressed in terms of a natural, but non-physical parameter $x$, with $-1 < x < 1$. In terms of this parameter, the following product form was derived for the order-parameter $R$ :

$$R(x) = \prod_{n=1}^{\infty} \frac{(1-x^n)(1-x^{5n})}{(1-x^{3n})^2}. \tag{4.5}$$

Subsequently Joyce [25] showed that, when expressed in terms of another product form that defined the reciprocal activity $z'$, $R(z')$ satisfied an *algebraic* equation of degree 4 in $\mathbb{R}^3$.

An example with a different flavour is provided by the generating function for the number of staircase polygons given in terms of the area $(q)$, horizontal semi-perimeter $(x)$ and vertical semi-perimeter $(y)$, equivalent up to a translation. It is [6]

$$G(x,y,q) = y\frac{J_1}{J_0} \text{ where} \tag{4.6}$$

$$J_1(x,y,q) = \sum_{n\geq 1} \frac{(-1)^{n-1}x^n q^{\binom{n+1}{2}}}{(q)_{n-1}(yq)_n} \text{ and} \tag{4.7}$$

$$J_0(x,y,q) = \sum_{n\geq 0} \frac{(-1)^n x^n q^{\binom{n+1}{2}}}{(q)_n(yq)_n}, \tag{4.8}$$

where $(a)_n = \prod_{i=0}^{n-1}(1 - aq^i)$.

In this case, it is clear that if we look at $G(x,1,q)$ in the complex $q$-plane with $x$ held fixed, the solution possesses a natural boundary on the unit circle.

The procedure which we have outlined in the discussion around eqn. (2) is a particularly useful first step in the study of such problems. One anisotropises, generates enough terms in the generating function to be able to construct the first few functions $H_n$, then studies the denominator pattern. If it appears that the zeros are becoming dense on the unit circle, one has good reason to suspect that the problem is D-unsolvable. If on the other hand there are only one or two zeros, one is in an excellent position to seek the solution in terms of the D-finite or *CDA* functions of mathematical physics—many of which are defined in [1]. In some cases one may be able to *prove* that the observed denominator pattern persists. In that case, one has proved the observed results.

The construction of the functions $H_n$ deserves some comment. At very low order this can often be done exactly, by combinatorial arguments based on the allowed graphs. Beyond this, our method is to generate the coefficients in the expansion, *assume* it is rational, then by essentially constructing the Padé approximant one conjectures the solution. Typically, one might generate 50–100 terms in the expansion and find a rational function with numerator and denominator of perhaps degree 5 or 10. Thus the first 10 or 20 terms of the series are used to identify the rational function, the remainder are used to confirm it. Hence while this is not a derivation that proves that this is the required rational function, the chance of it not being so is extraordinarily small.

It should be said explicitly that this technique is computationally demanding. That is to say, the generation of sufficient terms in the generating function is usually quite difficult. Only with improved algorithms—most notably the combination of the finite lattice method [13, 16] with a transfer matrix formulation, as discussed in Chapter 7 —and computers with large physical memory that are needed for the efficient implementation of such algorithms, has it been possible to obtain expansions of the required length in a reasonable time. The technique is still far from routine, with each problem requiring a significant calculational effort. .

An additional, and exceptionally valuable feature of the method comes when the numerical work, described above, is combined with certain functional relations that the anisotropised generating functions must satisfy. In the language of statistical mechanics, these key functional relations are called *inversion relations* and imply a connection between the generating function and its analytic continuation, usually involving the reciprocal of one or more of the expansion variable(s). As we show below, the existence of these inversion relations, coupled with any obvious symmetries (usually a symmetry with respect to the interchange of $x$ and $y$), coupled with the *observed* behaviour of the functions $H_n$ (described above) can yield an *implicit* solution to the underlying problem with no further calculation. An example of this is the solution [2] of the zero-field free energy of the two-dimensional Ising model. A more detailed discussion of this aspect, and its extension to problems in combinatorics, is given in [5].

In the remainder of this chapter, we describe the method in some detail in a few cases, then go on to apply it to a range of problems in statistical mechanics and combinatorics. We also take the first steps in extending the inversion relation idea from its natural home in statistical mechanics to the arena of combinatorics—where it sits less naturally due to the absence of an underlying Hamiltonian, the symmetries of which give rise to the inversion relation. It is in one sense comforting to discover that, without exception, the long-standing unsolved problems of statistical mechanics that we discuss are all found to be D-unsolvable.

Other important aspects of the method, such as the connection of these ideas with concepts of integrability, and with the existence of a Yang-Baxter equation, are not explored here.

## 4.2 Staircase Polygons

The enumeration of staircase polygons by perimeter is one of the simpler combinatorial exercises, and is addressed elsewhere in Chapter 3, but is nevertheless useful pedagogically, as so many distinct methods can be demonstrated by its solution. To this long list we add the experimental approach of studying the early terms of the two variable series expansion of the perimeter generating function and *observing* a functional relation, in this case called an *inversion relation*, it expresses the generating function in terms of the generating function with the arguments inverted.

We first write the perimeter generating function as

$$P(x,y) = \frac{1 - x^2 - y^2}{2} - \frac{\sqrt{x^4 - 2x^2y^2 - 2x^2 + y^4 - 2y^2 + 1}}{2} \tag{4.9}$$

$$= \sum_{m,n} p_{m,n} x^{2m} y^{2n} = \sum_n H_n(x^2) y^{2n} \tag{4.10}$$

where $p_{m,n}$ is the number of staircase polygons with horizontal perimeter $2m$ and vertical perimeter $2n$, defined up to a translation. Then $H_n(x^2)$ is the generating function for staircase polygons with $2n$ vertical bonds.

From *observation* of the early terms, it is clear that

$$H_n(x^2) = x^2 S_n(x^2)/(1 - x^2)^{2n-1}$$

for $n > 1$, where $S_n(x^2)$ is a symmetric, unimodal polynomial with non-negative coefficients, of degree $(n - 2)$. This observed symmetry can be expressed formally as

$$x^{2n} H_n(x^2) + x^2 H_n(1/x^2) = 0, \ n > 1.$$

This in turn translates into the functional relation

$$P(x,y) + x^2 P(1/x, y/x) = -y^2.$$

There is also an obvious symmetry relation $P(x,y) = P(y,x)$, and these observations are sufficient to implicitly solve the problem by calculating the functions $H_n$ order by order in polynomial time.

Of course, this must rank as one of the least impressive ways of solving this fairly simple model. However the purpose of this example is twofold. Firstly to show that this essentially experimental method can be applied to combinatorial structures in order to discover an inversion relation. Secondly, to show that once one has such an inversion relation, then this, coupled with symmetry and the structure of the functions $H_n$ (plus certain analyticity assumptions), provides an alternative method for obtaining a solution (albeit experimentally). Once one has such a conjectured solution, it is a comparatively easy task to prove that it is correct.

Numerous other polygon problems can also be tackled similarly [5].

## 4.3 Three-Choice Polygons

The problem of three-choice polygons [12] is an interesting one, as for nearly 10 years we knew everything about this model except a closed form solution! We had a polynomial time algorithm to generate the coefficients in its series expansion—which is tantamount to a solution—and detailed knowledge of its asymptotic behaviour. Then in 2006 Guttmann and Jensen [21] found the Fuchsian ODE satisfied by the generating function. It was found to be an 8th order ODE with polynomial coefficients of degree 35 to 37, so it is not surprising that the solution was not easily found!

They are self-avoiding polygons on a square lattice, defined up to a translation, and constructed according to the following rules: After a step in the $y$ direction, one may take a step in either the same direction or in the $\pm x$ direction. However after a step in the $+x$ direction, one may only make steps $+x$ or $+y$, while after a step in the $-x$ direction, one may only make steps $-x$ or $-y$. In unpublished work, A R Conway and Rechnitzer anisotropised the model in order to see whether the methods discussed here give insight into the solution.

Let $P_3(x,y) = \sum_{m,n} a_{m,n} x^m y^n$ be the perimeter generating function, where $a_{m,n}$ gives the number of 3-choice polygons, distinct up to translation, with $2m$ horizontal bonds and $2n$ vertical bonds. Then

$$P_3(x,y) = \sum_n H_n(x) y^n,$$

where

$$H_n(x) = P_n(x)/Q_n(x)$$

is a rational function of $x$. The degree of the numerator polynomial increases like $3n$ while the denominators are observed to be

$$Q_n(x) = (1-x)^{2n-1}(1+x)^{2n-7}, \ n \text{ even},$$
$$= (1-x)^{2n-1}(1+x)^{2n-8}, \ n \text{ odd},$$

where there are no terms in $(1+x)$ for $n < 5$. It is not difficult to construct a combinatorial argument, based on the way the polygon can "grow", that is consistent with this behaviour. This argument has recently been sharpened to a proof [4]. It has also been proved [4] that the solution is D-finite, and it clearly cannot be algebraic as the asymptotic behaviour of the number of coefficients [12] includes a logarithmic term.

An inversion relation for this model can be found experimentally [5], and the solution possesses $(x,y)$ symmetry. Nevertheless, because the degree of the numerator polynomial grows like $3n$ we do not have enough constraints to implicitly solve the model, as we did for staircase polygons above.

## 4.4 Hexagonal Directed Animals

A directed site animal $\mathscr{A}$ on an acyclic lattice is defined to be a set of vertices such that all vertices $p \in \mathscr{A}$ are either the (unique) origin vertex or may be reached from the origin by a connected path, containing bonds only in the allowed lattice directions, through sites of $\mathscr{A}$.

In [14, 10] it was found that the number of such animals of perimeter $n$ grew asymptotically like $\mu^n/\sqrt{n}$, where $\mu = 4$ for the triangular lattice, and $\mu = 3$ for the square lattice. Furthermore, the generating function was given by the solution of a simple algebraic equation. For the hexagonal lattice however we [10] found similar asymptotic growth but with $\mu = 2.025131 \pm 0.000005$, and we were unable to solve for the generating function.

In order to gain more insight into this seemingly anomalous situation, the model was anisotropised [20]. Let $A_h(x,s) = \sum_{m,n} a_{m,n} x^m s^n$ be the site generating function, where $a_{m,n}$ gives the number of hexagonal lattice site animals, with $n$ sites supported [8] one particular way and $m$ sites in total. Then

$$A_h(x,s) = \sum_n H_n(x)s^n,$$

where

$$H_n(x) = P_n(x)/Q_n(x)$$

is a rational function of $x$.

For the square (and triangular) lattices, the corresponding result has been obtained exactly [8]. For the square lattice, it is

$$A_{sq}(x,s) = \frac{1}{2}\left((1 - \frac{4x}{(1+x)(1+x-sx)})^{-\frac{1}{2}} - 1\right). \tag{4.11}$$

Writing this as

$$A_{sq}(x,s) = \sum_n H_n(x)s^n, \tag{4.12}$$

expansion readily yields

$$
\begin{aligned}
H_0(x) &= x/(1-x),\\
H_1(x) &= x^2/(1-x)^3,\\
H_2(x) &= x^3(1+x+x^2)/(1-x)^5(1+x),\\
H_3(x) &= x^4(1+2x+4x^2x+2x^3+x^4)/(1-x)^7(1+x)^2,\\
H_4(x) &= x^5(1+3x+9x^2+9x^3+9x^4+3x^5+x^6)/(1-x)^9(1+x)^3,\\
H_5(x) &= x^6[1,4,16,24,36,24,16,4,1]/(1-x)^{11}(1+x)^4.
\end{aligned}
$$

Here it can be seen that the functions $H_n(x)$ have just two denominator zeros, at $x = 1$ and $x = -1$. As discussed above, this is the hallmark of a solvable model.

However for the hexagonal lattice generating function, the denominator pattern, while regular, contains terms of the form $(1 - x^k)$ where $k$ is an increasing function of $n$. In fact, the first occurrence of the factor $(1 - x^{2k})$ is in $H_k$. The first few functions $H_n(x)$ for the hexagonal lattice are:

$$H_0(x) = x/(1-x),$$
$$H_1(x) = x/(1-x)^3(1+x),$$
$$H_2(x) = x^2(1+x+x^3)/(1-x)^5(1+x)^2(1+x^2),$$
$$H_3(x) = x^3(1+x)(1+x+3x^3-x^4+x^5)/(1-x)^7(1+x)^3(1+x^2)^2,$$
$$H_4(x) = x^4[1,3,4,10,12,14,16,13,14,7,6,4,0,1]/$$
$$(1-x)^9(1+x)^4(1+x^2)^3(1-x-x^2)(1+x+x^2).$$

The enumerations in [20] are complete up to $H_9(x)$.

The degree of the numerator also increases faster than linearly, so using any inversion relation (and we don't yet have one), would not provide a solution. We conclude that this is evidence for the existence of a natural boundary in the appropriate complex plane, and hence that the solution is likely to be D-unsolvable.

This is then consistent with the seemingly anomalous value of the growth constant $\mu$.

## 4.5 Self-Avoiding Walks and Polygons

We turn now to our main topic, that of self-avoiding polygons and walks.

A study of anisotropic square lattice SAW has been reported in [11]. Writing the SAW generating function $C(x)$ in the now familiar form as

$$C(x,y) = \sum_{m,n\geq0} c_{m,n}x^m y^n = \sum_{n\geq0} H_n(x)y^n, \tag{4.13}$$

the first eleven functions, $H_0(x), \cdots, H_{10}(x)$ were found [11].

The first few are:

$$H_0(x) = (1+x)/(1-x),$$
$$H_1(x) = 2(1+x)^2/(1-x)^2,$$
$$H_2(x) = 2(1+7x+14x^2+16x^3+9x^4+3x^5)/(1-x)^3(1+x)^2 \text{ and}$$
$$H_3(x) = 2(1+10x+29x^2+44x^3+41x^4+22x^5+7x^6)/(1-x)^4(1+x)^2.$$

The first occurrence of the term $(1-x^3)$ appears in $H_5(x)$ and the term $(1+x^2)$ first appears in $H_7$. Higher order cyclotomic polynomials then systematically occur as $n$ increases. The denominator pattern appears to be predictable, though this has not been proved. The degree of the numerator is equal to the degree of the denominator in all cases observed.

Again we see the characteristic hallmark of a D-unsolvable problem. Similar behaviour is observed for SAP. We can write the SAP generating function $P(x)$ as

$$P(x,y) = \sum_{m,n\geq 1} p_{m,n} x^{2m} y^{2n} = \sum_{n\geq 1} H_n(x) y^{2n},$$

where $p_{m,n}$ is the number of square lattice polygons, equivalent up to a translation, with $2n$ horizontal steps and $2m$ vertical steps. In [15] the first nine functions, $H_1(x), \cdots, H_9(x)$, were calculated, and these were found to behave in a manner characteristic of D-unsolvable problems—that is, the zeros appear to build up on the unit circle. The first few are:

$$H_1(x) = x/(1-x),$$
$$H_2(x) = x(1+x)^2/(1-x)^3,$$
$$H_3(x) = x(1+8x+17x^2+12x^3+3x^4)/(1-x)^5,$$
$$H_4(x) = x(1+18x+98x^2+204x^3+178x^4+70x^5+11x^6)/(1-x)^7,$$
$$H_5(x) = xP_9(x)/(1-x)^9(1+x)^2,$$
$$H_6(x) = xP_{15}(x)/(1-x)^{11}(1+x)^4,$$
$$H_7(x) = xP_{20}(x)/(1-x)^{13}(1+x)^6(1+x^2+x^4).$$

In the above equations, $P_k(x)$ denotes a polynomial of degree $k$. As was the case for 3-choice polygons, a combinatorial argument can be given for the form of the denominators, and this can be found in the next chapter. In the case of 3-choice polygons there were only two roots of unity in the denominator, whereas here the degree of the roots of unity steadily increases. The occurrence of new terms in the denominator, corresponding to higher roots of unity, can be identified with the first occurrence of specific graphs. In this way the denominator pattern can be predicted (as shown in the next chapter) though rather more work is required to refine this observation into a proof.

A similar study of hexagonal lattice polygons, in unpublished work by Enting and Guttmann, led to similar conclusions. Furthermore, it was observed that the denominators of the functions $H_n$ for the square and hexagonal lattice polygons are simply related.

## 4.6 The 8-Vertex Model

As a test of the idea that "solvable" models should, when anisotropised, have functions $H_n$ with only one or two denominator zeros, Tsukahara and Inami [29] studied the 8-vertex model—which is one of the most difficult statistical mechanics models that has been exactly solved [2]. While it might be thought straightforward to expand the solution in the desired form, this turns out not to be so. According to Tsukahara and Inami [29], the exact solution in terms of elliptic parameters is very implicit, and they have been unable to obtain an expansion directly from the solution.

The model can be described as two interpenetrating planar Ising models, coupled by a four-spin coupling, two spins in each of the sub-lattices. Let the coupling in one sub-lattice be $L$, that in the other be $K$, and the four-spin coupling be $M$. Then the usual high-temperature expansion variables are $t_1 = \tanh K$, $t_2 = \tanh L$, $t_3 = \tanh M$. New high temperature variables may be defined as follows:

$$z_1 = \frac{t_1 + t_2 t_3}{1 + t_1 t_2 t_3}, \tag{4.14}$$

$$z_2 = \frac{t_2 + t_1 t_3}{1 + t_1 t_2 t_3}, \tag{4.15}$$

$$z_3 = \frac{t_3 + t_1 t_2}{t_1 + t_2 t_3}. \tag{4.16}$$

Then it has recently been shown [29] that the logarithm of the reduced partition function per face

$$\log \Lambda(z_1, z_2, z_3) = \sum_{l,m,n} a_{l,m,n} z_1^{2l} z_2^{2m} z_3^{2n}$$

satisfies

$$\log \Lambda(z_1, z_2, z_3) + \log \Lambda\left(\frac{1 - z_2^2}{z_1(1 - z_3^2)}, -z_2, -z_3\right) = \log(1 - z_2^2). \tag{4.17}$$

A summation over $l$ allows the reduced partition function to be written as

$$\log \Lambda(z_1, z_2, z_3) = \sum_{m,n} R_{m,n}(z_1^2) z_2^{2m} z_3^{2n}. \tag{4.18}$$

After a complicated graphical calculation [29], it was found that

$$R_{1,0}(z^2) = z^2/(1 - z^2), \tag{4.19}$$
$$R_{1,1}(z^2) = 2z^4/(1 - z^2)^3, \tag{4.20}$$
$$R_{2,0}(z^2) = z^2(2 - 5z^2 + z^4)/(1 - z^2) \tag{4.21}$$
$$R_{1,2}(z^2) = 3z^6(1 + z^2)/(1 - z^2)^5. \tag{4.22}$$

It is then argued [29] that the general form of the coefficients is

$$R_{m,n}(z^2) = P_{m,n}(z^2)/(1 - z^2)^{2m+2n-1}.$$

This behaviour then accords with the expected behaviour of solvable models. That is to say, there is only a finite number—in this case 1—of denominator singularities.

## 4.7 Conclusion

In this chapter we have described a powerful numerical technique capable of indicating whether a problem is likely to be readily D-solvable or not. The hallmark of unsolvability, which is the build up of zeros on the unit circle in the complex $x$-plane in the functions $H_n(x)$ of the anisotropised models, can, in favourable cases, be refined into a proof.

In the most favourable cases, where in addition an inversion relation can be obtained—as in the case of staircase polygons— and if in addition the functions $H_n$ are sufficiently simple, and with only a pole at 1 on the unit circle, an exact solution can be implicitly obtained.

The prospect of solving hitherto unsolved problems by predicting the numerator and denominator of the functions $H_n$ by a combination of combinatorial and symmetry based arguments remains open.

Other methods for conjecturing solutions from the available terms in a series expansion include the computer program *NEWGRQD* [19], the Maple package *GFUN* and its multivariate generalisation *MGFUN* [22], which all search for D-finite solutions.

The concept of a natural boundary an indicator or proof of unsolvability in some sense has been seen earlier in other areas. Flajolet [17] has shown that certain context-free languages are ambiguous because their generating function has the unit circle as a natural boundary. In a study of the ice model, which includes various models of ice and ferro-electrics [18] it was found that the parameterised solution had the entire negative real axis as a natural boundary, except for two special values of the parameter, which coincided with the two cases, KDP and IKDP, that had been solved.

Much work remains to be done in classifying precisely what class of functions is excluded by certain observed behaviour, and in developing methods to solve problems which are identifiable as D-unsolvable.

## References

1. M. Abramowitz and I. A. Stegun (Editors), *Handbook of Mathematical Functions*, (Dover, New York) (1972).
2. R. J. Baxter, in *Fundamental Problems in Statistical Mechanics*, **5**, (1981) 109-41. E. G. D. Cohen ed. (North Holl. Amsterdam).
3. R. J. Baxter, Hard hexagons: exact solution, J. Phys. A. **13**, (1980) L61-70.
4. M. Bousquet-Mélou, (Private communication).
5. M. Bousquet-Mélou, A. J. Guttmann, W. P. Orrick and A. Rechnitzer, Inversion relations, reciprocity and polyominoes, Annals of Combinatorics, **3** (1999), 225-30.
6. M. Bousquet-Mélou and X. Viennot, Empilements de segments et $q$-énumération de polyominos convexes dirigés. J. Combin. Theory Ser. A **60**, (1992) 196-224.
7. F. Bergeron and C. Reutenauer, Combinatorial resolution of systems of differential equations III: a special class of differentiably algebraic series. Europ. J. Combinatorics, **11** (1990) 501-12.

8. M. Bousquet-Mélou, New enumerative results on two-dimensional directed animals, Disc. Math **180** (1998) 73-106.

9. C. H. Chang, The spontaneous magnetisation of a two-dimensional rectangular Ising model, Phys. Rev. **88**, (1952) 1422-6.

10. A. R. Conway, R. Brak and A. J. Guttmann, Directed animals on two-dimensional lattices, J. Phys. A. **26**, (1993), 3085-91.

11. A. R. Conway and A. J. Guttmann, Square Lattice Self-Avoiding Walks and Corrections to Scaling, Phys. Rev. Letts. **77**, (1996) 5284-7.

12. A. R. Conway, A. J. Guttmann and M. P. Delest, The number of three-choice polygons, Mathl. Comput. Modelling **26** (1997) 51-8.

13. T. de Neef and I. G. Enting, Series expansions from the finite lattice method, J. Phys. A: Math. Gen. **10**, (1977) 801-5.

14. D. Dhar, M. H. Phani and M. Barma, Enumeration of directed site animals on two-dimensional lattices, J. Phys. A. **15**, (1982) L279-84.

15. I. G. Enting and A. J. Guttmann, In preparation.

16. I. G. Enting, Series Expansions from the Finite Lattice Method, Nucl. Phys. B (Proc. Suppl.) **47**, (1996) 180-7.

17. P. Flajolet, Analytic models and ambiguity of context-free languages, Theor. Comp. Sci., **49**, (1987), 283-309.

18. M. L. Glasser, D. B. Abraham and E. H. Lieb, Analytic properties of the free energy of the Ice models, J. Math. Phys, **13**, (1972), 887-900.

19. A. J. Guttmann, Asymptotic Analysis of Power Series Expansions, in *Phase Transitions and Critical Phenomena,* **13**, eds. C. Domb and J. Lebowitz, Academic Press, (1989) 1-234. (Available from www.ms.unimelb.edu.au/~tonyg)

20. A. J. Guttmann and A. R. Conway, *Statistical Physics on the Eve of the Twenty-First Century.* ed. M. T. Batchelor, (World Scientific), (1999).

21. A. J. Guttmann and I. Jensen, Fuchsian differential equation for the perimeter generating function of three-choice polygons, *Séminaire Lotharingien de combinatoire,* **54** 1-14.

22. Programs developed by B. Salvy, P. Zimmerman, F. Chyzak and colleagues at INRIA, France. Available from http://pauillac.inria.fr/algo

23. M. T. Jaekel and J. M. Maillard, Symmetry relations in exactly soluble models, J. Phys. A: Math. Gen. **15**, (1982) 1309-25.

24. M. T. Jaekel and J. M. Maillard, Inverse functional relation on the Potts model, J. Phys. A: Math. Gen. **15**, (1982) 2241-57.

25. G. S. Joyce, On the hard hexagon model and the theory of modular functions. Phil. Trans. of the Roy. Soc. Lond., **325**, (1988), 643-706.

26. L. Lipshitz, D-finite power series. J. Algebra, **122** (1989) 353-73.

27. R. P. Stanley, Differentiably finite power series, Europ. J. Comb., **1**, (1980), 175-88.

28. R. P. Stanley, Enumerative Combinatorics., **2**,

29. H. Tsukahara and T. Inami, Test of Guttmann and Enting's conjecture in the Eight Vertex Model, J. Phys. Soc. Japan **67**, (1998) 1067-1070.

# Chapter 5
# The Anisotropic Generating Function of Self-Avoiding Polygons is not D-Finite

Andrew Rechnitzer

## 5.1 Introduction

The enumeration of self-avoiding polygons, and other families of lattice animals, is one of the most famous problems in enumerative combinatorics, and despite many years of intensive study these problems remain completely open.

Let $p_n$ be the number of self-avoiding polygons on the square lattice of perimeter $2n$ and let $G(z) = \sum p_n z^n$ be the corresponding generating function. Neither an explicit nor a useful implicit expression is known for either the $G(z)$ or $p_n$. The most efficient means of computing $p_n$ is the finite-lattice method (see Chapter 7). This method is exponentially faster than brute-force enumeration (and considerably so!), but still requires exponential time and space.

One of the most fruitful approaches has been the study of simpler combinatorial models in which extra conditions such as directedness or convexity are imposed (see Chapter 3). Almost all these models when enumerated by their number of bonds, however, share the property that their generating functions are the solutions of ordinary linear differential equations with polynomial coefficients—*differentiably finite* or *D-finite* functions.

**Definition 1.** Let $F(z)$ be a formal power series in $z$ with coefficients in $\mathbb{C}$. It is said to be *differentiably finite* or *D-finite* if there exists a non-trivial differential equation:

$$P_d(z)\frac{d^d}{dz^d}F(z) + \cdots + P_1(z)\frac{d}{dz}F(z) + P_0(z)F(z) = 0, \qquad (5.1)$$

with $P_j$ a polynomial in $z$ with complex coefficients [12].

D-finite functions have many nice properties including having a finite number of singularities [22]. Additionally a knowledge of the differential equation is sufficient

Andrew Rechnitzer
Department of Mathematics, University of British Columbia, Vancouver, Canada, e-mail: andrewr@math.ubc.ca

to compute the coefficients of the generating function in linear time and also their asymptotic behaviour.

In this work we seek to show that the generating function of self-avoiding polygons is distinctly different from those of models that have been solved to date, in that it is not D-finite, and so give some explanation why the problem remains unsolved. One way of doing this would be show that it has an infinite number of singularities; while there is strong numerical data for the location of the dominant singularity (see [2, 10] for example), very little is known about subdominant singularities.

Guttmann and Enting [8] devised a numerical method for examining the singularity structure of solved and unsolved lattice models based on their *anisotropic* generating function. Their survey of these generating functions demonstrated a distinct difference between solved and unsolved bond animal problems. They observed a similar difference for thermodynamic functions of lattice models of magnets such as the Ising model free-energy and susceptibility. They proposed that this could be used as a test of "solvability"; it provides compelling evidence that the anisotropic generating functions of many unsolved problems are not D-finite. In this article we prove that this is indeed the case for self-avoiding polygons.

To form this generating function we distinguish between vertical and horizontal bonds, and so count according to the vertical and horizontal half-perimeters

$$G(x,y) = \sum_{P \in \mathcal{G}} x^{|P|_\leftrightarrow} y^{|P|_\updownarrow}, \tag{5.2}$$

where $\mathcal{G}$ is the set of all self-avoiding polygons, $|P|_\leftrightarrow$ and $|P|_\updownarrow$ respectively denote the horizontal and vertical half-perimeters of a polygon $P$. By partitioning $\mathcal{G}$ according to the vertical half-perimeter we may resum the above generating function as

$$G(x,y) = \sum_{n \geq 1} y^n \sum_{P \in \mathcal{G}_n} x^{|P|_\leftrightarrow} = \sum_{n \geq 1} H_n(x) y^n, \tag{5.3}$$

where $\mathcal{G}_n$ is the set of SAPs with $2n$ vertical bonds, and $H_n(x)$ is its horizontal half-perimeter generating function.

In some sense, the anisotropic generating function is a more manageable object than the isotropic. Splitting the set of animals, $\mathcal{G}$, into separate simpler subsets, $\mathcal{G}_n$, breaks the problem into smaller pieces, each of which is easier to study than the whole. If one seeks to compute the *isotropic* generating function then one must examine *all* possible configurations that can occur in $\mathcal{G}$. Arguably, this is the reason that the only families of bond animals that have been solved are those with severe topological restrictions (such as directedness or convexity). On the other hand, if we examine the generating function of $\mathcal{G}_n$, then the number of different shapes that can occur is always finite (this idea will be made more precise below). Similarly, instead of trying to study the properties of the whole (possibly unknown) generating function, the anisotropy separates the generating function into separate simpler pieces, $H_n(x)$, that can be calculated exactly (for small $n$). By studying the properties of these coefficients, particularly their singularities, we can obtain some idea of the properties of the generating function as a whole.

While no solution is known for $G(x,y)$, the first few coefficients of $y$ may be computed exactly [1]. This was done for self-avoiding polygons (and a range of other problems) by Guttmann and Enting [8] and they observed the following:

- $H_n(x)$ is a rational function of $x$,
- the degree of the numerator of $H_n(x)$ is equal to the degree of its denominator,
- the denominators of $H_n(x)$ (we denote them $D_n(x)$) are products of cyclotomic polynomials [2] and the first few are:

$$D_1(x) = (1-x)$$
$$D_2(x) = (1-x)^3$$
$$D_3(x) = (1-x)^5$$
$$D_4(x) = (1-x)^7$$
$$D_5(x) = (1-x)^9(1+x)^2$$
$$D_6(x) = (1-x)^{11}(1+x)^4$$
$$D_7(x) = (1-x)^{13}(1+x)^6(1+x+x^2)$$
$$D_8(x) = (1-x)^{15}(1+x)^8(1+x+x^2)^3$$
$$D_9(x) = (1-x)^{17}(1+x)^{10}(1+x+x^2)^5$$
$$D_{10}(x) = (1-x)^{19}(1+x)^{12}(1+x+x^2)^7(1+x^2). \tag{5.4}$$

The singularities of the $H_n(x)$, if this pattern persists will become dense on $|x| = 1$. A similar pattern was observed for many unsolved models and is absent in solved models such as staircase or convex polygons. Guttmann and Enting suggested that this pattern of singularities becoming dense on $|x| = 1$ was the hallmark of an unsolvable problem, and that it could be used as a test of *solvability*; D-finite functions of two variables do not display this behaviour.

Then definition of D-finite can be extended to encompass multivariate functions [12] and so cover the anisotropic generating functions considered here.

**Definition 2.** Let $G(x,y)$ be a formal power series in $y$ with coefficients that are rational functions of $x$. Such a series is said to be D-finite if there exists a non-trivial differential equation:

---

[1] More precisely, the first hundred (or so) terms of the expansion of $H_n(x)$ can be computed using either brute force or the finite-lattice method. The first few tens of these can then be fitted using Padé approximants. and the remaining terms can be used to "verify" the conjectured form. We also note that one can show that $H_n(x)$ is rational using transfer matrix arguments and bounds are given for the numerator and denominator degrees in [15, 16], and so the conjectured forms are exact.

[2] The cyclotomic polynomials $\Psi_k(x)$ are the factors of the polynomials $(1-x^n)$. More precisely

$$(1-x^n) = \prod_{k|n} \Psi_k(x).$$

If $k$ is a prime number then $\Psi_k(x) = 1 + x + x^2 + \cdots x^{k-1}$.

$$Q_d(x,y)\frac{\partial^d}{\partial y^d}G(x,y) + \cdots + Q_1(x,y)\frac{\partial}{\partial y}G(x,y) + Q_0(x,y)G(x,y) = 0, \qquad (5.5)$$

with $Q_j$ a polynomial in $x$ and $y$ with complex coefficients

One can show that such functions cannot have a dense set of singularities.

**Theorem 1 (from [5]).** *Let $f(x,y) = \sum_{n\geq 0} H_n(x)y^n$ be a D-finite series in $y$ with coefficients $H_n(x)$ that are rational functions of $x$. Let $S_n$ be the set of poles of $H_n(x)$ and $S = \cup_n S_n$. Then $S$ has only a finite number of singularities.*

Ideally we would like to determine all of the singularities of the coefficients, $H_n(x)$, but unfortunately we have not been able to do so. Instead, we are able to prove that the first occurrence of each cyclotomic factor in the denominators of $H_n(x)$ does not cancel with the corresponding numerator.

**Theorem 2.** *Write $G(x,y) = \sum H_n(x)y^n$. The function $H_{3k-2}(x)$ has simple poles at the zeros of $\Psi_k(x)$ except when $k = 2$.*

This then implies that the set of singularities of the $H_n(x)$ form a dense set on $|x| = 1$ and so we have the corollary and our main result:

**Corollary 1.** *Let $S_n$ be the set of singularities of the coefficient $H_n(x)$. The set $S = \bigcup_{n\geq 1} S_n$ is dense on the unit circle $|x| = 1$. Consequently the self-avoiding polygon anisotropic half-perimeter generating function is not a D-finite function of $y$.*

In Section 5.2 we describe the "haruspicy" techniques that are used to define equivalence classes on the set of polygons. Each equivalence class has a simple rational generating function. Adding these generating functions together proves that $H_n(x)$ is rational and that its denominator is a product of cyclotomic polynomials (see Theorem 3). In Section 5.3 we determine which of these equivalence classes cause the first appearance of $\Psi_k(x)$ as a denominator factor. It turns out that these equivalence classes have a simple description and we can use this to find a functional equation satisfied by their generating function. Analysing this functional equation then completes the proof of Theorem 2. Finally in Section 5.4 we describe extensions of this work to other problems.

## 5.2 Haruspicy

In [15], the author developed techniques which allow us to determine properties of the coefficients, $H_n(x)$, whether or not they are known in some nice form. The central idea is to reduce the set of polygons to some sort of minimal set; various properties of the $H_n(x)$ may be inferred by examining the bond configurations of those minimal polygons. Since polygons are a type of bond animal, we refer to this approach as *haruspicy*; the word refers to techniques of divination based on the examination of the forms and shapes of the organs of animals.

## 5.2.1 Sections, Squashing and Posets

We start by describing how polygons may be cut up into simpler pieces that can be reduced in a consistent way. In particular we cannot violate self-avoidance, and we also need to conserve enough information to recover the original polygon. We cut the polygon into *pages* each of which may be expanded or reduced independently from the rest of the polygon.

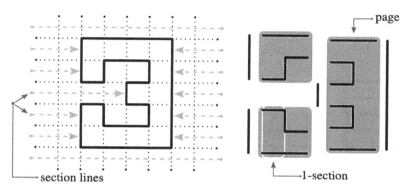

**Fig. 5.1** *Section lines*, indicated by the grey dashed lines in the left-hand figure, split the polygon into *pages*. The pages are shown in right-hand figure. Each column in a page is a *section*. This polygon is split into three pages, each containing two sections; a 1-section is highlighted. Ten vertical bonds lie between pages and four vertical bonds lie within the pages.

**Definition 3.** We construct the *section lines* of a polygon in the following way. Draw horizontal lines from the extreme left and the extreme right of the lattice towards the polygon so that the lines run through the middle of each lattice cell. The lines are terminated when they first touch a vertical bond (see Fig. 5.1).

Cut the lattice along each section line from infinity until it terminates at a vertical bond. Then from this vertical bond cut vertically in both directions until another section line is reached. In this way the polygon is split into *pages* (see Fig. 5.1); we consider the vertical bonds along these vertical cuts to lie *between* pages, while the other vertical bonds lie *within* the pages.

We cannot stretch or expand the horizontal bonds within a page independently of each other without violating self-avoidance. Instead we can expand or delete the horizontal bonds in a given column of a page together.

**Definition 4.** We call a *section* the set of horizontal bonds within a single column of a given page. Equivalently, it is the set of horizontal bonds of a column of a polygon between two neighbouring section lines. A section with $2k$ horizontal bonds is a $k$-section. The number of $k$-sections in a polygon, $A$, is denoted by $\sigma_k(A)$.

A polygon can now be encoded as a list of pages and sections within those pages. Many of these sections, however, are not really needed to encode the shape (in some loose sense of the word) of the polygon. If two neighbouring sections are the same we can remove one of them and still leave the shape of the polygon unchanged.

**Definition 5.** We say that a section is a *duplicate section* if the section immediately on its left (without loss of generality) is identical and there are no vertical bonds between them (see Fig. 5.2).

One can squash or reduce polygons by *deletion* of duplicate sections by slicing the polygon on either side of the duplicate section, removing the section and re-combining the polygon, as illustrated in Fig. 5.2. By reversing the section deletion process we define *duplication* of a section.

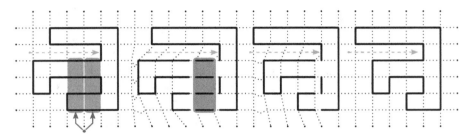

**Fig. 5.2** The two indicated sections are duplicates. We can delete the duplicate by slicing on either side separating the polygon into three pieces. The middle piece, being the duplicate, is removed and the remainder of the polygon is recombined. Reversing the steps duplicates the section duplication. Also indicated is a section line which separates the duplicate sections from the rest of the columns in which they lie.

Section-deletion defines a partial order, $\preceq$, on the set of polygons (see Fig. 5.2). Hence polygons together with this partial order form a partially ordered set, or poset.

**Lemma 1.** *Let $P$ and $Q$ be two polygons in $\mathcal{G}_n$. Write $P \preceq Q$ if $P$ can be obtained from $Q$ by a sequence of section-deletions. This relation is a partial order on the set of polygons.*

*Proof.* Let $A$, $B$ and $C$ be polygons. A partial order must be reflexive, anti-symmetric and transitive.

- reflexive—By definition $A \preceq A$.
- anti-symmetric—If $A \preceq B$, then either $A = B$ or $A$ can be obtained from $B$ by a sequence of deletions. This implies that either $A = B$ or $|A|_{\hookleftarrow} < |B|_{\hookleftarrow}$. Similarly if $B \preceq A$ then either $B = A$ or $|B|_{\hookleftarrow} < |A|_{\hookleftarrow}$. Hence if $A \preceq B$ and $B \preceq A$ then $A = B$.
- transitive—If $A \preceq B$ then there exists a sequence of section-deletions that takes $B$ to $A$. Similarly if $B \preceq C$, then there exists another sequence of section-deletions that takes $C$ to $B$. Concatenating these gives a sequence of deletions that takes $C$ to $A$, and hence $A \preceq C$.

$\square$

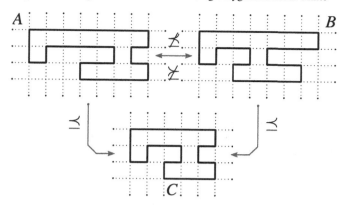

**Fig. 5.3** Polygons $A$ and $B$ can be reduced by section-deletions to $C$, so $C \preceq A, B$. However $A$ cannot be reduced to $B$ or vice-versa. Also $C$ does not contain any duplicate sections and so is section-minimal. Since all three polygons reduce to the same minimal polygon, they are all section-equivalent.

### 5.2.2 Minimal Polygons, Equivalence Classes and Generating Functions

It is clear that a polygon cannot be reduced to nothing. We quickly reach a polygon without duplicate sections—an example is given in Fig. 5.3. These minimal elements of the self-avoiding polygon poset can be used to reconstruct any polygon by duplicating sections; a knowledge of the minimal polygons is sufficient to reconstruct the entire set and its generating function.

**Definition 6.** A *section-minimal polygon*, $A$, is a polygon such that for all polygons, $B$, satisfying $B \preceq A$, then $B = A$. *i.e.* $A$ cannot be reduced any further.

**Lemma 2.** *Every polygon $C$ reduces by section-deletions to a unique section-minimal polygon.*

*Proof.* Number the pages of a given polygon $B$ from $1, 2, \ldots$ (from left to right, top to bottom). Consider, without loss of generality, the first page. We can encode the sections that lie within this page as a sequence $(s_1^{\alpha_1}, s_2^{\alpha_2}, \ldots s_j^{\alpha_j})$, where $s_i^{\alpha_i}$ denotes $\alpha_i$ repetitions of the section $s_i$. If we enforce the condition that $s_i \neq s_{i+1}$ then the $\alpha_i$ are unique. Deleting all the duplicate sections within this page reduces it to the unique sequence $(s_1^1, s_2^1, \ldots, s_j^1)$.

Note that section-deletion does not delete pages, nor does it move sections between pages, and so repeating this process for each page will reduce $B$ to a unique minimal polygon. □

If two polygons reduce to the same minimal polygon then they have (roughly speaking) similar shapes (see Fig. 5.3). We use this idea to define an equivalence relation.

**Lemma 3.** *If two polygons, A and B reduce to the same minimal polygon then we say that they are section-equivalent and write $A \approx B$. Section-equivalence is an equivalence relation.*

*Proof.* It follows almost directly from the definition that section-equivalence is reflexive, symmetric and transitive.                                                              □

This equivalence relation induces equivalence classes each of which has a simple rational generating function whose denominator is a product of cyclotomic polynomials. This shows the link between the minimal polygons and the structure of $H_n(x)$.

**Definition 7.** Section-equivalence partitions the set of polygons into equivalence classes each of which can be characterised by the minimal polygon within the class. We refer to the equivalence class of a section-minimal polygon, $A$, as the section-expansion of $A$. We write:

$$\mathscr{X}(A) = \{B \in \mathscr{G} \mid A \preceq B\}. \tag{5.6}$$

Note that all the elements in such an expansion must have the same number of vertical bonds. We write the horizontal bond generating function of the expansion of a minimal element, $A$, as

$$G(A) = \sum_{B \in \mathscr{X}(A)} x^{|B|_{\leftrightarrow}} \qquad \text{if } A \text{ is section-minimal.} \tag{5.7}$$

**Lemma 4.** *Let P be a section-minimal polygon; its expansion has the following generating function:*

$$G(P) = \prod_k \left( \frac{x^k}{1 - x^k} \right)^{\sigma_k(P)} \tag{5.8}$$

*Proof.* Let $P$ be a section-minimal polygon. Each page of the polygon can be encoded as a sequence of sections $(s_1, \ldots, s_j)$, with $s_i \neq s_{i+1}$. Since we can duplicate any section in $P$ any number of times, given any $(\alpha_1, \ldots, \alpha_j) \in \mathbb{Z}^{+j}$ there exists a polygon $Q$ whose corresponding page is encoded by a sequence of sections $(c_1^{\alpha_1}, \ldots c_j^{\alpha_j})$. So

$$G(P) = \prod_{\text{pages}} \prod_i \sum_{\alpha_i} |x_k|_{\leftrightarrow} (x^{|s_i|_{\leftrightarrow}})^{\alpha_i}$$

$$= \prod_{\text{pages}} \prod_i \frac{x^{|s_i|_{\leftrightarrow}}}{1 - x^{|s_i|_{\leftrightarrow}}} \tag{5.9}$$

where $|s_i|_{\leftrightarrow}$ is the number of horizontal bonds in $s_i$. The result follows.      □

**Lemma 5.** *If $\mathscr{G}_n$ is a set of polygons with n vertical bonds, then the set of section-minimal elements in $\mathscr{G}_n$ is finite.*

*Proof.* Let $A$ be a section-minimal polygon in $\mathscr{G}_n$. It is clear that $A$ cannot contain more than $n$ rows. Between any two columns of $A$ there must be at least a single vertical bond. If there is no vertical bond between two columns, then the horizontal bond configuration in each column must be the same and so they will be duplicates of each other and so $A$ is not minimal. Hence $A$ contains at most $2n + 1$ columns. Since there are a finite number of bond configurations containing at most $n$ rows and $2n + 1$ columns there are only a finite number of section-minimal polygons.     □

We can now prove two theorems about the coefficients $y^n$ in the polygon generating function.

**Theorem 3.** *If $G(x,y) = \sum_{n\geq 0} H_n(x)y^n$ is the anisotropic generating function of self-avoiding polygons, $\mathscr{G}$, then*

- *$H_n(x)$ is a rational function,*
- *the degree of the numerator of $H_n(x)$ cannot be greater than the degree of its denominator, and*
- *the denominator of $H_n(x)$ is a product of cyclotomic polynomials.*

*Proof.* Let $\mathscr{M}$ be the set of section-minimal polygons of $\mathscr{G}_n$. Since each polygon in $\mathscr{G}_n$ is an element in the expansion of exactly one element in $\mathscr{M}$ we can write

$$H_n(x) = \sum_{B\in\mathscr{G}_n} x^{|B|_{\leftrightarrow}} = \sum_{A\in\mathscr{M}} G(A) \tag{5.10}$$

Lemmas 5 and 4 imply that this sum is a finite sum of rational functions with the desired properties. The result follows.     □

Looking a little more carefully at the number of $k$-sections present in minimal polygons gives the following theorem.

**Theorem 4.** *If $H_n(x)$ has a denominator factor $\Psi_k(x)$, then $\mathscr{G}_n$ must contain a section-minimal polygon containing a $K$-section for some $K \in \mathbb{Z}^+$ divisible by $k$. Further if $H_n(x)$ has a denominator factor $\Psi_k(x)^\alpha$, then $\mathscr{G}_n$ must contain a section-minimal polygon that contains $\alpha$ sections that are $K$-sections for some (possibly different) $K \in \mathbb{Z}^+$ divisible by $k$.*

*Proof.* Let $\mathscr{M} = \{M_i\}$ be the set of section-minimal polygons $\in \mathscr{G}_n$.

$$H_n(x) = \sum_i GM_i$$

$$= \sum_i \prod_K \left(\frac{x^K}{1-x^K}\right)^{\sigma_K(M_i)}$$

$$= \sum_i x^{|M_i|_{\leftrightarrow}} \prod_k \Psi_k(x)^{-\sum_d \sigma_{kd}(M_i)}$$

$$= \frac{< \text{some polynomial in } x >}{\prod_k \Psi_k(x)^{\mu_k}} \tag{5.11}$$

where $\mu_k \leq \max_i \{\sum_d \sigma_{kd}(M_i)\}$—this is an inequality since the numerator and denominator could share common cyclotomic factors. Consequently if there is no minimal element $M_i$ containing a $K$-section (for some $K$ divisible by $k$) then $\mu_k = 0$, and the denominator cannot contain $\Psi_k(x)$.                                                      □

The above theorems can be generalised to most interesting sets of bond animals—such as self-avoiding walks, bond animals, bond trees and directed bond animals (see [15, 16]).

By determining how many vertical bonds are required to construct a section-minimal polygon with $\alpha$ $k$-sections, one can use the above theorems to show that

$$\text{The denominator of } H_n(x) \text{ divides } \prod_{k=1}^{\lceil n/3 \rceil} \Psi_k(x)^{2n-6k+5}. \tag{5.12}$$

In fact the denominator of $H_n(x)$ appears (as far as available data permits us to observe) to be exactly the right hand side of the above expression divided by a single power of $\Psi_2(x)$.

There is a similar result [15] for the corresponding generating function of general bond animals (in which $x$ is conjugate to the total number of horizontal bonds), namely

$$\text{The denominator of } H_n(x) \text{ divides } \Psi_1(x)^{3n+1} \prod_{k=2}^{\lfloor n/2 \rfloor} \Psi_k(x)^{2n-3k+4}. \tag{5.13}$$

The denominator of $H_n(x)$ appears to be exactly equal to the right-hand side of the above expression.

These results can be considered upper bounds on the exponents of $\Psi_k(x)$ in the denominator of $H_n(x)$. These are bounds, rather than equalities, since denominator factors might cancel with terms in the numerator. Demonstrating that a given factor does or does not cancel is considerably more difficult and we have only been able to do so in the case of the first occurrence of $\Psi_k(x)$. This is what we do below.

## 5.3 Analysing 2-4-2 Polygons

In this section we study the first occurrence of a given cyclotomic factor in the denominators of the $H_n(x)$. We start by characterising the section-minimal polygons that give rise to them. These polygons turn out to be significantly easier to construct than general self-avoiding polygons and we can find a functional equation for their generating function. The singularities of the solution of this equation give the singularities of the $H_n(x)$ and so lead us to Theorem 2.

## 5.3.1 The First k-Section

Examining the denominators (see (5.4)) of the first few $H_n(x)$ we see that $\Psi_k(x)$ first appears in the denominator of $H_{3k-2}(x)$ (with the exception of $\Psi_2$ which first appears in $H_5(x)$). We start by showing that it takes $6k - 4$ vertical bonds to build a polygon that contains a k-section, and so $\Psi_k(x)$ cannot occur in the denominators of $H_n(x)$ for $n < 3k - 2$.

**Lemma 6.** *To the left (without loss of generality) of a k-section there are at least $3k - 2$ vertical bonds, of which at least $2k - 1$ obstruct section lines. Hence no polygon with fewer than $6k - 4$ vertical bonds may contain a k-section. Further, it is always possible to construct a polygon with $6k - 4$ vertical bonds and a single k-section.*

*Proof.* Consider a vertical line drawn through a k-section (as depicted in the left-hand side of Fig. 5.4). The line starts outside the polygon and then as it crosses horizontal bonds it alternates between the inside and outside of the polygon. More precisely, there are $k + 1$ segments of the line that lie outside the polygon and $k$ segments that lie inside the polygon. Call the segments that lie within the polygon "*inside gaps*" and those that lie outside "*outside gaps*".

Draw a horizontal line through an inside gap (as depicted in the top-right of Fig. 5.4). This line must cross at least one vertical bond to the left of the gap (since it is inside the polygon) and then another to the right of the gap. Hence to the left of any inside gap there must be at least one vertical bond. Similarly there must be at least one vertical bond to the right of any inside gap.

Draw a horizontal line through the topmost of the $k + 1$ outside gaps. Since the line need not intersect the polygon it need not cross any vertical bonds at all. Similarly for the bottommost outside gap.

Now consider a horizontal line through one of the other outside gaps (as depicted in the bottom-right of Fig. 5.4). Traverse this line from the left towards the outside gap. If no vertical bonds are crossed then a section line may be drawn from the left into the outside gap. This splits the k-section into two smaller sections and so contradicts our assumptions. Hence the line must cross at least one vertical bond to block section lines. If only a single vertical bond is crossed before reaching the gap then the gap would lie inside the polygon. Hence the line must cross at least two (or any even number) vertical bonds before reaching the gap. Similar reasoning shows that it must also cross an even number of vertical bonds to the right of the gap.

Since any k-section contains $k$ inside gaps, a topmost outside gap, a bottommost outside gap and $k - 1$ other outside gaps, there must be at least $k \times 1 + 2 \times 0 + 2 \times (k - 1) = 3k - 2$ vertical bonds to its left and $3k - 2$ vertical bonds to its right. The polygons depicted in Fig. 5.5 are constructed by adding "*hooks*". In this way it is possible to construct a section-minimal polygon with $(6k - 4)$ vertical bonds and exactly one k-section.                                                                                    □

Now that we have established that $\Psi_k$ cannot occur before $H_{3k-2}$, we bound the exponent with which it occurs in $H_{3k-2}$ by bounding the number of k-sections that a section-minimal polygon can have.

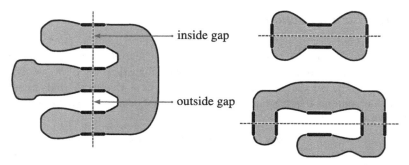

**Fig. 5.4** Vertical and horizontal lines drawn through a $k$-section show the minimum number of vertical bonds required in their construction.

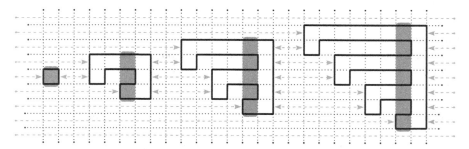

**Fig. 5.5** Section-minimal polygons with $6k - 4$ vertical bonds and a single $k$-section may be constructed by concatenating such "*hook*" configurations.

**Lemma 7.** *A section-minimal polygon with $6k - 4$ vertical bonds cannot contain more than one $k$-section. Hence the factor $\Psi_k(x)$ in the denominator of $H_{3k-2}(x)$ cannot occur with exponent greater than 1.*

*Proof.* Let $A$ be a section-minimal polygon with $6k - 4$ vertical bonds and more than one $k$-section. We show that $A$ cannot exist. The second statement of the lemma then follows from the first by Theorem 4.

Assume that $A$ has only a single $k$-section in each column. By Lemma 6, there are $3k - 2$ vertical bonds to the left of the leftmost $k$-section and $3k - 2$ vertical bonds to the right of the rightmost $k$-section. Between any two $k$-sections there must be at least one vertical bond (or they would be duplicates). Hence $A$ contains more than $6k - 4$ vertical bonds. If, on the other hand, $A$ contains a column with two or more $k$-sections, then to the left of this column there must be at least $6k - 4$ vertical bonds and similarly to its right. Hence $A$ contains at least $12k - 8$ vertical bonds. Hence $A$ cannot exist. $\qquad\square$

In order to proceed we need to split the set of polygons with $6k - 4$ vertical bonds into those that contain a $k$-section and those which do not. While we can, in principle define these sets of polygons, it is much easier to define a superset of those that contain a $k$-section, and this does not significantly alter the subsequent analysis.

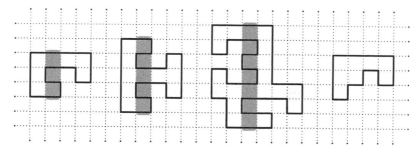

**Fig. 5.6** Four section-minimal *2-4-2* polygons. The first three contain a 2-, 3- and 4-section respectively, while the rightmost only contains 1-sections.

**Definition 8.** Number the rows of a polygon $P$ starting from the topmost row (row 1) to the bottommost (row r). Let $v_i(P)$ be the number of vertical bonds in the $i^{th}$ row of $P$. If $(v_1(P), \ldots, v_r(P)) = (2, 4, 2, \ldots, 4, 2)$ then we call $P$ a *2-4-2* polygon. We denote the set of such *2-4-2* polygons with $2n$ vertical bonds by $\mathscr{P}_n^{242}$. Note that this set is empty unless $2n = 6k - 4$.

**Lemma 8.** *A section-minimal polygon with* $(6k - 4)$ *vertical bonds that contains one k-section must be a 2-4-2 polygon. On the other hand, a section-minimal 2-4-2 polygon need not contain a k-section.*

*Proof.* The first statement follows by arguments given in the proof of Lemma 6. The rightmost polygon in Fig. 5.6 shows that a *2-4-2* polygon need not contain a $k$-section. □

Now that we have isolated the polygons that contain a $k$-section, the following lemma shows that we can ignore the effect of the remaining polygons.

**Lemma 9.** *The factor* $\Psi_k(x)$ *appears in the denominator of the generating function* $\sum_{P \in \mathscr{P}_{3k-2}^{242}} x^{|P|_{\leftrightarrow}}$ *with exponent exactly equal to 1 if and only if it appears in the denominator of* $H_{3k-2}(x)$ *with exponent exactly equal to one.*

*Proof.* The set of *2-4-2* polygons is closed under section-deletion (since it does not move vertical bonds between rows). Similarly the complement of this set is closed under section-deletion. One can then prove that similar results to Theorems 3 and 4 hold for these sets.

Hence the horizontal half-perimeter generating functions of these sets are rational and their denominators are products of cyclotomic factors. Since $\mathscr{P}_{3k-2} \setminus \mathscr{P}_{3k-2}^{242}$ does not contain a polygon with $k$-section (or indeed, by Lemma 6, any section with more than $2k$ horizontal bonds), it follows (by similar results to Theorem 4) that the denominator of the horizontal half-perimeter generating function of this set is a product of cyclotomic polynomials $\Psi_j(x)$ for $j < k$. Consequently this generating function is not singular at the zeros of $\Psi_k(x)$.

By Lemma 7 every section-minimal polygon in $\mathscr{P}_{3k-2}^{242}$ contains at most one $k$-section, and so the exponent of $\Psi_k(x)$ in the denominator of the horizontal half-perimeter generating function of $\mathscr{P}_{3k-2}^{242}$ is either one or zero (due to cancellations with the numerator). The result follows since this denominator factor may not be cancelled by adding the other generating function.                                    □

## 5.3.2 Hadamard Products and Functional Equations

Lemma 9 tells us that in order to prove that the denominator of $H_{3k-2}(x)$ has a factor of $\Psi_k(x)$ it suffices to examine the generating function of $P_{3k-2}^{242}$. Since 2-4-2 polygons have simpler structure than general self-avoiding polygons, this task is much easier. The technique we use is a variation of the Temperley method [21] and leads to functional equations very similar to those in [3]. It also appears in [4].

We construct 2-4-2 polygons by cutting them into smaller 2-4-2 polygons (see Fig. 5.7). In particular we decompose them into a rectangle of unit height and a sequence of 2-4-2 polygons each of height 3. Call these 2-4-2 polygons of height three "building blocks". We then glue these pieces back together. A functional equation for the generating function of all 2-4-2 polygons can then be obtained from the generating function of the building blocks.

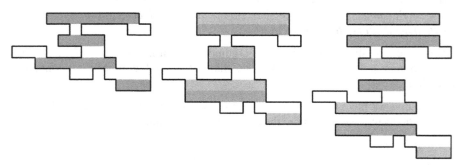

**Fig. 5.7** Decomposing 2-4-2 polygons into a sequence of building blocks (2-4-2 polygons of height three). Highlight each row with 2 vertical bonds. Then duplicate each of these rows excepting the bottommost. By cutting along each of the duplicated rows, the polygon is then uniquely decomposed into a rectangle of unit height and a sequence of building blocks.

**Lemma 10.** *Let $T(t,s;x,y)$ be the generating function of 2-4-2 polygon building blocks, where t and s are conjugate to the length of top and bottom rows (respectively). Then T may be expressed as*

$$T(t,s;x,y) = 2\left(\hat{T}(t,s;x,y) + \hat{T}(s,t;x,y)\right), \tag{5.14}$$

*where the generating function $\hat{T}(t,s;x,y)$ is given by*

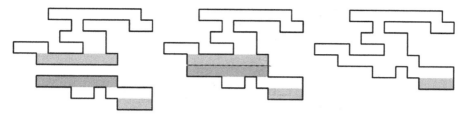

**Fig. 5.8** Constructing a *2-4-2* polygon from a (shorter) *2-4-2* polygon and a building block. When the building block and the polygon are squashed together, the total vertical perimeter is reduced by 2, and the total horizontal perimeter is reduced by twice the width of the joining row.

$$\hat{T}(t,s;x,y) = y^4 \left( A(s,t;x) \cdot [\![stx]\!][\![tx]\!]^2 \cdot B(s,t;x) \right.$$
$$+ A(s,t;x) \cdot [\![stx]\!][\![stx^2]\!][\![tx]\!]^2 \cdot B(s,t;x)$$
$$+ A(s,t;x) \cdot [\![stx]\!][\![tx]\!]^3 \cdot B(s,t;x)$$
$$+ C(s,t;x) \cdot [\![sx]\!][\![tx]\!]^3 \cdot B(s,t;x)$$
$$\left. + C(s,t;x) \cdot [\![sx]\!][\![x]\!][\![tx]\!]^3 \cdot B(s,t;x) \right). \tag{5.15}$$

*We have used $[\![f]\!]$ as shorthand for $\frac{f}{1-f}$, and the generating functions A, B and C are:*

$$A(s,t;x) = 1 + [\![x]\!] + 2[\![sx]\!] + 2[\![tx]\!] + [\![sx]\!][\![tx]\!] +$$
$$[\![sx]\!]^2 + [\![sx]\!][\![x]\!] + [\![tx]\!]^2 + [\![tx]\!][\![x]\!] \tag{5.16a}$$
$$B(s,t;x) = 1 + [\![tx]\!] + [\![x]\!] \tag{5.16b}$$
$$C(s,t;x) = 1 + [\![sx]\!] + [\![x]\!]. \tag{5.16c}$$

*Proof.* Figures 5.9 and 5.10 show how to construct the generating function $\hat{T}$ of building blocks in one orientation; each building block can be placed in one of four orientations. To obtain all building blocks we must reflect the blocks counted by $\hat{T}$ about both horizontal and vertical lines. Reflecting about a vertical line multiplies $\hat{T}$ by 2. Reflecting about a horizontal line interchanges the roles of $s$ and $t$. This proves the first equation.

We compute $\hat{T}$ by considering all the section-minimal polygons that contribute to it. This is done in detail in [16]. One can decompose each section-minimal polygon into one of the five polygons given in Fig. 5.9. The left and right ends of these polygons are made up of the *frills* given in Fig. 5.10. This calculation can be (and has been) verified using the *P*-partition techniques in [20].

The above equation for $\hat{T}(t,s;x,y)$ follows by expanding each of the sections in the minimal polygons. □

Larger *2-4-2* polygons can be constructed by gluing a building block onto a smaller *2-4-2* polygon as illustrated in Fig. 5.8. Gluing combinatorial objects together usually corresponds to multiplying their generating functions. However, when we glue together these objects we require that the top row of the building

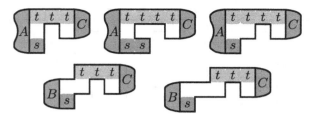

**Fig. 5.9** The section-minimal building blocks of *2-4-2* polygons. The *"frills"*, denoted *A*, *B* and *C* are given in Fig. 5.10.

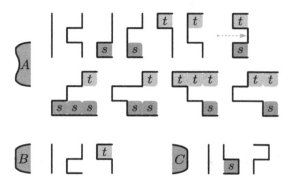

**Fig. 5.10** The *"frills"* of the building blocks in Fig. 5.9.

block has the same length as the bottom row of the *2-4-2* polygon. The corresponding operation on their generating functions is a type of Hadamard product.

**Definition 9.** Let $f(t) = \sum_{t \geq 0} f_n t^n$ and $g(t) = \sum_{t \geq 0} g_n t^n$ be two power series in $t$. We define the (restricted) Hadamard product $f(t) \odot_t g(t)$ to be

$$f(t) \odot_t g(t) = \sum_{n \geq 0} f_n g_n. \tag{5.17}$$

It is generally quite difficult to (explicitly) calculate the Hadamard product of two functions. However when one of the functions is rational the problem is much simpler.

**Lemma 11.** *Let* $f(t) = \sum_{t \geq 0} f_n t^n$ *be a power series, then*

$$f(t) \odot_t \frac{1}{1 - \alpha t} = f(\alpha) \tag{5.18a}$$

$$f(t) \odot_t \frac{k! t^k}{(1 - \alpha t)^{k+1}} = \left. \frac{\partial^k f}{\partial t^k} \right|_{t=\alpha}. \tag{5.18b}$$

*Proof.* The second equation follows from the first by differentiating with respect to $\alpha$. The first equation follows because

$$f(t) \odot_t \frac{1}{1 - \alpha t} = f(t) \odot_t \sum_{n \geq 0} \alpha^n t^n = \sum_{n \geq 0} f_n \alpha^n = f(\alpha). \tag{5.19}$$

$\square$

We can now use the building block generating function and the above Hadamard product to find an equation (though not yet in a usable form) for the generating function of *2-4-2* polygons.

**Lemma 12.** *Let $f(s; x, y)$ be the generating function of 2-4-2 polygons, where $s$ is conjugate to the length of the bottom row of the polygon. This generating function satisfies the following equation*

$$f(s; x, y) = \frac{ysx}{1 - sx} + f(t; x, y) \odot_t \left( \frac{1}{y} T(t/x, s; x, y) \right), \tag{5.20}$$

*where $T(t, s; x, y)$ is the generating function of the 2-4-2 building blocks.*

*Proof.* Write $f(s; x, y) = \sum_n f_n(x, y) s^n$ and $T(t, s; x, y) = \sum_n T_n(s; x, y)$, where $f_n(x, y)$ is the generating function of *2-4-2* polygons whose *bottom* row has length $n$, and $T_n(s; x, y)$ is the generating function of *2-4-2* building blocks, whose *top* row has length $n$. The above recurrence becomes:

$$f(s; x, y) = \frac{ysx}{1 - sx} + \sum_{n \geq 1} f_n(x, y) T_n(s; x, y) / (yx^n). \tag{5.21}$$

This follows because a *2-4-2* polygon is either a rectangle of unit height (counted by $\frac{ysx}{1-sx}$) or may be constructed by gluing a *2-4-2* polygon, whose last row is of length $n$ (counted by $f_n(x, y)$) to a *2-4-2* polygon whose top row is of length $n$ (counted by $T_n(s; x, y)$).

To explain the factor of $1/(yx^n)$ see Fig. 5.8; when the building block is joined to the polygon (centre) and the duplicated row is "*squashed*" (right), the total vertical half-perimeter is reduced by 1 (two vertical bonds are removed) and the total horizontal half-perimeter is reduced by the length of the join (two horizontal bonds are removed for each cell in the join). Hence if the join is of length $n$, the perimeter weight needs to be reduced by a factor of $(yx^n)$. $\square$

In order to turn the Hadamard equation in the above lemma into a more standard functional equation we use Lemma 11, and rewrite $T(t/x, s; x, y)/y$ in (a nonstandard) partial fraction form:

$$y^3 \left[ c_0 \cdot t^0 + \sum_{k=0}^{5} c_{k+1} \frac{k! t^k}{(1 - t)^{k+1}} + c_7 \frac{1}{1 - st} + c_8 \frac{1}{1 - stx} \right], \tag{5.22}$$

where the $c_i$ are (large and ugly) rational functions of $s$ and $x$. We will need $c_8$:

$$c_8 = -\frac{2sx^2(s^2x^2 + sx - s + 1)}{(1 - sx)^4(1 - x)^2}. \tag{5.23}$$

We do not require the other coefficients in the analysis that follows. We note that their denominators are products of $(1-x), (1-s)$ and $(1-sx)$. When $s=1$ some singularities of $T$ coalesce and we have

$$T(t/x,1;x,y)/y = y^3 \left[ \hat{c}_0 \cdot t^0 + \sum_{k=0}^{6} \hat{c}_{k+1} \frac{k! t^k}{(1-t)^{k+1}} + \hat{c}_8 \frac{1}{1-tx} \right], \tag{5.24}$$

where the $\hat{c}_i$ are (slightly simpler) rational functions of $x$. Again, we will need $\hat{c}_8$:

$$\hat{c}_8 = -2 \frac{x^3(1+x)}{(1-x)^6} = c_8|_{s=1}. \tag{5.25}$$

We note that the denominators of the $\hat{c}_i$ are products of $(1-x)$. Applying Lemma 11, we find:

$$f(t;x,y) \odot_t T(t/x,s;x,y)/y =$$
$$y^3 \left[ \sum_{k=0}^{5} c_{k+1} \frac{\partial^k f}{\partial t^k}(1;x,y) + c_7 f(s;x,y) + c_8 f(sx;x,y) \right], \tag{5.26}$$

where we have made use of the fact that $[t^0]f(t;x,y) = 0$ (there are no rows of zero length). When $s=1$ the coalescing poles change equation (5.26) to:

$$f(t;x,y) \odot_t T(t/x,1;x,y)/y =$$
$$y^3 \left[ \sum_{k=0}^{6} \hat{c}_{k+1} \frac{\partial^k f}{\partial t^k}(1;x,y) + \hat{c}_8 f(x;x,y) \right] \tag{5.27}$$

These equations give the following lemma:

**Lemma 13.** *Let $f(s;x,y)$ be the generating function for 2-4-2 polygons enumerated by bottom row-width, half-horizontal perimeter and half-vertical perimeter ($s,x$ and $y$ respectively). Write $f(s;x,y) = \sum_{n\geq 1} f_n(s;x)y^{3n-2}$, where the coefficient $f_n(s;x)$ is the generating function for $\mathscr{P}_{3n-2}^{242}$. These coefficients satisfy the following equations:*

$$f_1(s;x) = \frac{sx}{1-sx} \tag{5.28a}$$

$$f_{n+1}(s;x) = \sum_{k=0}^{5} c_{k+1} \frac{\partial^k f_n}{\partial s^k}(1;x) + c_7 f_n(s;x) + c_8 f_n(sx;x) \tag{5.28b}$$

$$f_{n+1}(1;x) = \sum_{k=0}^{6} \hat{c}_{k+1} \frac{\partial^k f_n}{\partial s^k}(1;x) + \hat{c}_8 f_n(x;x). \tag{5.28c}$$

*The second of these is only valid when $s \neq 1$; when $s=1$ it reduces to the last.*

*Proof.* Apply Lemma 11 to the partial fraction form of $T(t,s;x,y)$ for general $s$, and when $s = 1$. Extracting the coefficients of $y^{3n+1}$ from these equations gives the above recurrences.     □

### 5.3.3 Proof of Theorem 2

We complete the proof of Theorem 2 by showing that $f_n(1;x)$ is singular at the zeros of $\Psi_n(x)$. We are able to do this by induction on the recurrences in the previous lemma. It turns out that we are able to disregard most of these recurrences except for the terms involving $f_n(sx;x)$ and $f_n(x;x)$; these are the only terms that introduce new denominator factors. We will require the following lemma to show that the coefficients $c_8$ and $\hat{c}_8$ cannot cancel these factors since they do not contain cyclotomic factors (except $\Psi_2(x)$).

**Lemma 14.** *Consider the coefficient $c_8(s;x)$ defined above. When $s = x^k$, $c_8(x^k,x)$ has a single zero on the unit circle at $x = -1$ when $k$ is even. When $k$ is odd $c_8(x^k,x)$ has no zeros on the unit circle.*

*Proof.* When $s = x^k$, the coefficient $c_8$ is

$$c_8(x^k,x) = \frac{2x^{k+2}(k^{2k+2} + x^{k+1} - x^k + 1)}{(1 - x^{k+1})^4(1 - x)^2}. \tag{5.29}$$

Let $\xi$ be a zero of $c_8(x^k,x)$ that lies on the unit circle; $\xi$ must be a solution of $x^{2k+2} + x^{k+1} - x^k + 1 = 0$. Hence:

$$\xi^k - \xi^{k+1} = \xi^{2k+2} + 1$$
$$1/\xi - 1 = \xi^{k+1} + \xi^{-k-1}. \tag{5.30}$$

Since $\xi$ lies on the unit circle we may write $\xi = e^{i\theta}$:

$$e^{-i\theta} - 1 = e^{i(k+1)\theta} + e^{-i(k+1)\theta}$$
$$= 2\cos((k+1)\theta). \tag{5.31}$$

Since the right hand-side of the above expression is real the left-hand side must also be real. Therefore $\theta = 0, \pi$ and $\xi = \pm 1$. If $\xi = 1$ then $p_k(\xi) = 2$. On the other hand, if $\xi = -1$ then $p_k(\xi) = 4$ if $k$ is odd and is zero if $k$ is even.

Since the denominator of $c_8(x^k,x)$ is not zero when $k$ is even and $x = -1$ the result follows. One can verify that there are no multiple zeros at $x = -1$ by examining the derivative of the numerator.

□

**Proof of Theorem 2** :
Consider the recurrence given in Lemma 13. This implies that $f_n(s;x)$ is a rational

function of $s$ and $x$. Further, since $f_n(1;x)$ is a well-defined (and rational) function, the denominator of $f_n(s;x)$ does not contain any factors of $(1-s)$.

Let $\mathbb{C}_n(s;x)$ be the set of polynomials of the form

$$\prod_{k=1}^{n} \Psi_k(x)^{a_k}(1 - sx^k)^{b_k}, \tag{5.32}$$

where $a_k$ and $b_k$ are non-negative integers. We define $\mathbb{C}_n(x) = \mathbb{C}_n(0;x)$ (i.e. polynomials which are products of cyclotomic polynomials). We first prove by induction on $n$ that $f_n$ may be written as

$$f_n(s;x) = \frac{N_n(s;x)}{(1 - sx^n)D_n(s;x)}, \tag{5.33}$$

where $N_n(s;x)$ and $D_n(s;x)$ are polynomials in $s$ and $x$ with the restriction that $D_n(s;x) \in \mathbb{C}_{n-1}(s;x)$. Then we consider what happens when $s = 1$ and $x$ is a zero of $\Psi_k$.

For $n = 1$, equation (5.33) is true, since $f_1(s;x) = \frac{sx}{1-sx}$. Now assume equation (5.33) is true up to $n$ and apply the recurrence. The only term that may introduce a new zero into the denominator is $c_8(s;x)f_n(sx;x)$. By assumption $f_n(s;x) = \frac{N_n(sx;x)}{(1-sx^{n+1})D_n(sx;x)}$, and $D_n(sx;x) \in C_n(s;x)$. Hence equation (5.33) is true for $n+1$, and so is also true for all $n \geq 1$.

Let $\xi$ be a zero of $\Psi_k(x)$. We wish to prove that $f_n(1;x)$ is singular at $x = \xi$ and we do so by proving that for $k = 1,\ldots,n$, the generating function $f_k(x^{n-k};x)$ is singular at $x = \xi$, and then setting $k = n$. We proceed by induction on $k$ for fixed $n$.

If we set $k = 1$, then we see that

$$f_1(x^{n-1};x) = \frac{x^n}{1 - x^n}, \tag{5.34}$$

and so the result is true. Now let $k \geq 2$ and assume that the result is true for $k - 1$, i.e. $f_{k-1}(x^{n-k+1};x)$ is singular at $x = \xi$. The recurrence relation and equation (5.33) together imply

$$f_k(s;x) = \frac{N(s;x)}{D(s;x)} + c_8(s;x)f_{k-1}(sx;x), \tag{5.35}$$

where $N$ and $D$ are polynomials in $s$ and $x$ and $D(s;x) \in \mathbb{C}_{k-1}(s;x)$. Setting $s = x^{n-k}$ yields

$$f_k(x^{n-k};x) = \frac{N(x^{n-k};x)}{D(x^{n-k};x)} + c_8(x^{n-k};x)f_{k-1}(x^{n-k+1};x), \tag{5.36}$$

and we note that $D(x^{n-k};x) \in \mathbb{C}_{n-1}(x)$. In the case $k = n$ the above equation is still true, since $\hat{c}_8 = c_8|_{s=1}$.

Equation (5.36) shows that $f_k(x^{n-k})$ is singular at $x = \xi$ only if the contribution from $c_8(x^{n-k};x)f_{k-1}(x^{n-k+1};x)$ is singular at $x = \xi$. This is true (by assumption) unless $c_8(x^{n-k};x) = 0$ at $x = \xi$. By Lemma 14, $c_8(x^{n-k};x)$ is non-zero at $x = \xi$, except when $n = k = 2$.

In the case $n = k = 2$ this proof breaks down, and indeed we see that $H_4(x)$ is not singular at $x = -1$. Excluding this case, $f_k(x^{n-k};x)$ is singular at $x = \xi$ and so $f_n(1;x)$ is also singular at $x = \xi$. By Lemma 9, $H_{3k-2}(x)$ is singular at $x = \xi$. $\quad\square$

This theorem then allows us to prove the main aim of this chapter; the anisotropic generating function of self-avoiding polygons is not a D-finite function.

**Corollary 1.** *Let $S_n$ be the set of singularities of the coefficient $H_n(x)$. The set $S = \bigcup_{n \geq 1} S_n$ is dense on the unit circle $|x| = 1$. Consequently the self-avoiding polygon anisotropic half-perimeter generating function is not a D-finite function of $y$.*

*Proof.* For any $q \in \mathbb{Q}$, there exists $k$, such that $\Psi_k(e^{2\pi i q}) = 0$. By Theorem 2, $H_{3k-2}(x)$ is singular at $x = e^{2\pi i q}$, excepting $x = -1$. The set $S$ is dense on $|x| = 1$ and so has an infinite number of accumulation points. By Theorem 1 $G(x,y) = \sum H_n(x)y^n$ is not a D-finite power series in $y$. $\quad\square$

Since the specialisation of a D-finite power series is D-finite, the above result extends to self-avoiding polygons on hypercubic lattices.

**Corollary 2.** *Let $G_d$ be the generating function of self-avoiding polygons on the $d$-dimensional hyper-cubic lattice defined by:*

$$G_d(x_1,\ldots,x_{d-1},y) = \sum_P y^{|P|_d} \prod_{i=1}^{d-1} x_i^{|P|_i},$$

*where $|P|_i$ is half the number of bonds in parallel to the unit vector $\tilde{e}_i$. If $d = 1$, then this generating function is zero, and otherwise is a non-D-finite power series in $y$.*

*Proof.* When $d = 1$ then there are no self-avoiding polygons and so the generating function is zero. Now consider $d \geq 2$. The square lattice generating function $G(x,y)$ can be recovered from $G_d$ by setting $x_2 = \cdots = x_{d-1} = 0$. Since any well-defined specialisation of a D-finite power series is itself D-finite [12], it follows that if $G_d$ were D-finite, then so would $G(x,y)$. This contradicts Corollary 1 and so $G_d$ is not D-finite. $\quad\square$

## 5.4 Discussion

We have shown above that the anisotropic generating function of self-avoiding polygons on the square lattice, $G(x,y)$, is not a D-finite function of $y$. This result then extends to prove that the anisotropic generating function of self-avoiding polygons on any hypercubic lattice is either trivial (in one dimension) or a non-D-finite function (in dimensions 2 and higher). Similar results hold for directed-bond animals

[17], general bond animals and bond trees [14]. Unfortunately, work on a similar result for self-avoiding walks appears to be beyond the scope of these techniques. The self-avoiding walk analogue of 2-4-2 polygons appear to be quite complicated [18] and it is at all not clear that one can find recurrences such as those in Lemma 13.

There are several non-D-finiteness results for generating functions of other combinatorial problems, such as bargraphs enumerated by their site-perimeter [6], a number of lattice animal models related to heaps of dimers [5] and certain types of matchings [11]; these results rely upon a knowledge of the generating function—either in closed form or via some sort of recurrence. The result for self-avoiding polygons is, as far as we are aware, the first result on the D-finiteness of a completely unsolved model.

Unfortunately we are not able to use this result to obtain information about the nature of the isotropic generating function $G(z,z)$; one can easily construct a two-variable function that is not D-finite, that reduces to a single variable D-finite function. Consider, for example, the following function

$$F(x,y) = \sum_{n \geq 1} \frac{y^n}{(1 - x^n)(1 - x^{n+1})} \tag{5.37}$$

By Theorem 1 this is not a D-finite function of $y$. However, setting $x = y = z$ reduces $F$ to a simple rational, and hence D-finite, function:

$$F(z,z) = \frac{1}{1 - z} \sum_{n \geq 1} \left( \frac{z^n}{1 - z^n} - \frac{z^{n+1}}{1 - z^{n+1}} \right)$$
$$= \frac{z}{(1 - z)^2}. \tag{5.38}$$

On the other hand, the anisotropisation of *solvable* lattice models does not alter the nature of the generating function. Unfortunately we are unable to determine how far this phenomenon extends since we know so little about the nature of the generating functions of unsolved models.

We note that if the isotropic generating function is indeed not D-finite then it will not be found using computer packages such as GFUN [19] or differential approximants [7] which can only find D-finite solutions. At best one might hope that the solution may satisfy some sort of $q$-linear equation.

As noted above, the techniques developed for self-avoiding polygons have been successfully applied to other families of bond animals. Recent series expansion work by I. Jensen [9] shows that there is some possibility that these techniques can be extended to families of site animals (such as self-avoiding polygons enumerated by their area).

It would also be very interesting to apply these ideas to pattern-avoiding permutations—though it is not entirely clear how to "anisotropise" a permutation. Noonan and Zeilberger conjecture that the generating function of permutations avoiding a given pattern is D-finite [13]. This conjecture has helped drive developments in this field and any progress towards its resolution would constitute a major advance.

# References

1. A Maple package developed by Bruno Salvy, Paul Zimmermann and Eithne Murray at INRIA, France. Available from `http://algo.inria.fr/libraries/` at time of printing.
2. D. Bennett-Wood, J.L. Cardy, I.G. Enting, A.J. Guttmann, and A.L. Owczarek. On the non-universality of a critical exponent for self-avoiding walks. *Nuclear Physics B*, 528(3):533–552, 1998.
3. M. Bousquet-Mélou. A method for the enumeration of various classes of column-convex polygons. *Discrete Math*, 154(1–3):1–25, 1996.
4. M. Bousquet-Mélou, A.J. Guttmann, W.P. Orrick, and A. Rechnitzer. Inversion relations, reciprocity and polyominoes. *Annals of Combinatorics*, 3(2):223–249, 1999.
5. M. Bousquet-Mélou and A. Rechnitzer. Lattice animals and heaps ofdimers. *Discrete Mathematics*, 258:235–274, 2002.
6. M. Bousquet-Mélou and A. Rechnitzer. The site-perimeter of bargraphs. *Advances in Applied Mathematics*, 31(1):86–112, 2003.
7. A.J. Guttmann. Asymptotic analysis of power-series expansions. In C. Domb and J.L. Lebowitz, editors, *Phase Transitions and Critical Phenomena*, volume 13, pages 1–234. Academic, New York, 1989.
8. A.J. Guttmann and I.G. Enting. Solvability of Some Statistical Mechanical Systems. *Physical Review Letters*, 76(3):344–347, 1996.
9. I. Jensen. Anisotropic series for site-animals. Personal communication with the author.
10. I. Jensen and A.J. Guttmann. Self-avoiding polygons on the square lattice. *J. Phys. A: Math. Gen*, 32:4867–76, 1999.
11. M. Klazar. Non-P-recursiveness of numbers of matchings or linear chord diagrams with many crossings. *Advances in Applied Mathematics*, 30(1–2):126–136, 2003.
12. L. Lipshitz. D-finite power series. *J. Algebra*, 122(2):353–373, 1989.
13. J. Noonan and D. Zeilberger. The Enumeration of Permutations with a Prescribed Number of Forbidden Patterns. *Advances in Applied Mathematics*, 17(4):381–407, 1996.
14. A. Rechnitzer. The anisotropic generating functions of bond-animals and bond trees are not D-finite. Unpublished.
15. A. Rechnitzer. Haruspicy and anisotropic generating functions. *Advances in Applied Mathematics*, 30(1–2):228–257, 2003.
16. A. Rechnitzer. Haruspicy 2: The anisotropic generating function of self-avoiding polygons is not D-finite. *Journal of Combinatorial Theory Series A*, 113(3):520–546, 2006.
17. A. Rechnitzer. Haruspicy 3: the anisotropic generating function of directed bond-animals is not D-finite. *Journal of Combinatorial Theory Series A*, 113(6):1031–1049, 2006.
18. A. N. Rogers. On the anisotropic generating function of self-avoiding walks. Personal communcation with author—work formed part of ANR's PhD thesis submitted to The University of Melbourne, 2004.
19. B. Salvy and P. Zimmerman. GFUN: a Maple package for the manipulation of generating and holonomic functions in one variable. *ACM Transactions on Mathematical Software*, 20(2):163–177, 1994.
20. R.P. Stanley. *Enumerative Combinatorics*, volume 1. Cambridge University Press, Cambridge, 1996.
21. H.N. Temperley. Combinatorial Problems Suggested by the Statistical Mechanics of Domains and of Rubber-Like Molecules. *Physical Review*, 103(1):1–16, 1956.
22. E.T. Whittaker and G.N. Watson. *A Course of Modern Analysis*. Cambridge University Press, Cambridge, 1996.

# Chapter 6
# Polygons and the Lace Expansion

Nathan Clisby and Gordon Slade

## 6.1 Introduction

The lace expansion was introduced by Brydges and Spencer in 1985 [7] to analyse weakly self-avoiding walks in dimensions $d > 4$. Subsequently it has been generalised and greatly extended, so that it now applies to a variety of problems of interest in probability theory, statistical physics, and combinatorics, including the strictly self-avoiding walk, lattice trees, lattice animals, percolation, oriented percolation, the contact process, random graphs, and the Ising model. A recent survey is [42].

In this chapter, we give an introduction to the lace expansion for self-avoiding walks, with emphasis on self-avoiding polygons. We focus on combinatorial rather than analytical aspects.

The chapter is organised as follows. In Sec. 6.2, we briefly introduce the random walk model underlying our self-avoiding walk models. In Sec. 6.3, we discuss several examples of taking the reciprocal of a generating function, as this is what the lace expansion succeeds in doing for the self-avoiding walk. The lace expansion for self-avoiding walks is derived in Sec. 6.4. Some of the rigorous results for self-avoiding walks and polygons in dimensions $d > 4$, obtained using the lace expansion, are stated without proof in Sec. 6.5. In Sec. 6.6, we indicate how the lace expansion can be used to enumerate self-avoiding walks in all dimensions, as well as to compute coefficients in the $1/d$ expansion for the connective constant $\mu$ and certain critical amplitudes. Some heuristic ideas and numerical results concerning

Nathan Clisby
Department of Mathematics and Statistics, The University of Melbourne, Victoria, Australia, e-mail: N.Clisby@ms.unimelb.edu.au

Gordon Slade
Department of Mathematics, University of British Columbia, Vancouver, Canada, e-mail: slade@math.ubc.ca

the series analysis of the lace expansion and its relevance for the antiferromagnetic singularity of the susceptibility are provided in Sec. 6.7. Finally, in Sec. 6.8, we give a brief indication of an extension to a different model, by discussing some of the results for high-dimensional lattice trees that have been obtained using the lace expansion.

## 6.2 Preliminaries

The self-avoiding walk models we study are based on underlying random walk models. To define the latter, we fix a finite set $\mathcal{N} \subset \mathbb{Z}^d$ that is invariant under the symmetry group of $\mathbb{Z}^d$, i.e., under permutation of coordinates or replacement of any coordinate $x_i$ by $-x_i$. Our two basic examples are the *nearest-neighbour model*

$$\mathcal{N} = \{x \in \mathbb{Z}^d : \|x\|_1 = 1\} \tag{6.1}$$

and the *spread-out model*

$$\mathcal{N} = \{x \in \mathbb{Z}^d : 0 < \|x\|_\infty \le L\}, \tag{6.2}$$

where $L$ is a fixed (usually large) constant. The norms are defined, for $x = (x_1, \ldots, x_d) \in \mathbb{Z}^d$, by $\|x\|_1 = \sum_{j=1}^d |x_j|$ and $\|x\|_\infty = \max_{1 \le j \le d} |x_j|$.

An $n$-step walk with steps in $\mathcal{N}$ is a sequence $\omega = (\omega(0), \omega(1), \ldots, \omega(n))$ of points in $\mathbb{Z}^d$, with $\omega(j+1) - \omega(j) \in \mathcal{N}$ for $j = 0, 1, \ldots, n-1$. The walk $\omega$ is a self-avoiding walk if $\omega(i) \ne \omega(j)$ for all $i \ne j$. We will be interested in generating functions for certain classes of walks. These generating functions have the form $G(z) = \sum_{\omega \in \mathcal{C}} z^{|\omega|}$, where $\mathcal{C}$ is some specific class of walks (e.g., all walks with $\omega(0) = 0$), $z \in \mathbb{C}$ is a parameter, and $|\omega|$ denotes the number of steps in the walk $\omega$.

We denote the cardinality of $\mathcal{N}$ by $\Omega$, so that $\Omega = 2d$ for the nearest-neighbour model and $\Omega = (2L+1)^d - 1$ for the spread-out model. We also define

$$D(x) = \begin{cases} 1/\Omega & (x \in \mathcal{N}) \\ 0 & (x \notin \mathcal{N}). \end{cases} \tag{6.3}$$

Thus $D(x)$ is the probability for a random walk on $\mathbb{Z}^d$, which chooses steps uniformly from $\mathcal{N}$, to move from 0 to $x$ in a single step.

The Fourier transform of an absolutely summable function $f : \mathbb{Z}^d \to \mathbb{R}$ is defined by

$$\hat{f}(k) = \sum_{x \in \mathbb{Z}^d} f(x) e^{ik \cdot x}, \quad (k \in [-\pi, \pi]^d). \tag{6.4}$$

For the nearest-neighbour model, direct calculation gives

$$\hat{D}(k) = d^{-1} \sum_{j=1}^d \cos k_j \quad \text{(nearest-neighbour model)}.$$

The *convolution* of two absolutely summable functions on $\mathbb{Z}^d$ is defined by

$$(f * g)(x) = \sum_{y \in \mathbb{Z}^d} f(y)g(x-y). \tag{6.5}$$

The Fourier transform has the convenient property that $\widehat{f * g} = \hat{f}\hat{g}$. The original function $f(x)$ can be recovered from its Fourier transform via the inversion formula

$$f(x) = \int_{[-\pi,\pi]^d} \hat{f}(k) e^{-ik\cdot x} \frac{d^d k}{(2\pi)^d}. \tag{6.6}$$

## 6.3 Generating Functions and Their Reciprocals

In its simplest setting, the lace expansion can be understood as a way to take the reciprocal of the generating function for the number of self-avoiding walks. In this section, we consider four examples of generating functions and their reciprocals.

Suppose that the power series $G(z) = \sum_{n=0}^{\infty} g_n z^n$ has a non-zero radius of convergence, and suppose for simplicity that $g_0 = 1$. Then $G(z)$ is non-zero in a neighbourhood of the origin, so its reciprocal $F(z) = 1/G(z)$ has a power series expansion $F(z) = \sum_{m=0}^{\infty} f_m z^m$ with a non-zero radius of convergence. Knowledge of $f_0, \ldots, f_n$ uniquely determines $g_0, \ldots, g_n$, and vice versa. The identity $F(z)G(z) = 1$ implies that $f_0 = 1$, and, for $n \geq 1$,

$$\sum_{m=0}^{n} f_m g_{n-m} = 0. \tag{6.7}$$

This can be regarded as the recursion relation

$$g_n = -\left(f_1 g_{n-1} + f_2 g_{n-2} + \cdots + f_n\right). \tag{6.8}$$

**Example 1.** Let $G(z) = \sum_{\omega:0\to\cdot} z^{|\omega|}$ be the generating function for all random walks on $\mathbb{Z}^d$ that take steps in the set $\mathcal{N}$ and start at 0. There are $\Omega^n$ $n$-step walks, so

$$G(z) = \sum_{n=0}^{\infty} \Omega^n z^n = \frac{1}{1 - \Omega z} = \frac{1}{F(z)}. \tag{6.9}$$

In this example, the reciprocal $F(z) = 1 - \Omega z$ takes the very simple form of a linear function.

**Example 2.** Let $g_0(x) = \delta_{0,x}$, and, for $n \geq 1$, let $g_n(x)$ be the number of $n$-step walks that take steps in $\mathcal{N}$, start at $\omega(0) = 0$, and end at $\omega(n) = x$. Let $G(x;z) = \sum_{n=0}^{\infty} g_n(x) z^n$ be the generating function for such walks. By conditioning on the first step, we see that for $n \geq 1$,

$$g_n(x) = \sum_{y \in \mathbb{Z}^d} \Omega D(y) g_{n-1}(x-y) = (\Omega D * g_{n-1})(x), \tag{6.10}$$

where we have used the convolution defined in (6.5). The Fourier transform of (6.10) is

$$\hat{g}_n(k) = \Omega \hat{D}(k)\hat{g}_{n-1}(k). \tag{6.11}$$

It follows that

$$\hat{g}_n(k) = (\Omega \hat{D}(k))^n, \tag{6.12}$$

and hence the Fourier transform of $G(x;z)$ is given by

$$\hat{G}(k;z) = \sum_{n=0}^{\infty} (\Omega \hat{D}(k))^n z^n = \frac{1}{1 - \Omega z \hat{D}(k)}. \tag{6.13}$$

Thus the reciprocal of the Fourier transform of the generating function takes the simple form of a linear function. The generating function $G(x;z)$ can then be recovered as an integral using (6.6). Note that the generating function $1/(1 - \Omega z)$ of Example 1 is just $\hat{G}(0;z)$ since setting $k = 0$ corresponds to summation over $x$ in (6.4), which counts all $n$-step walks regardless of their endpoint.

**Example 3.** Let $G(z)$ be the generating function for nearest-neighbour self-avoiding walks started from the root of an infinite regular tree of degree $\Omega \geq 2$. The self-avoidance constraint merely eliminates immediate reversals, so there are $\Omega(\Omega - 1)^{n-1}$ $n$-step walks when $n \geq 1$, and therefore

$$G(z) = 1 + \sum_{n=1}^{\infty} \Omega(\Omega - 1)^{n-1} z^n = \frac{1+z}{1 - (\Omega - 1)z} = \frac{1}{1 - \Omega z - \Pi(z)} \tag{6.14}$$

with

$$\Pi(z) = \frac{-\Omega z^2}{1+z}. \tag{6.15}$$

For $|z| < 1$, $\Pi(z)$ has the power series representation

$$\Pi(z) = -\Omega \sum_{m=2}^{\infty} (-z)^m. \tag{6.16}$$

In particular, the case $\Omega = 2$ is just the nearest-neighbour self-avoiding walk on the 1-dimensional lattice $\mathbb{Z}$, for which

$$G(z) = \frac{1+z}{1-z}. \tag{6.17}$$

In a manner that is not immediately apparent from the formula, the subtracted term $\Pi(z) = \frac{\Omega z^2}{1+z}$ in the reciprocal $F(z) = 1 - \Omega z - \Pi(z)$ serves to eliminate the immediate reversals that are allowed in the generating function $1/(1 - \Omega z)$ for simple random walks on the tree. We will see how the lace expansion leads to (6.16) in Sec. 6.4.4.

**Example 4.** Now comes the example of most interest. Fix $d \geq 2$, and let $c_n(x)$ denote the number of $n$-step self-avoiding walks that take steps in $\mathcal{N}$, start at 0, and end at $x$. Let $c_n = \sum_{x \in \mathbb{Z}^d} c_n(x)$ denote the number of $n$-step self-avoiding walks that take steps in $\mathcal{N}$, start at 0, and end anywhere. It is a well-known consequence of the simple inequality $c_{m+n} \leq c_m c_n$ that the limit $\mu = \lim_{n \to \infty} c_n^{1/n}$ exists, as discussed at greater length in Chapter 2. The limit $\mu$ is known as the *connective constant*.

Let $\chi(z) = \sum_{n=0}^{\infty} c_n z^n$ be the generating function for self-avoiding walks that start at 0, and let $G(x;z) = \sum_{n=0}^{\infty} c_n(x) z^n$ be the generating function for those which end at $x$. It is clear that $\chi(z)$ has radius of convergence $z_c = 1/\mu$. The radius of convergence of $G(x;z)$ cannot be smaller, and it was shown by Hammersley [17] also to be $z_c$, for any $x$.

The first two terms of $\chi(z)$ are $\chi(z) = 1 + \Omega z + \cdots$, and hence its reciprocal is of the form

$$\chi(z) = \frac{1}{1 - \Omega z - \Pi(z)}, \tag{6.18}$$

with $\Pi(z) = \sum_{m=2}^{\infty} \pi_m z^m$ for some coefficients $\pi_m$. Similarly,

$$\hat{G}(k;z) = \frac{1}{1 - \Omega z \hat{D}(k) - \hat{\Pi}(k;z)}, \tag{6.19}$$

with $\hat{\Pi}(k;z) = \sum_{m=2}^{\infty} \hat{\pi}_m(k) z^m$ for some coefficients $\hat{\pi}_m(k)$. By definition, $\hat{G}(0;z) = \chi(z)$, so (6.18) is a special case of (6.19).

In this example, the problem of determining the coefficients $\hat{\pi}_m(k)$ or $\pi_m = \hat{\pi}_m(0)$ is more difficult. The purpose of the lace expansion is to find a convenient representation for $\pi_m$ and $\hat{\pi}_m(k)$, which can then be used to better understand $\chi(z)$ and $\hat{G}(k;z)$. As we explain in the next section, $\pi_m$ can be expressed in terms of the number of self-avoiding polygons and other so-called *lace graphs*. According to (6.8),

$$\hat{c}_n(k) = \Omega \hat{D}(k) \hat{c}_{n-1}(k) + \sum_{m=2}^{n} \hat{\pi}_m(k) \hat{c}_{n-m}(k), \tag{6.20}$$

or, equivalently,

$$c_n(x) = (\Omega D * c_{n-1})(x) + \sum_{m=2}^{n} (\pi_m * c_{n-m})(x). \tag{6.21}$$

The first term on the right-hand side of (6.20) is familiar from (6.11), and by itself would give all random walks, not just the self-avoiding ones. The second term on the right-hand side serves to eliminate self-intersecting walks from the count.

Setting $k = 0$ in (6.20) gives

$$c_n = \Omega c_{n-1} + \sum_{m=2}^{n} \pi_m c_{n-m}, \tag{6.22}$$

and knowledge of the coefficients $\pi_m$ for $2 \leq m \leq n$ would allow for the recursive determination of $c_n$ (and vice-versa). This approach to the enumeration of self-avoiding walks has proved fruitful, and is discussed further in Sec. 6.6.

## 6.4 The Lace Expansion

In this section, we give a quick sketch of the derivation of the lace expansion. We follow the original approach of Brydges and Spencer [7]; an alternate approach based on inclusion-exclusion is discussed e.g. in [42]. Further details can be found in [7] or, for a more recent account, [42]. Our presentation is closely based on [42].

The derivation is essentially unchanged when the setting is generalised to self-avoiding walks on an arbitrary graph $\mathbb{G} = (\mathbb{V}, \mathbb{E})$ with vertex set $\mathbb{V}$ and edge set $\mathbb{E}$, and we work in this more general setting. For example, $\mathbb{G}$ might be the hypercubic lattice or the honeycomb lattice, but we make no assumption that $\mathbb{G}$ is regular, and it could be finite or infinite. For simplicity, we assume that $\mathbb{G}$ does not contain loops (edges of the form $\{x, x\}$), but it would be easy to relax this assumption.

We set $c_0(x, y) = \delta_{x,y}$, and, for $n \geq 1$, let $c_n(x, y)$ denote the number of $n$-step self-avoiding walks in $\mathbb{G}$ that take steps in $\mathbb{E}$, begin at $x \in \mathbb{V}$, and end at $y \in \mathbb{V}$. With a little notational effort, it is also possible to include the case of walks which are weighted according to the specific steps they take. We do not work in such generality here, although in the literature it is common to consider weighted steps (see e.g. [23]).

### 6.4.1 The Recursion Relation

The lace expansion gives rise to a function $\pi_m(x, y)$, defined below, such that for $n \geq 1$,

$$c_n(x, y) = \sum_{v \in \mathbb{V}} c_1(x, v) c_{n-1}(v, y) + \sum_{m=2}^{n} \sum_{v \in \mathbb{V}} \pi_m(x, v) c_{n-m}(v, y). \tag{6.23}$$

In the translation invariant case, e.g. $\mathbb{V} = \mathbb{Z}^d$, we have $c_n(x, y) = c_n(0, y - x) \equiv c_n(y - x)$ and similarly for $\pi_m$, the sums over $v$ on the right-hand side reduce to convolutions, and we recover (6.21). All the formulae of Example 4 then also apply. In particular, the lace expansion gives an expression for the reciprocal of the generating functions $\chi(z)$ and $\hat{G}(k; z)$ via (6.18) and (6.19).

### 6.4.2 Definition of $\pi_m(x, y)$

In this section, we define $\pi_m(x, y)$ and sketch the derivation of (6.23). Let $\mathscr{W}_m(x, y)$ denote the set of all $m$-step random walk paths (possibly self-intersecting) that start at $x \in \mathbb{V}$ and end at $y \in \mathbb{V}$. Given $\omega \in \mathscr{W}_m(x, y)$, let

$$U_{st}(\omega) = \begin{cases} -1 & \text{if } \omega(s) = \omega(t) \\ 0 & \text{if } \omega(s) \neq \omega(t). \end{cases} \tag{6.24}$$

Then

$$c_n(x,y) = \sum_{\omega \in \mathscr{W}_n(x,y)} \prod_{0 \le s < t \le n} (1 + U_{st}(\omega)), \qquad (6.25)$$

since the product is equal to 1 if $\omega$ is a self-avoiding walk and is equal to 0 otherwise. We call any set of pairs $st$, with $s < t$ chosen from $\{0, 1, 2, \ldots, n\}$, a *graph*. Let $\mathscr{B}_n$ denote the set of all graphs. Expansion of the product in (6.25) gives

$$c_n(x,y) = \sum_{\omega \in \mathscr{W}_n(x,y)} \sum_{\Gamma \in \mathscr{B}_n} \prod_{st \in \Gamma} U_{st}(\omega). \qquad (6.26)$$

We call a graph $\Gamma \in \mathscr{B}_n$ *connected*[1] if both 0 and $n$ are endpoints of edges in $\Gamma$, and if in addition, for any integer $c \in (0,n)$, there are $s, t \in [0,n]$ such that $s < c < t$ and $st \in \Gamma$. In other words, $\Gamma$ is connected if, as intervals of real numbers, $\cup_{st \in \Gamma}(s,t)$ is equal to the connected interval $(0,n)$. The set of all connected graphs on $[0,n]$ is denoted $\mathscr{G}_n$. See Fig. 6.1.

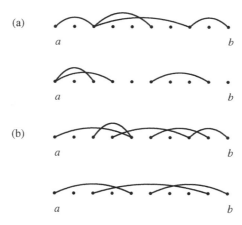

**Fig. 6.1** Graphs in which an edge $st$ is represented by an arc joining $s$ and $t$. The graphs in (a) are not connected, whereas the graphs in (b) are connected.

If we partition the sum over all graphs according to whether: (a) 0 does not occur in an edge in the graph, or (b) 0 does occur in an edge, then we are led to the identity (6.23) with

$$\pi_m(x,y) = \sum_{\omega \in \mathscr{W}_m(x,y)} \sum_{\Gamma \in \mathscr{G}_m} \prod_{st \in \Gamma} U_{st}(\omega). \qquad (6.27)$$

Case (a) gives rise to the first term on the right-hand side of (6.23): the graphs not containing 0 produce a self-avoidance constraint that omits the requirement that

---

[1] This is not the standard graph-theory definition of a connected graph.

the initial vertex at the origin be avoided subsequently. Case (b) gives rise to the second term on the right-hand side of (6.23), with $[0, m]$ the extent of the connected component containing 0: the lack of an edge that passes over $m$ means that the walk segments before and after time $m$ are independent, and the arbitrary graphs on the interval $[m, n]$ produce a self-avoidance constraint during that interval.

### 6.4.3 Representation of $\pi_m(x, y)$ via Laces

An important alternate representation for $\pi_m(x, y)$ can be obtained in terms of laces. A *lace* is a minimally connected graph, i.e., a connected graph for which the removal of any edge would result in a disconnected graph. The set of laces on $[0, m]$ is denoted by $\mathscr{L}_m$, and the set of laces in $\mathscr{L}_m$ which consist of exactly $N$ edges is denoted $\mathscr{L}_m^{(N)}$. See Fig. 6.2.

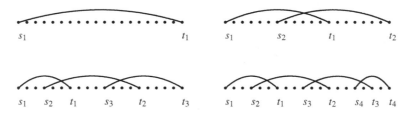

**Fig. 6.2** Laces in $\mathscr{L}_m^{(N)}$ for $N = 1, 2, 3, 4$, with $s_1 = 0$ and $t_N = m$.

Given a connected graph $\Gamma \in \mathscr{G}_m$, the following prescription associates to $\Gamma$ a unique lace $\mathsf{L}_\Gamma \subset \Gamma$: The lace $\mathsf{L}_\Gamma$ consists of edges $s_1 t_1, s_2 t_2, \ldots$, with $t_1, s_1, t_2, s_2, \ldots$ determined, in that order, by

$$t_1 = \max\{t : 0t \in \Gamma\}, \quad s_1 = 0,$$

$$t_{i+1} = \max\{t : \exists s < t_i \text{ such that } st \in \Gamma\}, \quad s_{i+1} = \min\{s : st_{i+1} \in \Gamma\}.$$

Given a lace $L$, the set of all edges $st \notin L$ such that $\mathsf{L}_{L \cup \{st\}} = L$ is denoted $\mathscr{C}(L)$. Edges in $\mathscr{C}(L)$ are said to be *compatible* with $L$. See Fig. 6.3.

We write $L \in \mathscr{L}_m^{(N)}$ as $L = \{s_1 t_1, \ldots, s_N t_N\}$, with $s_l < t_l$ for each $l$. The fact that $L$ is a lace is equivalent to a certain ordering of the $s_l$ and $t_l$. For $N = 1$, we simply have $0 = s_1 < t_1 = m$. For $N \geq 2$, $L \in \mathscr{L}_m^{(N)}$ if and only if

$$0 = s_1 < s_2, \quad s_{l+1} < t_l \leq s_{l+2} \quad (l = 1, \ldots, N-2), \quad s_N < t_{N-1} < t_N = m$$

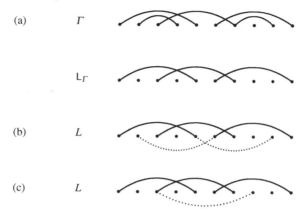

**Fig. 6.3** (a) A connected graph $\Gamma$ and its associated lace $L = L_\Gamma$. (b) The dotted edges are compatible with the lace $L$. (c) The dotted edge is not compatible with the lace $L$.

(for $N = 2$ the vacuous middle inequalities play no role); see Fig. 6.2. Thus $L$ divides $[0, m]$ into $2N - 1$ subintervals:

$$[s_1, s_2], \ [s_2, t_1], \ [t_1, s_3], \ [s_3, t_2], \ \dots, [s_N, t_{N-1}], \ [t_{N-1}, t_N]. \qquad (6.28)$$

Of these, intervals of the form $[t_i, s_{i+2}]$ can have zero length, whereas all others have length at least 1.

The sum over connected graphs in (6.27) can be converted to a double sum, first over all laces $L$, and then over connected graphs for which the above prescription produces $L$. This gives

$$\sum_{\Gamma \in \mathcal{G}_m} \prod_{st \in \Gamma} U_{st} = \sum_{L \in \mathcal{L}_m} \prod_{st \in L} U_{st} \sum_{\Gamma \in \mathcal{G}_m : L_\Gamma = L} \prod_{s't' \in \Gamma \setminus L} U_{s't'}. \qquad (6.29)$$

The sum over $\Gamma$ on the right-hand side can then be resummed explicitly (for details, see [7] or [42]) to obtain the formula

$$\pi_m(x, y) = \sum_{\omega \in \mathcal{W}_m(x,y)} \sum_{L \in \mathcal{L}_m} \prod_{st \in L} U_{st}(\omega) \prod_{s't' \in \mathcal{C}(L)} (1 + U_{s't'}(\omega)). \qquad (6.30)$$

We restrict the sum in (6.30) to laces with $N$ edges, and introduce a minus sign to obtain a non-negative integer, to define

$$\pi_m^{(N)}(x, y) = \sum_{\omega \in \mathcal{W}_m(x,y)} \sum_{L \in \mathcal{L}_m^{(N)}} \prod_{st \in L} (-U_{st}(\omega)) \prod_{s't' \in \mathcal{C}(L)} (1 + U_{s't'}(\omega)). \qquad (6.31)$$

The right hand side of (6.31) is zero unless $N < m$ (since otherwise $\mathcal{L}_m^{(N)}$ is empty), and hence

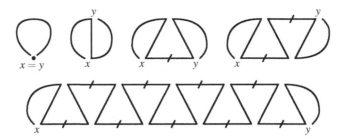

**Fig. 6.4** Self-intersections required for a walk $\omega$ with $\prod_{st \in L} U_{st}(\omega) \neq 0$, for the laces with $N = 1, 2, 3, 4$ bonds depicted in Fig. 6.2. The picture for $N = 11$ is also shown. A slashed subwalk may have length zero .

$$\pi_m(x, y) = \sum_{N=1}^{m-1} (-1)^N \pi_m^{(N)}(x, y). \tag{6.32}$$

Note that each term in the sum (6.31) is either 0 or 1. The first product in (6.31) is equal to 1 precisely when $\omega(s) = \omega(t)$ for each edge $st \in L$. The second product is equal to 1 precisely when $\omega(s') \neq \omega(t')$ for each $s't' \in \mathscr{C}(L)$. Thus the edges in the lace require $\omega$ to have certain self-intersections, while the compatible edges enforce certain self-avoidance conditions. The self-intersections required are illustrated in Fig. 6.4. We refer to the walk configurations of Fig. 6.4 as *lace graphs*.

The simplest term is $\pi_m^{(1)}(x, y)$, which is zero if $y \neq x$, and which is the number of $m$-step self-avoiding returns to $x$ when $y = x$. In the translation invariant case, $\pi_m^{(1)}(x, y)$ can be expressed in terms of the number $p_m$ of $m$-step unrooted unoriented self-avoiding polygons, by $\pi_m^{(1)}(x, y) = 2mp_m\delta_{x,y}$ when $m > 2$ (by convention, $p_2 = 0$).

For $N \geq 2$, $\pi_m^{(N)}(x, y)$ counts $m$-step walk configurations as indicated in Fig. 6.4. The number of loops in a diagram is equal to the number of edges in the corresponding lace. In these diagrams, each line represents a self-avoiding walk, and the overall walk begins at $x$ and ends at $y$. The lines which are slashed correspond to subwalks which may consist of zero steps, but the others correspond to subwalks consisting of at least one step. The combined number of steps taken by all the subwalks is $m$. If the $2N - 1$ subwalks in the $N$-loop diagram are sequentially labeled $1, 2, \ldots, 2N - 1$, then the subwalks are mutually avoiding (apart from the required intersections) due to the effect of the compatible edges, with the following patterns: [123] for $N = 2$; [1234], [345] for $N = 3$; [1234], [3456], [567] for $N = 4$; [1234], [3456], [5678], [789] for $N = 5$; and so on for larger $N$. In the above, e.g., for $N = 4$, the meaning is that subwalks $1, 2, 3, 4$ are mutually avoiding apart from the enforced intersections explicitly depicted, as are subwalks $3, 4, 5, 6$ and subwalks $5, 6, 7$. However, subwalks not grouped together are permitted to freely intersect, e.g., for $N = 4$, subwalks $1, 2$ are permitted to intersect subwalks $5, 6, 7$, and subwalks $3$ and $4$ can intersect subwalk $7$.

### 6.4.4 Walks Without Immediate Reversals

The algebra used in deriving the lace expansion does not depend on the precise form of the interaction $U_{st}(\omega)$, and other choices are possible. For example, we could instead take

$$U_{st}(\omega) = \begin{cases} -1 \text{ if } \omega(s) = \omega(t) \text{ and } t = s+2 \\ 0 \text{ otherwise.} \end{cases} \tag{6.33}$$

With this choice, $c_n(x,y)$ of (6.25) simply counts the number of walks from $x$ to $y$ that do not make any immediate reversals. For a non-zero contribution to $\pi_m(x,y)$ in (6.30), laces must have all edges of length 2, and thus there is a unique lace on the interval $[0,m]$ and this lace contains $m-1$ edges. In this case, the lace graphs consist of successive immediate reversals. For a vertex transitive (translation invariant) graph with vertices of degree $\Omega$, an examination of (6.31) shows that

$$\sum_{v \in \mathbb{V}} \pi_m^{(N)}(v) = \begin{cases} \Omega \text{ if } N = m-1 \\ 0 \quad \text{otherwise.} \end{cases} \tag{6.34}$$

This reproduces the formula (6.16) of Example 3, with $\Pi(z) = \sum_{m=2}^{\infty} \pi_m z^m$ and $\pi_m = \sum_{N=1}^{m-1} (-1)^N \sum_{u \in \mathbb{V}} \pi_m^{(N)}(v) = (-1)^{m-1}\Omega$, as it leads to

$$\Pi(z) = \sum_{m=2}^{\infty} \pi_m z^m = -\Omega \sum_{m=2}^{\infty} (-z)^m. \tag{6.35}$$

## 6.5 Self-Avoiding Walks and Polygons in Dimensions $d > 4$

The major mathematical problem for self-avoiding walks on $\mathbb{Z}^d$ is to prove the existence and compute the values of the universal critical exponents $\gamma, \nu, \alpha$ which appear in the predicted asymptotic formulas

$$c_n \sim A\mu^n n^{\gamma-1}, \quad \langle |\omega(n)|^2 \rangle_n \sim Dn^{2\nu}, \quad c_n(e) \sim B\mu^n n^{\alpha-2}. \tag{6.36}$$

Here $\langle \cdot \rangle_n$ denotes expectation with respect to the uniform measure on the set of $n$-step self-avoiding walks started from the origin. In the last formula, $n$ is required to be odd for the nearest-neighbour model, and $e$ represents a neighbour of the origin.

In this section, we describe some of the results that have been obtained in this direction using the lace expansion, for self-avoiding walks in dimensions $d > 4$. The hypothesis $d > 4$ is used to ensure convergence of the lace expansion, e.g., in the sense that $\sum_{m=2}^{\infty} m|\pi_m|z_c^m < \infty$, where $z_c = 1/\mu$. There are now several different approaches to proving convergence of the lace expansion, and we make no attempt here to explain them. Perhaps the simplest approach, and many references to other approaches, can be found in [42].

All known convergence proofs require a small parameter to ensure convergence. To prove that the critical exponent $\gamma$ is equal to 1 amounts to proving that $\frac{d}{dz}[1/\chi(z)]|_{z=z_c}$ is finite and non-zero, since $c_n \sim A\mu^n$ corresponds to $\chi(z) \sim A(1-\mu z)^{-1}$ as $z \nearrow z_c$. This, in turn, amounts to analysing $z_c \frac{d}{dz}\Pi(z_c)$ or, equivalently, $\sum_{m=2}^{\infty} m\pi_m z_c^m$. The $m = 2$ term in this sum is equal to $2\Omega z_c^2$, and assuming that $z_c$ is close to $(\Omega - 1)^{-1}$ and that $\Omega$ is large, this is close to $2\Omega^{-1}$. For the nearest-neighbour model with $d$ large, or for the spread-out model with $d > 4$ and $L$ large, this approximation turns out to be reasonably accurate and $\sum_{m=2}^{\infty} m|\pi_m| z_c^m$ is $O(d^{-1})$ or $O(L^{-d})$. In [19, 21], Hara and Slade used a computer assisted proof to show that even for the nearest-neighbour model when $d = 5$ the small parameter is small enough to allow for a proof of convergence of the lace expansion.

In particular, the following three theorems were proved in [19, 21].

**Theorem 1.** *For the nearest-neighbour model in dimensions $d \geq 5$, there are positive constants $A, D$ such that the following hold:*

*(a)*   $c_n = A\mu^n[1 + O(n^{-\varepsilon})]$ *as $n \to \infty$, for any $\varepsilon < 1/2$.*
*(b)*   $\langle |\omega(n)|^2 \rangle_n = Dn[1 + O(n^{-\varepsilon})]$ *as $n \to \infty$, for any $\varepsilon < 1/4$.*

*For $d = 5$, $A \in [1, 1.493]$ and $D \in [1.098, 1.803]$.*

Theorem 1 is alluded to in the general discussion of the asymptotic behavior of $c_n$ in Chapter 1, and additionally provides explicit bounds on the error terms. A corollary of $(a)$ is that $\lim_{n\to\infty} c_{n+1}/c_n = \mu$. This is believed to be true in all dimensions, but remains unproved for $d = 2, 3, 4$. It was proved by Kesten [32] that $\lim_{n\to\infty} c_{n+2}/c_n = \mu^2$ in all dimensions.

Let $C_d[0, 1]$ denote the set of continuous functions $f : [0, 1] \to \mathbb{R}^d$, equipped with the supremum norm. Given an $n$-step self-avoiding walk $\omega$, we define a rescaled version $X_n \in C_d[0, 1]$ of the self-avoiding walk by setting $X_n(k/n) = (Dn)^{-1/2}\omega(k)$ for $k = 0, 1, 2, \ldots, n$, and taking $X_n(t)$ to be the linear interpolation of this. We denote by $dW$ the Wiener measure on $C_d[0, 1]$. The following theorem shows that for $d \geq 5$ the scaling limit of the self-avoiding walk is Brownian motion.

**Theorem 2.** *For the nearest-neighbour model in dimensions $d \geq 5$, $X_n$ converges in distribution to Brownian motion, i.e., for any bounded continuous function $f : C_d[0, 1] \to \mathbb{C}$,*

$$\lim_{n\to\infty} \langle f(X_n) \rangle_n = \int f dW.$$

Perhaps the most basic application of the above theorem is to the case $f(X) = e^{ik \cdot X(1)}$. In this case, Theorem 2 gives

$$\lim_{n\to\infty} \langle e^{ik \cdot \omega(n)/\sqrt{Dn}} \rangle_n = e^{-|k|^2/2d}, \qquad (6.37)$$

i.e., the scaling limit of the endpoint of the self-avoiding walk has a Gaussian distribution. Note that the expression under the limit in the above equation can also be written as $\hat{c}_n(k/\sqrt{Dn})/c_n$; this shows the relevance of the Fourier transform of $c_n(x)$ in understanding the scaling limit.

The results for self-avoiding polygons in dimensions $d \geq 5$ are less complete than those above. Since $p_n = \frac{1}{2n} \sum_{e \in \mathcal{N}} c_{n-1}(e)$, the study of $c_n(x)$ is more general than the study of self-avoiding polygons. Ideally one would like a result which states that $c_{n-1}(e) \sim B\mu^n n^{-d/2}$ for $d > 4$, but this has not been proved. The following theorem from [21] proves an upper bound on the generating function for $c_n(x)$ which is consistent with this asymptotic behaviour. Madras [36] has proved bounds valid for general $d$, believed not to be sharp.

**Theorem 3.** *For the nearest-neighbour model in dimensions $d \geq 5$, for any $a < (d-2)/2$,*

$$\sup_{x \in \mathbb{Z}^d} \sum_{n=0}^{\infty} n^a c_n(x) \mu^{-n} < \infty.$$

Stronger results for $c_n(x)$ have been obtained for the spread-out model in dimensions $d > 4$. The best results can be found in [24], and include the following as a very special case. In the statement of the theorem, $\mu$ denotes the connective constant for the spread-out model.

**Theorem 4.** *Consider the spread-out model in dimensions $d > 4$. Let $0 < \delta < \min\{1, \frac{d-4}{2}\}$. There is an $L_0$ such that for $L \geq L_0$ the following statements hold:*
*(a) There exist positive constants $a$ and $b$ (depending on $d$ and $L$), such that for all $k \in \mathbb{R}^d$ with $|k|^2$ bounded by a constant, as $n \to \infty$,*

$$\hat{c}_n(k/\sqrt{bn}) = a\mu^n e^{-|k|^2/2d} \left[1 + O(n^{-(d-4)/2}) + O(|k|^2 n^{-\delta})\right]. \tag{6.38}$$

*(b) There are constants $C_1, C_2$ (depending on $d$ but not $L$) such that*

$$C_1 \mu^n L^{-d} n^{-d/2} \leq \sup_{x \in \mathbb{Z}^d} c_n(x) \leq C_2 \mu^n L^{-d} n^{-d/2}. \tag{6.39}$$

Note that for $k = 0$, Theorem 4(a) gives

$$c_n = a\mu^n \left[1 + O(n^{-(d-4)/2})\right], \tag{6.40}$$

which is a better error bound than that proved for the nearest-neighbour model in Theorem 1. It was predicted in [15] that the $O(n^{-(d-4)/2})$ is sharp, and by universality we expect it to hold also for the nearest-neighbour model (this is confirmed numerically in [8]). In [24], a version of Theorem 4 is obtained in the much more general setting of cycle-free networks of mutually-avoiding self-avoiding walks. For arbitrary networks, possibly with cycles, see [26].

Theorem 4(a) provides a central limit theorem for the endpoint of the spread-out self-avoiding walk in dimensions $d > 4$. It is natural to wonder if Theorem 4(b) extends to a local central limit theorem, i.e., a statement that $c_n^{-1} c_n(x\sqrt{bn})$ is asymptotically Gaussian (when $x$ has the same parity as $n$). Such an extension follows from the results of [23, 24]. Some care is needed with such a statement, since $c_n(0) = 0$ for all $n \geq 1$, and to eliminate this local effect we average over a region that grows with $n$. For the averaging, we denote the cube of radius $R$ centred at $x \in \mathbb{Z}^d$ by

$$C_R(x) = \{y \in \mathbb{Z}^d : \|x - y\|_\infty \leq R\}, \tag{6.41}$$

with cardinality $|C_R(x)|$. Let $\lfloor x \rfloor$ denote the closest lattice point in $\mathbb{Z}^d$ to $x \in \mathbb{R}^d$ (with some arbitrary rule to break ties).

**Theorem 5.** *Consider the spread-out model in dimensions $d > 4$. Let $R_n$ be any sequence with $\lim_{n \to \infty} R_n = \infty$ and $\lim_{n \to \infty} n^{-1/2} R_n = 0$. There is an $L_0$ such that for $L \geq L_0$, and for any $x \in \mathbb{R}^d$ with $|x|^2 [\log R_n]^{-1}$ sufficiently small, as $n \to \infty$,*

$$\frac{1}{|C_{R_n}(\lfloor x\sqrt{bn} \rfloor)|} \sum_{y \in C_{R_n}(\lfloor x\sqrt{bn} \rfloor)} \frac{c_n(y)}{c_n} \sim \left(\frac{d}{2\pi bn}\right)^{d/2} e^{-d|x|^2/2}, \tag{6.42}$$

*in the sense that the limit of the ratio of the two sides is 1.*

The Gaussian limit (6.42) does not follow directly from the convergence of the Fourier transform in Theorem 4. The latter implies that sums over cubes of side $\sqrt{n}$ converge to integrals of the Gaussian density, whereas (6.42) permits arbitrarily slow growth of $R_n$.

Finally, we mention that rigorous results have been proved for the scaling of the weakly self-avoiding walk on a 4-dimensional *hierarchical* lattice, using renormalisation group methods [5, 6]. The 3-dimensional cubic lattice appears to be well beyond the reach of any currently known methods. For $d = 2$, there is very strong evidence [34] that the scaling limit is given by $SLE_{8/3}$, but the problem of proving existence of the scaling limit remains open.

## 6.6 Self-Avoiding Walk Enumeration and $1/d$ Expansions

### 6.6.1 Self-Avoiding Walk Enumeration via the Lace Expansion

For the nearest-neighbour model on $\mathbb{Z}^d$, (6.22) states that

$$c_n = 2d c_{n-1} + \sum_{m=2}^{n} \pi_m c_{n-m}. \tag{6.43}$$

Let

$$\pi_m^{(N)} = \sum_{x \in \mathbb{Z}^d} \pi_m^{(N)}(x) \tag{6.44}$$

denote the number of $N$-loop lace graphs of length $m$, so that $\pi_m = \sum_{N=1}^{m-1} (-1)^N \pi_m^{(N)}$. Equation (6.43) recursively expresses the number of self-avoiding walks of length $n$ in terms of $\pi_m$, and thus allows for the determination of $c_n$ from the number of lace graphs with $m \leq n$ and $N \leq n - 1$. The lace graph trajectories, shown in Fig. 6.4, are less spatially extended than SAWs of the same length, and are hence less numerous.

In [8], nearest-neighbour self-avoiding walks on $\mathbb{Z}^d$ were enumerated in dimensions $d \geq 3$ by enumeration of lace graphs together with (6.43). The value of $c_n$ was determined for $n \leq 30$ for $d = 3$ and for $n \leq 24$ for *all* $d \geq 4$ (knowledge of $\pi_m$ for $m \leq 24$ and $d \leq 12$ determines $\pi_m$ also for $d > 12$, since lace graphs with at most 24 steps can occupy at most 12 dimensions). In practice, for the cubic lattice it was found that there are approximately 525 times as many 30-step self-avoiding walks as compared to 30-step lace graphs. This factor was found to get much larger as the dimension is increased: the factor for $d = 4$, $n = 24$ is approximately 1700, for $d = 5$, $n = 24$, it is approximately 6200, while for $d = 6$, $n = 24$, it is approximately 20000.

The simplest lace graphs are the self-avoiding returns counted by $\pi_m^{(1)}$, and their enumeration is equivalent to the enumeration of self-avoiding polygons. Polygons were counted in [8] using the so-called *two-step method*. The two-step method is an innovation for the direct enumeration of self-avoiding walks and reduces the exponential complexity of the enumeration problem. For details, we refer to [8]. In [8], the two-step method was used to count polygons and, more generally, to enumerate lace graphs needed to determine $\pi_m$. For polygons, the results of [8] are slightly better than those for general lace graphs: $p_{32}$ is determined for $d = 3$, and $p_{26}$ is determined for $d = 4$.

### 6.6.2 $1/d$ *Expansions via the Lace Expansion*

Next, we indicate how knowledge of the values of $\pi_m$ for the $d$-dimensional nearest-neighbour model can be combined with estimates on the lace expansion to derive $1/d$ expansions for the connective constant $\mu$ and for the amplitudes $A$ and $D$ of Theorem 1.

Let $z_c = 1/\mu$ denote the radius of convergence of the susceptibility $\chi(z) = \sum_{n=0}^{\infty} c_n z^n$. It is a consequence of the simple inequality $c_{n+m} \leq c_n c_m$ that $c_n \geq \mu^n$ for all $n$, and from this it follows that $\chi(z) \nearrow \infty$ as $z \nearrow z_c$. Therefore $1/\chi(z) = 1 - 2dz - \Pi(z) \searrow 0$ as $z \nearrow z_c$. In particular, $\lim_{z \nearrow z_c} \Pi(z) = 1 - 2dz_c$. It is proved in [21] that, for $d \geq 5$, $\lim_{z \nearrow z_c} \Pi(z) = \Pi(z_c)$, and therefore $z_c$ obeys the equation

$$z_c = \frac{1}{2d}\left[1 - \sum_{m=2}^{\infty} \pi_m z_c^m\right]. \tag{6.45}$$

This equation was used recursively in [22] to prove that $z_c$ has an asymptotic expansion $z_c \sim \sum_{i=1}^{\infty} a_i (2d)^{-i}$, to all orders, with integer coefficients $a_i$.

Well-developed lace expansion methods (see [8]) show that for each $N \geq 1$ and $j \geq 2$ there are constants $C_N, C_{N,j}$, independent of sufficiently large $d$, such that

$$\sum_{m=2}^{\infty} \sum_{M=N}^{m-1} \pi_m^{(M)} z_c^m \leq C_N d^{-N}, \qquad \sum_{m=j}^{\infty} \pi_m^{(N)} z_c^m \leq C_{N,j} d^{-j/2}. \tag{6.46}$$

It then follows from (6.45) and (6.46) that

$$z_c = \frac{1}{2d}\left[1 - \sum_{m=2N}^{2N}\sum_{M=1}^{N}(-1)^M \pi_m^{(M)} z_c^m\right] + O(d^{-N-2}), \qquad (6.47)$$

where we have used the fact that $z_c$ has an asymptotic expansion in powers of $d^{-1}$, to replace an error term of order $d^{-N-3/2}$ by one of order $d^{-N-2}$. Knowledge of the coefficients $\pi_m^{(M)}$ (as polynomials in $d$) for $m \le 2N$ and $M \le N$ permits the recursive calculation of the terms in the $1/d$ expansion for $z_c$ up to and including order $d^{-N-1}$. The enumerations of [8] with $m \le 24$, $M \le 12$ give

$$z_c = \frac{1}{2d} + \frac{1}{(2d)^2} + \frac{2}{(2d)^3} + \frac{6}{(2d)^4} + \frac{27}{(2d)^5} + \frac{157}{(2d)^6} + \frac{1065}{(2d)^7} + \frac{7865}{(2d)^8} + \frac{59665}{(2d)^9}$$
$$+ \frac{422421}{(2d)^{10}} + \frac{1991163}{(2d)^{11}} - \frac{16122550}{(2d)^{12}} - \frac{805887918}{(2d)^{13}} + O\left(\frac{1}{(2d)^{14}}\right). \qquad (6.48)$$

Taking the reciprocal gives

$$\mu = 2d - 1 - \frac{1}{2d} - \frac{3}{(2d)^2} - \frac{16}{(2d)^3} - \frac{102}{(2d)^4} - \frac{729}{(2d)^5} - \frac{5533}{(2d)^6} - \frac{42229}{(2d)^7}$$
$$- \frac{288761}{(2d)^8} - \frac{1026328}{(2d)^9} + \frac{21070667}{(2d)^{10}} + \frac{780280468}{(2d)^{11}} + O\left(\frac{1}{(2d)^{12}}\right). \qquad (6.49)$$

Equations (6.48) and (6.49) more than double the length of the previously known series [10, 22, 39], which were known up to and including the term $-102(2d)^{-4}$ in (6.49). The error estimates are rigorous. The above expansions would appear to have radius of convergence zero, but there is no proof of this; it would be of interest to study their Borel summability. The critical temperature of the spherical model is known to have an asymptotic $1/d$ expansion with radius of convergence zero [13], and the suggestion that this is true rather generally for $1/d$ expansions of critical points was made in [11]. Note the change in sign at the term $(2d)^{-10}$; a similar sign change is observed in [13] for the critical temperature of the spherical model.

It is proved in [21] that for $d \ge 5$ the amplitudes $A$ and $D$ of Theorem 1 are given by the formulas

$$\frac{1}{A} = 2dz_c + \sum_{m=2}^{\infty} m\pi_m z_c^m, \qquad D = A\left[2dz_c + \sum_{m=2}^{\infty} r_m z_c^m\right], \qquad (6.50)$$

where $r_m = \sum_{x \in \mathbb{Z}^d} |x|^2 \pi_m(x)$. The formula for $A$ can be understood from the fact that $\gamma = 1$ for $d \ge 5$ and hence the susceptibility $\chi(z) = 1/F(z)$ should be given approximately by $[F'(z_c)(z - z_c)]^{-1}$, according to Taylor's theorem. The coefficient $c_n$ of $z^n$ is given in this approximation to be $[-z_c F'(z_c)]^{-1} z_c^{-n}$. This gives $A^{-1} = -z_c F'(z_c)$, and using the formula $F(z) = 1 - 2dz - \Pi(z)$ from (6.18) gives the above formula for $A$. The formula for $D$ can be found similarly, using the fact that $\sum_x |x|^2 c_n(x)$ is the coefficient of $z^n$ in $-\nabla^2 \hat{G}(k;z)|_{k=0}$, where $\nabla$ represents the gradient with respect to the vector $k$. To leading order, if we write $\hat{G}(k;z) = 1/\hat{F}(k;z)$, this is given by

$$-\nabla^2 \hat{G}(0;z) = \frac{\nabla^2 \hat{F}(0;z)}{\hat{F}(0;z)^2} \approx \frac{\nabla^2 \hat{F}(0;z_c)}{[F'(z_c)(z-z_c)]^2}, \tag{6.51}$$

where we have used $\hat{F}(0;z) = F(z)$. Expansion of the right-hand side in powers of $z$ then gives the desired formula for $D$.

It can be argued from (6.50) using an extension of (6.46) (see [8]) that

$$\frac{1}{A} = 2dz_c + \sum_{m=2}^{2N} \sum_{M=1}^{N} (-1)^M m \pi_m^{(M)} z_c^m + O(d^{-N-1}) \tag{6.52}$$

and

$$D = A \left[ 2dz_c + \sum_{m=2}^{2N} \sum_{M=1}^{N} (-1)^M r_m^{(M)} z_c^m \right] + O(d^{-N-1}), \tag{6.53}$$

with $r_m^{(M)} = \sum_{x \in \mathbb{Z}^d} |x|^2 \pi_m^{(M)}(x)$. Insertion of (6.48) and the enumerations of [8] for $m \le 24$, $M \le 12$ into (6.52) and (6.53) then give

$$A = 1 + \frac{1}{2d} + \frac{4}{(2d)^2} + \frac{23}{(2d)^3} + \frac{178}{(2d)^4} + \frac{1591}{(2d)^5} + \frac{15647}{(2d)^6} + \frac{164766}{(2d)^7} + \frac{1825071}{(2d)^8}$$
$$+ \frac{20875838}{(2d)^9} + \frac{240634600}{(2d)^{10}} + \frac{2684759873}{(2d)^{11}} + \frac{26450261391}{(2d)^{12}} + O\left(\frac{1}{(2d)^{13}}\right), \tag{6.54}$$

$$D = 1 + \frac{2}{2d} + \frac{8}{(2d)^2} + \frac{42}{(2d)^3} + \frac{284}{(2d)^4} + \frac{2296}{(2d)^5} + \frac{21024}{(2d)^6} + \frac{210306}{(2d)^7} + \frac{2242084}{(2d)^8}$$
$$+ \frac{24909542}{(2d)^9} + \frac{280764914}{(2d)^{10}} + \frac{3079111998}{(2d)^{11}} + \frac{29964810674}{(2d)^{12}} + O\left(\frac{1}{(2d)^{13}}\right). \tag{6.55}$$

This extends the series up to and including order $(2d)^{-5}$ that were reported in [12, 39] and [38] for $A$ and $D$, respectively, and also provides rigorous error estimates.

## 6.7 The Antiferromagnetic Singularity

There is strong numerical evidence that the number of self-avoiding walks on the lattice $\mathbb{Z}^d$ for $d = 2$ is given asymptotically by

$$c_n \sim \mu^n n^{\gamma-1} \left( A + \frac{a_1}{n} + \frac{a_2}{n^{3/2}} \cdots \right) + (-\mu)^n n^{\alpha-2} \left( b_0 + \frac{b_1}{n} + \cdots \right) \tag{6.56}$$

and for $d \ge 3$ by

$$c_n \sim \mu^n n^{\gamma-1} \left( A + \frac{a_1}{n^\theta} + \cdots \right) + (-\mu)^n n^{\alpha-2} \left( b_0 + \frac{b_1}{n^\theta} + \cdots \right) \tag{6.57}$$

(with a log correction when $d = 4$); see [28] for $d = 2$ and [8] for $d \geq 3$. Similar asymptotic behaviour applies also for the honeycomb lattice [30]. The $\mu^n n^{\gamma - 1}$ term is the familiar leading asymptotic form for $c_n$, but the $(-\mu)^n n^{\alpha - 2}$ term also appears, with the polygon exponent $\alpha$. These two terms are reflections of singularities of the generating function $\chi(z)$: one of the form $(1 - z/z_c)^{-\gamma}$ at $z = z_c$, and another of the form $(1 + z/z_c)^{1 - \alpha}$ at $z = -z_c$. The latter is referred to as the *antiferromagnetic singularity*.

The direct theoretical evidence for existence of the antiferromagnetic singularity seems to be rather thin, despite the fact that its existence has been recognised for decades [14]. The bipartite nature of $\mathbb{Z}^d$ and the honeycomb lattice plays an important role in the existence of the antiferromagnetic singularity: for example, numerical evidence shows that the antiferromagnetic singularity does not occur for the triangular lattice [29].

The hyperscaling relation $2 - \alpha = d\nu$ allows the leading behaviour in the second term of (6.57) to be rewritten as $(-\mu)^n n^{-d\nu}$, and the predicted value of $\nu$ implies that $d\nu > 1$ for all $d \geq 2$. This corresponds to a finite value for $\chi(-z_c)$, but indicates that derivatives of $\chi$ of order $d\nu - 1 = 1 - \alpha$ or higher will diverge at $-z_c$.

In this section, we first make a connection between the numerically observed sign alternation in the sequence $\pi_m$ and the existence of the antiferromagnetic singularity. We then draw parallels between the role of the polygon exponent $\alpha$ in the asymptotic behaviour of $\pi_m$ and in the asymptotic behaviour of the susceptibility near the antiferromagnetic singularity. Finally, we report the results of series analysis of $1/\chi(z)$ (equivalent to an analysis of $\Pi(z)$) via differential approximants, which provides the locations of the zeroes of the susceptibility.

This section is not devoted to rigorous results, but is a combination of heuristic arguments and numerical observations.

### 6.7.1 Sign Alternation of $\pi_m$

There is now significant numerical evidence that $\pi_m$ alternates in sign for nearest-neighbour self-avoiding walks on $\mathbb{Z}^d$ ($d \geq 1$) and on the honeycomb lattice, which are all bipartite lattices. For $d = 1$, the sign alternation is immediate from Example 3 of Section 6.3, where $\Pi(z) = -2 \sum_{m=2}^{\infty} (-z)^m$. The values of $\pi_m$ can be computed for $m \leq 71$ on the square lattice and for $m \leq 105$ on the honeycomb lattice using (6.22) and the enumeration of $c_n$ given in [28, 30]. In both cases, the signs are strictly alternating. For $d \geq 3$, the values of $\pi_m$ are known for $m \leq 30$ when $d = 3$ and for $m \leq 24$ for *all* $d \geq 4$, due to the lace graph enumerations of [8]. In all these cases, the signs are strictly alternating.

On the other hand, for the triangular lattice direct enumeration shows that $\pi_2 = \pi_3 = -6$, and the strict sign alternation fails. A similar result is observed for the fcc lattice. The triangular and fcc lattices are not bipartite.

A bipartite graph is characterised by the absence of odd cycles, which in the translation invariant case is equivalent to the vanishing of $\pi_m^{(1)}$ for all odd $m$. This means that we can write (recall (6.32) and (6.44)) for integer $m$

$$\pi_{2m} = -\sum_{M=1}^{m} \left( \pi_{2m}^{(2M-1)} - \pi_{2m}^{(2M)} \right) \qquad (6.58)$$

$$\pi_{2m-1} = \sum_{M=1}^{m-1} \left( \pi_{2m-1}^{(2M)} - \pi_{2m-1}^{(2M+1)} \right). \qquad (6.59)$$

For $\mathbb{Z}^d$, the values of $\pi_m^{(N)}$ are enumerated in [8] for $m \leq 30$, $d = 2,3$, and $m \leq 24$, $d \geq 4$. For this range of parameters, we have explicitly verified that $\pi_m^{(2M-1)} - \pi_m^{(2M)} > 0$ for even $m$, and that $\pi_m^{(2M)} - \pi_m^{(2M+1)} > 0$ for odd $m$. With (6.58) and (6.59), this gives rise to sign alternation for $\pi_m$. (It also raises the question of whether $\pi_m^{(N)} - \pi_m^{(N+1)}$ has a combinatorial interpretation.) The enumerations of [8] actually give a more refined version of the above two inequalities: they are obeyed also if $\pi_m^{(N)}$ is replaced by $\pi_{m,\delta}^{(N)}$, which counts the number of lace graphs which occupy exactly $\delta$ dimensions.

If the sequence $\pi_m$ does indeed strictly alternate in sign for sufficiently large $m$, then its generating function $\Pi(z) = \sum_{m=2}^{\infty} \pi_m z^m$ will have its dominant singularity on the *negative* real axis. We expect that on lattices such as $\mathbb{Z}^d$, the honeycomb lattice and the triangular lattice, the singularity of $\chi(z)$ at $z_c$ is not merely a pole, so that $\Pi(z)$ will also be singular at $z_c$. (Note, however, that for self-avoiding walks on a tree of degree $\Omega \geq 3$, Example 3 shows that $\chi(z)$ has a simple pole at $z_c = 1/(\Omega - 1) < 1$ while $\Pi(z)$ has its closest singularity at $-1$.) A singularity of $\Pi(z)$ at $-x_c$ with $0 < x_c < z_c$ is possible a priori, but since $\chi(z)$ is analytic in the disk of radius $z_c$ such a singularity of $\Pi(z)$ could only be a pole, corresponding to a zero of $\chi(z)$. We find no numerical evidence for this possibility on $\mathbb{Z}^d$ or the honeycomb lattice, as we discuss below. Assuming the absence of such a pole, we conclude that sign alternation of $\pi_m$ corresponds to a singularity of $\Pi(z)$ at $-z_c$, and that this is the origin of the antiferromagnetic singularity.

There is currently no proof that $\pi_m$ remains strictly alternating on the honeycomb lattice or on $\mathbb{Z}^d$ for any $d \geq 2$. It is an appealing though perhaps difficult problem to prove this.

### 6.7.2 $\pi_m$ and the Polygon Exponent $\alpha$

The asymptotic form (6.57) implies that $1/\chi(z)$ should vanish like $(1 - z/z_c)^\gamma$ near $+z_c$, while near $-z_c$ its singular part should behave like $(1 + z/z_c)^{1-\alpha} = (1 + z/z_c)^{d\nu-1}$. Consider first the dimensions $d = 2,3$. For $d = 2,3$, the numerical values of $\gamma$ and $\nu$ are such that the antiferromagnetic singularity is the dominant one, i.e., $d\nu - 1 < \gamma$. Since $\pi_m$ is the $m^{\text{th}}$ coefficient in the series for $1/\chi(z)$, this suggests that

$$\pi_m \sim c(-\mu)^m m^{\alpha-2} \quad (d = 2,3). \qquad (6.60)$$

Recall that $\pi_m = \sum_{N=1}^{\infty}(-1)^N \pi_m^{(N)}$ is the alternating sum of lace graph counts. The term $\pi_m^{(1)}$ just counts self-avoiding returns, whose asymptotic behaviour is predicted to be given by $\mu^m m^{\alpha-2}$, again with the polygon exponent. It is tempting to guess that $\pi_m^{(N)}$ has this same asymptotic form for all $N$, but we have insufficient data to verify this numerically. If it does, then the form (6.60) would arise from an alternating sum of counts of objects which are each governed by the polygon exponent $\alpha$.

The case $d = 4$ is delicate due to logarithmic corrections, but dimensions $d > 4$ are also more subtle. For $d > 4$, the error term in (6.40) can be expected to be sharp, and to be the source of a singularity $(1 - z/z_c)^{(d-2)/2}$ in the susceptibility. Since the power $1 - \alpha$ is also equal to $\frac{d-2}{2}$, this indicates that the ferromagnetic and antiferromagnetic singularities are of equal strength, so that (6.60) should be replaced by

$$\pi_m \sim [c_+ + c_-(-1)^m]\mu^m m^{-d/2} \quad (d > 4). \tag{6.61}$$

The observed fact that the sequence $\pi_m$ does alternate in sign for $m \leq 24$ suggests that the antiferromagnetic singularity still dominates, in the sense that $|c_-| > |c_+|$.

For dimensions $d > 4$, it is proved in [23, 24] that for spread-out models with $d > 4$ and $L$ large (see (6.2)), $\pi_m^{(N)} \leq (cL^{-d})^N \mu^m m^{-d/2}$ and therefore $|\pi_m| \leq \sum_{N=1}^{\infty} \pi_m^{(N)} \leq CL^{-d}\mu^m m^{-d/2}$. This verifies (6.61) as a one-sided bound, and supports the belief that $\pi_m^{(N)}$ may have the same asymptotic form for all $N$. In particular, it proves that $\Pi(z)$ cannot have a singularity inside the disk of radius $z_c$. Universality suggests that the same behaviour should apply also for the nearest-neighbour model.

For dimensions $d = 2, 3$, there are no rigorous results to help verify (6.60), so it is useful to turn to series analysis. The extensive methodology that has been developed for series analysis is discussed in Chapter 8 by Guttmann and Jensen (a classic earlier reference is [16]). An important method of series analysis is that of differential approximants, a powerful technique which is quite generally applicable and often (although not always) more effective than any other known method. The method of differential approximants can provide information about the singularities of a generating function for which we know a finite number of series coefficients, including poles, power law singularities and their confluent corrections.

We have applied the method of differential approximants to analyse the reciprocal series $1/\chi(z)$ for the square, honeycomb and triangular lattices, and for $\mathbb{Z}^d$ with $3 \leq d \leq 8$ (the susceptibility series itself was analysed in the original papers [8, 28, 29, 30]). Long series which are approaching the asymptotic regime are the most amenable to series analysis, and we obtain our most accurate results for the 2-dimensional lattices. Information for $\mathbb{Z}^d$ with $d \geq 3$ is far more difficult to extract owing to the availability of only relatively short series [8], and the existence of strong corrections to scaling for $d = 3$.

In all cases, we find that $\Pi(z)$ clearly has radius of convergence $z_c$ with leading singularity at the ferromagnetic singularity $-z_c$. This confirms the leading exponential growth $(-\mu)^m$ of (6.60). However, we find that information about critical exponents is degraded with reciprocal series, and more accurate estimates for the power law correction to the leading exponential behaviour $(-\mu)^n$ for $\pi_m$ are obtained from the prior analyses of $\chi(z)$ than we find via analysis of $1/\chi(z)$. (Dlog Padé approximants give the same results for $\chi(z)$ and $1/\chi(z)$, but the dlog Padé method is not as accurate as more general differential approximants, which give different results.)

### 6.7.3 Zeroes of the Susceptibility

Although the analysis of $1/\chi(z)$ via differential approximants did not prove to be fruitful for an accurate determination of the exponent $\alpha$, such analysis does yield the location of susceptibility zeroes, which correspond to poles of $1/\chi(z)$. In this section, we report the location of zeroes of $\chi(z)$ found in this manner.

*Square lattice.* For the square lattice, the analysis of $\chi(z)$ in [28] clearly confirms the existence of the antiferromagnetic singularity at $-z_c = -0.379052\ldots$ with the polygon exponent $\alpha = \frac{1}{2}$. We find that the differential approximants for $1/\chi(z)$, most clearly the first-order inhomogeneous approximants, detect a pole at $z^* \approx -0.3758$. This would seem to imply the unexpected result that the radius of convergence of $\Pi(z)$ is strictly less than $z_c$. However, direct integration of the differential equations of the differential approximant method to determine the amplitude of $\chi(z)$ at this point shows that $\chi(z^*) > 0$. The apparent contradiction can be resolved as follows:
Let us assume that

$$\chi(z) \sim A(z)(1 + z/z_c)^{1/2} + B(z) \tag{6.62}$$

near $z = -z_c$ (with the critical amplitude $A(z)$ and background amplitude $B(z)$ analytic at $-z_c$), and also assume that $\chi(z) > 0$ for $z \in (-z_c, 0]$. If we integrate the differential equation from the differential approximant for $\chi(z)$ along the negative axis to a point $z = -z_c + \varepsilon$, we obtain a function of the form (6.62), with approximate values for the exponent and $z_c$. If we then integrate the differential equation around the circle $|z + z_c| = \varepsilon$, starting and ending at $z = -z_c + \varepsilon$, the non-analytic part will pick up a factor of $\exp(\pi i) = -1$. We are now on a different Riemann sheet, and it is possible for $\chi(z)$ to have a zero provided that there is a solution of $-A(z)(1 + z/z_c)^{1/2} + B(z) = 0$. We confirm this numerically by using the procedure of Velgakis et al. [45] to find $A(z)$ and $B(z)$ in the vicinity of $z = -z_c$, and observe that there is a zero at a value close to $z^* = -0.3758$. Thus the pole observed in the analysis of $1/\chi(z)$ in fact corresponds to a zero of $\chi(z)$ in the Riemann sheet visited by circling around the singularity at $-z_c$. We do not detect any other poles in $1/\chi(z)$.

*Honeycomb lattice.* For the honeycomb lattice, the antiferromagnetic singularity is clearly seen in [30]. Without performing a careful error analysis we find, via differential approximant analysis, a complex conjugate pair of simple poles of $1/\chi(z)$ at $z = -0.262426 \pm 0.676916i$, with the error likely to be confined to the final digit. These poles lie outside the circle $|z| = z_c$. We confirm that these are genuine zeroes of $\chi(z)$ by integrating the differential equations obtained via the differential approx-

imant method to obtain a representation of $\chi(z)$, and confirming that the amplitude
of $\chi(z)$ is very small in the vicinity of these points.

*Triangular lattice.* For the triangular lattice there is, as expected, no sign of the antiferromagnetic singularity [28]. Analysis of $1/\chi(z)$ reveals two complex conjugate
pairs of poles at $z = -0.464 \pm 0.331i$, and $z = -0.204 \pm 0.611i$, well outside the radius of convergence determined by $z_c = 0.24091\ldots$. We have again confirmed that
these poles correspond to zeroes of the susceptibility.

*Dimensions* $3 \le d \le 8$. Using differential approximants and the enumerations of [8],
we clearly observe the antiferromagnetic singularity at $-z_c$, but we find no evidence
of any poles for $1/\chi(z)$ anywhere in the complex plane, and thus no evidence of
zeroes of the susceptibility.

As a technical point, we find that the first-order inhomogeneous approximants
seem to be much more effective than higher order approximants at pinpointing the
location of the poles of $1/\chi(z)$. This is probably due to the fact that the more restrictive functional form, which does not allow for confluent corrections, is more
appropriate for fitting the function in the immediate vicinity of the pole.

## 6.8 Lattice Trees

The lace expansion has been applied to a wide range of models [42]. The simplest
extension beyond self-avoiding walks is to lattice trees. In this section we discuss
some results for lattice trees in dimensions $d > 8$, without entering into details about
the methods of proof.

A *lattice tree* on $\mathbb{Z}^d$ is defined to be a finite connected[2] set of bonds which
contains no cycles (closed loops). Bonds are pairs $\{x,y\}$ of vertices of $\mathbb{Z}^d$, with
$y - x \in \mathcal{N}$, where $\mathcal{N}$ is given either by the nearest-neighbour set (6.1) or the spread-
out set (6.2). Although a lattice tree $T$ is defined as a set of bonds, we will write $x \in T$
if $x$ is an element of a bond in $T$. The number of bonds in $T$ is denoted $|T|$, and the
number of vertices in $T$ is thus $|T| + 1$.

A basic combinatorial problem is to count the number of lattice trees of fixed
size. Let $t_n^{(1)}$ denote the number of $n$-bond lattice trees that contain the origin. It is
customary to count lattice trees modulo translation, namely to consider $t_n$ defined
by

$$t_n = \frac{1}{n+1} t_n^{(1)}. \tag{6.63}$$

A sub-additivity argument [33] shows that there is a positive constant $\lambda$ such that
$\lim_{n \to \infty} t_n^{1/n} = \lambda$. The precise asymptotic behaviour of $t_n$ as $n \to \infty$ is believed to be
given by

$$t_n \sim A\lambda^n n^{-\theta}, \tag{6.64}$$

where $\theta$ is a universal critical exponent. The bounds

---

[2] This is the standard graph-theory definition of a connected graph.

$$c_1\lambda^n n^{-c_2\log n} \leq t_n \leq c_3\lambda^n n^{-(d-1)/d}, \tag{6.65}$$

were proved respectively in [27] and [37] for general dimensions $d \geq 2$. The upper bound does provide a power law correction, but it is predicted that $\theta > (d-1)/d$ for all $d \geq 2$.

Let $\bar{x}_T = (|T|+1)^{-1}\sum_{x\in T} x$ denote the centre of mass of $T$ (considered as a set of equal masses at the *vertices* of $T$), and let

$$R(T)^2 = \frac{1}{|T|+1}\sum_{x\in T}|x-\bar{x}_T|^2 \tag{6.66}$$

denote the squared radius of gyration of $T$. The typical length scale of a lattice tree is characterized by the average radius of gyration $R_n$, defined by

$$R_n^2 = \frac{1}{t_n^{(1)}}\sum_{T:|T|=n,T\ni 0} R(T)^2. \tag{6.67}$$

It is predicted that there is a universal critical exponent $v$ such that

$$R_n \sim Dn^v. \tag{6.68}$$

Based on a field theoretic representation, it was argued in [35] that the upper critical dimension for lattice trees is 8. Further evidence for this was given in [2, 43, 44]. The mean-field values of the exponents are $\theta = \frac{5}{2}$ and $v = \frac{1}{4}$. The value $v = \frac{1}{4}$ corresponds in (6.68) to $n$ being asymptotic to a multiple of $R_n^4$, which is a statement of 4-dimensionality. The fact that two 4-dimensional objects generically do not intersect above eight dimensions gives a quick prediction that $d = 8$ is the upper critical dimension for lattice trees.

The lace expansion has been used to prove a number of results for lattice trees in dimensions $d > 8$. The following theorem from [20] proves that $\theta = \frac{5}{2}$ and $v = \frac{1}{4}$ in high dimensions.

**Theorem 6.** *For nearest-neighbour lattice trees with $d$ sufficiently large, or for spread-out lattice trees with $d > 8$ and $L$ sufficiently large, there are positive constants $A$ and $D$ (depending on $d,L$) such that for every $\varepsilon < \min\{\frac{1}{2},\frac{d-8}{4}\}$,*

$$t_n = A\lambda^n n^{-5/2}[1+O(n^{-\varepsilon})], \tag{6.69}$$
$$R_n = Dn^{1/4}[1+O(n^{-\varepsilon})]. \tag{6.70}$$

A *lattice animal* is a finite connected set of bonds which may contain closed loops. It is believed that lattice animals belong to the same universality class as lattice trees, so that both models have the same critical exponents and scaling limits. Results related to Theorem 6 have been obtained for lattice animals, in terms of generating functions [18].

Information about the spatial distribution of lattice trees is contained in the number $t_n(x)$ of $n$-bond lattice trees containing the vertices $0, x \in \mathbb{Z}^d$. The scaling behaviour of the Fourier transform of $t_n(x)$ in high dimensions is given in the following theorem from [9].

**Theorem 7.** *For nearest-neighbour lattice trees with $d$ sufficiently large, or for spread-out lattice trees with $d > 8$ and $L$ sufficiently large, as $n \to \infty$,*

$$\hat{t}_n^{(2)}(kD_1^{-1}n^{-1/4}) \sim A\lambda^n n^{-1/2} \int_0^\infty dt\, t\, e^{-t^2/2} e^{-|k|^2 t/2d}, \qquad (6.71)$$

*where $D_1 = 2^{3/4}\pi^{-1/4}D$, and where $A$ and $D$ are the constants of Theorem 6.*

The scaling of the Fourier variable $k$ by $kn^{-1/4} = kn^{-\nu}$ in (6.71) corresponds to rescaling the lattice $\mathbb{Z}^d$ to $n^{-1/4}\mathbb{Z}^d$, and (6.71) is a statement about the scaling limit of $t_n^{(2)}(x)$ in Fourier language. Theorem 7 provides a first step in understanding the scaling limit of lattice trees in dimensions $d > 8$—the full scaling limit has been obtained by proving corresponding statements for the number of $n$-bond lattice trees containing vertices $x_1, \ldots, x_l$ for all $l \geq 1$. Under the hypotheses of Theorem 7, the scaling limit for $d > 8$ has been shown to be given by the random measure on $\mathbb{R}^d$ known as integrated super-Brownian excursion (ISE) [9, 42], as was first conjectured by Aldous [1]. In a somewhat different formulation, the scaling limit can also be interpreted as the measure-valued stochastic process known as the canonical measure of super-Brownian motion [25]. This is part of a larger story in which the scaling limit of various high-dimensional critical branching models can be understood in terms of super-Brownian motion [41, 42].

A mathematically rigorous analysis of critical exponents for lattice trees in low dimensions appears to be beyond the reach of current methods. However, there has been recent progress by Brydges and Imbrie [3, 4] on a very natural model of continuous branched polymers in $\mathbb{R}^d$, which is expected to be in the same universality class as lattice trees. Inspired by ideas of Parisi and Sourlas [40], in their remarkable paper [3] Brydges and Imbrie proved existence of critical exponents for their continuum model in dimensions $d = 2$ and 3 (with partial results for $d = 4$), with values $\theta = 1$ for $d = 2$, and $\theta = \frac{3}{2}$, $\nu = \frac{1}{2}$ for $d = 3$. An alternate approach to the results of Brydges and Imbrie has recently been obtained by Kenyon and Winkler [31].

# Acknowledgements

NC acknowledges support by the Australian Research Council and the Centre of Excellence for Mathematics and Statistics of Complex Systems. The work of GS was supported in part by NSERC of Canada.

# References

1. D. Aldous. Tree-based models for random distribution of mass. *J. Stat. Phys.*, **73**:625–641, (1993).
2. A. Bovier, J. Fröhlich, and U. Glaus. Branched polymers and dimensional reduction. In K. Osterwalder and R. Stora, editors, *Critical Phenomena, Random Systems, Gauge Theories*, Amsterdam, (1986). North-Holland.
3. D.C. Brydges and J.Z. Imbrie. Branched polymers and dimensional reduction. *Ann. Math.*, **158**:1019–1039, (2003).
4. D.C. Brydges and J.Z. Imbrie. Dimensional reduction formulas for branched polymer correlation functions. *J. Stat. Phys.*, **110**:503–518, (2003).
5. D.C. Brydges and J.Z. Imbrie. End to end distance from the Green's function for a hierarchical self-avoiding walk in four dimensions. *Commun. Math. Phys.*, **239**:523–547, (2003).
6. D.C. Brydges and J.Z. Imbrie. Green's function for a hierarchical self-avoiding walk in four dimensions. *Commun. Math. Phys.*, **239**:549–584, (2003).
7. D.C. Brydges and T. Spencer. Self-avoiding walk in 5 or more dimensions. *Commun. Math. Phys.*, **97**:125–148, (1985).
8. N. Clisby, R. Liang, and G. Slade. Self-avoiding walk enumeration via the lace expansion. *J. Phys. A: Math. Theor.*, **40**:10973–11017, (2007).
9. E. Derbez and G. Slade. The scaling limit of lattice trees in high dimensions. *Commun. Math. Phys.*, **193**:69–104, (1998).
10. M.E. Fisher and D.S. Gaunt. Ising model and self-avoiding walks on hypercubical lattices and "high-density" expansions. *Phys. Rev.*, **133**:A224–A239, (1964).
11. M.E. Fisher and R.R.P. Singh. Critical points, large-dimensionality expansions, and the Ising spin glass. In G.R. Grimmett and D.J.A. Welsh, editors, *Disorder in Physical Systems*. Clarendon Press, Oxford, (1990).
12. D.S. Gaunt. $1/d$ expansions for critical amplitudes. *J. Phys. A: Math. Gen.*, **19**:L149–L153, (1986).
13. P.R. Gerber and M.E. Fisher. Critical temperatures of classical $n$-vector models on hypercubic lattices. *Phys. Rev. B*, **10**:4697–4703, (1974).
14. A.J. Guttmann and S.G. Whittington. Two-dimensional lattice embeddings of connected graphs of cyclomatic index two. *J. Phys. A: Math. Gen.*, **11**:721–729, (1978).
15. A.J. Guttmann. Correction to scaling exponents and critical properties of the $n$-vector model with dimensionality $> 4$. *J. Phys. A: Math. Gen.*, **14**:233–239, (1981).
16. A.J. Guttmann. Asymptotic analysis of power-series expansions. In C. Domb and J.L. Lebowitz, editors, *Phase Transitions and Critical Phenomena,* Volume 13, pages 1–234. Academic Press, New York, (1989).
17. J.M. Hammersley. The number of polygons on a lattice. *Proc. Camb. Phil. Soc.*, **57**:516–523, (1961).
18. T. Hara and G. Slade. On the upper critical dimension of lattice trees and lattice animals. *J. Stat. Phys.*, **59**:1469–1510, (1990).
19. T. Hara and G. Slade. The lace expansion for self-avoiding walk in five or more dimensions. *Rev. Math. Phys.*, **4**:235–327, (1992).
20. T. Hara and G. Slade. The number and size of branched polymers in high dimensions. *J. Stat. Phys.*, **67**:1009–1038, (1992).
21. T. Hara and G. Slade. Self-avoiding walk in five or more dimensions. I. The critical behaviour. *Commun. Math. Phys.*, **147**:101–136, (1992).
22. T. Hara and G. Slade. The self-avoiding-walk and percolation critical points in high dimensions. *Combin. Probab. Comput.*, **4**:197–215, (1995).
23. R. van der Hofstad and G. Slade. A generalised inductive approach to the lace expansion. *Probab. Theory Related Fields*, **122**:389–430, (2002).
24. R. van der Hofstad and G. Slade. The lace expansion on a tree with application to networks of self-avoiding walks. *Adv. Appl. Math.*, **30**:471–528, (2003).

25. M. Holmes. Convergence of lattice trees to super-Brownian motion above the critical dimension. *Electr. J. Prob.*, **13**:671–755, (2008).

26. M. Holmes, A.A. Járai, A. Sakai, and G. Slade. High-dimensional graphical networks of self-avoiding walks. *Canad. J. Math.*, **56**:77–114, (2004).

27. E.J. Janse van Rensburg. On the number of trees in $\mathscr{Z}^d$. *J. Phys. A: Math. Gen.*, **25**:3523–3528, (1992).

28. I. Jensen. Enumeration of self-avoiding walks on the square lattice. *J. Phys. A: Math. Gen.*, **37**:5503–5524, (2004).

29. I. Jensen. Self-avoiding walks and polygons on the triangular lattice. *J. Stat. Mech.*, P10008, (2004).

30. I. Jensen. Honeycomb lattice polygons and walks as a test of series analysis techniques. *J. Phys.: Conf. Series*, **42**:163–178, (2006).

31. R. Kenyon and P. Winkler. Branched polymers in 2 and 3 dimensions. To appear in *Amer. Math. Monthly*.

32. H. Kesten. On the number of self-avoiding walks. *J. Math. Phys.*, **4**:960–969, (1963).

33. D.J. Klein. Rigorous results for branched polymer models with excluded volume. *J. Chem. Phys.*, **75**:5186–5189, (1981).

34. G.F. Lawler, O. Schramm, and W. Werner. On the scaling limit of planar self-avoiding walk. In *Fractal geometry and applications: a jubilee of Benoît Mandelbrot, Part 2*, pages 339–364, Proc. Sympos. Pure Math., 72, Part 2. Amer. Math. Soc., Providence, RI, (2004).

35. T.C. Lubensky and J. Isaacson. Statistics of lattice animals and dilute branched polymers. *Phys. Rev.*, **A20**:2130–2146, (1979).

36. N. Madras. Bounds on the critical exponent of self-avoiding polygons. In R. Durrett and H. Kesten, editors, *Random Walks, Brownian Motion and Interacting Particle Systems*, Boston, (1991). Birkhäuser.

37. N. Madras. A rigorous bound on the critical exponent for the number of lattice trees, animals and polygons. *J. Stat. Phys.*, **78**:681–699, (1995).

38. A.M. Nemirovsky, K.F. Freed, T. Ishinabe, and J.F. Douglas. End-to-end distance of a single self-interacting self-avoiding polymer chain: $d^{-1}$ expansion. *Phys. Lett. A*, **162**:469–474, (1992).

39. A.M. Nemirovsky, K.F. Freed, T. Ishinabe, and J.F. Douglas. Marriage of exact enumeration and $1/d$ expansion methods: Lattice model of dilute polymers. *J. Stat. Phys.*, **67**:1083–1108, (1992).

40. G. Parisi and N. Sourlas. Critical behavior of branched polymers and the Lee–Yang edge singularity. *Phys. Rev. Lett.*, **46**:871–874, (1981).

41. E. Perkins. Super-Brownian motion and critical spatial stochastic systems. *Canad. Math. Bull.*, **47**:280–297, (2004).

42. G. Slade. *The Lace Expansion and its Applications*. Springer, Berlin, (2006). Lecture Notes in Mathematics Vol. 1879. Ecole d'Eté de Probabilités de Saint–Flour XXXIV–2004.

43. H. Tasaki. *Stochastic geometric methods in Statistical Physics and Field Theories*. PhD thesis, University of Tokyo, (1986).

44. H. Tasaki and T. Hara. Critical behaviour in a system of branched polymers. *Prog. Theor. Phys. Suppl.*, **92**:14–25, (1987).

45. M.J. Velgakis, G.A. Baker, Jr., and J. Oitmaa. Integral approximants. *Comput. Phys. Commun.*, **99**:307–322, (1997).

# Chapter 7
# Exact Enumerations

Ian G. Enting and Iwan Jensen

## 7.1 Generating Functions and Enumeration

Apart from the intrinsic combinatorial interest of the problem, enumerations of polygons and polyominoes play an important role in enumerative approaches to lattice statistics systems such as percolation and the Ising/Potts models. Consequently, the history of enumeration techniques for polygons and polyominoes is intertwined with the more general developments in enumerative techniques in statistical mechanics. Two accounts of techniques [21, 23], almost 30 years apart, span much of this development. As with expansions in statistical physics, the development of combinatorial enumeration of polygons and polyominoes has reflected the trend, noted in statistical mechanics by Wortis [39], of replacing combinatorial complexity by algebraic complexity. In the later sections of this chapter we describe some key steps in the evolution of enumeration techniques. The bulk of this chapter describes transfer matrix techniques, which represent a continuation of the trend towards increasingly algebraic techniques. These are introduced using staircase and convex polygons as examples in Section 3. Sections 4 and 5 consider the application of transfer matrix techniques to self-avoiding polygons and polyominoes respectively. Technical issues of implementation are described in Section 6.

As noted in Chapter 1, early interest in self-avoiding polygons arose from their role in series expansions for the Ising model [38]. High-temperature expansions for the Ising model free energy, in terms of the variable $\tanh(J/k_B T)$, involve sums over graphs with all vertices of even degree. On planar lattices, duality relations imply that the same class of graphs (on the dual lattice) is also required for low-temperature expansions of Ising models.

The thermodynamic analogy, defined by restricting the Ising sum to include only polygons, immediately suggests the investigation of the generating function

Ian G. Enting and Iwan Jensen
Department of Mathematics and Statistics, The University of Melbourne, Victoria, Australia, e-mail: ienting@unimelb.edu.au, I.Jensen@ms.unimelb.edu.au

$$P(x) = \sum_n p_n x^n, \tag{7.1}$$

where $p_n$ is the number of polygons (per site) that can be embedded on a lattice. The Ising model sum for the free energy has a singularity with critical exponent $2 - \alpha$. The same exponent notation is used for the restricted sum defining the polygon generating function leading to the expression

$$P(x) \sim B(1 - x/x_c)^{2-\alpha} \qquad \text{with } 1/x_c = \mu, \tag{7.2}$$

where $B$ is a constant also known as a critical amplitude. The relation between the asymptotic behaviour of the $p_n$ and the singularity in $P(x)$ is discussed in Chapter 1. In a magnetic field the graphs contributing to the low-temperature expansion of the planar Ising model have a field dependent weighting proportional to the area enclosed by the graph. This analogy suggests an interest in area-weighting of planar polygons. Spatial moments for polygons also have analogues in spin systems and the notation for the exponents reflects this analogy. Polyomino enumerations are by contrast more closely related to expansions of the statistics of various percolation models.

Exact enumeration of polygons and polyominoes is limited by the rate of growth in computing requirements. In direct enumeration techniques that effectively enumerate each polygon individually, a growth of at least $\mu^n$ is inevitable. In contrast, the 'transfer matrix' techniques described below, perform sums over classes of polygon segments. For direct enumerations, the primary issue is the rate of growth in CPU time. For transfer matrix techniques, the growth in memory requirements can also be a significant constraint. Another constraint on indirect techniques arises when additional information is required—properties such as area, spatial moments and interaction counts. In direct enumerations that create a representation of each object, such information can generally be extracted with only a cost in time. For indirect enumerations there will be a similar time penalty for dealing with more information and maybe an additional penalty in both time and memory space, through having to deal with more classes.

While the main focus of this chapter is on the enumeration of self-avoiding polygons that are otherwise unrestricted, a number of special cases are of interest, including various forms of convex polygons. Many restricted cases turn out to be exactly solvable, as described in Chapter 3. Convex polygons are related to the work in the present chapter because they appear as 'correction terms' in the transfer matrix techniques described later in the present chapter. Indeed a number of the exact solutions were first obtained, without proof, by fitting differential approximants [11] to such 'correction series'. Consideration of these correction relations led to the development of the 'pruning algorithm' [18], described below, which has led to the most extensive enumerations of polygons. In the present chapter, we also use some of the special cases to provide tutorial examples for our description of transfer matrix methods.

Polygons have been enumerated for the various regular crystal lattices in two and three dimensions, by perimeter, area and both. Other enumerations have been performed for hyper-cubic lattices and two-dimensional directed lattices. In addition, enumerations weighted by properties such as spatial moments and interaction count have been performed in a number of cases, particularly in two dimensions. Such auxiliary information can be counted as part of the enumeration, with varying degrees of difficulty. Direct enumerations and indirect approaches such as transfer matrix techniques differ in the relative difficulty of particular generalisations.

## 7.2 Cluster-Based Enumeration Techniques

### 7.2.1 General Principles

Martin [22] has noted that the first few terms of a series such as polygon enumerations are typically calculated simply by drawing the graphs. Often one counts as a single case only graphs that differ under all translational and rotational symmetries—'space types' in the terminology of Domb [5]. The counts for each type must be multiplied by the appropriate symmetry factor in order to give a count for 'oriented space types'. Figure 7.1 uses the triangular lattice case to illustrate the issues showing that the sum over space types × orientations leads to the series $P_{TRI}(x) = 2x^3 + 3x^4 + 6x^5 + 15x^6 + \ldots$

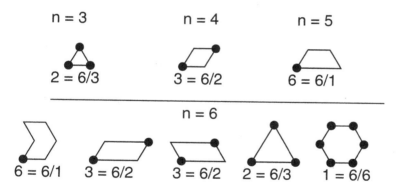

**Fig. 7.1** Space types for self-avoiding polygons of various sizes on the triangular lattice, with the integers giving the number of rotations that give translationally distinct cases. The number of distinct rotational cases is equal to 6 divided by the number of vertices in each class that is equivalent under rotations. The solid dots indicate one such equivalence class in each case shown.

The choice of enumerating such 'oriented space types' corresponds to analogous enumerations in lattice statistical mechanics. An alternative case, where the requisite statistic is the sum over space types, (i.e. without the rotational symmetry factor) occurs in studies of chemical isomers and has been considered by Vöge et al. [36].

### 7.2.2 Polyominoes

General issues of computer enumeration of clusters have been reviewed by Martin [22]. His account is more general than is mostly relevant for the polygon case. The most important point is the need for an easily-computed canonical labeling to ensure that each distinct object is counted exactly once.

In Ising model expansions, an important technique which is applicable to bipartite lattices is the method of partial generating functions [28]. This transforms the model by performing a summation over all possible states of one sublattice. The original series expansion is recovered by algebraic transformation of a multivariate series expansion on the remaining sublattice. This approach was extended to various forms of cluster expansion in a series of papers by Sykes and co-workers [29, 30, 31, 32, 33, 34]. An alternative approach to using sublattice expansions to enumerate percolation statistics was given by Enting [7]. This approach was based on the expression of bond percolation clusters as the $q \to 1$ limit of the $q$-state Potts model. In principle, a similar limit for loop expansions may be able to give a sublattice expansion for polygons, but it seems that this possibility has not been exploited.

### 7.2.3 Polygons

Rushbrooke and Eve [25] gave the enumeration of polygons on square and simple cubic lattices to 18 and 12 steps, respectively, all counted by hand, duplicating results of Wakefield [38] and unpublished results by Sykes. They also reported the count of 14-step simple cubic polygons obtained by 50 hours of machine time. They later [26] reported the count of 16-step simple cubic polygons, with a description of the algorithm:

- take an algorithm for counting self-avoiding walks, e.g. as described by Martin [21], and count the cases that return to the origin;
- exploit the symmetry by forcing the first step to be in the +X direction;
- terminate construction whenever the end-point gets too far from the origin to be able to return within the requisite number of steps;
- note that what is being counted is $2n\,p_n/z$ where $z$ is the lattice coordination number.

In addition, a number of 'counting theorems' were developed relating enumerations of various classes of graph. These were mainly applied to the enumeration of self-avoiding walks and are not considered here.

A 'dimerisation' approach, building paths from pairs of pre-computed walks, has been described by Torrie and Whittington [35]. Macdonald et al. [20] described an extension involving building paths from three pre-computed components.

An alternative form of enumeration was to extend walks and polygons by adding $k$ steps at a time. Algorithms of this type have recently been described by Clisby et al [2]. As with the finite lattice expansions described below, one of the key computational trade-offs is between execution time and storage requirements.

## 7.3 Transfer Matrix Techniques

We start by describing transfer matrix techniques for the enumeration of staircase polygons. We have already seen in Chapter 3 that this model can be solved exactly by both perimeter and area. However, the model serves perfectly as a gentle introduction to transfer matrix techniques.

### 7.3.1 Staircase Polygons

Recall that a staircase polygon can be seen as consisting of two directed walks starting at the origin, moving only to the right and up, and terminating once the walks join at a vertex. If we look at a diagonal line $x + y = k + 1/2$ then for any integer $k$ this line will intersect a polygon at 0 (miss the polygon) or 2 edges (intersect the polygon), see Fig. 7.2. We start with $k = 0$ such that the line intersects the first two edges of the staircase polygon. We then move the line upward (increase $k$ by 1) and as we do this we add an edge to each walk. There are only four new configurations corresponding to the four possible steps. We need only keep track of the gap between the two walks, where the gap is the minimal number of iterations required in order to join the two walks. As we move the line, the gap is either increased by a unit (the upper walk moves up and the lower walk moves right), decreased by a unit (the upper walk moves right and the lower walk moves up) or remains constant in two possible ways (both walks move up or right). These moves are illustrated in the right-most panel of Fig. 7.2.

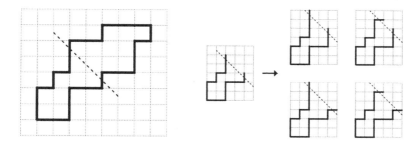

**Fig. 7.2** The left panel shows a typical staircase polygon bisected in two place by a boundary line (dashed line). On the right we show the four possible new configuration arising as the boundary line is moved one step forwards.

Let $C_j^{(k)}$ be the number of configurations with a gap of $j$ after $k$ iterations. We then have the following very simple algorithm: Set $C_1^{(0)} = x$ (where $x$ is a variable conjugate to the half-perimeter of the polygon). Run through all the possible values of the new gap $j = 1, \ldots, k+1$ and do the following updates: $C_{j+1}^{(k+1)} := xC_j^{(k)}$, $C_{j-1}^{(k+1)} := xC_j^{(k)}$ and $C_j^{(k+1)} := 2xC_j^{(k)}$. Here $a := b$ is short-hand for assign to $a$ the value $a + b$. The coefficient of the term $xC_1^{(k)}$ is the number of staircase polygons of half-perimeter $k+2$. Note that the use of the variable $x$ is somewhat superfluous in the case of staircase polygons since the generating function at iteration $k$ is just $x^{k+1}C_j^{(k)}$, but it is included here for reasons of generality and in most other cases the generating function will be a (non-trivial) polynomial in $x$. Naturally we need not actually keep all the entries $C_j^{(k)}$ since only the current and subsequent values are needed for the calculation so we can replace $C_j^{(k)}$ with $C_j^{(k')}$, where $k' = k \bmod 2$. We just have to initially set to zero all entries in the next step and keep a running total $S(k)$ of the number of staircase polygons. In almost all applications of enumeration algorithms we are only interested in calculation up to a certain pre-set maximal order $n$. This means that we can ignore some further entries in $C_j^{(k)}$ because they correspond to counting staircase polygons larger than $n$. More precisely we have that the half-perimeter is increased by 1 in each iteration and a gap of size $j$ requires a further $j$ iterations to close so that the entry in $C_j^{(k)}$ contributes to staircase polygons of size at least $k + j$. Thus if $k + j > n$ we can ignore this entry (in particular we always require that the gap $j \leq n/2$). This is an example of what we shall refer to as *pruning*, that is to say the discarding of any superfluous configurations.

Formally we can view the transformation from the set of states $C_j^{(k)}$ to $C_j^{(k+1)}$ as a matrix multiplication (hence our use of the nomenclature *transfer matrix algorithm*) with $k$ counting the number of iterations of the transfer matrix algorithm. However, as can be readily seen from the algorithm the transfer matrix is extremely sparse and there is generally no reason to list it explicitly (it is given *implicitly* by the updating rules). However, to illustrate the point we shall list the transition (or transfer) matrix $T$, which has entries

$$t_{i,j} = \begin{cases} 2 & i = j \\ 1 & |i - j| = 1, \\ 0 & \text{otherwise}, \end{cases}$$

explicitly

$$T = \begin{pmatrix} 2 & 1 & 0 & 0 & 0 & 0 & 0 & \ldots \\ 1 & 2 & 1 & 0 & 0 & 0 & 0 & \ldots \\ 0 & 1 & 2 & 1 & 0 & 0 & 0 & \ldots \\ 0 & 0 & 1 & 2 & 1 & 0 & 0 & \ldots \\ 0 & 0 & 0 & 1 & 2 & 1 & 0 & \ldots \\ \vdots & \vdots & \vdots & & \ddots \end{pmatrix}$$

The number of staircase polygons of perimeter $2n$ is the $(1,1)$ element $t_{1,1}^{(n-1)}$ in $T^{n-1}$. In this way we find the series $P_{\text{stair}}(x) = x^2 + 2x^3 + 5x^4 + 14x^5 + \cdots$.

## 7.3.1.1 Enumeration by Area

The use of generating functions when counting staircase polygons by perimeter is trivial. A less trivial and therefore more illuminating case is the enumeration of staircase polygons by area. A configuration with gap $j$ contributes an addition of $j$ units of area. So with $q$ the variable conjugate to area we get the following algorithm: Set $C_1^{(0)} = 1$. Run through all possible gaps $j = 1, \ldots, k+1$ and do the following updates: $C_{j+1}^{(k+1)} :\overset{+}{=} q^j C_j^{(k)}$, $C_{j-1}^{(k+1)} :\overset{+}{=} q^j C_j^{(k)}$ and $C_j^{(k+1)} :\overset{+}{=} 2q^j C_j^{(k)}$. The result after the first few iterations is:

$$C_1^{(0)} = 1$$
$$C_1^{(1)} = 2q C_1^{(0)} = 2q$$
$$C_2^{(1)} = q C_1^{(0)} = q$$
$$C_1^{(2)} = 2q C_1^{(1)} + q^2 C_2^{(1)} = 4q^2 + q^3$$
$$C_2^{(2)} = q C_1^{(1)} + 2q^2 C_2^{(1)} = 2q^2 + 2q^3$$
$$C_3^{(2)} = q^2 C_2^{(1)} = q^3$$
$$C_1^{(3)} = 2q C_1^{(2)} + q^2 C_2^{(2)} = 8q^3 + 4q^4 + 2q^5$$
$$C_2^{(3)} = q C_1^{(2)} + 2q^2 C_2^{(2)} + q^3 C_3^{(2)} = 4q^3 + 5q^4 + 4q^5 + q^6$$
$$C_3^{(3)} = q^2 C_2^{(2)} + 2q^3 C_3^{(2)} = 2q^4 + 2q^5 + 2q^6$$
$$C_4^{(3)} = q^3 C_3^{(2)} = q^6$$

Notice that in this case (unlike in the enumeration by perimeter) the entries $C_j^{(k)}$ are non-trivial polynomials in $q$. The term $q C_1^{(k)}$ counts the contribution to the area generating function of staircase polygons of half-perimeter $k+1$

$$\mathscr{A}(q) = \sum_k q C_1^{(k)}.$$

In matrix form we have

$$T = \begin{pmatrix} 2q & q & 0 & 0 & 0 & 0 & 0 & \ldots \\ q^2 & 2q^2 & q^2 & 0 & 0 & 0 & 0 & \ldots \\ 0 & q^3 & 2q^3 & q^3 & 0 & 0 & 0 & \ldots \\ 0 & 0 & q^4 & 2q^4 & q^4 & 0 & 0 & \ldots \\ 0 & 0 & 0 & q^5 & 2q^5 & q^5 & 0 & \ldots \\ \vdots & \vdots & \vdots & & \ddots & & & \end{pmatrix}$$

In this case pruning is a little more interesting. The creation of a gap $j$ involves inserting at least $\sum_{i=1}^{j-1} i$ units of area and closing the gap takes at least $\sum_{i=1}^{j} i$ units of area for a total of at least $j^2$. If $k \geq j$ a further $k + 1 - j$ must have been used. So if $j^2 + k + 1 - j > n$ we can discard this entry.

In this way we find the series $\mathscr{A}(q) = q + 2q^3 + 4q^3 + \cdots$.

### 7.3.2 Convex Polygons

Convex polygons have been enumerated by Guttmann and Enting using transfer matrix techniques [12]. Here we briefly describe their algorithm. The number of convex polygons can be counted by considering bounding rectangles of width $w$ and length $l$. Let $C_{w,l}$ denote the number of convex polygons with minimal bounding rectangle $w \times l$. The convexity constraint implies that all such convex polygons have half-perimeter $w + l$. Due to the symmetry of the square lattice $C_{w,l} = C_{l,w}$ we need consider only rectangles with $l \geq w$. The half-perimeter generating function for convex polygons is

$$C(x) = \sum_{w=1}^{\infty} C_{w,w} x^{2w} + 2 \sum_{w=1}^{\infty} \sum_{l=w+1}^{\infty} C_{w,l} x^{w+l}$$

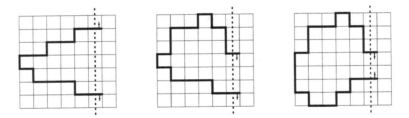

**Fig. 7.3** Examples of the types of configurations counted by $N_{i,j}^l$, $U_{i,j}^l$ and $B_{i,j}^l$ respectively. The arrows indicate the directions in which the edges must move.

A vertical boundary line at position $k - \frac{1}{2}$, $k > 0$, along the $x$-axis intersects a partially completed convex polygon at 2 edges, as indicated by the dashed lines in Fig. 7.3. The edges are connected forming a loop which extends to the left border (column 0) of the rectangle. The upper edge can be extended upwards and to the right until it hits the upper border (at row $w$) of the rectangle. It then turns around and can be extended downwards and to the right. Similarly, the lower edge is extended downwards (and to the right) until it hits the lower border (at row 0) where it turns before being extended upwards and to the right. By moving the boundary to the right we can 'build' up the rectangle column by column. It suffices to keep track of the positions $i$ and $j$ of the two edges. Due to the nature of the problem we have that $0 \leq i < j \leq w$. For each pair $(i, j)$ there are four possible scenarios: The polygon has touched *none* of the borders, only the *upper* border, only the *lower*

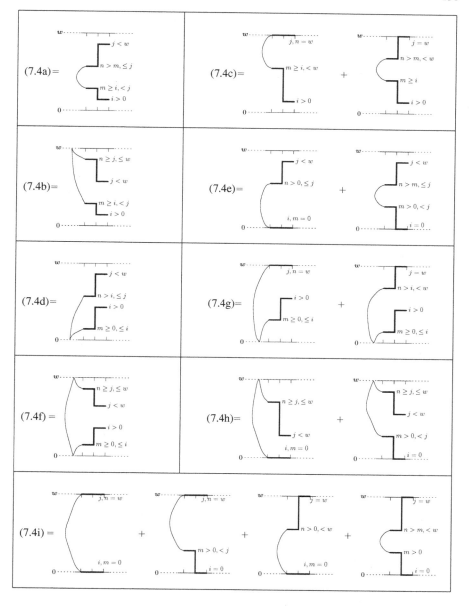

**Fig. 7.4** Graphical representation of the transfer matrix equations for convex polygons of width $w$. The thick solid lines shows the moves of the edges associated with the sums in Eqs. (7.4a)–(7.4i) as we move the boundary line (not shown) from position $k$ to position $k+1$. The thin lines between the edges shows (schematically) how they are connected to the left of column $k$ and whether or not the upper and/or lower border has been touched. The horizontal steps are indexed $i, j, m, n$ as in the equations and the corresponding limits of summation are also indicated.

border or *both* borders. To enumerate convex polygons of width $w$ we therefore introduce four families of matrices $N_{i,j}^k$, $U_{i,j}^k$, $L_{i,j}^k$, and $B_{i,j}^k$, corresponding to the four scenarios above. Here $N_{i,j}^k$ is the number of partially completed convex polygons after $k$ iterations with edges at positions $i$ and $j$ such that neither the lower nor the upper border of the rectangle has been touched. Similarly for the other three matrices. The number of convex polygons in a $w \times l$ rectangle is then $C_{w,l} = \sum_{i,j} B_{i,j}^l$.

Initially all entries in the matrices are set to 0 except:

$$N_{i,j}^1 = 1 \qquad 0 < i < j < w \tag{7.3a}$$

$$U_{i,w}^1 = 1 \qquad 0 < i < w \tag{7.3b}$$

$$L_{0,j}^1 = 1 \qquad 0 < j < w \tag{7.3c}$$

$$B_{0,w}^1 = 1 \tag{7.3d}$$

The updating rules for the matrices are:

$$N_{i,j}^{k+1} = \sum_{m=i}^{j-1} \sum_{n=m+1}^{j} N_{m,n}^k \qquad 0 < i < j < w \tag{7.4a}$$

$$U_{i,j}^{k+1} = \sum_{m=i}^{j-1} \sum_{n=j}^{w} U_{m,n}^k \qquad 0 < i < j < w \tag{7.4b}$$

$$U_{i,w}^{k+1} = \sum_{m=i}^{w-1} U_{m,w}^k + \sum_{m=i}^{w-2} \sum_{n=m+1}^{w-1} N_{m,n}^k \qquad 0 < i < w \tag{7.4c}$$

$$L_{i,j}^{k+1} = \sum_{m=0}^{i} \sum_{n=i+1}^{j} L_{m,n}^k \qquad 0 < i < j < w \tag{7.4d}$$

$$L_{0,j}^{k+1} = \sum_{n=1}^{j} L_{0,n}^k + \sum_{m=1}^{j-1} \sum_{n=m+1}^{j} N_{m,n}^k \qquad 0 < j < w \tag{7.4e}$$

$$B_{i,j}^{k+1} = \sum_{m=0}^{i} \sum_{n=j}^{w} B_{m,n}^k \qquad 0 < i < j < w \tag{7.4f}$$

$$B_{i,w}^{k+1} = \sum_{m=0}^{i} B_{m,w}^k + \sum_{m=0}^{i} \sum_{n=i+1}^{w-1} L_{m,n}^k \qquad 0 < i < w \tag{7.4g}$$

$$B_{0,j}^{k+1} = \sum_{n=j}^{w} B_{0,n}^k + \sum_{m=1}^{j-1} \sum_{n=j}^{w} U_{m,n}^k \qquad 0 < j < w \tag{7.4h}$$

$$B_{0,w}^{k+1} = B_{0,w}^k + \sum_{m=1}^{w-1} U_{m,w}^k + \sum_{n=1}^{w-1} L_{0,n}^k + \sum_{m=1}^{w-2} \sum_{n=m+1}^{w-1} N_{m,n}^k \tag{7.4i}$$

The updating of the matrix $N_{i,j}^k$ is given in eq. (7.4a) and illustrated graphically in Fig. 7.4. In this case neither the top nor the bottom border of the rectangle has been touched yet. The final positions $i$ and $j$ of the lower and upper edges must therefore satisfy $0 < i < j < w$. Let $m$ and $n$ be the position of the lower and upper edge, respectively, before moving the boundary a step forwards. The lower edge must move downwards while the upper edge must move upwards (before taking a step to the right). Hence $m$ can range from $i$ to $j - 1$ while $n$ can range from $m + 1$ to $j$. This ensures that the edges move in the prescribed direction and don't cross ($n > m$) and of course they are not allowed to 'over-shoot' and hence $m \geq i$ and $n \leq j$. The updating rules for the other matrices can be derived similarly. In general there is a formula for how the matrices are updated in the 'interior' of the rectangle when $0 < i < j < w$, see equations (7.4b), (7.4d) and (7.4f). In addition for $\boldsymbol{U}^k_{i,w}$, eq. (7.4c), (and $\boldsymbol{L}^k_{0,j}$, eq. (7.4e)) there is a formula for updating at the upper (lower) border where contributions from $\boldsymbol{N}^k_{m,n}$ have to be added. Finally for $\boldsymbol{B}^k_{i,j}$ there are the three special cases: firstly $\boldsymbol{B}^k_{i,w}$, eq. (7.4g), where the top border is touched (but the bottom is not), secondly $\boldsymbol{B}^k_{0,j}$, eq. (7.4h), where the bottom border is touched, and thirdly $\boldsymbol{B}^k_{0,w}$, eq. (7.4i), where both borders are touched.

## 7.4 Transfer Matrix Algorithms for Self-Avoiding Polygons

Enting [6] was the first to use transfer matrix techniques to enumerate self-avoiding polygons. The original application was to square lattice SAPs counted by perimeter. Following this initial work the method was generalised to other lattices and to enumeration by area. The next qualitative advance was the use of *pruning* by Jensen and Guttmann [18] to produce an exponentially faster algorithm. Jensen [16] has also implemented efficient parallel versions of the algorithm.

Below we briefly review the common features of the finite-lattice method and transfer matrix techniques. We then give a detailed description of the square lattice algorithm and a brief outline of the honeycomb and triangular lattice algorithms.

The first terms in the series for the polygon generating function can be calculated by using transfer matrix techniques to count the number of polygons in finite sub-lattices. The generating function for the number of SAPs per vertex of the infinite lattice is obtained by combining the contributions from these sub-lattices. On the square lattice the obvious and natural choice is rectangles $w$ cells wide and $l$ cells long. Due to the symmetry of the square lattice one need consider only rectangles with $l \geq w$. For other lattices there may be many 'natural' choices for the finite sub-lattices. However, for regular lattices one can generally always choose an implementation using $w \times l$ rectangles and in practice this is what we have always chosen to do.

In implementations of the finite lattice method two types of finite lattice generating functions have been widely used. The first of these is $L_{wl}(x)$ which counts the number of polygons of length exactly $l$ fitting within a $w \times l$ rectangle, that is polygons which touch the left and right side of the rectangle. The second type of generation function is $G_{wl}(x)$ which counts the number of polygons of length exactly $l$ and width exactly $w$, that is polygons which touch all the sides of the rectangle. $L_{wl}(x)$ counts polygons of width $w' < w$ several times, $w - w' + 1$ times to be precise, so that

$$L_{wl}(x) = \sum_{k=1}^{w} (w - k + 1)G_{wl}(x)$$

In applying the transfer matrix technique to the enumeration of polygons we regard them as sets of edges on the finite lattice with the properties:

(1)  A weight $x$ is associated with each occupied edge.
(2)  All vertices are of degree 0 or 2.
(3)  Apart from isolated sites, the graph has a single connected component.
(4a) In implementations using $L_{wl}(x)$ each graph must span the rectangle from left to right.
(4b) In implementations using $G_{wl}(x)$ each graph must span the rectangle from left to right and from bottom to top.

The most important change to the transfer matrix algorithm from the implementations in the previous section is that rather than adding a whole column at a time, we build up each column by adding a single lattice cell at a time (details will be given below). We shall refer to the boundary line configuration prior to a move as the 'source' and after the move as the 'target'. There are two major advantages to this approach. Firstly, the boundary line can intersect a generic SAP in many places (not just a few as in the previous simple examples) and the loops bisected by the boundary line can be nested within one another in a complicated fashion. This would make adding a column at a time quite complicated. By building up each column cell by cell the possible updates and transformations of the configurations along the boundary line become much simpler. Secondly, and more importantly, this implementation of the transfer matrix algorithm is much more efficient. The number of possible configurations along the boundary line grows exponentially with $w$, say as $N_{\text{Conf}} \propto \lambda^w$. If we add a whole column at once then each configuration can give rise to a sub-set of the total number of configurations of a size proportional to $N_{\text{Conf}}$. This means that the total number of operations required to add a column grows like $N_{\text{Conf}}^2 \propto \lambda^{2w}$. However, when adding a single cell at a time a given source configuration gives rise to only a few (one or two on the square lattice) target configurations. So the total number of operations grows like $wN_{\text{Conf}} \propto w\lambda^w$.

### 7.4.1 Square Lattice

In the original application [6], valid polygons were required to span the enclosing rectangle in the lengthwise direction. Clearly polygons with projection on the $y$-axis $< w$, that is polygons which are narrower than the width of the rectangle, are counted many times. As described above, it is easy to obtain the polygons of width exactly $w$ and length exactly $l$ from this enumeration. Any polygon spanning such a rectangle has a perimeter of length at least $2(w + l)$. By adding the contributions from all rectangles of width $w \leq w_{\max}$ (where the choice of $w_{\max}$ depends on available computational resources, as discussed below) and length $w \leq l \leq 2w_{\max} - w + 1$ (with contributions from rectangles with $l > w$ counted twice) the number of polygons per vertex of an infinite lattice is obtained correctly up to perimeter $4w_{\max} + 2$.

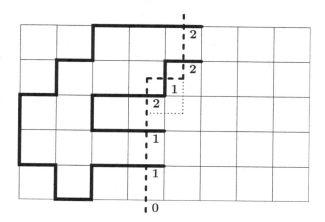

**Fig. 7.5** A snapshot of the boundary line (dashed line) during the transfer matrix calculation on the square lattice. Polygons are enumerated by successive moves of the kink in the boundary line, as exemplified by the position given by the dotted line, so that one vertex at a time is added to the rectangle. To the left of the boundary line we have drawn an example of a partially completed polygon. The numbers along the boundary line is the encoding of the edge states of the loops intersected by the boundary line.

The transfer matrix technique involves drawing a boundary line through the rectangle intersecting a set of up to $w + 2$ edges. Polygons in a given rectangle are enumerated by moving the boundary line so as to add one vertex or lattice cell at a time, as shown in Fig. 7.5. In this fashion we build up the rectangle column by column with each column built up vertex by vertex. As we move the boundary line it intersects partially completed polygons consisting of disjoint loops. Eventually all the loops must be connected to form a single polygon. For each configuration of occupied or empty edges along the intersection we maintain a (perimeter) generating function for open loops to the left of the line cutting the intersection in that particular pattern. The updating of the generating functions depends primarily on the states of the two edges at the kink in the boundary line prior to the move (we shall refer to these edges as the kink edges). As the boundary line is moved the two new edges intersected by the boundary line can be either empty or occupied. We shall now briefly outline how the constraints (1)–(4) can be satisfied.

Constraint (1) is trivial to satisfy. We just multiply the generating function of the source configuration by $x^j$, where $j$ is the number of newly added steps, and add the result to the generating function of the target configuration.

Constraint (2) is easy to satisfy. If both kink edges were empty we can leave both new edges empty or insert a partial new loop by occupying both of the new edges. If one of the kink edges was occupied then one of the new edges must also be occupied. If both of the kink edges are occupied both of the new edges must be empty. It is easy to see that these rules leads to graphs satisfying constraint (2).

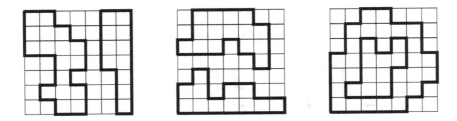

**Fig. 7.6** Three ways in which graphs with separate components could occur.

Constraint (3) is the most difficult to satisfy. We have shown some examples of two-component graphs in Fig. 7.6. Graphs of the type shown in the left-most panel, where separate components occur side by side, are quite easy to avoid by never allowing the insertion of a new loop into the totally empty configuration except while building up the first column. This also ensures that all polygons touch the left-most border of the rectangle. There are only two distinct ways in which a pair of loops can be placed relative to one another—side by side or nested—as shown in the left-most panels of Fig. 7.7.

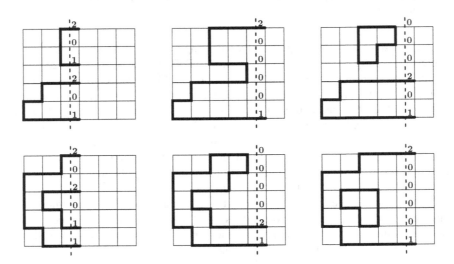

**Fig. 7.7** Illustration of how two partial loops can be placed relative to one another (left-most panels), how they can be connected to form a single loop (middle panels), and connections leading to graphs with more than one component (right-most panels). The numbers along the boundary line show how the configurations can be uniquely encoded in a computer program (see text for details).

The way to avoid situations leading to graphs with more than a single component, such as shown at the top right of Fig. 7.7, is to forbid a loop to close on itself if the boundary line intersects any other loops. So two loop ends can only be joined if they belong to different loops or all other edges are empty. To exclude loops which close on themselves we need to label the occupied edges in such a way that we can easily determine whether or not two loop ends belong to the same loop. The most obvious choice would be to give each loop a unique label. However, on two-dimensional lattices there is a more compact scheme relying on the fact that two loops can never intertwine. Each end of a loop is assigned one of two labels depending on whether it is the lower end or the upper end of a loop. Each configuration along the boundary line can thus be represented by a set of edge states $\{\sigma_i\}$, where

$$\sigma_i = \begin{cases} 0 & \text{empty edge,} \\ 1 & \text{lower end of a loop,} \\ 2 & \text{upper end of a loop.} \end{cases} \tag{7.5}$$

Configurations are read from the bottom to the top. The configuration along the intersection of the partially completed polygon in Fig. 7.5 is $\{0112122\}$ and further examples of this encoding are given in Fig. 7.7. It is easy to see that this encoding uniquely describes which loop-ends are connected. In order to find the upper loop-end, matching a given lower end, we start at the lower end and work upwards in the configuration counting the number of '1's and '2's we pass (the '1' of the initial lower end is *not* included in the count). We stop when the number of '2's exceeds the number of '1's. This '2' marks the matching upper end of the loop. It is worth noting that there are some restrictions on the possible configurations. Firstly, every lower loop-end must have a corresponding upper end, and it is therefore clear that the total number of '1's is equal to the total number of '2's. Secondly, as we look through the configuration starting from the bottom, the number of '1's is never smaller than the number of '2's. Ignoring the '0's, the '1's and '2's can be viewed as perfectly balanced parentheses. Those familiar with algebraic languages will immediately recognize that each configuration of labeled loop-ends forms a Motzkin word [4].

Constraint (4a) is automatically satisfied by the rules described above to satisfy constraint (3). In order to satisfy constraint (4b) we need to add more information to a configuration. In addition to the usual labeling of the intersection with the boundary line, we also have to indicate whether the partially completed polygon has reached neither, the lower, the upper or both borders of the rectangle.

In order to represent a given partial polygon we have to add some information to the usual set of edge states $\{\sigma_i\}$. We add two extra 'virtual' edge states $\sigma_b$ and $\sigma_t$, where $\sigma_b$ is 0 or 1 if the bottom of the rectangle hasn't or has been touched and similarly $\sigma_t$ is 0 or 1 if the top of rectangle hasn't or has been touched. A boundary state is denoted as $\{\sigma_0 \ldots \sigma_{w+1}; \sigma_b \sigma_t\}$. Note that when the boundary line is completely vertical it intersects only $w$ edges (rather than the usual $w+1$ edges when it has a kink) and $\sigma_{w+1}$ is thus unassigned. We mark this by a $*$ (though we encode it by a 0 in the actual program). Thus the set of edge states of the partial polygon in Fig. 7.5 is $\{0112122; 11\}$, since the polygon has touched both borders of the rectangle.

At first glance it would appear to be inefficient to calculate $G_{wl}(x)$ directly rather than use $L_{wl}(x)$ since for many boundary line configurations we now have to keep 4 distinct generating functions depending on which borders have been touched. However, as demonstrated in practice [18], it actually leads to an algorithm which is both exponentially faster and whose memory requirement is exponentially smaller. Realizing the full savings in time and memory usage require enhancements to the original algorithm. The most important is that of *pruning*. This procedure, details of which are given below, allows us to discard most of the possible configurations for large $w$ because they contribute only to polygons of length greater than $4w_{max} + 2$. Briefly this works as follows. Firstly, for each configuration we keep track of the current minimum number of steps $n_{cur}$ already inserted to the left of the boundary line in order to build up that particular configuration. Secondly, we calculate the minimum number of additional steps $n_{add}$ required to produce a valid polygon. There are three contributions, namely the number of steps required to close the polygon, the number of steps needed (if any) to ensure that the polygon touches both the lower and upper border, and finally the number of steps needed (if any) to extend at least $w$ edges in the length-wise direction (remember we only need rectangles with $l \geq w$). If the sum $n_{cur} + n_{add} > 4w_{max} + 2$ we can discard the partial generating function for that configuration, and of course the configuration itself, because it won't make a contribution to the polygon count up to the perimeter lengths we are trying to obtain. For instance, polygons spanning a rectangle with a width close to $w_{max}$ have to be almost convex, so very convoluted polygons are not possible. Thus configurations with many loop ends (non-zero entries) make no contribution at perimeter length $\leq 4w_{max} + 2$.

### 7.4.1.1 Derivation of Updating Rules

In this section we give a detailed description of how one derives the updating rules for the generating functions for partially completed polygons. Table 7.1 lists the possible local 'input' states as well as the 'output' states which arise as the kink in the boundary is propagated by one step, so as to include a vertex not situated at the top- or bottom-most borders of the rectangle. The most important boundary edges are the vertical edge intersecting the horizontal part of the boundary line and the horizontal edges immediately below and to the left. This is the position in which the lattice is being extended and the state of these edges determines the possible states of the newly added edges (the horizontal edge to the right and the vertical edge below the kink) intersected in Fig. 7.5 by the dotted line. In addition the state of an edge further afield may have to be changed if two loops of a partially completed polygon are joined at the kink.

00: Both kink edges in the input state are empty. We can leave the new edges empty or occupy both by inserting a new loop leading to the two output states '00' (weight 1) and '12' (weight $x^2$), respectively. These are the only possibilities

**Table 7.1** The various 'input' states and the 'output' states which arise as the boundary line (dashed line) is moved in order to include a vertex. Full lines indicate the four local edges involved in the update. Thick edges are occupied by a part of the polygon and the loops indicate how these edges are connected to other edges intersected by the boundary line.

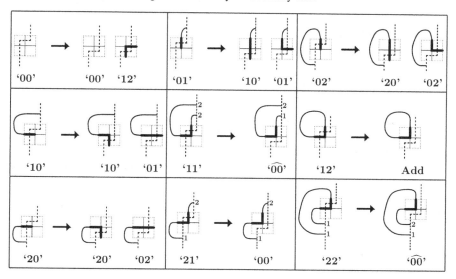

because the added vertex must have degree 0 or 2 according to constraint (2). Only the boundary states in the kink are changed in this update and all other edge states remain fixed.

01: The vertical edge is occupied by a lower loop end while the horizontal edge is empty. Constraint (2) forces us to occupy one and only one of the new edges. So we either continue straight down to output state '10' or make a 90° turn to output state '01'. Each of these updates picks up a weight $x$.

02: Same as above except the vertical edge intersects an upper loop end.

10: Similar to input '01' except it is the horizontal edge which is occupied.

20: Similar to input '02' except it is the horizontal edge which is occupied.

11: Both edges in the input state are occupied by lower loop ends. Constraint (2) forces us to leave both new edges empty. We are joining two loops so in addition to setting the output states to '0' we have to change the matching upper end of the inner-most loop to a lower end in the new boundary state. This relabeling is indicated in table 7.1 by putting a 'hat' over the output states.

22: Same as for '11' except we are joining two upper loop ends and we have to relabel the matching lower loop end as an upper end (see the middle lower panel of Fig. 7.7).

21: The horizontal edge in the input state is occupied by an upper loop end while the vertical edge is a lower loop end. Constraint (2) forces us to leave both new edges empty. We set the output states to '0' and no further changes are required.

12: The horizontal edge in the input state is occupied by a lower loop end while the vertical edge is an upper loop end. This means that we are closing a loop on itself (note that the two loop ends must belong to the same loop because a polygon has no self-intersections). This is only allowed if all other boundary states are empty. The generating function is accumulated into the running total for the given length.

The updating at the top border uses the same rules as above but with the additional restriction that $\sigma_{w+1} = 0$ while $\sigma_w \neq 1$. The only additional rule is that if we are calculating $G_{wl}(x)$ we need to set $\sigma_t = 1$ when inserting a new loop in the '00' input state.

The updating at the bottom border also uses the same rules as above but these are restricted in that $\sigma_0 = 0$ in the *output* state. Since the first non-zero state must be a '1' we get the updating rules: '00' → '00'; '10' or '01' → '01'; '11' → '$\widehat{00}$'; and '12' → 'accum'. After applying these rules we shift the whole boundary line configuration by one unit, e.g., we set $\sigma_i = \sigma_{i+1}$ for $i$ from 0 to $w$ and we set $\sigma_{w+1} = 0$. In this case, if we are calculating $G_{wl}(x)$, we need to set $\sigma_b = 1$ if the input state had any non-zero entries. Finally, we may apply the symmetry transformation described below.

### 7.4.1.2 Symmetry

Symmetries of the underlying lattice can be used to reduce the number of configurations we need to retain in any given calculation. We have already seen how the basic symmetry of the square lattice allows us to reduce the computational complexity, since we need only consider rectangles with $l \geq w$ because $G_{wl}(x) = G_{lw}(x)$. We note here that there is a further symmetry which we can use. After a column has been completed (and the boundary line is completely vertical) configurations are symmetric with respect to reflections. That is, configurations such as $\{010200*; 11\}$ and $\{001020*; 11\}$ have the same generating function. The first configuration is a single loop whose lower end is 1 unit from the bottom of the rectangle with the upper end 2 units from the top of the rectangle while both borders have been touched. The second configuration is a single loop with the lower end 2 units from the bottom of the rectangle, the upper end 1 unit from the top of the rectangle while having touched both borders. Given the symmetry of the square lattice it is clear that any partially completed polygon resulting in the first configuration must have a matching symmetric polygon leading to the second configuration. So their generating functions must be identical and we can discard one of them while multiplying the other generating function by 2. Similarly the configurations $\{010200*; 10\}$ and $\{001020*; 01\}$ have the same loop ends as before but the first has touched only the bottom of the rectangle while the second has touched only the top of the rectangle. Again symmetry dictates that they must have the same generating function.

### 7.4.1.3 Pruning

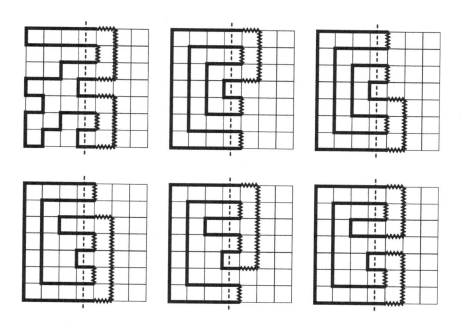

**Fig. 7.8** Examples of partially generated polygons (thick solid lines) to the left of the intersection (dashed line) and how to close them in a valid way (thick wavy line). Upper left panel shows how to close the configuration {12112212}. The upper middle and right panels show the two possible closures of the configuration {11112222}. The lower panels show the three possible closures of the configuration {11121222}.

The number of steps needed to ensure a spanning polygon is straightforward to calculate. The complicated part of the pruning approach is the algorithm to calculate the number of steps required to close the polygon. There are very many special cases depending on the position of the kink in the intersection and whether or not the partially completed polygon has reached the bottom and/or top of the bounding rectangle. So in the following we will only briefly describe some of the simple contributions to the closing of a polygon.

Firstly, if the partial polygon contains separate pieces these have to be connected as illustrated in Fig. 7.8. Separate pieces are easy to locate since all we have to do is start at the bottom of the intersection and as we move upwards we count the number of '1's and '2's in the configuration. Whenever these numbers are equal a separate piece has been found and (provided one is not at the last edge in the configuration) the currently encountered 2-edge can be connected to the next 1-edge above. $n_{add}$ is incremented by the number of steps (the distance) between the edges and the two edge-states are removed from the configuration before further processing. It is a little less obvious that if the configuration starts (ends) as {112...2} ({1...122}) the two lower (upper) edges can safely be connected (note that there can be any number of '0's interspersed between the occupied edges). Again $n_{add}$ is incremented by the number of steps between the edges, and the two edge-states are removed from the configuration—leading to the new configuration $\{001\ldots2\}$ ($\{1\ldots200\}$)—before further processing. After these operations we may be left with a configuration which has just a single 1-edge and a single 2-edge. We are almost done since these two edges can be connected to form a valid polygon. This is illustrated in Fig. 7.8 where the upper left panel shows how to close the partial polygon with the intersection $\{12112212\}$, which contain three separate pieces. After connecting these pieces we are left with the configuration $\{10012002\}$. We now connect the two 1-edges and note that the first two-edge is relabeled to a 1-edge (it has become the new lower end of the loop). Thus we get the configuration $\{00001002\}$ and we can now connect the remaining two edges and end up with a valid completed polygon. Note that in the last two cases, in addition to the steps spanning the distance between the edges, an additional two horizontal steps had to be added in order to form a valid loop around the intervening edges. If the transformation above doesn't result in a closed polygon we must have a configuration of the form $\{111\ldots222\}$. The difficulty lies in finding the way to close such configurations with the smallest possible number of additional steps. Suffice to say that if the number of non-zero entries is small one can easily devise an algorithm to try all possible valid ways of closing a polygon and thus find the minimum number of additional steps.

In Fig. 7.8 we show all possible ways of closing polygons with 8 non-zero entries. Note that we have shown the generic cases here. In actual cases there could be any number of 0-edges interspersed in the configurations and this would determine which way of closing would require the least number of additional steps.

### 7.4.1.4 Computational Complexity

The time required to obtain the number of polygons on $w \times l$ rectangles grows exponentially with $w$. Time and memory requirements are basically proportional to the maximal number of distinct configurations along the boundary line. When there is no kink in the intersection (a column has just been completed) we can calculate this number, $N_{conf}(w)$, exactly. Each boundary line configuration consists of '0's and an equal number of '1's and '2's with the latter forming a perfectly balanced parenthesis system. This corresponds to a Motzkin path [27, Ch. 6] (just map 0 to a horizontal step, 1 to a north-east step, and 2 to a south-east step). The number of Motzkin paths $M_n$ with $n$ steps is easily derived from the generating function $\mathcal{M}(x) = \sum_n M_n x^n$, which satisfies, $\mathcal{M}(x) = 1 + x\mathcal{M}(x) + x^2\mathcal{M}^2(x)$, so that

$$\mathcal{M}(x) = [1 - x - \sqrt{(1+x)(1-3x)}]/2x^2. \tag{7.6}$$

When the boundary line has a kink the number of configurations exceeds $N_{conf}(w)$ but clearly is less than $N_{conf}(w+1)$. From (7.6) we see that asymptotically $N_{conf}(w)$ grows like $3^w$ (up to a power of $w$). So the same is true for the maximal number of boundary line configurations and hence for the computational complexity of the algorithm. Note that the total number of SAPs grows like $\mu^{2n}$ (where $\mu \simeq 2.638$ on the square lattice), while the complexity of the transfer matrix algorithm grows as $3^{n/4}$. Since $\sqrt[4]{3} \simeq 1.316$ we see that even the basic algorithm without pruning leads to a very substantial exponential improvement over direct enumeration.

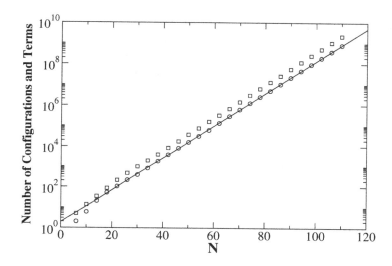

**Fig. 7.9** The maximal number of configurations (circles) and terms (squares) required in order to count the number of SAPs up to perimeter $n$. The solid line is drawn as a guide to the eye and would correspond to a growth rate of exactly $2.06^{1/4} = 1.198\ldots$.

Pruning results in a further exponential improvement to the algorithm. In this case the algorithm is too complicated to be analysed exactly. Instead we simply run the algorithm with different cut-offs $n$ and monitor the maximal number of configurations required during the calculation of the number of polygons of size up to $n$. In Fig. 7.9 we plot the maximal number of configurations and terms required as a function of $n$. We clearly see an exponential growth with $n$ and we find a growth constant $\lambda \approx 2.06^{1/4} \simeq 1.198\ldots$ which is a vast improvement on the unpruned algorithm where $\lambda = 3^{1/4} \simeq 1.316\ldots$.

### 7.4.2 Honeycomb Lattice

Enting and Guttmann [9] were the first to use transfer matrix techniques to enumerate SAPs on the honeycomb lattice. They implemented the honeycomb lattice as a square lattice with some edges removed, resulting in a brickwork lattice (Fig. 7.10). The basic transfer matrix algorithm is essentially identical to the square lattice case and we shall provide no further details here. The only major point of difference to notice is that the honeycomb lattice lacks the usual rotational symmetry of the square lattice so the generating functions of $w \times l$ and $l \times w$ rectangles are no longer identical. Efficiency therefore dictates that the cases $w \leq l$ and $w > l$ be treated separately and an algorithm has to be written for each case. However, the differences between the two cases are minor and relate predominantly to effects at the borders of the rectangle.

Enting and Guttmann [9] obtained the perimeter generating function to perimeter 82. Jensen [17] implemented pruning and used a parallel version to enumerate SAPs to length 158. Vöge, Guttmann and Jensen [36] enumerated honeycomb SAPs by area up to 35. As mentioned in Chapter 1 they are studied in theoretical chemistry where they are know as models of benzenoid hydrocarbons.

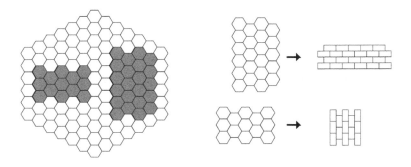

**Fig. 7.10** In the left-most panel we show a section of the honeycomb lattice and two rectangular shaped regions used in the finite lattice method. The right-most panels show how these regions are mapped to the brickwork lattice.

### 7.4.3 Triangular Lattice

Self-avoiding polygons on the triangular lattice were first enumerated using the finite-lattice method by Enting and Guttmann [10]. They implemented the triangular lattice as a square lattice with additional edges connecting the top-left and bottom-right vertices of each unit cell (see Fig. 7.11). Rectangles of size $w \times l$ can be used as our finite lattices. Due to the symmetry of the triangular lattice we need consider only rectangles with $l \geq w$. As usual with the transfer matrix technique we intersect the rectangle with a boundary line. In most cases it is most efficient to let the boundary line cut through the edges of the lattice. However, on the triangular lattice it is more efficient to let the boundary line cut through the vertices [10]. Essentially this variation leads to only half as many intersected vertices (as opposed to edges) along the boundary line. As we show below there is a small price to pay since we have to introduce a new type of vertex state (in addition to the usual three edge states empty, lower and upper). But obviously $4^w < 3^{2w}$ so we are clearly better off using this approach. SAPs in rectangles of a given width $w$ are enumerated by moving the intersection so as to add one vertex at a time, as shown in Fig. 7.11. If we draw a SAP and then cut it by a line we observe that the partial SAP to the left of this line consists of a number of loops connecting two vertices. In addition it is possible that the SAP touches a vertex (that is the SAP comes in along one edge and exits along another edge but without crossing the boundary line). All these cases are illustrated in Fig. 7.11.

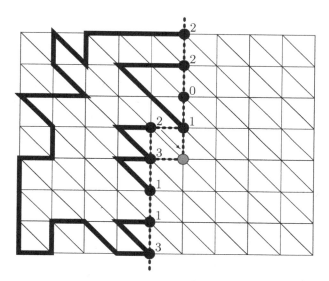

**Fig. 7.11** A snapshot of the boundary line (dashed line) during the transfer matrix calculation on the triangular lattice. SAPs are enumerated by successive moves of the kink in the boundary line so that one vertex (shaded) at a time is added to the rectangle. To the left of the boundary line we have drawn an example of a partially completed SAP.

Each configuration along the boundary line can thus be represented by a set of states $\{\sigma_i\}$, where

$$\sigma_i = \begin{cases} 0 & \text{empty vertex,} \\ 1 & \text{vertex is a lower loop-end,} \\ 2 & \text{vertex is an upper loop-end,} \\ 3 & \text{touched (degree 2) vertex,} \end{cases} \tag{7.7}$$

If we read from the bottom to the top, the configuration along the intersection of the partial SAP in Fig. 7.11 is $\{311321022\}$.

### 7.4.3.1 Updating Rules

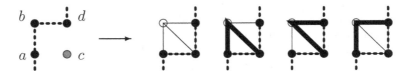

**Fig. 7.12** The four possible outputs from a single iteration of the TM algorithm. Depending on the states of the three vertices $a$, $b$, and $d$ in the input, some of the outputs cannot occur.

In Fig. 7.12 we have illustrated what can happen locally as the boundary line is moved. Before the move, the boundary line intersects the vertices $a$, $b$ and $d$ and after the move the vertices $a$, $c$ and $d$ are intersected by the boundary line. In a basic iteration step we can insert steps along the edges emanating from vertex $b$. Since vertex $b$ can't have degree greater than 2 we can insert at most two new steps. However, depending on the states of vertices $a$ and $d$ in the source, some of the edge configurations in Fig. 7.12 may be forbidden. The updating of the partial generating function depends most crucially on the state of vertex $b$ and to a somewhat lesser extent on the states of the vertices $a$ and $d$. The basic limitation on the allowed outputs are that conditions (2)–(4) must be enforced. In the following we shall briefly describe how the updating rules are derived.

**State of vertex $b$ is 0.** Since vertex $b$ is empty, all the outputs in Fig. 7.12 are possible. In the first output we insert no steps. This is always allowed and no changes are made to the configuration.

In the other three outputs we insert a partial loop. There are restrictions on the insertion of steps to vertices $a$ and $d$. We cannot insert a step to a vertex in state 3. Otherwise the first two outputs are always allowed. The last output is a little more complicated. If vertex $a$ is in state 1 and vertex $d$ is in state 2 we cannot join the two vertices since this would result in a closed loop.

After the insertion of new steps we have to assign a state to vertex $c$ and quite possibly change the states of vertices $a$ and $d$ (and perhaps the states of some other vertices in the target configuration). The state of vertex $c$ will be 0 (no step), 1 (lower loop-end) or 2 (upper loop-end). Next we consider what happens to vertices $a$ and $d$. When these vertices are empty in the source they can take the values just listed above in the target. If they are occupied in the source they either retain their state in the target (no steps inserted) or change to state 3 (a step is inserted). In the latter case we may have to change the state of other vertices in the target. We may join two lower (upper) loop-ends and then we must change the matching upper (lower) loop-end of the inner-most loop to the lower (upper) loop-end of the new joined loop.

**State of vertex $b$ is 1.** A lower end of a loop enters vertex $b$ and it has to be continued by inserting a single step (partial loops cannot be inserted since this would make vertex $b$ of degree 3) either to vertex $c$ which becomes a state 1 vertex; to vertex $a$ if not in state 3 or state 2 (a closed loop would be formed); or to vertex $d$ if not in state 3. Again we have to change the states of vertices $a$ and $d$ when a step is inserted on these vertices. If the source state of the vertices was 0 the target state becomes 1, otherwise the target state becomes 3 and as above we may need to change the state of other vertices as well.

**State of vertex $b$ is 2.** An upper end of a loop enters vertex $b$. The upper end can always be continued to vertex $c$; to vertex $d$ if it is not in state 3; and to vertex $a$ provided it is not in state 3 or 1 (this would result in a closed loop). The states of the target vertices are changed as described above.

**State of vertex $b$ is 3.** This is the simplest situation. Vertex $b$ is of degree 2 so no steps can be inserted and only the output with all empty edges is allowed. The state of vertex $c$ is 0 and the states of all other vertices are unchanged.

## 7.4.4 Enumerations by Area

Enumeration by area uses essentially the same transfer matrix algorithm. The updating rules as described above are practically unchanged. The only difference is of course that we need to count the number of enclosed unit cells rather that the perimeter. This is easily accomplished. When the kink in the boundary line is moved we simply need to determine whether of not the new cell lies inside or outside the polygon. Notice that as we run along the boundary line and pass an occupied edge we change from the inside to the outside and vice versa. We naturally start by being outside the polygon. As we move from row 0 to row $w$ and encounter the first occupied edge we move to the inside, passing the second occupied edge takes us to the outside again and so on. In other words if there is an *odd* number of occupied edges below the new cell, it is on the inside of the polygon and we must multiply the source generating function by $q$ before adding it to the target generating function. If on the other hand there is an *even* number of edges below the new cell it lies outside the polygon and we merely add the source generating function to the target generating function.

### 7.4.4.1 Area-Weighted Moments

The area-weighted moments are quantities of particular interest and they can be calculated easily from the full perimeter and area generating function

$$\mathscr{P}(x, q) = \sum_{n,m} p_{n,m} x^n q^m, \tag{7.8}$$

where $p_{n,m}$ is the number of polygons with perimeter $n$ and area $m$. From this we get the area-weighted generating functions,

$$\mathscr{P}_k(x) = (q\frac{\partial}{\partial q})^k \mathscr{P}(x,q)\Big|_{q=1} = \sum_n \sum_m m^k p_{n,m} x^n = \sum_n p_n^{(k)} x^n, \qquad (7.9)$$

and we define the average moments of area for a polygon with perimeter $n$

$$\langle a^k \rangle_n = p_n^{(k)}/p_n = \sum_m m^k p_{n,m}/p_n. \qquad (7.10)$$

In order to calculate the moments of area through this approach we need to calculate a full two-parameter generating function, which generally will require a lot of computer memory. If we are only interested in the first few moments there is a much more efficient approach [3]. We simply replace the variable $q$ by $1+v$ thus obtaining the function

$$F(x,v) = \sum_{n,m} p_{n,m} x^n (1+v)^m = \sum_{n,m} \sum_{k=0}^m \binom{m}{k} p_{n,m} x^n v^m. \qquad (7.11)$$

Let $F_i(x)$ be the coefficient of $v^i$ in $F(x,v)$. Then we see that

$$F_0(x) = \sum_{n,m} p_{n,m} x^n = \mathscr{P}(x),$$

$$F_1(x) = \sum_{n,m} m p_{n,m} x^n = \mathscr{P}_1(x),$$

$$F_2(x) = \sum_{n,m} m(m-1)/2 p_{n,m} x^n = [\mathscr{P}_2(x) - \mathscr{P}_1(x)]/2,$$

and so on. Thus if we are only interested in the first and second moments of area we can truncate the series $F(x,v)$ at second order in $v$ and find the relevant moments as $\mathscr{P}_1(x) = F_1(x)$ and $\mathscr{P}_2(x) = 2F_2(x) + F_1(x)$. The growth in memory requirements is still dominated by the exponential growth in the number of configurations. However, we have managed to turn the calculation of these moments from a problem with a prefactor cubic in $W_{max}$ (the area is proportional to $W_{max}^2$) into a problem with a prefactor linear in $W_{max}$.

## 7.4.5 Metric Properties

Here we shall briefly describe how one can obtain information about the metric properties of SAPs, that is how one can calculate quantities describing the typical size of SAPs of a given length.

### 7.4.5.1 Caliper Size

The simplest size measure for a SAP is the *caliper size*. The caliper size is simply [24] the average sum of the spans of the polygons in a given direction. First let

$d_n = \sum_{w,l} w[x^n] G_{w,l}(x)$, where as usual $G_{w,l}(x)$ is the generating function of SAPs with minimal bounding rectangle $w \times l$. One can also look at higher moments $d_n^k = \sum_{w,l} w^k [x^n] G_{w,l}(x)$. Then the caliper size is simply $\langle D_n \rangle = d_n/p_n$, and likewise for the higher moments. One expects that $\langle D_n^k \rangle \simeq n^{kv}$. Guttmann and Enting [13] studied the first two moments of the caliper size for SAPs on the square lattice for $n$ up to 56.

### 7.4.5.2  Radius of Gyration

The caliper size is trivial to compute and comes essentially for free in our transfer matrix calculations. However, the series are often not very well behaved and the estimates for the size exponent $v$ are therefore not very accurate. It is therefore important to study other size measures. The obvious choice is to study the mean-squared radius of gyration. When counting by perimeter this is measured using the *vertices* along the perimeter, while in enumeration by area it is more natural to use the centre-points of the enclosed lattice cells.

In the following we show how the definition of the radius of gyration can be expressed in a form suitable for a transfer matrix calculation. We use the enumeration by perimeter as our example. Note again that we define the radius of gyration according to the *vertices* of the SAP and that the number of vertices equals the perimeter length. The radius of gyration of $n$ points at positions $\mathbf{r}_i$ is

$$n^2 R_n^2 = \sum_{i>j} (\mathbf{r}_i - \mathbf{r}_j)^2 = (n-1) \sum_i (x_i^2 + y_i^2) - 2 \sum_{i>j} (x_i x_j + y_i y_j). \qquad (7.12)$$

This last expression is suitable for a transfer matrix calculation. As usual [24] we actually calculate the generating function, $\mathscr{R}_g^2(x) = \sum_n p_n \langle R^2 \rangle_n n^2 x^n$, because the coefficients in this series are integer valued. Note that $\langle R^2 \rangle_n$ is the average radius of gyration of SAPs with perimeter $n$. In order to do this we have to maintain five partial generating functions for each possible boundary configuration $\sigma$, namely

- $P(x)$, the number of (partially completed) polygons.
- $R^2(x)$, the sum of the squared components of the distance vectors.
- $X(x)$, the sum of the $x$-component of the distance vectors.
- $Y(x)$, the sum of the $y$-component of the distance vectors.
- $XY(x)$, the sum of the 'cross' product of the components of the distance vectors, e.g., $\sum_{i>j} (x_i x_j + y_i y_j)$.

As the boundary line is moved to a new position each target configuration $\sigma$ might be generated from source configurations $\sigma'$ in the previous boundary position. If we again regard our lattice as embedded in a square lattice then the position of a new added vertex is given by the column number $i$ and row number $j$. The partial generation functions are updated as follows

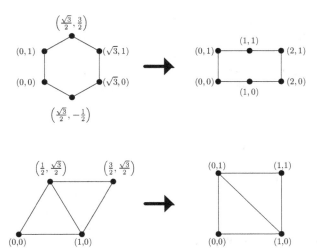

**Fig. 7.13** The transformation from the co-ordinates of the vertices on the honeycomb and triangular lattices to their brickwork and square lattice embeddings.

$$P(x,\sigma) = \sum_{\sigma'} x^{\delta(\sigma')} P(x,\sigma'),$$

$$R^2(x,\sigma) = \sum_{\sigma'} x^{\delta(\sigma')} [R^2(x,\sigma') + W_R(i,j)P(x,\sigma')],$$

$$X(x,\sigma) = \sum_{\sigma'} x^{\delta(\sigma')} [X(x,\sigma') + W_X(i,j)P(x,\sigma')], \qquad (7.13)$$

$$Y(x,\sigma) = \sum_{\sigma'} x^{\delta(\sigma')} [Y(x,\sigma') + W_Y(i,j)P(x,\sigma')],$$

$$XY(x,\sigma) = \sum_{\sigma'} x^{\delta(\sigma')} [XY(x,\sigma') + U_X(i,j)X(x,\sigma') + U_Y(i,j)Y(x,\sigma')]$$

where $\delta(\sigma')$ is the number of steps added to the source configuration $\sigma'$ in order to produce the target configuration $\sigma$. The $W$'s and $U$'s are lattice dependent weights. In all cases these weights are 0 if the newly added vertex isn't part of the perimeter of the polygon. In the case of the square lattice the non-zero weights are quite obvious. The weights for the honeycomb and triangular lattices can be worked out from Fig. 7.13.

If the new vertex is part of the perimeter the weights are:

$$\text{Square}: \quad \begin{cases} W_X = i & W_Y = j & W_R = i^2 + j^2 \\ U_X = i & U_Y = j \end{cases}$$

$$\text{Triangular}: \quad \begin{cases} W_X = 2i + j & W_Y = j & W_R = (2i + j)^2 + 3j^2 \\ U_X = 2i + j & U_Y = 3j \end{cases}$$

$$\text{Honeycomb}: \quad \begin{cases} W_X = i & W_Y = 3j - d_{i,j} & W_R = 3i^2 + (3j - d_{i,j})^2 \\ U_X = 3i & U_Y = 3j - d_{i,j} & d_{i,j} = \mathrm{mod}(i+j, 2) \end{cases}$$

## 7.5 Polyominoes or Lattice Animals

### 7.5.1 The Square Lattice

The method we use to enumerate polyominoes on the square lattice is based on the method used by Conway [3] for the calculation of series expansions for percolation problems, and is similar to the methods for the enumeration of self-avoiding polygons. In the following we give a brief description of the algorithm used to count polyominoes.

As for SAPs, the number of fixed polyominoes that span rectangles of width $w$ and length $l$ are counted using a transfer matrix algorithm. By combining the results for all $w \times l$ rectangles with $w \leq w_{max}$ and $w + l \leq 2w_{max} + 1$ we can count all polyominoes up to $n_{max} = 2w_{max}$. Due to symmetry we consider only rectangles with $l \geq w$ while counting the contributions from rectangles with $l > w$ twice.

We draw our boundary line through a set of $W$ cells. Polyominoes in rectangles of a given width are counted by moving the intersection so as to add one cell at a time, as shown in Fig. 7.14. Each configuration can be represented by a set of states $S = \{\sigma_i\}$, where the value of the state $\sigma_i$ at position $i$ must indicate first of all if the cell is occupied or empty. An empty cell is simply indicated by $\sigma_i = 0$. Since we have to ensure that we count only connected graphs more information is required if a cell is occupied. We need a way of describing which occupied cells along the intersection are connected to one another via a set of occupied cells to the left of the intersection. The most compact encoding of this connectivity is [3]

$$\sigma_i = \begin{cases} 0 & \text{empty cell,} \\ 1 & \text{occupied cell not connected to others,} \\ 2 & \text{first among a set of connected cells,} \\ 3 & \text{intermediate among a set of connected cells,} \\ 4 & \text{last among a set of connected cells.} \end{cases} \quad (7.14)$$

Configurations are read from the bottom to the top. As an example, the configuration along the intersection of the partially completed polyomino in Fig. 7.14 is $S = \{201023404\}$.

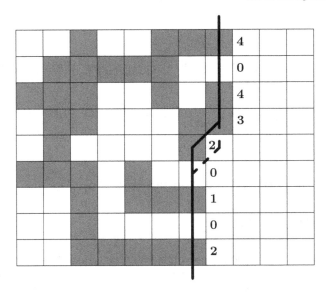

**Fig. 7.14** A snapshot of the intersection (solid line) during the transfer matrix calculation on the square lattice. Polyominoes are enumerated by successive moves of the kink in the intersection, as exemplified by the position given by the dashed line, so that one cell at a time is added to the rectangle. To the left of the intersection we have drawn, using shaded squares, an example of a partially completed polyomino. Numbers along the intersection indicate the encoding of this particular configuration.

### 7.5.1.1 The Updating Rules

In Table 7.2 we have listed the possible local 'input' states and the 'output' states which arise as the kink in the intersection is propagated by one step. The most important cell on the intersection is the 'lower' one situated at the bottom of the kink (the cell marked with the second '2' (counting from the bottom) in Fig. 7.14). This is the position in which the lattice is being extended. Obviously the new cell can be either empty or occupied. The state of the upper cell (the cell marked '3' in Fig. 7.14) is likely to be changed as a result of the move. In addition the state of a cell further afield may have to be changed if a branch of a partially completed polyomino terminates at the new cell or if two independent sections of a partially completed polyomino join at the new cell.

Details of how these updating rules are derived can be found in [15]. Here a few comments will have to suffice.

10: The lower cell is an isolated occupied cell and the new cell can be empty only if there are no other occupied cells on the intersection (otherwise we generate graphs with separate components) and if both the lower and upper borders have been touched. The result enumerates valid polyominoes and the partial generating function is added to the total polyomino generating function.

**Table 7.2** The various 'input' states and the 'output' states which arise as the intersection is moved in order to include one more cell of the lattice. Each panel contains two 'output' states where the left (right) most is the configuration in which the new cell is empty (occupied).

| Lower \ Upper | 0 | 1 | 2 | 3 | 4 |
|---|---|---|---|---|---|
| 0 | 00  10 | 01  24 | 02  23 | 03  33 | 04  34 |
| 1 | add  10 | –  24 | –  23 | –  33 | |
| 2 | $\overline{00}$  20 | $\overline{01}$  23 | $\overline{02}$ $\widehat{23}$ | 02  23 | 01  24 |
| 3 | 00  30 | 01  33 | 02  33 | 03  33 | 04  34 |
| 4 | $\overline{00}$  40 | $\overline{01}$  34 | $\overline{02}$  33 | $\overline{03}$ $\widehat{33}$ | |

14: This situation never occurs. The upper cell is last among a set of occupied cells, so the cell immediately to its left is also occupied. This in turn is connected to the lower cell, which therefore cannot be an isolated cell.

20: The lower cell is first among a set of occupied cells, so if the new cell is empty, another cell in the set changes its state. Either the *first* intermediate cell becomes the new first cell, and its state is changed from 3 to 2, or, if there are no intermediate cells, the last cell becomes an isolated cell, and its state is changed from 4 to 1. This relabeling of a matching cell is indicated in Table 7.2 by over-lining.

22: When the new cell is occupied, two separate pieces of the polyomino are joined. The new cell remains the first cell in the joined piece while the upper cell becomes an intermediate cell. The last cell in the innermost set of connected cells also becomes an intermediate cell in the joined piece. We indicate this type of transformation by putting a hat over the string.

## 7.5.2 The Honeycomb Lattice

Polyominoes on the honeycomb lattice have been enumerated by transfer matrix methods [37]. The algorithm is based on the observation that since the honeycomb lattice has vertices of degree 3 a polyomino is just a SAP (enumerated of course by area) with an arbitrary number of punctures.

## 7.5.3 The Triangular Lattice

Polyominoes on the triangular lattice (triominoes) have been enumerated by transfer matrix methods. In this case it is convenient to regard triominoes as site lattice animals on the hexagonal lattice.

## 7.6 Implementation Details

### 7.6.1 Modular Arithmetic

The integer coefficients of the generating functions can be very large and in particular much larger than the limit allowed by most computers ($2^{64} - 1$ on 64-bit computers). Furthermore some calculations may involve multiplication further reducing the size of the largest coefficients one can handle without overflow. The solution is to do the calculations using modular arithmetic [19]. This involves performing the calculation modulo various integers $m_i$, so that in a run using $m_i$ we calculate the residues $r_{i,n} = p_n \pmod{m_i}$. The Chinese remainder theorem ensures that any integer has a unique representation in terms of residues. The $m_i$ are called moduli and must be chosen so they are mutually prime, e.g., none of the $m_i$ have a common divisor. If the largest absolute values occurring in the final expansion is $N$, then we have to use a number of moduli $k$ such that $p_1 p_2 \cdots p_k / 2 > N$.

The calculations for the perimeter (or area) generating functions involve only additions so we can use moduli close to the upper limit permitted by the specific computer, say prime numbers of the form $2^{62} - t_i$ on 64-bit machines. The calculation of the area-weighted moments and the radius of gyration require a lot more memory for the generating functions (plus the radius of gyration calculation involves multiplication with quite large integers) so in this case we normally use prime numbers of the form $2^{30} - t_i$ for the moduli $m_i$.

### 7.6.2 Data Management

For fast access the generating functions must be stored in main memory using a data array. The main task is to find an efficient way of storing this data, that is to say, we don't want to waste precious resources so we want to store only what we absolutely need, and in addition require that access to this stored data must be easy and fast. Fortunately this is a well-developed area of programming and suitable approaches are readily available in the literature.

The representation, organisation and management of the data structures required to perform the calculations efficiently are often matters of personal preference and should be left to the individual. For this reason we shall only very briefly touch on two issues. One, that of hashing, is indispensable for efficiency and the other, related to storage, is useful.

#### 7.6.2.1 Hashing

First we need to transform a state into a 'key' which uniquely identifies the state. Recall that on the square lattice a boundary state is denoted as $\sigma = \{\sigma_0 \ldots \sigma_{w+1}; \sigma_b \sigma_t\}$, where $\sigma_i = 0$, 1 or 2 while $\sigma_b$ and $\sigma_t$ equal 0 or 1. The integer key, $S = \sum_{j=0}^{w+1} \sigma_j 3^j + \sigma_b 3^{w+2} + \sigma_t 3^{w+3}$, is unique. We could use this key as the entry in a large array of generating functions. However, not every integer in the range $3^{w+3} + 2 \times 3^{w+2}$ of all possible key values encodes a configuration. We have already seen that the number of encoding configurations is bounded between the number of Motzkin paths of length $w + 1$ and $w + 2$. At $w = 25$ this would mean that less than 0.2% of the integer keys in the range are actually encoding and with pruning this becomes fewer still. So using an array over the possible key values would be extremely wasteful and using some sort of lexicographically ordered list of Motzkin paths would also be inefficient when pruning is used.

The solution to this problem is the use of hashing. The general idea is that in cases such as this where the number of realised keys is sparse within the range of possible key values, we should map the keys to the entries of an array only a little larger than the number of realised keys. This is done using a so-called hash function. Ideally one would like the hash function to be injective so that no two occurring keys are mapped to the same entry. This is known as perfect hashing. In practice this is very difficult to achieve and we thus generally have to settle for a hash function which can lead to a small number of keys being mapped to the same entry. These "collisions" are then treated separately. The most efficient way is generally through the use of linked lists. Probably the most commonly used type of hash functions (and the ones we use) are essentially a kind of random number generator, i.e., the keys are mapped (uniformly) onto the array entries. This can be achieved by using hash functions of the type:

$$f_{hash}(S) = ((S \times h_m \bmod h_p) \bmod M)$$

where $h_m$ is an integer called the hash multiplier, $h_p$ is a prime number and $M$ is size of the array. Generally $h_p$ exceeds the largest possible key value. By varying $h_m$ and $h_p$ it should almost always be possible to find a well behaved hash function, i.e., almost uniform and with few collisions.

Notice that by using linked lists collisions can be managed quite efficiently and one need not really have $M$ greater than the number of occurring keys.

### 7.6.2.2 Storage of Generating Functions

In a calculation of the number of polygons up to size $n$, the terms in the generating functions for partially completed polygons associated with the configuration $\sigma$ could simply be stored in arrays of length $n$ (say using an $M \times n$ array). However, this is quite wasteful. Notice first of all that often (such as on the square or honeycomb lattices) polygons are of even length. This means that in any occurring generating function half the terms (either all even or all odd terms) are zero. Furthermore, since we use pruning for any configuration $\sigma$, we know that the smallest polygon we can construct from $\sigma$ has perimeter $n_c = n_{cur} + n_{add}$ so we only need store the $n - n_c$ non-zero terms (on the square or honeycomb lattice this number is further reduced by a factor of 2). The average number of terms one needs to store grows much more slowly than $n$.

In order to manage this data, we use an array $H$ of size $M$ onto which we map the keys $S$ using our hash function. $H$ is simply pointing to entries in an array $C$, with a size greater than the number of occurring keys, where we store information about the configurations $\sigma$. Essentially we keep track of the keys $S$ using linked lists and $H[f_{\text{hash}}(S)]$ points to the entry in the array $C$ where we start the linked list of keys mapped to $f_{\text{hash}}(S)$. Stored in $C$ is: the key $S$; information about the generating function (namely $n_{\text{cur}}$ and $n_{\text{s}} = n - n_{\text{c}}$, that is the minimum number of steps already inserted and the number of terms stored in the generating function); a pointer to the next key entry in the linked list; and a pointer to the entry in an array $G$ at which we start the storage of the terms in the generating function.

## 7.6.3 Parallel Algorithms

The computational complexity of the FLM grows exponentially with the number of terms one wishes to calculate. It is therefore little wonder that implementations of the algorithms have always been geared towards using the most powerful computers available. In the past decade or so parallel computing has become the paradigm for high performance computing. The early machines were largely dedicated massively parallel processing machines which more recently have been superseded by clusters.

The transfer matrix algorithms used in the calculations of the finite lattice contributions are perfectly suited to parallel computations.

The most basic concerns in any efficient parallel algorithm is to minimise the communication between processors and to ensure that each processor does the same amount of work and uses the same amount of memory. In practice one naturally has to strike some compromise and accept a certain degree of variation across the processors.

One of the main ways of achieving a good parallel algorithm using data decomposition is to try to find an invariant under the operation of the updating rules. That is, we seek to find some property about the configurations along the boundary line which does not alter in a single iteration. The algorithm for the enumeration of polygons is quite complicated since not all possible configurations occur due to pruning, and an update at a given set of edges might change the state of an edge far removed, e.g., when two lower loop-ends are joined we have to relabel one of the associated upper loop-ends as a lower loop-end in the new configuration. However, there still is an invariant since any edge not directly involved in the update cannot change from being empty to being occupied and vice versa. That is, only the kink edges can change their occupation status. This invariant allows us to parallelise the algorithm in such a way that we can do the calculation completely independently on each processor with just two redistributions of the data set each time an extra column is added to the lattice.

The main points of the algorithm are summarized below:

1. With the boundary line straight (having no kinks), distribute the data across processors so that configurations with the same occupation pattern along the *lower* half of the boundary line are placed on the same processor.
2. Do the TM update inserting the top half of a new column. This can be done *independently* by each processor because the occupation pattern in the lower half remains unchanged.
3. Upon reaching the half-way mark redistribute the data so that configurations with the same occupation pattern along the *upper* half of the boundary line are placed on the same processor.
4. Do the TM update inserting the bottom-half of a new column.
5. Go back to step 1.

The redistribution among processors can be done as follows:

1. On each processor run through the configurations to establish the configuration pattern $c$ of each configuration and calculate $n(c)$, the number of configurations with a given pattern.
2. On processor 0 calculate the *global* sum of $n(c)$.
3. Sort the array $n(c)$ in descending order on processor 0.
4. On processor 0 assign each pattern to a processor $p(c)$ such that:
   a. Set $p_{id} = 0$.
   b. Assign the *most* frequent unassigned pattern $c$ to processor $p_{id}$.
   c. If the number of configurations assigned to $p_{id}$ is less than the number of configurations assigned to processor 0 then assign the *least* frequent unassigned patterns to $p_{id}$ until the desired inequality is achieved.
   d. set $p_{id} = (p_{id} + 1) \bmod N_p$, where $N_p$ is the number of processors.
   e. Repeat from (b) until all patterns have been assigned.
5. Send $p(c)$ to all processors.
6. On each processor run through the configurations sending each configuration to its assigned processor.

A simple timing of the various subroutines of the parallel algorithm shows that the typical time to do a redistribution is the same as the average time taken per iteration in order to move the kink once. Since the maximal time use at $n_{max} = 110$, when enumerating square lattice SAPs by perimeter, occurs at $W = 24$ there are 24 iterations and just 2 redistributions per added column, so the overall cost of parallel execution is less than 10%.

## 7.7 Concluding Remarks

As noted in the introduction, techniques of polygon enumeration have shown a general trend towards replacing geometric combinatorial complexity with algebraic complexity. In two dimensions, this is particularly true of the transfer matrix methods. As well as the generalisations that we have described in detail, a number of other cases have been considered. These include interacting polygons [1], punctured polygons and polyominoes [14], and directed lattices [8]. Furthermore, a number of these special cases, directed lattices, interactions and 'holes' can in principle be combined with each other and with other generalisations such as areas and spatial moments described above. Current indications are that there is little to be learnt from such complications.

Hopefully, the detailed descriptions of this chapter will allow the readers to devise their own FLM algorithms for other enumeration problems, or devise further improvements to existing FLM algorithms.

# References

1. Bennett-Wood D, Enting I G, Gaunt D S, Guttmann A J, Leask J L, Owczarek A L and Whittington S G 1998 Exact enumeration study of free energies of interacting polygons and walks in two dimensions *J. Phys. A* **31** 4725–4741
2. Clisby N, Liang R and Slade G 2007 Self-avoiding walk enumeration via the lace expansion *J. Phys. A* **40** 10973–11017
3. Conway A R 1995 Enumerating 2D percolation series by the finite lattice method *J. Phys. A* **28** 335–349
4. Delest M P and Viennot G 1984 Algebraic languages and polyominoes enumeration *Theor. Comput. Scie.* **34** 169–206
5. Domb C 1960 On the theory of cooperative phenomena in crystals *Advances in Physics* **9** 149–361
6. Enting I G 1980 Generating functions for enumerating self-avoiding rings on the square lattice *J. Phys. A* **13** 3713–3722
7. Enting I G 1986 Lattice models of firn closure: I. Percolation on the interstices of the BCC lattice *J. Phys. A* **19** 2841–2854
8. Enting I G and Guttmann A J 1985 Self-avoiding polygons on the square, L and Manhattan lattices *J. Phys. A* **18** 1007–1017
9. Enting I G and Guttmann A J 1989 Polygons on the honeycomb lattice *J. Phys. A* **22** 1371–1384
10. Enting I and Guttmann A 1992 Self-avoiding rings on the triangular lattice *J. Phys. A: Math. Gen.* **25** 2791–2807
11. Guttmann A J 1989 Asymptotic analysis of power-series expansions in *Phase Transitions and Critical Phenomena* (eds. C Domb and J L Lebowitz) (New York: Academic) vol. 13 1–234
12. Guttmann A J and Enting I G 1988 The number of convex polygons on the square and honeycomb lattices *J. Phys. A* **21** L467–474
13. Guttmann A J and Enting I G 1988 The size and number of rings on the square lattice *J. Phys. A* **21** L165–172
14. Guttmann A J, Jensen I, Wong L H and Enting I G 2000 Punctured polygons and polyominoes on the square lattice *J. Phys. A* **33** 1735–1764
15. Jensen I 2001 Enumerations of lattice animals and trees *J. Stat. Phys.* **102** 865–881
16. Jensen I 2003 A parallel algorithm for the enumeration of self-avoiding polygons on the square lattice *J. Phys. A* **36** 5731–5745
17. Jensen I 2006 Honeycomb lattice polygons and walks as a test of series analysis techniques *J. Phys.: Conf. Ser.* **42** 163–178
18. Jensen I and Guttmann A J 1999 Self-avoiding polygons on the square lattice *J. Phys. A* **32** 4867–4876

19. Knuth D E 1969 *Seminumerical Algorithms. The Art of Computer Programming, Vol 2.* (Reading, Mass: Addison Wesley)

20. MacDonald D, Hunter D L, Kelly K and Jan N 1992 Self-avoiding walks in two to five dimensions: exact enumerations and series study *J. Phys. A* **25** 1429–1440

21. Martin J L 1962 The exact enumeration of self-avoiding walks on a lattice *Proc. Camb. Phil. Soc.* **58** 92–101

22. Martin J L 1974 Computer techniques for evaluating lattice constants in *Phase Transitions and Critical Phenomena* (eds. C Domb and M Green) (New York: Academic) vol. 3 97–112

23. Martin J L 1990 The impact of large-scale computing on lattice statistics *J. Stat. Phys.* **58** 774–779

24. Privman V and Rudnick J 1985 Size of rings in two dimensions *J. Phys. A* **18** L789–L793

25. Rushbrooke G S and Eve J 1959 On noncrossing lattice polygons *J. Chem Phys.* **31** 1333–1334

26. Rushbrooke G S and Eve J 1962 High-temperature Ising partition function and related noncrossing polygons for the simple cubic lattice *J. Math. Phys.* **3** 185–189

27. Stanley R P 1999 *Enumerative Combinatorics* vol. 2 (Cambridge: Cambridge University Press)

28. Sykes M F, Essam J and Gaunt D S 1965 Derivation of low-temperature expansions for the Ising model of a ferromagnet and an antiferromagnet *J. Math. Phys.* **6** 283–298

29. Sykes M F 1986 Generating functions for connected embeddings in a lattice: I. Strong embeddings *J. Phys. A* **19** 1007–1025

30. Sykes M F 1986 Generating functions for connected embeddings in a lattice: II. Weak embeddings in a lattice: II. Weak embeddings *J. Phys. A* **19** 1027–1032

31. Sykes M F 1986 Generating functions for connected embeddings in a lattice: III. Bond percolation in a lattice: III. Bond percolation *J. Phys. A* **19** 2425–2429

32. Sykes M F 1986 Generating functions for connected embeddings in a lattice: IV. Site percolation *J. Phys. A* **19** 2431–2437

33. Sykes M F and Wilkinson M K 1986 Generating functions for connected embeddings in a lattice: V. Application to the simple cubic and body-centred cubic lattice *J. Phys. A* **19** 3407–3414

34. Sykes M F and Wilkinson M K 1986 Derivation of series expansions for a study of percolation processes *J. Phys. A* **19** 3415–3424

35. Torrie G and Whittington S G 1975 Exact enumeration of neighbour-avoiding walks on the tetrahedral and body-centred cubic lattices *J. Phys. A* **8** 1178–1184

36. Vöge M, Guttmann A J and Jensen I 2002 On the number of benzenoid hydrocarbons *J. Chem. Inf. Comput. Sci.* **42** 456–466

37. Vöge M and Guttmann A 2003 On the number of hexagonal polyominoes *Theo. Comp. Sci.* **307** 433–453

38. Wakefield A J 1951 Statistics of the simple cubic lattice *Proc. Camb. Phil. Soc.* **47** 419–435 and 799–810

39. Wortis M F 1974 Linked cluster expansions in *Phase Transitions and Critical Phenomena* (eds. C Domb and M Green) (New York: Academic) vol. 3 114–180

# Chapter 8
# Series Analysis

Anthony J. Guttmann and Iwan Jensen

## 8.1 Objective and General Principles

As we have seen in earlier chapters, the problem of determining the critical behaviour of various generating functions, such as that for SAP and polyominoes is an
unsolved problem. One is thus forced to resort to numerical methods, of which the
most successful for determining the precise behaviour of a given model on a given
lattice, is the method of exact series expansions. In this method, one generates as
many terms as possible in the generating function, so that if the generating function
is written

$$F(x) = \sum_{n \geq 0} f_n x^n,$$

the coefficient $f_n$ which counts the number of objects with some measure of size
indexed by $n$—typically the perimeter or area—is known for $n \leq N$.

The fundamental problem of series analysis is this: Given a *finite* number of terms
in the series expansion of a function $F(x)$ what can one say about the asymptotic
and in general singular behaviour of $F(x)$ or $f_n$? This after all is a property of the
*infinite* series. The problem is thus mathematically ill-posed, as given the first $N$
coefficients of a power series expansion, one can add to it the function $x^{N+1} H(x)$,
for any function $H(x)$. The behaviour of the modified function is then not reflected
in the known series coefficients.

It is thus a (usually) unstated assumption that the coefficients to hand are indeed
representative of the underlying function, so that a careful analysis can then reveal
something of the true large-$n$ behaviour. For the same reason, quoting error bars
in any method of series analysis is fraught with difficulty. It must be understood
that quoted error bars are in no sense rigorous. Often unfortunately they reflect the
optimism of the investigator in the quality of his/her investigations! Usually the best

Anthony Guttmann and Iwan Jensen
Department of Mathematics and Statistics, The University of Melbourne, Victoria, Australia, e-
mail: tonyg@ms.unimelb.edu.au e-mail: iwan@ms.unimelb.edu.au

one can do is to calculate some mean and variance of a range of estimates, and try and present evidence that there is no systematic drift of estimates. If there is drift, one can try and estimate that too.

We will show examples of this type of error analysis, which can, in favourable cases, give rise to surprisingly accurate critical parameters. Most of the generating functions pertaining to polygons, polyominoes and polyhedra are believed to have algebraic singularities, though in most cases this has not been proved. That is to say, the generating function above is believed to behave as

$$F(x) \sim A(1 - x/x_c)^\theta \text{ as } x \to x_c^-.$$  (8.1)

Hence it follows that

$$f_n = [x^n]F(x) \sim \frac{An^{-\theta-1}}{\Gamma(\theta)x_c^n}.$$  (8.2)

Here $A$ is referred to as the *critical amplitude,* $x_c$ as the *critical point,* and $\theta$ as the *critical exponent.*

A more comprehensive review of various methods used to analyse and estimate the asymptotic behaviour of series can be found in [7].

## 8.2 Ratio Method

The ratio method was perhaps the earliest systematic method of series analysis employed, and is still a useful starting point, prior to the application of more sophisticated methods. It was first used by M F Sykes in his 1951 D Phil studies, under the supervision of C Domb. From equation (8.2), it follows that the *ratio* of successive terms

$$r_n = \frac{f_n}{f_{n-1}} = \frac{1}{x_c}\left(1 - \frac{\theta+1}{n} + O(\frac{1}{n})\right).$$  (8.3)

From this idea, it is then natural to plot the successive ratios $\{r_n\}$ against $1/n$. If the correction term $O(\frac{1}{n})$ can be ignored, such a plot will be linear, with gradient $-\frac{\theta+1}{x_c}$, and intercept $1/x_c$ at $1/n = 0$.

We show this method in action by considering the application of the ratio method to the polygon generating function for SAP on the triangular lattice. The first few terms in the generating function (in fact from $p_3$ to $p_{26}$) are: 2, 3, 6, 15, 42, 123, 380, 1212, 3966, 13265, 45144, 155955, 545690, 1930635, 6897210, 24852576, 90237582, 329896569, 1213528736, 4489041219, 16690581534, 62346895571, 233893503330, 880918093866. Plotting successive ratios against $1/n$ results in the plot shown in Fig. 8.1. The critical point is known [16] to be at $x_c \approx 0.240917574\ldots$ $= 1/4.15079722\ldots$. From Fig. 8.1, one sees that the locus of points, after some initial (low $n$) curvature becomes linear to the naked eye for $n > 15$ or so, (corresponding to $1/n < 0.067$). Visual extrapolation to $1/x_c$ is quite obvious. A straight line drawn through the last $4-6$ data points intercepts the horizontal axis around $1/n \approx 0.13$. Thus the gradient is approximately $\frac{4.1508-2.8}{-0.13} \approx -10.39$, from which we conclude that the exponent $\theta + 1 = -2.50$. It is known [19] that the exact value is $\theta = -7/2$, which is in complete agreement with this simple graphical analysis.

Various refinements of the method can be readily derived. If the critical point is known exactly, it follows from equation (8.3) that estimators of the exponent $\theta$ are given by

$$\theta = n(1 - x_c \cdot r_n) - 1 + O(1).$$

Similarly, if the exponent $\theta$ is known, estimators of the critical point $x_c$ are given by

$$x_c = \frac{1}{r_n}\left(1 - \frac{\theta + 1}{n} + O(\frac{1}{n})\right).$$

One problem with the ratio method is that if the singularity closest to the origin is not the singularity of interest (the so-called *physical singularity*), then the ratio method will not give information about the physical singularity. Worse still, if the closest singularity to the origin is a conjugate pair of singularities, lying in the complex plane and off the real axis, the ratios will vary dramatically in both sign and magnitude. To overcome this difficulty G A Baker Jr [1] proposed the use of Padé approximants applied to the logarithmic derivative of the series expansion.

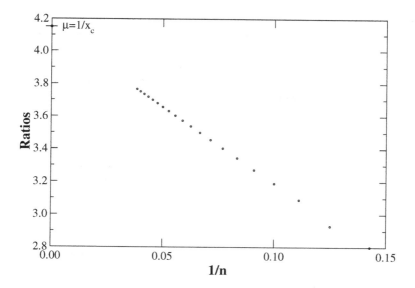

**Fig. 8.1** Plot of ratios against $1/n$ for triangular lattice polygons. A straight line through the last few data points intercepts the Ratios axis at $1/x_c$.

## 8.3 Padé Approximants

The basic idea of Padé approximation is very simple. Given a function $F(x)$ with a simple pole at some point $x_c$ we then use the series expansion of $F(x)$ to form an approximation to $F(x)$ as a ratio of two polynomials,

$$F(x) = \frac{P_i(x)}{Q_j(x)} \qquad (8.4)$$

where $P_i(x)$ and $Q_j(x)$ are polynomials of degree $i$ and $j$, respectively, whose coefficients are chosen such that the first $i + j + 1$ terms in the series expansion for $F(x)$ are identical to those of the expansion for $P_i(x)/Q_j(x)$. It is a convention to impose the normalisation condition $Q_j(0) = 1$.

In order to use this basic Padé approximation scheme for polygon problems we must first transform the series into a suitable form, which brings us to the classic method called Dlog-Padé approximation [1]. If we have a function with the expected critical behaviour typical of regular singular points, as given by equation (8.1), then taking the derivative of the log of $F(x)$ gives

$$\widehat{F}(x) = \frac{\mathrm{d}}{\mathrm{d}x} \log F(x) \simeq \frac{\theta}{x - x_c} + C. \qquad (8.5)$$

This form is perfectly suited for Padé analysis and we see that an estimate $x_c^*$ of the critical point $x_c$ can be obtained from the roots of the denominator polynomial $Q_j(x)$, while an estimate of the critical exponent $\theta$ is given by the residue at the pole found at $x = x_c^*$. Such an estimate of the exponent is known as an *unbiased* estimate. If $x_c$ is exactly known, as is sometimes the case, a *biased* estimate of the critical exponent $\theta$ can be obtained from the residue of the Padé approximant to $\widehat{F}(x)$ at $x_c$, that is

$$\theta = \lim_{x \to x_c} (x - x_c) \frac{P_i(x)}{Q_j(x)}. \qquad (8.6)$$

Finally, if $F(x) \sim A(1 - x/x_c)^\theta$ as $x \to x_c^-$, then once estimates $x_c^*$ and $\theta^*$ of the critical point and critical exponent, respectively, have been obtained, one can then estimate the critical amplitude $A$ by forming Padé approximants to

$$(x_c^* - x)F(x)^{1/\theta^*}|_{x=x_c^*},$$

which should approximate $x_c^* A^{1/\theta^*}$, from which estimates of $A$ follow.

Noting that the Dlog-Padé function $\widehat{F}(x) = F'(x)/F(x)$, we see that forming a Dlog-Padé approximant is simply equivalent to seeking an approximation to $F(x)$ by solving the first order homogeneous differential equation.

$$F'(x)Q_j(x) - F(x)P_i(x) = 0.$$

This observation leads us straight into the more powerful and more general method of differential approximants by noting that we can approximate $F(x)$ by a solution to a higher order ODE (possibly inhomogeneous). This method was first proposed and developed by Guttmann and Joyce [11] in 1972, and was subsequently extended to the inhomogeneous case by Au-Yang and Fisher [5] and Hunter and Baker [13] in 1979.

## 8.4 Differential Approximants

As we have seen in earlier chapters the majority of polygon and polyomino models in statistical mechanics and combinatorics have generating functions with regular singular points. From the known exact solutions it is clear that the generating functions are often algebraic, or otherwise are given by the solution of simple linear ordinary differential equations. This observation (originally made in the context of the Ising model) forms the nucleus of the method of *differential approximants*. The basic idea is to approximate the function $F(x)$ by solutions to differential equations with polynomial coefficients. The singular behaviour of such ODEs is a well-known classical mathematics problem (see e.g. [6, 14]) and the singular points and exponents are easily calculated. Even if the function *globally* is not a solution of such a linear ODE (as is the case for SAP, as proved in Chapter 6) one hopes that *locally* in the vicinity of the (physical) critical points the generating function can still be well approximated by a solution to a linear ODE.

An $M^{th}$-order differential approximant (DA) to a function $F(x)$ is formed by matching the coefficients in the polynomials $Q_k(x)$ and $P(x)$ of degree $N_k$ and $L$, respectively, so that (one) of the formal solutions to the inhomogeneous differential equation

$$\sum_{k=0}^{M} Q_k(x) \left( x \frac{\mathrm{d}}{\mathrm{d}x} \right)^k \tilde{F}(x) = P(x) \tag{8.7}$$

agrees with the first $N = L + \sum_k (N_k + 1)$ series coefficients of $F(x)$. The function $\tilde{F}(x)$ thus agrees with the power series expansion of the (generally unknown) function $F(x)$ up to the first $N$ series expansion coefficients. We normalise the DA by setting $Q_M(0) = 1$ thus leaving us with $N$ rather than $N + 1$ unknown coefficients to find, in order to specify the ODE. From the theory of ODEs, the singularities of $F(x)$ are approximated by zeros $x_i$, $i = 1, \ldots, N_M$ of $Q_M(x)$, and the associated critical exponent $\lambda_i$ is estimated from the indicial equation. If there is only a single root at $x_i$ this is just

$$\lambda_i = M - 1 - \frac{Q_{M-1}(x_i)}{x_i Q'_M(x_i)}. \tag{8.8}$$

The physical critical point is the first singularity on the positive real axis.

In order to locate the singularities of the series in a systematic fashion we often use the following procedure: We calculate all $[L; N_0, N_1, N_2]$ and $[L; N_0, N_1, N_2, N_3]$ second- and third-order inhomogeneous differential approximants with $|N_i - N_j| \leq 2$, that is the degrees of the polynomials $Q_k$ differ by at most 2. In addition we demand that the total number of terms used by the DA is at least $N_{\max} - 10$, where

$N_{\max}$ is the total number of terms available in the series. Each approximant yields $N_M$ possible singularities and associated exponents from the $N_M$ zeroes of $Q_M(x)$ (most of these are not singularities of the series but merely spurious zeros). Next these zeros are sorted into equivalence classes by the requirement that they lie at most a distance $1/2^j$ apart, where we typically start with $j = 35$. An equivalence class is accepted as a singularity if an associated zero appears in more than 75% of the total number of approximants, and an estimate for the singularity and exponent is obtained by averaging over the included approximants (the spread among the approximants is also calculated). The calculation is then repeated for $j - 1, j - 2, \ldots$ until a minimum value of 8 or 10. To avoid outputting well-converged singularities at every level, once an equivalence class has been accepted, the data used in the estimate is discarded, and the subsequent analysis is carried out on the remaining data only.

One advantage of this method is that spurious outliers, some of which will almost always be present when so many approximants are generated, are discarded systematically and automatically. Unfortunately, it is not possible to provide rigorous error bounds for differential approximant estimates. In quoting errors we have adopted the following general procedure: For typical individual estimates with a fixed value of $L$ the error is calculated from the spread (basically one or two standard deviations) among the approximants used in obtaining the estimate. Note that these error bounds should *not* be viewed as a measure of the true error as they cannot include possible systematic sources of error. The final estimates (and error bounds) take into account the individual estimates and their error bounds. Note that DA estimates *are not* statistically independent so the true error may exceed the estimated error-bars. This is frequently accommodated by doubling or tripling the calculated error.

### 8.4.1 The Honeycomb SAP Generating Function

As a first example we apply the differential approximant analysis to the generating function for SAP on the honeycomb lattice. On this lattice the critical point, critical exponent and some universal amplitude ratios are known exactly, so this model provides us with a perfect test-bed for series analysis. In Table 8.1 we have listed the estimates for the critical point $x_c^2$ and exponent $2 - \alpha$ obtained from second- and third-order DAs. We note that all the estimates are in perfect agreement (surely a best case scenario) in that within 'error-bars' they take the same value. From this we arrive at the estimate $x_c^2 = 0.2928932186(5)$ and $2 - \alpha = 1.5000004(10)$. The final estimates are in perfect agreement with the conjectured [19, 20] exact values $x_c^2 = 1/\mu^2 = 1/(2 + \sqrt{2}) = 0.292893218813\ldots$ and $2 - \alpha = 3/2$.

Before proceeding we will consider possible sources of systematic errors. First and foremost is the possibility that the estimates might display a systematic drift as the number of terms used is increased, and secondly there is the possibility of numerical errors. The latter possibility is quickly dismissed. The calculations were performed using 128-bit real numbers. The estimates from a few approximants were

**Table 8.1** Critical point and exponent estimates for self-avoiding polygons.

| $L$ | Second order DA | | Third order DA | |
|---|---|---|---|---|
| | $x_c^2$ | $2 - \alpha$ | $x_c^2$ | $2 - \alpha$ |
| 0 | 0.29289321854(19) | 1.50000065(41) | 0.29289321865(12) | 1.50000040(28) |
| 5 | 0.29289321875(21) | 1.50000010(59) | 0.29289321852(48) | 1.50000041(99) |
| 10 | 0.29289321855(23) | 1.50000060(48) | 0.29289321878(32) | 1.49999999(97) |
| 15 | 0.29289321859(19) | 1.50000054(43) | 0.29289321861(37) | 1.50000035(67) |
| 20 | 0.29289321866(15) | 1.50000038(33) | 0.29289321860(21) | 1.50000049(43) |

compared to values obtained using MAPLE with 100 digits accuracy and this clearly showed that the program was numerically stable and rounding errors were negligible. In order to address the possibility of systematic drift and lack of convergence to the true critical values we refer to Fig. 8.2 (this is probably not really necessary in this case but we include the analysis here in order to present the general method).

In the left panel of Fig. 8.2 we have plotted the estimates from third-order DAs for $x_c^2$ vs. the highest order coefficient index $N < N_{max}$ used by the DA. Each dot in the figure is an estimate obtained from a specific approximant. As can be seen the estimates clearly settle down to the conjectured exact value (solid line) as $N$ is increased and there is little to no evidence of any systematic drift at large $N$. One curious aspect though is the widening of the spread in the estimates around $N = 140$. We have no explanation for this behaviour but it could quite possibly be caused by just a few 'spurious' approximants. In the right panel we show the variation in the exponent estimates with the critical point estimates. The 'curve' traced out by the estimates passes through the intersection of the lines given by the exact values. We have not been able to determine the reason for the apparent branching into two parts. However, the lower 'branch' contains many more approximants than the upper one, and is therefore the selected branch.

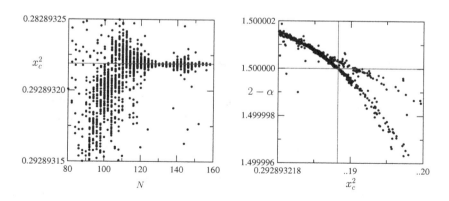

**Fig. 8.2** Plot of estimates from third order differential approximants for $x_c^2$ vs. the highest order term used, and the right panel shows $2 - \alpha$ vs. $x_c^2$. The straight lines are the exact predictions.

The differential approximant analysis can also be used to find possible non-physical singularities of the generating function. Averaging over the estimates from the DAs shows that there is an additional non-physical singularity on the negative $x$-axis at $x = x_- = -1/\mu_-^2 = -0.41230(2)$, where the estimates of the associated critical exponent $\alpha_-$ are consistent with the exact value $\alpha_- = 3/2$. In the left panel of Fig. 8.3 we have plotted $\alpha_-$ vs. the highest order term used by the DAs and we clearly see the convergence to $\alpha_- = 3/2$. If we take this value as being exact we can get a refined estimate of $x_-$ from the plot in the right panel of Fig. 8.3, where we notice that the estimates for $\alpha_-$ cross the value $3/2$ for $x_- = -0.412305(5)$ which we take as our final estimate. From this we then get $\mu_- = 1.557366(10)$.

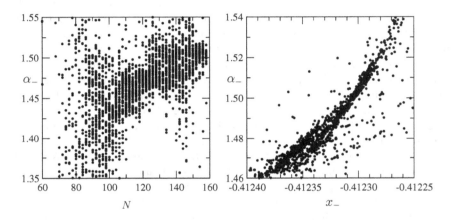

**Fig. 8.3** Plot of estimates from third order differential approximants for the location $x_-$ of the non-physical singularity and the associated exponent $\alpha_-$. The left panel shows $\alpha_-$ vs. the highest order term used, and the right panel shows $\alpha_-$ vs. $x_-$.

## 8.5 Amplitude Estimates

Now that the exact values of $\mu$ and the exponents have been confirmed we turn our attention to the "fine structure" of the asymptotic form of the coefficients. In particular we are interested in obtaining accurate estimates for the leading critical amplitudes. The method of analysis consists in fitting the coefficients to an assumed asymptotic form. Generally one must include a number of asymptotic terms in order to account for the behaviour of the generating function at both the physical singularity and the non-physical singularities as well as accounting for sub-dominant corrections to the leading order behaviour. As we hope to demonstrate, this method of analysis can not only yield accurate amplitude estimates, but it is often possible to clearly demonstrate which corrections to scaling are present.

Before proceeding with the analysis we briefly consider the kind of terms which occur in the generating functions, and how they influence the asymptotic behaviour of the series coefficients. At the most basic level a function $G(x)$ with a power-law singularity[1]

$$G(x) = \sum_n g_n x^n \sim A(x)(1 - \mu x)^{-\xi},$$ (8.9)

where $A(x)$ is analytic in the vicinity of $x = x_c = 1/\mu$, gives rise to the following asymptotic form of the coefficients:

$$g_n \sim \mu^n n^{\xi-1} \left[ \tilde{A} + \sum_{i \geq 1} a_i / n^i \right],$$ (8.10)

that is, we get the dominant exponential growth given by the term $\mu^n$, modified by a sub-dominant term given by the term $n^{\xi-1}$, involving the critical exponent $\xi$, followed by analytic corrections. The amplitude $\tilde{A}$ is related to the function $A(x)$ in (8.9) via the relation $\tilde{A} = A(1/\mu)/\Gamma(\xi)$. If $G(x)$ has a non-analytic correction to scaling such as

$$G(x) = \sum_n g_n x^n \sim (1 - \mu x)^{-\xi} \left[ A(x) + B(x)(1 - \mu x)^\Delta \right],$$ (8.11)

we get the more complicated form,

$$g_n \sim \mu^n n^{\xi-1} \left[ \tilde{A} + \sum_{i \geq 1} a_i / n^i + \sum_{i \geq 0} b_i / n^{\Delta+i} \right].$$ (8.12)

A singularity on the negative $x$-axis $\propto (1 + \mu_- x)^{-\eta}$ leads to additional corrections of the form

$$\sim (-1)^n \mu_-^n n^{\eta-1} \sum_{i \geq 0} c_i / n^i.$$ (8.13)

Singularities in the complex plane are still more complicated. However, a pair of singularities on the imaginary axis at $\pm i/\tau$, that is a term of the form $D(x)(1 + \tau^2 x^2)^{-\eta}$, generally results in coefficients that change sign according to a $+ + - -$ pattern. This can be accommodated by terms of the form

$$\sim (-1)^{\lfloor n/2 \rfloor} \tau^n n^{\eta-1} \sum_{i \geq 0} d_i / n^i.$$ (8.14)

All of these possible contributions must then be put together in an assumed asymptotic expansion for the coefficients $g_n$ and we obtain estimates for the unknown amplitudes by directly fitting $g_n$ to the assumed form. That is, we take a sub-sequence of terms $\{g_n, g_{n-1}, \ldots, g_{n-k}\}$, plug into the assumed form and solve

---

[1] We have rewritten equation (8.1) in a more convenient form for this analysis.

the $k + 1$ linear equations to obtain estimates for the first few amplitudes. As we shall demonstrate below this allows us to probe the asymptotic form.

### 8.5.1 Estimating the Polygon Amplitude $\tilde{A}$

Here we illustrate the method by analysing the coefficients of the generating function for honeycomb lattice polygons,

$$P(x) = \sum_{n=0} p_{2n} x^n.$$

As well as the physical singularity of interest at $x = x_c^2$, there is a non-physical singularity at $x = x_-$, where $|x_-| > x_c^2$. The asymptotic form of the coefficients $p_n$ of the generating function of square and triangular lattice SAP has been previously studied in detail [3, 17, 15]. There is now clear numerical evidence that the leading correction-to-scaling exponent for SAPs is $\Delta_1 = 3/2$, as predicted by Nienhuis [19, 20]. As argued in [3] this leading correction term combined with the $2 - \alpha = 3/2$ term of the SAP generating function produces an *analytic* background term as can be seen from equation (8.11). Indeed, in the previous analysis of SAPs there was no sign of non-analytic corrections-to-scaling to the generating function (a strong indirect argument that the leading correction-to-scaling exponent must be half-integer valued). At first we ignore the singularity at $x_-$ (since $|x_-| > x_c^2$ it is exponentially suppressed) and obtain estimates for $\tilde{A}$ by fitting $p_n$ to the form

$$p_n = \mu^n n^{-5/2} \left[ \tilde{A} + \sum_{i=1}^{k} a_i/n^i \right]. \tag{8.15}$$

That is, we take a sub-sequence of terms $\{p_n, p_{n-2}, \ldots, p_{n-2k}\}$ ($n$ even), plug into the formula above and solve the $k + 1$ linear equations to obtain estimates for the amplitudes. It is then advantageous to plot estimates for the leading amplitude $\tilde{A}$ against $1/n$ for several values of $k$. The results are plotted in the left panel of Fig. 8.4. Obviously the amplitude estimates are not well behaved and display clear parity effects. So clearly we can't just ignore the singularity at $x_-$ (which gives rise to such effects) and we thus try fitting to the more general form

$$p_n = \mu^n n^{-5/2} \left[ \tilde{A} + \sum_{i=1}^{k} a_i/n^i \right] + (-1)^{n/2} \mu_-^n n^{-5/2} \sum_{i=0}^{k} b_i/n^i. \tag{8.16}$$

The results from these fits are shown in the middle panel of Fig. 8.4. Now we clearly have very well-behaved estimates (note the significant change of scale along the $y$-axis from the left to the middle panel). In the right panel we take a more detailed look at the data and from the plot we estimate that $\tilde{A} = 1.2719299(1)$. We notice that as more and more correction terms are added ($k$ is increased) the plots of the amplitude estimates exhibit less curvature and the slope become less steep. This is very strong evidence that (8.16) indeed is the correct asymptotic form of $p_n$.

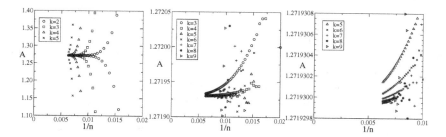

**Fig. 8.4** Plots of fits for the self-avoiding polygon amplitude $\tilde{A}$ using in the left panel the asymptotic form (8.15) which ignores the singularity at $x = x_-$, and in the middle panel the asymptotic form (8.16) which includes the singularity at $x = x_-$. The right panel gives a closer look at the data from the middle panel.

## 8.5.2 The Correction-to-Scaling Exponent

In this section we shall briefly show how the method of direct fitting can be used to differentiate between various possible values for the leading correction-to-scaling exponent $\Delta_1$. There are two competing theoretical predictions, $\Delta_1 = 3/2$ by Nienhuis [19] and $\Delta_1 = 11/16$ by Saleur [21]. As already stated there is now firm evidence from previous work that the Nienhuis result is correct. Here we shall present further evidence. Different values for $\Delta_1$ lead to different assumed asymptotic forms for the coefficients. For the SAP series we argued that a value $\Delta_1 = 3/2$ (or indeed any half-integer value) would result only in *analytic* corrections to the generating function and thus that $p_n$ asymptotically would be given by (8.16). If we have a generic value for $\Delta_1$ we would get

$$p_n = \mu^n n^{-5/2} \left[ \tilde{A} + \sum_{i=1}^{k} a_i/n^i + \sum_{i=0}^{k} b_i/n^{\Delta_1+i} \right] + (-1)^{n/2} \mu_-^n n^{-5/2} \sum_{i=0}^{k} c_i/n^i. \quad (8.17)$$

Fitting to this form we can then estimate the amplitude $b_0$ of the term $1/n^{\Delta_1}$. We would expect that if we used a manifestly incorrect value for $\Delta_1$ then $b_0$ should vanish asymptotically thus demonstrating that this term is really absent from (8.17). So let us fit to this form using the value $\Delta_1 = 11/16$. More precisely we fit to the generic form

$$p_n = \mu^n n^{-5/2} \sum_{i=0}^{k} a_i/n^{\alpha_i} + (-1)^{n/2} \mu_-^n n^{-5/2} \sum_{i=0}^{k} b_i/n^i. \quad (8.18)$$

First we include only the leading term arising from $\Delta_1$ using the sequence of exponents $\alpha_i = \{0, 11/16, 1, 2, 3, \ldots\}$. Next we fit to a form including additional analytical corrections arising from $\Delta_1$ leading to the sequence of exponents $\alpha_i = \{0, 11/16, 1, 27/16, 2, 33/16, 3, 49/16, \ldots\}$. More generally one also expects terms of the form $1/n^{m\Delta_1 + i}$ with $m$ a non-negative integer. This leads to fits to the form above but with $\alpha_i = \{0, 11/16, 1, 11/8, 27/16, 2, 33/16, 19/8, 43/16, 11/4, 3 \ldots\}$. The estimates of the amplitude of the term $1/n^{\Delta_1}$ obtained from fits to these forms are shown in Fig. 8.5. As can be seen from the left panel, where we fit to the first scenario, the amplitude clearly seems to converge to 0, which would indicate the absence of this term in the asymptotic expansion for $p_n$. In the middle and right panels we show the results from fits to the more general forms. The estimates are consistent with the amplitude being identically zero, though the evidence is not quite as convincing. This is however not really surprising given that the incorrect value $\Delta_1 = 11/16$ gives rise to a plethora of absent terms which will tend to greatly obscure the true asymptotic behaviour.

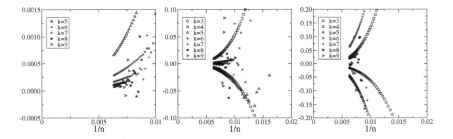

**Fig. 8.5** Plots of estimates for the amplitude of the term $1/n^{\Delta_1}$. The left panel shows results from fits to the form (8.18) where only the leading order term $1/n^{\Delta_1}$ is included (as well as analytical corrections). In the middle panel additional terms of the form $1/n^{\Delta_1 + i}$ are included and in the right panel terms like $1/n^{m\Delta_1 + i}$ are included.

## 8.6 Exact Fuchsian ODEs for Polygon Models

In recent work Zenine et al. [22, 23, 24] obtained, by experimental computer search, the linear differential equations whose solutions give some quantities of interest in the study of the Ising model of ferromagnetism. Adopting their methods, Guttmann and Jensen [8, 9] used the same ideas to find linear differential equations which have as a solution the generating function $\mathscr{T}(x)$ for three-choice polygons and $\mathscr{P}(x)$ for punctured staircase polygons.

Punctured staircase polygons [10] are staircase polygons with internal holes which are also staircase polygons (the polygons are mutually- as well as self-avoiding). Here we will study only the case with a *single* hole (see Fig. 8.6), and

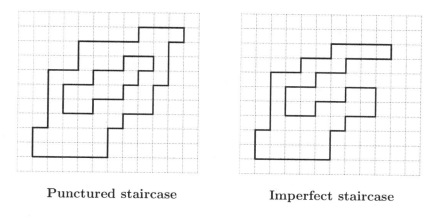

Punctured staircase                    Imperfect staircase

**Fig. 8.6** A punctured staircase polygon and an imperfect staircase polygon.

we will refer to these objects as punctured staircase polygons. The perimeter length of staircase polygons is even and thus the total perimeter (the outer perimeter plus the perimeter of the hole) is also even. We denote by $p_n$ the number of punctured staircase polygons of perimeter $2n$.

Three-choice self-avoiding walks on the square lattice were introduced by Manna [18] and can be defined as follows: Starting from the origin one can step in any direction; after a step upward or downward one can head in any direction (except backward); after a step to the left one can only step forward or head downward, and similarly after a step to the right one can continue forward or turn upward. As usual one can define a polygon version of the walk model by requiring the walk to return to the origin. So a three-choice polygon [12] is simply a three-choice self-avoiding walk which returns to the origin, but has no other self-intersections. There are two distinct classes of three-choice polygons. The three-choice rule either leads to staircase polygons or *imperfect staircase polygons* [4] (see Fig. 8.6). The three-choice rules produce imperfect staircase polygons in two ways and staircase polygons of perimeter $n$ in $n$ ways. We denote by $t_n$ the number of three-choice polygons of perimeter $2n$.

Here we briefly outline the method used to find the exact ODE, which we will illustrate by looking at the perimeter generating function of three-choice polygons. Assume we have a function $F(x)$ with a singularity at $x = x_c = 1/\mu$. Starting from a (long) series expansion for the function $F(x)$ we look for a linear differential equation of order $M$ of the form

$$\sum_{k=0}^{M} P_k(x) \frac{d^k}{dx^k} F(x) = 0, \tag{8.19}$$

such that $F(x)$ is a solution of this homogeneous linear differential equation, where the $P_k(x)$ are polynomials. In order to make it as simple as possible we start by searching for a Fuchsian [14] equation. Such equations have only regular singular points. There are several reasons for searching for a Fuchsian equation, rather than a more general D-finite equation. Computationally the Fuchsian assumption simplifies the search for a solution. One may also argue, less precisely, that for "sensible" combinatorial models one would expect Fuchsian equations, as non-Fuchsian equations are characterized by explosive, super-exponential behaviour. Such behaviour is not normally characteristic of combinatorial problems. (The point at infinity may be an exception to this somewhat imprecise observation.) One may also ask the question whether most of the problems in combinatorics with D-finite solutions have Fuchsian solutions? While we have not made an exhaustive study, we know of no counter-example to this suggestion.

From the general theory of Fuchsian [14] equations it follows that the degree of $P_k(x)$ is at most $N_M - M + k$ where $N_M$ is the degree of $P_M(x)$. To simplify matters (reduce the order of the unknown polynomials) it is often advantageous to explicitly assume that the origin and $x = x_c$ are regular singular points and to set $P_k(x) = Q_k(x)S(x)^k$, where $S(x) = xR(x)$ and $R(x)$ is a polynomial of minimal degree having $x_c$ as a root (in our case we have $R(x) = 1 - 4x$). $S(x)$ could be generalised to include more regular singular points if some were known from other methods of analysis, but we have not found this to be particularly advantageous. Thus when searching for a solution of Fuchsian type there are only two parameters: namely the order $M$ of the differential equation and the degree $q_M$ of the polynomial $Q_M(x)$. Let $\rho$ be the degree of $S(x)$ (2 in our case), then for given $M$ and $q_M$ there are $L = (M+1)(q_M+1) + \rho M(M+1)/2 - 1$ unknown coefficients, where we have assumed without loss of generality that the leading order coefficient in $P_M(x) = Q_M(x)S(x)^M$ is 1. We can then search systematically for solutions by varying $M$ and $q_M$.

In this way we first found a solution with $M = 10$ and $q_M = 12$, which required the determination of $L = 206$ unknown coefficients. We have 260 terms in the half-perimeter series and thus have more than 50 additional terms with which to check the correctness of our solution. Having found this conjectured solution we then turned the ODE into a recurrence relation and used this to generate more series terms in order to search for a lower order Fuchsian equation. The lowest order equation we found was eighth order ($M = 8$) and with $q_M = 30$, which requires the determination of $L = 321$ unknown coefficients. Thus from our original 260 term series we could not have found this $8^{th}$ order solution since we did not have enough terms to determine all the unknown coefficients in the ODE. This raises the question as to whether perhaps there is an ODE of lower order than 8 that generates the coefficients? The short answer to this is no. Further study of our differential operator revealed that it can be factorised. In fact we found a factorization into three first-order linear operators, a second order and a third order. The generating function is a solution of the $8^{th}$ order operator, not of any of the smaller factors.

So the (half)-perimeter generating function $\mathscr{T}(x)$ for three-choice polygons is conjectured to be a solution of the linear differential equation of order 8,

$$\sum_{k=0}^{8} P_k(x) \frac{d^k}{dx^k} F(x) = 0 \qquad (8.20)$$

with

$$P_8(x) = x^3(1-4x)^4(1+4x)(1+4x^2)(1+x+7x^2)Q_8(x),$$
$$P_7(x) = x^2(1-4x)^3 Q_7(x), \quad P_6(x) = 2x(1-4x)^2 Q_6(x),$$
$$P_5(x) = 6(1-4x)Q_5(x), \quad P_4(x) = 24Q_4(x),$$
$$P_3(x) = 24Q_3(x), \quad P_2(x) = 144x(1-2x)Q_2(x),$$
$$P_1(x) = 144(1-4x)Q_1(x), \quad P_0(x) = 576Q_0(x),$$

(8.21)

where $Q_8(x), Q_7(x), \ldots, Q_0(x)$, are polynomials of degree $25, 31, 32, 33, 33, 32, 29,$ 29, and 29, respectively. See [8] for further details.

The singular points of the differential equation are given by the roots of $P_8(x)$. One can easily check that all the singularities (including $x = \infty$) are *regular singular points* so equation (8.20) is indeed of the Fuchsian type. It is thus possible, using the method of Frobenius, to obtain from the indicial equation the critical exponents at the singular points. These are listed in Table 8.2.

**Table 8.2** Critical exponents for the regular singular points of the Fuchsian differential equation satisfied by $\mathcal{T}(x)$.

| Singularity | Exponents |
|---|---|
| $x = 0$ | $-1, 0, 0, 0, 1, 2, 3, 4$ |
| $x = 1/4$ | $-1/2, -1/2, 0, 1/2, 1, 3/2, 2, 3$ |
| $x = -1/4$ | $0, 1, 2, 3, 4, 5, 6, 13/2$ |
| $x = \pm i/2$ | $0, 1, 2, 3, 4, 5, 6, 13/2$ |
| $1 + x + 7x^2 = 0$ | $0, 1, 2, 2, 3, 4, 5, 6$ |
| $x = \infty$ | $-2, -3/2, -1, -1, -1/2, 1/2, 3/2, 5/2$ |
| $Q_8(x) = 0$ | $0, 1, 2, 3, 4, 5, 6, 8$ |

We shall now consider the local solutions of the differential equation around each singularity. Recall that in general it is known [6, 14] that if the indicial equation yields $k$ critical exponents which differ by an integer, then the local solutions *may* contain logarithmic terms up to $\log^{k-1}$. However, for the Fuchsian equation (8.20) *only* multiple roots of the indicial equation give rise to logarithmic terms in the local solution around a given singularity, so that a root of multiplicity $k$ gives rise to logarithmic terms up to $\log^{k-1}$.

In particular this means that near any of the 25 roots of $Q_8(x)$ the local solutions have no logarithmic terms and the solutions are thus *analytic* since all the exponents are positive integers. The roots of $Q_8(x)$ are thus *apparent singularities* [6, 14] of the Fuchsian equation (8.20). There are methods for distinguishing real and apparent singularities (see, e.g, [6] §45) and in principle one should check that the roots of $Q_8(x)$ satisfy the conditions for being apparent singularities. However, this theoretical method is quite cumbersome. An easier numerical way to see that the roots of $Q_8(x)$ must be apparent singularities is as follows: We already found a 10th order

Fuchsian equation for which the polynomial $P_{10}(x)$ was of a form similar to $P_8(x)$ as listed in equation (8.21), but with the degree of $Q_{10}(x)$ being only 7. That is all the singularities as tabulated in Table 8.2 also appear in this higher order equation with the exception of the 25 roots of $Q_8(x)$ (at most 7 of these could appear in the order 10 Fuchsian equation). In fact we can find a solution of order 14 of the same form as above but with $Q_{14}(x)$ being just a constant. So at this order none of the roots of $Q_8(x)$ appear. Clearly any real singularity of the system cannot be made to vanish and we conclude that the 25 roots of $Q_8(x)$ must indeed be apparent singularities.

Assuming that only repeated roots give rise to logarithmic terms, and thus that a sequence of positive integers give rise to *analytic* terms, then near the physical critical point $x = x_c = 1/4$ we expect the singular behaviour

$$\mathscr{T}(x) \sim A(x)(1-4x)^{-1/2} + B(x)(1-4x)^{-1/2}\log(1-4x), \qquad (8.22)$$

where $A(x)$ and $B(x)$ are analytic in the neighbourhood of $x_c$. Note that the terms associated with the exponents $1/2$ and $3/2$ become part of the analytic correction to the $(1-4x)^{-1/2}$ term. Near the singularity on the negative $x$-axis, $x = x_- = -1/4$ we expect the singular behaviour

$$\mathscr{T}(x) \sim C(x)(1+4x)^{13/2}, \qquad (8.23)$$

where again $C(x)$ is analytic near $x_-$. We expect similar behaviour near the pair of singularities $x = \pm i/2$, and finally at the roots of $1+x+7x^2$ we expect the behaviour $\mathscr{T}(x) \sim D(x)(1+x+7x^2)^2\log(1+x+7x^2)$.

Next we turn our attention to the asymptotic behaviour of the coefficients of $\mathscr{T}(x)$. To standardise our analysis, we assume that the critical point is at 1. The growth constant of three-choice polygons is 4, so we normalise the series by considering a new series with coefficients $r_n$, defined by $r_n = t_{n+2}/4^n$. Thus the generating function we study is $\mathscr{R}(y) = \sum_{n\geq 0} r_n y^n = 4 + 3y + 2.625y^2 + \cdots$. From equations (8.22) and (8.23) it follows that the asymptotic form of the coefficients is

$$[y^n]\mathscr{R}(y) = r_n = \frac{1}{\sqrt{n}} \sum_{i\geq 0} \left( \frac{a_i \log n + b_i}{n^i} + (-1)^n \left( \frac{c_i}{n^{7+i}} \right) \right) + \mathrm{O}(\lambda^{-n}). \qquad (8.24)$$

The last term includes the effect of other singularities, further from the origin than the dominant singularities. These will decay exponentially since $\lambda > 1$ in the scaled variable $y = x/4$.

Using the recurrence relations for $t_n$ (derived from the ODE) it is easy and fast to generate many more terms $r_n$. We generated the first 100000 terms and saved them as floats with 500 digit accuracy (this calculation took less than 15 minutes). With such a long series it is possible to obtain accurate numerical estimates of the first 20 amplitudes $a_i$, $b_i$, $c_i$ for $i \leq 19$ with precision of more than 100 digits for the dominant amplitudes, shrinking to 10–20 digits for the the case when $i = 18$ or 19. In making these estimates we have ignored the exponentially decaying term, which is the last term in equation (8.23). In this way we confirmed an earlier conjecture [4] that $a_0 = \frac{3\sqrt{3}}{\pi^{3/2}}$, (where we have taken into account the different normalisation used in that paper). We also find that $b_0 = 3.173275384589898481765\ldots$ and $c_0 = \frac{-24}{\pi^{3/2}}$, though we have not been able to identify $b_0$. However, we have successfully identified further sub-dominant amplitudes, and find $a_1 = \frac{-89}{8\sqrt{3}\pi^{3/2}}$, $a_2 = \frac{1019}{384\sqrt{3}\pi^{3/2}}$, and $a_3 = \frac{-10484935}{248832\sqrt{3}\pi^{3/2}}$, and $c_1 = \frac{225}{\pi^{3/2}}$, $c_2 = \frac{-16575}{16\pi^{3/2}}$, and $c_3 = \frac{389295}{128\pi^{3/2}}$. It seems possible that the amplitudes $\pi^{3/2}\sqrt{3}a_i$ and $\pi^{3/2}c_i$ are rational.

Estimates for the amplitudes were obtained by fitting $r_n$ to the form given above using an increasing number of amplitudes. 'Experimentally' we find we need about the same total number of terms at $x_c$ and $-x_c = x_-$.

So in the fits we used the terms with amplitudes $a_i$, and $b_i$, $i = 0,\ldots,K$ and $c_i$, $i = 0,\ldots,2K$. Going only to $i = K$ with the $c_i$ amplitudes results in much poorer convergence and going beyond $2K$ leads to no improvement. For a given $K$ we thus have to estimate $4K + 3$ unknown amplitudes. So we use the last $4K + 3$ terms $r_n$ with $n$ ranging from $100000$ to $100000 - 4K - 2$ and solve the resulting system of $4K + 3$ linear equations. We find that the amplitudes are fairly stable up to around $2K/3$. We observed this by doing the calculation with $K = 30$ and $K = 40$ and then looking at the difference in the amplitude estimates. For $a_0$ and $b_0$ the difference is less than $10^{-131}$, while for $c_0$ the difference is less than $10^{-123}$. Each time we increase the amplitude index by 1 we lose around $10^6$ in accuracy. With $i = 20$ the differences are respectively around $10^{-16}$ and $10^{-8}$.

The excellent convergence is solid evidence (though naturally not a proof) that the assumptions leading to equation (8.24) are correct. Further evidence was obtained as follows: We can add extra terms to the asymptotic form and check what happens to the amplitudes of the new terms. If the amplitudes are very small it is highly likely that the terms are not truly present (if the calculation could be done exactly these amplitudes would be zero). One possibility is that our assumption about integer exponents leading only to analytic terms is incorrect. To test this we fitted to the form

$$\frac{1}{\sqrt{n}}\sum_{i\geq 0}\left(\frac{\tilde{a}_i\log n + \tilde{b}_i}{n^{i/2}} + (-1)^n\left(\frac{\tilde{c}_i}{n^{7+i}}\right)\right) + O(\lambda^{-n})$$

(as above, in making these estimates we have ignored the exponentially decaying term, which is the last term in the above equation). With $K = 30$ we found that the amplitudes $\tilde{a}_1$ and $\tilde{b}_1$ of the terms $\log n/n$ and $1/n$, respectively, were less than $10^{-60}$, while the amplitudes $\tilde{a}_3$ and $\tilde{b}_3$ were less than $10^{-50}$. We think we can safely say that all the additional terms we just added are *not* present. We found similar results if we added terms like $\log^2 n$ or additional $\log n$ terms at $y = -1$. That is, we found that those terms were not present. So this fitting procedure provides convincing evidence that the asymptotic form (8.24), and thus the assumption leading to this formula, is correct.

### 8.6.1 *Exact ODEs Modulo a Prime*

In the above calculations we searched for the ODE using the full series. However, if the size of the ODE is large then this is very time consuming both in terms of generating the actual series and then searching through values of $M$ and $D$ looking for the ODE. In very recent work [2] we have adopted a different and much more efficient strategy. Rather than perform the search on the full series we search only for a solution *modulo* a specific prime (in practice we used the prime $p_0 = 32749 = 2^{15} - 19$). The advantages of this approach are obvious. Firstly we only need generate a long series for a single prime (at least initially) and secondly solving the system determined by (8.25) amounts to finding whether or not the system of linear equations has a zero determinant. This is easily done using Gaussian elimination, and if a zero-determinant is found one can then proceed to solve the system, which yields the ODE *modulo* the prime $p_0$. In theory one has to worry about possible false positive results, but in widespread use we have never encountered this situation in practice (and in most cases one can check the result using a different prime). Below we give a brief outline of the procedure developed in [2]. Further details can be found there.

Here we illustrate the method by looking at a different generating function for punctured and imperfect staircase polygons. We study the case where the counting variable is the 'length' (extent along the main diagonal) of the polygons. The 'length' is equal to the sum of the coordinates of the point of the polygon furthest from the origin. For punctured staircase polygons this is also equivalent to counting according to the half-perimeter of the outer staircase polygon (rather than the total perimeter of the outer and inner polygons combined). We denote the associated generating functions as $\mathscr{L}(x)$ in the punctured case and $\mathscr{I}(x)$ in the imperfect case.

Starting from a (long) series expansion for some function $F(x)$ we look for a linear differential equation of order $M$ satisfied by $F(x)$. An essential constraint on the ODE of the type we shall consider here is that it be Fuchsian. In particular this means that $x = 0$ and $x = \infty$ are regular singular points. A form for the ODE that automatically satisfies this constraint is

$$\sum_{k=0}^{M} Q_k(x) \left( x \frac{\mathrm{d}}{\mathrm{d}x} \right)^k F(x) = 0, \tag{8.25}$$

where the $Q_k(x) = \sum_{j=0}^{D} q_{k,j} x^j$ are polynomials of degree $D$. The condition $a_{M,0} \neq 0$ makes $x = 0$ a regular singular point and the use of the operator $(x\mathrm{d}/\mathrm{d}x)$ rather than just $\mathrm{d}/\mathrm{d}x$ makes analysis around $x = \infty$ simple. Finding the ODE (if it exists) then essentially amounts to solving a system of $(M+1) \times (D+1)$ linear equations.

To determine the coefficients $q_{k,j}$ of the polynomials in (8.25) we arrange the set of linear equations in a well-defined order. There exists a non-trivial solution if the determinant of the matrix of the system of $(M+1) \times (D+1)$ linear equations vanishes. We test this by standard Gaussian elimination, creating an upper triangular matrix $U$ in the process. If we find that a diagonal element $U(N,N) = 0$ for some $N$, then a non-trivial solution exists. If $N < N_{MD} = (M+1) \times (D+1)$ we set to zero all $q_{k,j}$ in the ordered list beyond $N$. Of the remaining $q_{k,j}$ we set $q_{M,0} = 1$,

**Table 8.3** $M$ is the order of the ODE, $D$ is the degree of each polynomial multiplying each derivative, $N_{MD} = (M+1)(D+1)$, $N$ is the actual number of terms predicted by (8.26) as necessary to find an ODE of the given order $M$, and $\Delta$ is the difference $N_{MD} - N$. The first five columns gives this data for $\mathscr{L}(x)$ while the next five columns gives this data for $\mathscr{I}(x)$.

| Terms needed to find $\mathscr{L}(x)$ | | | | | Terms needed to find $\mathscr{I}(x)$ | | | | |
|---|---|---|---|---|---|---|---|---|---|
| $M$ | $D$ | $N_{MD}$ | $N$ | $\Delta$ | $M$ | $D$ | $N_{MD}$ | $N$ | $\Delta$ |
| 11 | 53 | 648 | 648 | 0 | 14 | 92 | 1395 | 1395 | 0 |
| 12 | 31 | 416 | 415 | 1 | 15 | 52 | 848 | 848 | 0 |
| 13 | 23 | 336 | 336 | 0 | 16 | 39 | 680 | 679 | 1 |
| 14 | 20 | 315 | 312 | 3 | 17 | 32 | 594 | 594 | 0 |
| 15 | 17 | 288 | 288 | 0 | 18 | 28 | 551 | 551 | 0 |
| 16 | 16 | 289 | 286 | 3 | 19 | 26 | 540 | 536 | 4 |
| 17 | 15 | 288 | 284 | 4 | 20 | 24 | 525 | 521 | 4 |
| 18 | 14 | 285 | 280 | 5 | 21 | 22 | 506 | 506 | 0 |
| 19 | 13 | 280 | 280 | 0 | 22 | 21 | 506 | 505 | 1 |
| 20 | 13 | 294 | 289 | 5 | 23 | 20 | 504 | 504 | 0 |
| 21 | 13 | 308 | 298 | 10 | 24 | 20 | 525 | 517 | 8 |
| 22 | 12 | 299 | 296 | 3 | 25 | 19 | 520 | 516 | 4 |

thus guaranteeing that $x = 0$ is a regular singular point and determine the remaining coefficients by back substitution. The $N$ for which $U(N,N) = 0$ is the minimum number of series coefficients needed to find the ODE within the constraint of a given $M$ and $D$. Obviously, $N \leq N_{MD} = (M+1) \cdot (D+1)$. Henceforth, $D$ will always refer to the minimum $D$ for which a solution can be found for a given $M$. Then, for example, we can define a unique non-negative deviation $\Delta$ by $N = N_{MD} - \Delta = (M+1) \cdot (D+1) - \Delta$. Examples of such constants are given in Table 8.3 based on our analysis of $\mathscr{L}(x)$ and $\mathscr{I}(x)$. A very striking empirical observation was made in [2], namely that the numbers $N$ in Table 8.3 are given by a simple linear relation

$$N = A \cdot M + B \cdot D - C = (M+1) \cdot (D+1) - \Delta \qquad (8.26)$$

where $A$, $B$ and $C$ are constants depending on the particular series. For $\mathscr{L}(x)$ they are $A = 9$, $B = 11$, $C = -34$ and for $\mathscr{I}(x)$ they are $A = 13$, $B = 14$, $C = -75$ as can be verified from Table 8.3. Note that (8.26) has no (positive) solution for $D$ if $M < B$. Thus $B = M_0$ is the minimum order possible for the ODE. Similarly, $A = D_0$ is the minimum possible degree and thus we can rewrite (8.26) in the more definitive form

$$N = D_0 \cdot M + M_0 \cdot D - C = (M+1) \cdot (D+1) - \Delta. \qquad (8.27)$$

The minimum order $M_0$ and degree $D_0$ can be inferred directly from the ODE independently of (8.27). The head polynomial $Q_M(x)$ in (8.25) can be factored *modulo* a prime and the greatest common divisor of these from several different orders $M$ is the polynomial $Q(x)$ whose zeros are the true singularities of the linear ODE. In all cases we have tested, the degree of this polynomial factor is the $D_0$ in (8.27).

## 8.6.2 Reconstructing the Exact ODE from Modular Results

Finding the minimal exact ODE for $\mathscr{I}(x)$ using the exact series coefficients would be a difficult task since the size of the coefficients grow as $2^{4n}$, so we would have to handle integers of some 1700 digits using an array of size $1396^2$ in order to solve the set of linear equations arising out of equation (8.25). This would be stretching the capacity of our current algorithms. So instead we decided to use a different and as we shall see much more efficient approach. It is possible to reconstruct the exact ODE using the results from several *modulo* prime calculations (actually as we shall show only 10 primes are needed). Here we schematically outline the procedure for finding the exact minimum order ODE.

**Procedure for ODE reconstruction:**

1. Generate a long series modulo a single prime.
2. Find ODEs at different orders and identify the constants $A$, $B$, and $C$ of (8.26).
3. Then use this formula to identify both the minimal order ODE and the ODE requiring the least number of terms.
4. Generate series for more primes $p_i$ long enough to find the minimal term ODEs.
5. Turn these ODEs into recurrences and generate longer series.
6. Use these series to find the minimal order ODE mod $p_i$.
7. Combine to find the exact minimal order ODE:
   (a) Use the Chinese Remainder Theorem to get coefficients $a_{ij}$.
      This gives us $b_{ij} = a_{ij}$ *modulo* $P$, where $P = \prod p_i$.
   (b) Find the exact rational coefficients say by using the Maple call $a_{ij} =$ iratrecon($b_{ij}, P$).

We managed to reconstruct the exact ODE for $\mathscr{I}(x)$ using 18 primes of the form $p_j = 2^{30} - r_j$. Reconstructing the exact series coefficients up to the length needed to find the exact ODE by the more traditional approach would require at least 10 times as many primes. We note that it takes only a few minutes to find the ODEs mod the primes and then reconstruct the exact ODE coefficient. Even fewer primes are actually needed. In general the numerators are much smaller than the denominators so we can modify the call to read $r/s = a_{ij} =$ iratrecon($b_{ij}, P, R, S$), where $R$ and $S$ are positive integers such that $|r| \leq R$ and $0 < s \leq S$ with $2RS \leq P$. If we assume that $s < \sqrt{r}$ then we may choose $S = \sqrt[4]{P}$ and $R = P/(2S)$ and we can then find the $a_{ij}$ using only 12 primes.

A further refinement is possible by generating the $a_{ij}$ starting from $a_{MD}$. We then multiply all the residues by the denominator of $a_{MD}$ *modulo* the respective primes. We then run through the remaining coefficients by decreasing first $j$ so as to generate all $a_{Mj}$. Whenever a fraction is encountered we multiply all residues by its denominator (after this we found that the only remaining denominator was 9). We then repeat for $i = M - 1$ and so on until all $a_{ij}$ have been exhausted. After this the modified residues for the $a_{ij}$ are representations of integer coefficients which we then reconstruct. This procedure can generate the exact integer coefficients of the ODE using only 10 primes. We note that for this problem the new procedure is at least 1000 times faster than the original one described above in Section 8.6.1.

For completeness and comparison to the results for three-choice polygons we list in Table 8.4 the critical points and exponents of $\mathscr{I}(x)$.

**Table 8.4** Critical exponents at the regular singular points of the Fuchsian differential equation satisfied by $\mathscr{I}(x)$ as obtained from the exact ODE.

| Singularity | Exponents |
|---|---|
| $x = 0$ | $0, 1, 2, 2, 7/3, 8/3, 3, 3, 3, 4, 5, 6, 7, 8$ |
| $x = 1/16$ | $0, 1, 2, 3, 4, 4, 5, 6, 13/2, 7, 8, 9, 10, 11$ |
| $x = 1/5$ | $0, 1/2, 1, 2, 3, 4, 5, 6, 7, 8, 9, 10, 11, 12$ |
| $x = 1/4$ | $-1, -1/2, 0, 1/2, 1, 3/2, 2, 5/2, 3, 7/2, 4, 5, 6, 7$ |
| $x = 1$ | $-2, -3/2, -1, 0, 1, 2, 3, 4, 5, 6, 7, 8, 9, 10$ |
| $x = -1/4$ | $0, 1, 2, 3, 4, 5, 6, 13/2, 7, 8, 9, 10, 11, 12$ |
| $x = \infty$ | $-2, -3/2, -7/6, -1, -1, -5/6, -1/4, 0, 0, 1/4, 1, 2, 3, 4$ |
| $P_{14}(x) = 0$ | $0, 1, 2, 3, 4, 5, 6, 7, 8, 9, 10, 11, 12, 14$ |

## 8.7 Conclusion

In this chapter we have given an outline of the principal methods used for the analysis of series. The method of series analysis was originally developed as a numerical tool, designed to estimate the various constants and exponents that appear in asymptotic estimates. More recently, with the enhancement of both algorithms and computational hardware, it has been possible in some cases to obtain very long series expansions. Then by use of the techniques outlined in Section 8.6, it is sometimes possible to actually obtain the exact ODE whose solution gives the generating function. In such cases the method becomes not just an approximate tool, but an exact one. This development is quite recent, and is likely to enable us to solve hitherto unsolvable problems.

## References

1. Baker G A Jr 1961 Application of the Padé approximant method to the investigation of some magnetic properties of the Ising model *Phys Rev* **124**, 768-74
2. Boukraa S, Guttmann A J, Hassani S, Jensen I, Nickel B, Maillard J M and Zenine N 2008 Experimental mathematics on the magnetic susceptibility of the square lattice Ising model *J. Phys. A: Math. Theor.* **41** 455202

3.  Conway A R and Guttmann A J 1996 Square lattice self-avoiding walks and corrections to scaling *Phys. Rev. Lett.* **77** 5284–5287
4.  Conway A R, Guttmann A J, and Delest M 1997 The number of three-choice polygons. *Mathl. Comput. Modelling* **26** 51–58
5.  Fisher M E and Au-Yang H 1979 Inhomogeneous differential approximants for power series *J. Phys. A: Gen. Phys.* **12** 1677–92.
6.  Forsyth A R 1902 *Part III. Ordinary linear equations* vol. IV of *Theory of differential equations.* (Cambridge: Cambridge University Press)
7.  Guttmann A J 1989 Asymptotic analysis of coefficients in *Phase Transitions and Critical Phenomena* Vol. 13 pp. 1–234, eds C Domb and J L Lebowitz (Academic: London)
8.  Guttmann A J and Jensen I 2006 Fuchsian differential equation for the perimeter generating function of three-choice polygons *Séminaire Lotharingien de Combinatoire* **54** B54c.
9.  Guttmann A J and Jensen I 2006 The perimeter generating function of punctured staircase polygons *J. Phys. A: Math. Gen.* **39**, 3871–3882
10. Guttmann A J, Jensen I, Wong L H and Enting I G 2000 Punctured polygons and polyominoes on the square lattice *J. Phys. A: Math. Gen.* **33** 1735–1764
11. Guttmann A J and Joyce G S 1972 On a new method of series analysis in lattice statistics *J. Phys. A: Gen. Phys.* **5** L81–L84
12. Guttmann A J, Prellberg T and Owczarek A L 1993 On the symmetry classes of planar self-avoiding walks *J. Phys. A: Math. Gen.* **26** 6615–6623
13. Hunter D L and Baker G A Jr 1979 Methods of series analysis III. Integral approximant methods *Phys Rev B* **19**, 3808–21.
14. Ince E L 1927 *Ordinary differential equations* (London: Longmans, Green and Co. Ltd.)
15. Jensen I 2003 A parallel algorithm for the enumeration of self-avoiding polygons on the square lattice *J. Phys. A: Math. Gen.* **36** 5731–5745
16. Jensen I 2004 Self-avoiding walks and polygons on the triangular lattice *J. Stat. Mech: Theor. Exp.* P10008
17. Jensen I and Guttmann A J 1999 Self-avoiding polygons on the square lattice *J. Phys. A: Math. Gen.* **32** 4867–4876
18. Manna S S 1984 Critical behaviour of anisotropic spiral self-avoiding walks *J. Phys. A: Math. Gen.* **17** L899–L903
19. Nienhuis B G (1982) Exact critical point and exponents of the O($n$) model in two dimensions. *Phys Rev Lett* **49**, 1062–1065.
20. Nienhuis B 1984 Critical behavior of two-dimensional spin models and charge asymmetry in the coulomb gas *J. Stat. Phys.* **34** 731–761
21. Saleur H 1987 Conformal invariance for polymers and percolation *J. Phys. A: Math. Gen.* **20**, 455–470
22. Zenine N, Boukraa S, Hassani S and Maillard J M 2004 The Fuchsian differential equation of the square lattice Ising model $\chi^{(3)}$ susceptibility *J. Phys. A: Math. Gen.* **37** 9651–9668
23. Zenine N, Boukraa S, Hassani S and Maillard J M 2005 Square lattice Ising model susceptibility: series expansion method and differential equation for $\chi^{(3)}$ *J. Phys. A: Math. Gen.* **38** 1875–1899
24. Zenine N, Boukraa S, Hassani S and Maillard J M 2005 Ising model susceptibility: the Fuchsian differential equation for $\chi^{(4)}$ and its factorization properties *J. Phys. A: Math. Gen.* **38** 4149–4173

# Chapter 9
# Monte Carlo Methods for Lattice Polygons

E.J. Janse van Rensburg

## 9.1 Introduction

It is frequently claimed that Monte Carlo simulation is a method of last resort. This may be true in the most general sense, but it remains surprising that a fairly simple statistical technique could, when applied appropriately, produce high quality data by sampling states randomly in a given model.

Models of lattice polygons, which are related to the self-avoiding walk, are notoriously resistant to exact solutions or to rigorous analysis. These difficulties arise from the non-Markovian nature of polygon models, and significant rigorous progress has only been made in a number of special circumstances, such as in two dimensions where conformal invariance techniques led to the prediction of the numerical values of certain exponents associated with these models [13], as discussed in Chapter 14.

More generally, only scaling arguments are available for the analysis of these models [11, 14], and such arguments have produced scaling relations between critical exponents, or in some cases, to predicted values for scaling exponents. The calculation of critical exponents, and the testing of scaling relations, is a numerical affair, and this is where Monte Carlo methods play a key role.

The earliest incarnation of the Monte Carlo method was during the Manhattan project, and Ulam is usually given credit for its invention [49]. A popular implementation is due to Metropolis and others [50]. In this chapter, a Monte Carlo simulation is understood to consist essentially of the sampling of states in a given polymer model from a given distribution along a Markov chain. This implementation is often referred to as "Markov chain Monte Carlo" (for a review see reference [3]).

Perhaps the first implementation of Monte Carlo techniques for models of polymers is due to Rosenbluth and Rosenbluth, who sampled lattice self-avoiding walk models of polymers up to length 64 steps [62]. A number of alternative approaches

E.J. Janse van Rensburg
Mathematics and Statistics, York University, Toronto, Ontario, Canada, e-mail: rensburg@yorku.ca

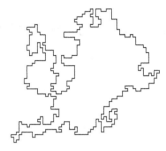

**Fig. 9.1** *A lattice polygon of length* 500. *This polygon was obtained by sampling uniformly in the set of polygons of length* 500 *edges using the pivot algorithm [45].*

have since been invented for the simulation of self-avoiding walks. Some of these methods can be implemented in a Metropolis-style simulation to sample self-avoiding walks from a given distribution (this is usually the uniform distribution on walks of given fixed length, or the Boltzmann distribution on walks of variable length). Examples of Monte Carlo algorithms for self-avoiding walk models of polymers include the Beretti-Sokal algorithm [2], the pivot algorithm [47, 48] as well as "static algorithms" such as dimerisation [60]. Generally, there are today a number of useful algorithms available for the sampling of walks, which includes PERM, an efficient generalization of the Rosenbluth algorithm [58].

In the case of lattice polygons, which are models of ring polymers, similar algorithms have been designed for efficient sampling, although the topological constraint imposed by ring closure has made it more difficult to design algorithms. Monte Carlo algorithms for fixed length polygons exist [6, 26] as well as an algorithm for sampling polygons of variable length [1, 4] which is called the BFACF-algorithm. The ergodicity properties of this last algorithm are rather interesting, and will be considered in this chapter.

The basic definitions for sampling of lattice polygons are as follows: The hypercubic lattice $\mathbb{Z}^d$ is composed of the points with integer coordinates in $\mathbb{R}^d$; these points are called *vertices* in $\mathbb{Z}^d$. Usually, $\mathbb{Z}^d$ is decorated by the addition of unit length line segments or *edges* between its vertices which are unit distance apart. This turns $\mathbb{Z}^d$ into an (undirected) graph, and polygons may be thought of as unrooted and unlabeled cycles in this graph. Two polygons are considered equivalent if one is a translate of the other in $\mathbb{Z}^d$. This equivalence defines a set of equivalence classes which we shall call *lattice polygons*, or simply *polygons*. An example of a lattice polygon is given in Fig. 9.1.

Define $p_n$ to be the number of lattice polygons of length $n$. Then, as we have seen in more detail in earlier chapters, $p_4 = 1$ and $p_6 = 2$ in $\mathbb{Z}^2$. The function $p_n$ is the most basic quantity in the study of lattice polygons.

Two polygons in $\mathbb{Z}^d$ can be concatenated (see figure 9.2) by translating and rotating the second polygon until its left-most bottom edge is parallel to the right-most top edges of the first polygon. As discussed in more detail in Chapter 2, by deleting

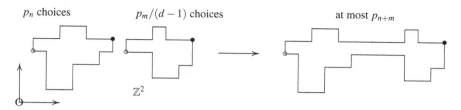

$p_n$ choices       $p_m/(d-1)$ choices       at most $p_{n+m}$

$\mathbb{Z}^2$

**Fig. 9.2** *Concatenating two polygons in the hypercubic lattice. Bottom vertices are denoted by ∘s, and top vertices are denoted by •s.*

these top and bottom edges, and inserting two edges to reconnect the two polygons into a single polygon, the polygons are concatenated to form a unique new polygon. If the first and second polygons had lengths $n$ and $m$, then there were $p_n$ choices for the first polygon, and $p_{m-1}/(d-1)$ choices for the second polygon (since the bottom edge of the second polygon must have the same orientation as the top edge of the first polygon), and the number of distinct polygons of lengths $n+m$ which can be obtained in this construction is at most $p_{n+m}$. Thus,

$$p_n p_m/(d-1) \leq p_{n+m}. \tag{9.1}$$

A basic theorem on sub-additive functions can be applied to this super-multiplicative inequality [20, 18] by taking logarithms. The result is that the limit

$$\lim_{n\to\infty} \frac{1}{n} \log p_n = \log \mu \tag{9.2}$$

exists. The *growth constant* $\mu$ determines the exponential rate at which $p_n$ increases with $n$. In particular, since $p_n \leq (2d)^n$, $\mu$ is finite, and existence of the above limit implies that $p_n = \mu^{n+o(n)}$ and $p_n \leq (d-1)\mu^n$. It can be shown that $\kappa = \log \mu$ is the connective constant of the self-avoiding walk [18].

More is known about $p_n$ and about lattice polygons. In particular, a pattern theorem for lattice polygons has been proved [38, 39] (see [23] for a proof for polygons using a method due to D J A Welsh). This result was a key ingredient in settling the Frisch-Wasserman-Delbruck conjecture [12, 15] in the cubic lattice [66]. This conjecture states that the probability that a ring polymer is non-trivially knotted increases to one as its length increases without bound.

The growth constant $\mu$ has been estimated from self-avoiding walk enumeration data in two dimensions by the finite lattice method [34], and in three dimensions by using lace expansions [10], see reference [19]. These best (current) estimates are

$$\mu = \begin{cases} 2.63815853031(2), & \text{in 2 dimensions;} \\ 4.684043(12), & \text{in 3 dimensions.} \end{cases} \tag{9.3}$$

and these estimates are many orders of magnitude better than can be obtained by current Markov chain Monte Carlo techniques. Corresponding results for other lattices can be found in the Appendix.

A theoretical (but not rigorous) interpretation of polygons as arising in the $N \to 0$ limit of an $O(N)$ model [54, 55], discussed in Chapter 1, suggests the asymptotic form

$$p_n = A n^{\alpha-3} \mu^n \left(1 + B n^{-\Delta} + C n^{-1} + \dots\right) \qquad (9.4)$$

for $p_n$, with a power law correction to the pure exponential growth term where $\alpha$ is often called the *entropic* or *specific heat* exponent. Corrections to this are both analytic (as in the terms $C n^{-1}$) and non-analytic (as in the terms $B n^{-\Delta}$), where the confluent correction exponent $\Delta$ is the first in a hierarchy of higher order corrections and there is strong evidence that $\Delta = 3/2$ in two dimensions [7, 5]. In the analysis of numerical data these corrections are often ignored or modified, and the assumption that $p_n \approx A n^{\alpha-3} \mu^n$ may be made for large values of $n$. Alternatively, the series of correction terms is truncated to include only the first non-analytic and analytic terms in linear least squares models for analysing collected data.

It is thought that $\alpha - 3 = -2.5$ in two dimensions [7, 5] (a numerical estimate gives $\alpha = 0.5000005 \pm 0.0000010$ [35, 36]) and that $\alpha - 3 \approx -2.763$ in three dimensions [43, 44]. Thus, the generating function of polygons, defined by

$$P(t) = \sum_{n=0}^{\infty} p_n t^n, \qquad (9.5)$$

is finite at its radius of convergence, which is given by $t = 1/\mu$ in equation (9.2). Simulations give the values $\alpha = 0.237 \pm 0.005$ and $\Delta = 0.56 \pm 0.03$ [41] in three dimensions. The singular behaviour of $P(t)$ suggested by equation (9.4) is

$$P_{sing}(t) \sim |\log(\mu t)|^{2-\alpha} \approx (1 - \mu t)^{2-\alpha} \qquad \text{as } t \nearrow 1/\mu, \qquad (9.6)$$

where $P_{sing}(t)$ is the singular part in $P(t)$. The singularity in $P(t)$ is at $t = 1/\mu$, and since $\alpha - 3 < -1$ [44], $P_{sing}(1/\mu)$ is finite. Hence, one would expect that

$$P(t) \approx P(1/\mu) \left(1 - C |\log(\mu t)|^{2-\alpha}\right) \qquad (9.7)$$

as $t \to 1/\mu$ from below.

The average length of polygons in this (grand canonical) ensemble (with $P(t)$ playing the role of a grand partition function) is given by

$$\langle n \rangle = \frac{d}{d \log t} \log P(t) \approx \text{Const.} + \frac{C(2-\alpha)|\log(t\mu)|^{1-\alpha}}{1 - C |\log(t\mu)|^{2-\alpha}}. \qquad (9.8)$$

By putting $t = 1/\mu$ in the above, one obtains the result that

$$\langle n \rangle |_{t=1/\mu} = \text{Const.} \qquad (9.9)$$

In other words, the expected length of polygons at the critical point $t = 1/\mu$ is finite, provided that $\alpha \leq 1$.

The mean-square radius of gyration of polygons of length, $\langle R^2 \rangle_n$, has also been computed. Since $C_1 n^{2/d} \leq \langle R^2 \rangle_n \leq C_2 n^2$ in $d$ dimensions, for some non-zero constants $C_1$ and $C_2$, it is assumed that

$$\langle R^2 \rangle_n = D n^{2\nu} (1 + o(1)) \tag{9.10}$$

where $\nu$ is the *metric exponent*. Conformal invariance arguments show that $\nu = 3/4$ in two dimensions [54], and in three dimensions the best estimate is $\nu = 0.5877 \pm 0.0006$ [41], obtained by analysing self-avoiding walk data obtained by Monte Carlo simulation using the pivot algorithm[1]. The combination of amplitudes in equations (9.4) and (9.10) gives the remarkable expression $AD = 5\sigma/32\pi^2$ where $\sigma = 2$ for the square lattice [8]. This has been tested using exact enumeration data: $A = 0.5623012(1)$ and $D = 0.05630944(1)$ for the square lattice [33, 34].

In this chapter a short review of the Monte Carlo simulation of lattice polygons is presented. The general framework of Markov chain Monte Carlo in the context of sampling lattice polygons is examined in section 2. The general implementation of Monte Carlo sampling, irreducibility, aperiodicity and detailed balance is explained. In addition, more advanced implementations of Monte Carlo algorithms via multiple Markov chain sampling and umbrella sampling are reviewed. The calculation of sample averages is explained, and in the context of umbrella sampling, the calculation of weighted averages. Statistical considerations are also briefly reviewed.

In section 3 a review of the two most common algorithms for sampling lattice polygons are given. These are the BFACF algorithm and the pivot algorithm and its variants. The free energies of polygon models of polymers can be estimated as well, and we explain how this can be done in section 4. We conclude this chapter in section 5 with a few final comments.

## 9.2 Monte Carlo Simulations and Polygons

Let $\mathbb{Z}^d$ be the $d$-dimensional hypercubic lattice consisting of all the vertices with integer Cartesian coordinates. A vertex $v \in \mathbb{Z}^d$ has coordinates $(X(v), Y(v), \ldots, Z(v))$, with $X(v)$ always denoting the first coordinate, and $Z(v)$ always denoting the last coordinate. Two vertices in $\mathbb{Z}^d$ are said to be *adjacent* if they are a unit distance apart.

A sequence of vertices $\{x_j\}_{j=0}^n$ is said to be an $n$-step *walk* if they are pairwise adjacent: $x_{i-1}$ is adjacent to $x_i$ for $i = 1, 2, \ldots, n$. If these vertices are also distinct, then the walk is said to be *self-avoiding*. An *edge* $e_i$ in a walk is the unit length line segment joining a pair of adjacent vertices $(x_{i-1}, x_i)$ in the lattice which are also adjacent in the walk. We say that $e_i$ is *incident* with $x_{i-1}$ and with $x_i$, the vertices

---

[1] See the end of section 9.3 for details of an improved algorithm, and see footnote 5 of Chapter 1 for more precise exponent estimates.

which are its endpoints. The definition of a walk is broadened by adding to a walk all the edges joining its adjacent vertices; this makes it possible to treat a walk as either a sequence of vertices, or a sequence of edges (or steps).

A walk $\omega = \{x_j\}_{j=0}^n$ is *closed* if its first vertex $x_0$ and its last vertex $x_n$ coincide. The closed walk is self-avoiding if the vertices $x_1, x_2, \ldots, x_n$ are distinct, and in this case it is a *polygon*.

Define *equivalence classes of polygons* by saying that two polygons are equivalent if (1) a permutation of the vertices will make them identical, or (2) if one can be translated in the lattice to become identical to the other, or (3) both these operations. The resulting equivalence classes of polygons will simply be refered to as *polygons*, and we may think of these as undirected and unlabeled sequences of distinct adjacent vertices which are equivalent up to translations in the lattice.

Consider a set $S$ of lattice polygons. $S$ may be endowed with a probability distribution $\Pi$, where for each $\omega \in S$ we call $\Pi(\omega)$ the *weight* of $\omega$. A Markov chain Monte Carlo algorithm consists of a scheme which samples polygons from $(S, \Pi)$ in such a way that a Markov chain of polygons $\{\omega_i\}$ is obtained. The polygons in the Markov Chain must have the correct weight in a sense that will be explained below. The set $S$ is the *state space* of the algorithm.

Suppose $\omega \in S$ is a polygon, and suppose that a procedure or construction has been defined which operates on $\omega$ to produce a state $v$, also in $S$. If the probability that $v$ is obtained from $\omega$ is denoted by $p(\omega \to v)$, then $p(\omega \to v)$ is the *transition probability matrix* of the algorithm. The procedures or constructions which update a polygon $\omega$ to propose a new state $v$ are the *elementary moves* of the algorithm.

The Monte Carlo algorithm defined by the transition probability matrix $p(\omega \to v)$ is said to be *reversible* if $p(\omega \to v)$ satisfies the condition

$$\Pi(\omega)\, p(\omega \to v) = \Pi(v)\, p(v \to \omega) \tag{9.11}$$

if $\Pi$ is the uniform distribution. In general, $\Pi$ is not uniform, in which case this is *a condition of detailed balance*.

A Monte Carlo algorithm $A$ is defined by its state space and by its condition of detailed balance (there are numerous resources explaining the basics of Monte Carlo simulation, see for example reference [63]). We say that the algorithm $A$ is irreducible if there is a sequence of elementary moves which will take a given state $\omega$ to any other state $v$ in the state space. The algorithm is aperiodic if it can visit states in state space at any step with non-zero probability. Commonly, a Monte Carlo algorithm that is both aperiodic and irreducible is said to be *ergodic*.

By summing equation (9.11) over $v$, one obtains

$$\Pi(\omega) = \sum_{v \in S} \Pi(v)\, p(v \to \omega). \tag{9.12}$$

In other words, the transition probability matrix has eigenvalue 1, and associated eigenvector $\Pi(\omega)$. In these circumstances, it follows from the fundamental theorem of Markov chain Monte Carlo that the unique stationary distribution of the algorithm is $\Pi(\omega)$, provided that the chain is both ergodic and satisfies a condition of detailed balance [49, 50].

The implication of the fundamental theorem is that spatial averages of functions over $S$ may be estimated by time series averages taken over a realization of a Markov chain by the algorithm. In particular, for a function $f(\omega)$ the average is defined by

$$\langle f \rangle_S = \sum_{\omega \in S} f(\omega) \Pi(\omega). \tag{9.13}$$

On the other hand, if $\{\omega_i\}_{i=1}^{N}$ is a realization of a Markov chain in $S$ using a Monte Carlo algorithm with stationary distribution $\Pi(\omega)$, then

$$\langle f \rangle_S \approx \langle f(\omega_i) \rangle_N = \frac{1}{N} \sum_{i=1}^{N} f(\omega_i) \tag{9.14}$$

and this becomes an equality as $N \to \infty$.

Normally, a confidence interval can be estimated by examining the variance of $f$, defined by $\mathrm{Var}_N(f) = \langle f(\omega_i)^2 \rangle_N - \langle f(\omega_i) \rangle_N^2$. For large $N$, one may estimate a 67% confidence interval $\sigma_N$ on $\langle f(\omega_i) \rangle_N$ by computing $\sigma_N^2 = \frac{1}{N} \mathrm{Var}_N(f)$. Under normal circumstances, this would be sufficient. However, it may be the case that the states generated along the realization of a Markov chain are correlated, and $\sigma_N$ would be an underestimate of the confidence interval.

Statistical analysis of an observable $f$ along a realized Markov chain proceeds instead by the computation of the autocorrelation time. The covariance $C_f(k)$ of $f$ along a Markov chain is defined by

$$C_f(k) = \langle f(\omega_i) f(\omega_{i+k}) \rangle_N - \langle f(\omega_i) \rangle_N^2 \tag{9.15}$$

and the *autocorrelation function* of $f$ is defined by $C_f(k)/C_f(0)$. This function depends on the dynamics of the Monte Carlo algorithm, and is influenced by the choice of elementary moves and state space. Typically the autocorrelation function decreases at an exponential rate $2\tau_e$ with increasing $k$. $\tau_e$ is the *exponential autocorrelation time* of the algorithm. An alternative measure of the autocorrelation time is obtained by assuming that $C_f(k)$ decays exponentially with $k$, and to integrate it to determine the *integrated autocorrelation time* $\tau_{int}$:

$$\tau_{int} = \sum_{k=-\infty}^{\infty} \frac{C_f(k)}{2C_f(0)} = \frac{1}{2} + \sum_{k=1}^{\infty} \frac{C_f(k)}{C_f(0)}. \tag{9.16}$$

As the length $N$ of the realized Markov chain in the state space $S$ increases to infinity, measurements of the integrated autocorrelation time allows one to estimate the variance of $f$ along the Markov chain by the asymptotic result

$$\mathrm{Var}_N(f) \sim \frac{2\tau_{int}}{N} \left( \langle f(\omega_i)^2 \rangle_N - \langle f(\omega_i) \rangle^2 \right). \tag{9.17}$$

This modifies the estimate of the variance by taking into account correlations along the realized Markov Chain, and a 67% confidence interval $\sigma_N$ on $\langle f(\omega_i)\rangle_N$ is given by $\sqrt{\mathrm{Var}_N(f)}$, while $2\sigma_N$ is the 95% confidence interval. The effect of the auto-correlation time is to reduce the number of states sampled along the chain ($N$) to $N/2\tau_{int}$ "effectively independent" samples. For more details, see for example chapter 9 in [46], and in particular section 9.2.2.

### 9.2.1 Multiple Markov Chain Monte Carlo Simulations

In many applications the condition of detailed balance in equation (9.11) is a function of a parameter $\beta$. This implies that the probability distribution $\Pi_\beta$ on the state space $S$ is a function of $\beta$. In addition, the transition probabilities $p_\beta(\omega \to v)$ are functions of $\beta$, which is set at a certain value before the simulation. In this event the condition of detailed balance becomes

$$\Pi_\beta(\omega)\, p_\beta(\omega \to v) = \Pi_\beta(v)\, p_\beta(v \to \omega) \qquad (9.18)$$

and sampling along realizations of Markov chains is performed by fixing the value of $\beta$ and implementing the algorithm.

Computing averages at fixed values of $\beta$ may be an inefficient use of computational resources, since one is often interested in sampling states for values of $\beta$ varying over an interval $[\beta_m, \beta_M]$. This will require the calculation of averages of observables as a function of $\beta$. One effective way in which this can be done is to implement a multiple Markov chain Monte Carlo [17, 67] version of the (existing) Markov chain Monte Carlo algorithm.

The implementation is as follows: Realize $N$ Markov Chains in parallel at a sequence of values of the parameter $\beta$: say at $\beta_m = \beta_1 < \beta_2 < \beta_3 < \ldots < \beta_N = \beta_M$. Normally, these chains could be allowed to evolve in parallel with no interactions amongst them, and averages can be computed at values of $\beta \in [\beta_m, \beta_M]$ by reweighting the observations (see for example [53] for details) and using importance sampling. The expected value of an observable is then estimated by the ratio estimator

$$\langle f \rangle_{\beta,N} \approx \frac{\langle \Pi_\beta f / \Pi_{\beta'} \rangle_{\beta',N}}{\langle \Pi_\beta / \Pi_{\beta'} \rangle_{\beta',N}} \qquad (9.19)$$

where sampling is done at $\beta'$ and we estimate the observable $f$ at $\beta$, and where one may choose $\beta' = \beta_j$ as a convenient value of $j$ among the parallel Markov chains in the simulation. In applications, choosing $\beta' = \beta_j$ for $\beta_j$ close to $\beta$ is usually sufficient to provide a good estimate of the observable $f$ at $\beta$.

In a multiple Markov chain Monte Carlo implementation one allows the parallel Markov chains to interact with one another [16, 17]. The normal implementation is to attempt the switching of the current states in two selected chains at a given time step. If the current state in chain $i$ is $\omega_i$ and the current state in chain $j$ is $\omega_j$, then the two states are swapped with probability

$$q(\omega_i, \omega_j) = \min \left\{ 1, \frac{\Pi_{\beta_i}(\omega_j)\Pi_{\beta_j}(\omega_i)}{\Pi_{\beta_i}(\omega_i)\Pi_{\beta_j}(\omega_j)} \right\}. \tag{9.20}$$

This whole process, implemented on the underlying $N$ parallel chains, is itself a Markov chain which is called the *composite Markov chain*. It can be verified explicitly that the swapping of states in the composite Markov chain satisfies a condition of detailed balance

$$[\Pi_{\beta_1}(\omega_1) \dots \Pi_{\beta_i}(\omega_i) \dots \Pi_{\beta_j}(\omega_j) \dots \Pi_{\beta_N}(\omega_N)] \, q_\beta(\omega_i, \omega_j)$$
$$= [\Pi_{\beta_1}(\omega_1) \dots \Pi_{\beta_i}(\omega_j) \dots \Pi_{\beta_j}(\omega_i) \dots \Pi_{\beta_N}(\omega_N)] \, q_\beta(\omega_j, \omega_i) \tag{9.21}$$

so that the invariant stationary distribution of the composite chain is the product distribution given by $\Pi[\beta_1, \beta_2, \dots, \beta_N] = \Pi_{i=1}^{N} \Pi_{\beta_i}$, see for example reference [68].

Sample means can be computed using any of the chains in the simulation. If $\beta$ coincides with one of the $\beta_i$, then the sample mean is given by equation (9.14) using data only sampled along chain $i$. If $\beta$ does not coincide with one of the $\beta_i$, then the nearest chain can be selected, and a ratio estimator as in equation (9.19) will give the calculated mean.

## 9.2.2 Umbrella Sampling

This is a generalization of the idea of importance sampling [69], and may be useful in applications where the weights of polygons in the state space $S$ vary over many orders of magnitude. To implement this method, assume at first that an estimate of the observable $f$ over $S$ with a given probability distribution $\Pi_\beta$ is required. Instead of sampling directly from $(S, \Pi_\beta)$, Markov chain Monte Carlo is performed from a different distribution, say $P$, and the expected value of $f$ is estimated over $\Pi_\beta$ by reweighting the data

$$\langle f \rangle_{\Pi_\beta} = \frac{\langle f\Pi_\beta/P \rangle_P}{\langle \Pi_\beta/P \rangle_P}, \tag{9.22}$$

where $\langle \cdot \rangle_P$ is the time series average of the realized Markov chain with states sampled from distribution $P$.

Normally, $\Pi_\beta$ is a Boltzmann distribution, and if the energy of a polygon $\omega \in S$ is given by $E(\omega)$, then the distribution is given by

$$\Pi_\beta(\omega) = \frac{e^{-\beta E(\omega)}}{\Phi_\beta} \tag{9.23}$$

where $\Phi_\beta$ is a normalising factor given by

$$\Phi_\beta = \sum_{\omega \in S} e^{-\beta E(\omega)}. \tag{9.24}$$

Umbrella sampling can be implemented by choosing a suitable umbrella distribution $P$. Practical experience shows that $P$ may be chosen as a sum of Boltzmann factors

$$P(\omega) = \sum_{j=1}^{N} w_j e^{-\beta_j E(\omega)} \tag{9.25}$$

for a set $\beta_1 < \beta_2 < \ldots < \beta_N$. The coefficients $w_j$ can be chosen to fix a suitable umbrella over the interval $[\beta_1, \beta_N]$. There are several schemes for determining the $w_j$, the most important criterion being a "flatness criterion" which requires that the binning of polygons by energy should give a nearly flat histogram. This requires that the $\{\beta_j\}$ are spaced closely together to allow sufficient overlap between the adjacent distributions composing the umbrella.

The flatness criterion ensures sufficient mobility for the Markov chain to sample polygons over the entire range of energies, and this allows for accurate estimates from the ratio estimator in equation (9.22). If the algorithm realizes a Markov chain $C$ in $S$ by sampling from the umbrella distribution $P(\omega)$, then

$$\langle f \rangle_\beta \approx \frac{\sum_{\omega \in C} f(\omega) e^{-\beta E(\omega)} / P(\omega)}{\sum_{\omega \in C} e^{-\beta E(\omega)} / P(\omega)} \tag{9.26}$$

is the ratio estimator of $f$ with respect to the Boltzmann distribution at $\beta$.

Important considerations in this implementation are the following: (1) Determining a suitable umbrella distribution $P$ may be difficult. There are adaptive schemes which may be used to grow an umbrella [51]. (2) The weight factors $w_j$ and the exponential factors in equation (9.26) involve very large numbers, or involve calculations with sets of numbers which may be many orders of magnitude different from one another. Implementing this requires careful programming to avoid overflows or problems with rounding which may skew numerical results.

## 9.3 Algorithms for Lattice Polygons

There are only two widely used Metropolis-style Monte Carlo algorithms for sampling lattice polygons along a Markov Chain. These are the BFACF algorithm [1, 4], and the pivot algorithm [45, 48]. Any one of these algorithms may be implemented in a multiple Markov chain algorithm, or in an umbrella sampling implementation.

## 9.3.1 The BFACF Algorithm

This simple algorithm has interesting properties [4]. It operates by sampling polygons of variable length along a Markov chain, using a simple elementary move (see figure 9.3) to generate new states along the chain.

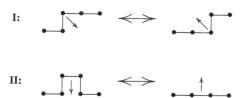

**Fig. 9.3** The elementary transitions of the BFACF algorithm.

Let $\omega_j$ be the current state of length $|\omega_j|$ sampled by the algorithm in a given realization of the Markov chain. To generate a next state $\omega_{j+1}$, the simple construction illustrated in Fig. 9.3 is implemented: Choose, with uniform probability, an edge in $\omega_j$, and translate this edge normal to itself in one of $2(d-1)$ possible directions a distance of one lattice step, while inserting or deleting adjacent edges to keep the polygon connected. Let $\omega'$ be the resulting object. We choose $\omega'$ as the next state $\omega_{j+1}$ in the Markov chain by implementing a (Metropolis-style) rejection technique.

If $\omega'$ is not self-avoiding (then it is also not a polygon), then the attempt fails, and $\omega_{j+1} = \omega_j$ is the next state in the Markov chain.

If $\omega'$ is a polygon, then there are three possible outcomes (as illustrated in Fig. 9.3). Either $\omega'$ has length $|\omega_j|+2$, or it has length $|\omega_j|$, or it has length $|\omega_j|-2$. Choose $\omega_{j+1} = \omega'$ with probabilities

$$\Pr(\omega_j \to \omega_{j+1}) = \begin{cases} \frac{|\omega_{j+1}|^{q-1}}{|\omega_j|^{q-1}}\beta^2, & \text{if } |\omega_{j+1}| = |\omega_j|+2; \\ 1, & \text{if } |\omega_{j+1}| \le |\omega_j|, \end{cases} \qquad (9.27)$$

where $q$ and $\beta$ are two parameters of this implementation. If the move is rejected, then put $\omega_{j+1} = \omega_j$.

Since the a priori probability of choosing an edge in $\omega_j$ to implement the elementary moves in Fig. 9.3 is $1/|\omega_j|$, it follows that the probability of obtaining a given $\omega_{j+1}$ as the next state is

$$p(\omega_j \to \omega_{j+1}) = \frac{1}{|\omega_j|}\left(\frac{|\omega_{j+1}|^{q-1}}{|\omega_j|^{q-1}}\beta^2\right) \qquad (9.28)$$

if $|\omega_{j+1}| = |\omega_j|+2$. In this case $p(\omega_{j+1} \to \omega_j) = \frac{1}{|\omega_{j+1}|}$ and this gives the condition of detailed balance

$$|\omega_j|^q \beta^{|\omega_j|} p(\omega_j \rightarrow \omega_{j+1}) = |\omega_{j+1}|^q \beta^{|\omega_{j+1}|} p(\omega_{j+1} \rightarrow \omega_j). \qquad (9.29)$$

One may explicitly check that the same condition holds if the proposed move conserves the length of the polygon.

By summing equation (9.29) over $\omega_{j+1}$, one obtains for any $\omega$

$$|\omega|^q \beta^{|\omega|} = \sum_v p(v \rightarrow \omega) |v|^q \beta^{|v|}, \qquad (9.30)$$

which is the specific case of equation (9.12) for this algorithm. Normally, this result would have been enough to determine the invariant stationary distribution. In this case, however, the algorithm is not known to be irreducible (it is aperiodic in this implementation because the rejection technique destroys any periodic cycles in the underlying Markov chain). The irreducibility of the algorithm has been determined in two dimensions [42, 47] (or in the square lattice). Thus, in the square lattice the algorithm is ergodic and obeys the condition of detailed balance in equation (9.29). The invariant stationary distribution of the algorithm in the square lattice is

$$\Pi_2(\omega; \beta, q) = \frac{|\omega|^q \beta^{|\omega|}}{\sum_v |v|^q \beta^{|v|}}. \qquad (9.31)$$

The state space of the algorithm in this case is the set of all *unrooted* polygons (which are equivalent under translations) on the square lattice. The parameters $q$ and $\beta$ can be chosen to simulate polygons of convenient average length. Observe that

$$\sum_v |v|^q \beta^{|v|} = \sum_{n>0} n^q p_n \beta^n. \qquad (9.32)$$

where $p_n$ is the number of unrooted polygons of length $n$, counted *modulo* translations. The expected length of polygons sampled along a Markov chain at $\beta$ is given by

$$\langle |\omega| \rangle = \frac{\sum_\omega |\omega|^{q+1} \beta^{|\omega|}}{\sum_v |v|^q \beta^{|v|}}. \qquad (9.33)$$

To leading order,

$$p_n = A n^{\alpha-3} \mu^n (1 + o(1)) \qquad (9.34)$$

where $\alpha = 1/2$ is the *specific heat* or *entropic* exponent of polygons. Substitution of equation (9.34) into equation (9.33) and estimating the series shows that $\beta_c = 1/\mu$ is the critical point of the parameter $\beta$ in this algorithm. If $\beta < \beta_c$, then the expected length of polygons sampled by the algorithm is

$$\langle |\omega| \rangle \sim \text{Const.} + \frac{C(2-\alpha-q)|\log(\beta\mu)|^{1-q-\alpha}}{1 - C|\log(\beta\mu)|^{2-q-\alpha}}. \qquad (9.35)$$

Since the algorithm is ergodic in two dimensions, this sample mean will estimate the expected value of $|\omega|$ over the state space of all lattice polygons in two dimensions. The algorithm is not ergodic in three dimensions since polygons have

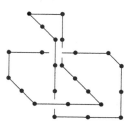

**Fig. 9.4** *A knotted lattice polygon with 24 steps in the cubic lattice.*

well-defined knot types here, and the elementary moves cannot change the knot type of a polygon. In what follows, the relation between lattice knots and the BFACF algorithm is examined.

### 9.3.1.1 The BFACF Algorithm and Lattice Knots

The ergodicity properties of the BFACF algorithm can only be understood in three dimensions in terms of lattice knots [31, 32]. A *knot* is an embedding of the circle in three space. Two knots are said to belong to the same *knot type* if there is an ambient orientation preserving homeomorphism of three space onto itself which takes one knot to the other. Since lattice polygons are (piecewise linear) embeddings of the circle in three space, they have well-defined knot types. *Lattice knots* are polygons in the cubic lattice, but with knot type determined by examining the polygon as an embedding of the circle in three space.

Define $p_n(K)$ to be the number of lattice polygons of length $n$ and of knot type $K$, counted up to equivalence under translations in the cubic lattice. Concatenation of two polygons of knot types $K$ and $L$ can still be done as illustrated in Fig. 9.2, and the result is a polygon with compound knot type $K\#L$. In particular, this implies that

$$p_n(K)\,p_m(L) \leq 2\,p_{n+m}(K\#L). \tag{9.36}$$

If $K = L = \emptyset$, where $\emptyset$ denote the unknot, then this implies that

$$p_n(\emptyset)\,p_m(\emptyset) \leq 2\,p_{n+m}(\emptyset) \tag{9.37}$$

with the result that the limit

$$\lim_{n\to\infty} \frac{1}{n}\log p_n(\emptyset) = \mu_\emptyset \tag{9.38}$$

exists, where $\mu_\emptyset$ is the growth constant of unknotted lattice polygons.

The Frisch-Wasserman-Delbruck conjecture in this context implies that $\mu_0 < \mu$. In polymer chemistry, this conjecture states that almost all ring polymers of sufficient length will be knotted. Numerical simulations [30, 70] indicate that

$$\log \mu - \log \mu_0 = (4.15 \pm 0.32) \times 10^{-6}. \tag{9.39}$$

In other words, $\mu = 4.6840\ldots$ and $\mu_0 = 4.6840\ldots$, and there is a difference in the next one or two digits. Current computational technology could determine the next couple of digits in $\mu$ and $\mu_0$ ($\mu$ is known at least to the accuracy in equation (9.3)), but this remains a tremendous computational challenge. The Frisch-Wasserman-Delbruck conjecture was settled by Sumners and Whittington in 1988 [66] and by Pippenger in 1989 [59] for a cubic lattice polygon model of ring polymers.

The existence of $\mu_0$ as a limit in equation (9.38) is a special case. For non-trivial knot types, it is not known that a similar limit exists. Instead one defines the lim sup

$$\log \mu_K = \limsup_{n \to \infty} \frac{1}{n} \log p_n(K). \tag{9.40}$$

The existence of $\mu_K$ as a limit is a major open problem in the study of lattice knots. It is known that [65]

$$\mu_0 \leq \mu_K < \mu. \tag{9.41}$$

Settling the first inequality is an open problem. Generally, it is believe that $\mu_0 = \mu_K$ for arbitrary and fixed non-trivial knot types $K$.

In analogy with $p_n$ it is expected that the asymptotic form for $p_n(K)$ should be

$$p_n(K) = A_K n^{\alpha_K - 3} \mu_K^n (1 + o(1)). \tag{9.42}$$

There is a tremendous amount of numerical data which supports this assumption [56, 57]. Numerical simulations in reference [57] indicate the rather surprising fact that $A_K$ is independent of $K$ (if the assumption that $\mu_0 = \mu_K$ is made). The entropic exponent of lattice knots, $\alpha_K$, appears to be related to the entropic exponent $\alpha_0$ of unknotted polygons by

$$\alpha_K = \alpha_0 + N_K \tag{9.43}$$

where $N_K$ is the number of prime components in the knot type $K$. For example, $\alpha_{3_1} = \alpha_0 + 1$, while $\alpha_{3_1 \# 3_1} = \alpha_0 + 2$. However, unlike $p_n$, the asymptotic formula for $p_n(K)$ in equation (9.42) does not rest on a theoretical foundation such as an $O(N)$ model. Settling this is unlikely to be a rigorous result, but this remains an open problem.

The ergodicity properties of the BFACF algorithm have also been resolved in three dimensions [31, 21]. In particular, the algorithm is not irreducible for lattice polygons on the cubic lattice. Sampling polygons along a Markov chain on the cubic lattice is not from a unique stationary distribution. Instead, the chain samples from a stationary distribution that depends on the initial state of the polygon. This fixes the ergodicity class of the algorithm for every simulation. The ergodicity classes are described by the following theorem:

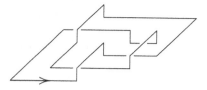

**Fig. 9.5** *A realization of the knot* $3_1^+$ *in* $L_1$.

**Theorem 1.** *The ergodicity classes of the BFACF algorithm, when applied to un-rooted lattice polygons in the cubic lattice, are the classes of polygons with the same knot types.*

In other words, the ergodicity classes coincide with the knot-types of lattice polygons. For the proof of this theorem, see reference [31], and in particular theorem 3.11 therein.

A *slab* $\mathbb{L}_w$ is the subset of the cubic lattice defined by

$$\mathbb{L}_w = \{\mathbf{x} \in \mathbb{Z}^3 \mid 0 \leq Z(\mathbf{x}) \leq w\}. \tag{9.44}$$

The BFACF algorithm has similarly been studied in a slab in the cubic lattice. In this case, the irreducibility of the algorithm is described by the following theorems [25]:

**Theorem 2.** *The ergodicity classes of the BFACF algorithm in the set of all unrooted lattice polygons in* $\mathbb{L}_2$ *are the classes of polygons with the same knot types.*

This theorem generalizes to other slabs:

**Theorem 3.** *Suppose that* $w \geq 2$. *Then the ergodicity classes of the BFACF algorithm in the set of all unrooted lattice polygons in* $\mathbb{L}_w$ *are the classes of polygons with the same knot types.*

There are several open problems which remain unresolved. In particular, the properties of the BFACF algorithm in $\mathbb{L}_1$ in three dimensions are not understood.

### 9.3.1.2 Implementation and Data Analysis

Suppose that the numerical values of an observable $A(\omega)$ are tracked along a Markov chain $\mathcal{M} = \{w_i\}$ realized by the BFACF algorithm. Then the sample average of $A(\omega)$ along the chain is

$$\ll A \gg_N = \frac{1}{N} \sum_{i=1}^{N} A(\omega_i) \tag{9.45}$$

if the chain has length $N$. Since the chain is ergodic in its ergodicity class $K$ (which is the set of polygons with knot type $K$, where $K$ is determined by the knot type of the first state $\omega_1$ in the Markov chain of sampled polygons) by theorem 3 in three dimensions, the sample average of the chain in three dimensions converges to

$$\lim_{N\to\infty} \ll A \gg_N = \sum_{\omega\in K} A(\omega)|\omega|^q t^{|\omega|} \qquad (9.46)$$

by the fundamental theorem of Markov chains and where it is assumed that the Markov chain is generated by setting the activity parameter of edges equal to $t$, and where the summation is over all polygons of knot type $K$. In two dimensions the algorithm is ergodic, and the sample average converges to expected values taken over all polygons.

The expected value of $A$ measured over a Boltzmann distribution is given by

$$\langle A \rangle = \sum_{\omega\in K} A(\omega) t^{|\omega|} / \sum_{\omega\in K} t^{|\omega|}, \qquad (9.47)$$

and comparison with equation (9.46) shows that the ratio-estimator

$$\langle A \rangle = \frac{\lim_{N\to\infty} \ll A/|\omega|^q \gg_N}{\lim_{N\to\infty} \ll 1/|\omega|^q \gg_N} \approx \frac{\ll A/|\omega|^q \gg_N}{\ll 1/|\omega|^q \gg_N} \qquad (9.48)$$

approximates $\langle A \rangle$ for large values of $N$.

In Fig. 9.6 the measured average lengths of cubic lattice polygons of fixed knot types are plotted against $t$ in a simulation of the BFACF algorithm. For small values of $t$, the observed average length is small, but it increases dramatically as $t$ approaches a critical point given by $t = 1/\mu_K$. The parameter $q$ was put equal to 3 in this simulation. This biased the simulation to longer polygons.

In analogy with equation (9.4) the number of polygons of knot type $K$ and length $n$ may be expected to be given by

$$p_n(K) = A_K n^{\alpha_K-3} \mu_K^n \left(1 + B_K n^{-\Delta_K} + \ldots \right). \qquad (9.49)$$

The expected length of polygons of knot type $K$ in the BFACF algorithm is given by

$$\langle n_K \rangle \approx \frac{[\alpha_K + q - 2]\mu_K t}{1 - t\mu_K} \left(1 - \frac{B_K \Delta_K [1 - t\mu_K]^{\Delta_K}}{\alpha_K + q - 2}\right). \qquad (9.50)$$

The growth constant $\mu_K$ can be estimated by considering the asymptotic behaviour of $\langle n_K \rangle$ as $z = (1 - t\mu_K) \to 0$ (or $t \to 1/\mu_K$). In particular, to leading order, we can use (9.50) to approximate $1/\langle n_K \rangle$:

$$\langle n_K \rangle^{-1} \approx \frac{1 - t\mu_K}{[\alpha_K + q - 2]\mu_K t} = \frac{1}{(\alpha_K + q - 2)\mu_K t} - \frac{1}{\alpha_K + q - 2}. \qquad (9.51)$$

An estimate of $\mu_K$ can be obtained by extrapolating to that value of $K$ for which $1/\langle n_K \rangle$ is zero. Such an analysis [57] gives the following estimates:

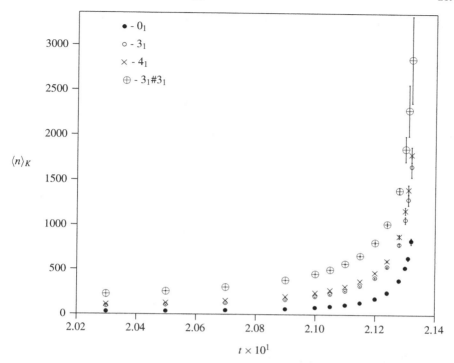

**Fig. 9.6** *Data generated by the BFACF algorithm in three dimensions for polygons of fixed knot types. These data are taken from [56, 57], and measured the mean length of knotted polygons as a function of t and of knot type. This data strongly supports the assumptions in equation (9.42) with $\alpha_K$ related to $\alpha_\emptyset$ by $\alpha_K = \alpha_\emptyset + N_K$ where $N_K$ is the number of prime components in K, see equation (9.43). In this simulation, the parameter q was set equal to 3, biasing the sampling to longer polygons.*

$$\mu(\emptyset) = 4.6852,$$
$$\mu(3_1) = 4.6832,$$
$$\mu(4_1) = 4.6833,$$
$$\mu(3_1 \# 3_1) = 4.6800,$$
$$\mu(3_1 \# 4_1) = 4.6841.$$

(9.52)

This may be interpreted as strong evidence that $\mu_K = \mu_\emptyset$ in equation (9.41).

To determine the entropic exponent, assume that $\mu_K = \mu_\emptyset$ and consider the approximation

$$\frac{\langle n_K \rangle}{\langle n_L \rangle} \approx \frac{\alpha_K + q - 2}{\alpha_L + q - 2} \left[ 1 + c(1 - t\mu_\emptyset)^\Delta \right]$$

(9.53)

for two knots of types K and L. A plot of $\langle n_K \rangle / \langle n_L \rangle$ against $(1 - t\mu_\emptyset)^\Delta$ should give a curve which will be linear for t close to $1/\mu_\emptyset$ and will have intercept $[\alpha_K + q - 2]/[\alpha_L + q - 2]$ from which we can estimate $\alpha_K$ in terms of $\alpha_L$. For the set of all polygons, the confluent exponent $\Delta$ has a value close to $1/2$. Putting $q = 3$ and assuming that the confluent exponent $\Delta_K = \Delta$ is independent of the knot type, then shows that

$$\frac{\langle n_K \rangle}{\langle n_L \rangle} = \frac{\alpha_K + 1}{\alpha_L + 1} + c_1(1 - t\mu_0)^{\Delta}. \tag{9.54}$$

If we assume that $\Delta = 1/2$ in each case, then these fits will produce a ratio $\rho(K,L) = (\alpha_K + 1)/(\alpha_L + 1)$ for $q = 3$. Analysing data collected in reference [57] gave the following estimates:

$$\frac{\alpha_{3_1} + 1}{\alpha_0 + 1} = 1.69 \pm 0.11 \tag{9.55}$$

$$\frac{\alpha_{4_1} + 1}{\alpha_0 + 1} = 1.67 \pm 0.11 \tag{9.56}$$

$$\frac{\alpha_{3_1} + 1}{\alpha_{4_1} + 1} = 1.01 \pm 0.11 \tag{9.57}$$

$$\frac{\alpha_{3_1 \# 3_1} + 1}{\alpha_{3_1} + 1} = 1.25 \pm 0.16 \tag{9.58}$$

If we assume that $\alpha_0 \approx 0.25$, then these results are consistent with equation (9.43).

## 9.3.2 The Pivot Algorithm

The BFACF algorithm samples polygons from the grand canonical ensemble by making small local and length-changing changes to a polygon, and accepting it as the next state using a probability distribution which is related to the Boltzmann distribution. The philosophy underlying the pivot algorithm is radically different. Instead, it samples polygons of fixed lengths (in the canonical ensemble) along a Markov Chain by making relatively large changes to it (these are called "pivot" moves).

The pivot algorithm was first invented for self-avoiding walks in the hypercubic lattice [40]; see for example also references [48, 64]. While first used as an algorithm for self-avoiding walks, the pivot algorithm was also shown to be useful in simulating polygons [45, 26].

Implementing the pivot algorithm involves a "global move" in the polygon: Select two distinct vertices or *pivots* $v_1$ and $v_2$ on a polygon $\omega$. These vertices cut $\omega$ into two self-avoiding walks, one of which is the shorter, $\omega_1$. The remaining part of the polygon is $\omega_2$, is also a self-avoiding walk, and has length at least equal to that of $\omega_1$. The shorter segment $\omega_1$ can be changed by the algorithm provided that its endpoints remains fixed in the lattice. This will leave open the possibility that the updated object is still a polygon.

The possible operations on $\omega_1$ can be chosen from the symmetry group of the underlying lattice. For example, a point reflection through the centre-of-mass of the pivots will leave the endpoints of $\omega_1$ unchanged. This operation is commonly called

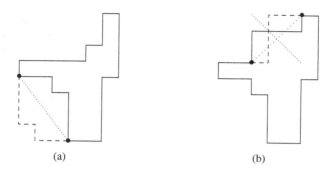

(a)                                          (b)

**Fig. 9.7** *Two examples of pivot moves. On the left hand side (a) is an inversion or a point reflection through the midpoint (centre of mass) of the line joining the pivots is shown. In (b) a reflection through the bisector of the line joining the pivots is illustrated. This move is only possible if the pivots lie on the same line which is in one lattice plane but inclined at 45° to the other lattice axes.*

an *inversion*, and it can be attempted on any polygon and any set of pivots. While it leaves the endpoints unchanged, the walk itself may be changed to a different conformation. In other words, for any two choices of the pivots, there is at least one move that can be made to attempt an update of the polygon.

More generally, the endpoints of $\omega_1$ (the pivots) may be on the same lattice axis, or in the same lattice plane, or may even be on the same diagonal line or plane. In each case there would be a selection of possible pivot moves that can be used to update the polygon and to select the next state. These moves are elements of the symmetry group of the underlying lattice. For example, if the two pivots $v_1$ and $v_2$ lie on the same lattice axis, then the following set of operations are possible: Rotations about the axis in multiples of 90°, a reflection through the midpoint (centre of mass) of the line joining $v_1$ and $v_2$, followed by a rotation about the axis in multiples of 90°. Further, first a reflection through the lattice planes containing the pivots, or through the lattice plane bisecting the axis, or through planes containing the lattice axis, but at a 45° angle to the lattice planes, among many other possibilities (the symmetry group of the cubic lattice has 48 elements). In the case of the square or cubic lattice, the algorithm is irreducible if enough of the group elements have been selected as possible pivot moves.

In normal implementations the entire symmetry group of the lattice is used to generate a new polygon. However, to show that the algorithm is irreducible requires only some of the elements in the symmetry group. The requirements are understood in the square and cubic lattice. In figure 9.7 some pivot moves are illustrated.

**Theorem 4.** *The pivot algorithm for lattice polygons in the square lattice is irreducible if inversions, and if reflections through bisectors of the lines joining the pivots, making a 45° angle with the lattice axes, are included in the set of elementary moves.*

In other words, the two moves illustrated in Fig. 9.7 are enough to generate an ergodic pivot algorithm in the square lattice. The proof of this fact is quite simple:

Select pivots in the convex hull of the polygon and perform inversions. Each such inversion increases the area included inside the polygon. Since the area is bounded, this must fail after a finite number of moves (each inversion increases the area by at least one half unit square). This occurs when the polygon is equal to its convex hull, that is, when it is a rectangle. Finally, it is possible to collapse any rectangular lattice polygon into a horizontally oriented rectangle with minimal area (vertical height equal to 1) using an inversion, and a move of the type illustrated in Fig. 9.7(b).

A representative polygon of length 5000 edges, sampled uniformly from the set of polygons of length 5000, is illustrated in Fig. 9.8.

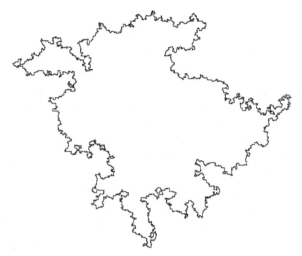

**Fig. 9.8** *A lattice polygon of length* 5000. *This polygon was obtained by sampling uniformly in the set of polygons of length* 5000 *edges using the pivot algorithm [45].*

The situation is somewhat more complex in three dimensions [45], but it has also been resolved. The following theorem states the necessary conditions for an irreducible pivot algorithm in three dimensions.

**Theorem 5.** *The pivot algorithm for lattice polygons in the cubic lattice is irreducible if inversions, $\pm\pi/2$ rotations about lines parallel to the lattice axes, and reflections through bisectors of lines joining the pivots and parallel to $Y = \pm X$, $Z = \pm Y$ and $X = \pm Z$, are included in the set of possible elementary Monte Carlo moves.*

For the methods of proof of theorems 4 and 5, see for example references [45, 26].

The pivot algorithm on the cubic lattice gives an ergodic algorithm with state space being the set of all lattice polygons of fixed length (sometimes refered to as the canonical ensemble). This is in contrast to the BFACF algorithm which samples

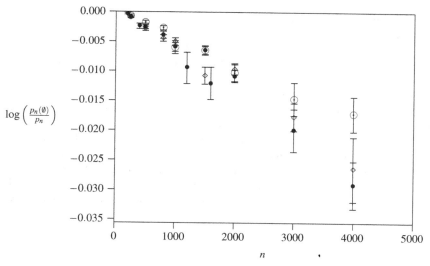

**Fig. 9.9** *Measurements of the probability that a polygon is the unknot as a function of n. These data were generated using the pivot algorithm in three dimensions on three lattices: the cubic lattice ($\odot$), the face-centered cubic lattice ($\bullet$) and the body-centered cubic lattice ($\diamond$).*

ergodically on the state space of lattice polygons with the same knot type in the grand canonical ensemble.

The pivot algorithm has been used extensively to examine polygon models of ring polymers. A particularly interesting example involves the detection of knots in lattice polygons. A study of this kind was first performed by Michels and Wiegels in 1982 [52]. The general idea here is to estimate the difference in equation (9.39), and that result was obtained by analysing the lattice data in figure 9.9. The face-centered cubic lattice and the body-centered cubic lattice data can similarly be analysed to estimate $\log \mu - \log \mu_0$. In the case of the face-centered cubic lattice the result is $(5.91 \pm 0.32) \times 10^{-6}$ and in case of the body-centered cubic lattice the result is $(5.82 \pm 0.37) \times 10^{-6}$. See references [30, 24] for more details.

Madras and Sokal [48] showed that the pivot algorithm was much faster than any local move algorithm in the simulation of self-avoiding walks. For walks of $n$ steps their implementation of the pivot algorithm had an integrated autocorrelation time of $O(n)$ in CPU units. They argued that this could not be improved upon as it takes time $O(n)$ to even write down an $n$-step self-avoiding walk. Kennedy [37] showed that it is possible to overcome the $O(n)$ barrier, by utilising geometric information to reduce the number of intersection tests, and only updating the data structure every so often. In CPU units, the integrated autocorrelation time was estimated to be of $O(n^{0.57})$ on the square lattice, and $O(n^{0.85})$ on the simple cubic lattice.

In recent work [9], a version of the pivot algorithm for self-avoiding walks has been implemented that is dramatically faster still. The basic pivot move, whether accepted or not, can be performed in CPU time $O(\log n)$, which leads to an integrated autocorrelation time of approximately $O(n^{0.19} \log n)$ on the square lattice and $O(n^{0.11} \log n)$ on the simple cubic lattice. For a walk of one million steps on the simple cubic lattice, this leads to an improvement of approximately a factor of 200 over Kennedy's method, and 1400 over the implementation of Madras and Sokal.[2] It is possible to obtain good statistics on walks of up to one billion steps on current generation machines using this technique![3] The key insight of [9] is the recognition that the self-avoiding walk can be represented as a binary tree, where each walk is represented as the concatenation of two sub-walks of roughly equal length. This version of the pivot algorithm can be straightforwardly implemented for self-avoiding polygons, and in the future will enable accurate simulations of extremely long self-avoiding polygons. Note, the CPU time per attempted pivot for the proposed implementation for self-avoiding polygons would be $O(\log n)$, as it is for walks, while the integrated autocorrelation time for self-avoiding polygons will certainly be different, as this depends on the autocorrelation function of the Markov chain.

## 9.4 Computing Free Energies of Lattice Polygons

It is possible to estimate the growth constant $\mu$ (defined by equation (9.2)) of a model of polygons, using a Monte Carlo simulation. In this section two methods are given, the first a "binning" method whereby relative free energies can be determined, and the second a method using atmospheres which is a generalisation of a method first developed for self-avoiding walks [61].

### 9.4.1 Free Energies by Binning

Traditionally estimates of the free energy were made by sampling polygons of variable length weighted by length from a Boltzmann distribution. Consider first the non-interacting case.

The *partition function* in these models is given by

$$Z_\beta = \sum_\omega e^{-\beta|\omega|} = \sum_{n=0}^{\infty} p_n e^{-\beta n} \tag{9.59}$$

where the summation runs over all polygons, and where $p_n$ is the number of polygons of length $n$ edges. The (intensive) free energy of this model is given by

$$F_\beta = \log Z_\beta. \tag{9.60}$$

---

[2] These numbers are highly dependent on compiler and machine details, and should be taken as a rough guide only.

[3] The upper bound is due to memory limitations in current generation machines.

The partition function is also the normalising factor of the expected value first encountered in equation (9.13). With a Boltzmann distribution this becomes explicitly

$$\langle f \rangle_\beta = \frac{\sum_\omega f(\omega) e^{-\beta|\omega|}}{\sum_\omega e^{-\beta|\omega|}} \tag{9.61}$$

In a simulation the data can be binned according to increasing length $|\omega|$ of the polygons. Since we are sampling from the distribution $\Pi_\beta(n) = p_n e^{-\beta n}/Z_\beta$, the number of samples of length $n$ will be proportional to $p_n e^{-\beta n}$. Thus, if $r_n$ is the observed number of samples in the bin for polygons of length $n$, then

$$\frac{r_{n+2}}{r_n} \approx \frac{p_{n+2} e^{-2\beta}}{p_n}. \tag{9.62}$$

By repeatedly estimating the ratio of the bins, an average can be computed and the ratio $p_{n+2}/p_n$ can be estimated:

$$\frac{p_{n+2}}{p_n} = \left\langle \frac{r_{n+2}}{r_n e^{-2\beta}} \right\rangle_\beta. \tag{9.63}$$

The ratio $p_{n+2}/p_n$ may be interpreted as the relative free energy of non-interacting polygons. If $n \to \infty$, this is believed to approach $\mu^2$, where $\mu$ is the growth constant of polygons (or self-avoiding walks) and where $\log\mu$ can be interpreted as the partition function.

The above can be generalized to weighted polygons in an interacting model with partition function

$$Z_\beta(\beta) = \sum_\omega e^{-\beta E(\omega)} = \sum_i p(E_i) e^{-\beta E_i} \tag{9.64}$$

where $E(\omega)$ is the energy of polygon $\omega$ and where $p(E_i)$ is the number of polygons of energy $E_i$. Binning data again, this time in bins for each $E(\omega)$, shows that the ratio of the observed number of samples at two different energies $E_1$ and $E_2$ will be

$$\frac{r(E_1)}{r(E_2)} = \frac{p(E_1) e^{-\beta E_1}}{p(E_2) e^{-\beta E_2}}. \tag{9.65}$$

From this, one may estimate the ratio

$$\frac{p(E_2)}{p(E_1)} = \left\langle \frac{r(E_1)}{r(E_2)} e^{-\beta(E_2 - E_1)} \right\rangle_\beta. \tag{9.66}$$

If these ratios are known to sufficient accuracy, then the sum in equation (9.64) can be estimated up to a constant multiplier, and thus the relative free energy can be estimated.

In general, the estimation of free energies by binning is a noisy and numerically difficult process, and free energy data obtained this way tend to be difficult to interpret.

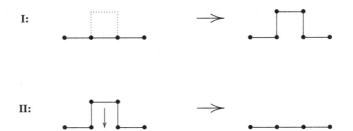

**Fig. 9.10** *Atmospheres of a polygon. In (I) a unit square is outlined on the polygon. By removing the edge on one side of the unit square, and adding three edges around its other three sides, the conformation on the right is obtained. If the resulting object is a polygon, then the unit square contributes to the positive atmosphere of the polygon. The size of the positive atmosphere is the total number of such unit squares on a polygon. In (II) a U-shaped conformation composed of three edges is removed and replaced by a single edge. The resulting object is always a polygon. The collection of U-shaped conformations on a polygon composes its negative atmosphere. The size of the negative atmosphere is the number of such U-conformations in the polygon. These definitions can obviously be generalised. Observe that the addition of edges on a positive atmospheric square creates a negative atmospheric U-shape conformation, and vice versa.*

### 9.4.2 Free Energies by Atmospheres

It is possible to estimate the ratio $p_{n+2}/p_n$ by using a technique inspired by the original method used to estimate the free energies of self-avoiding walk and tree models of branched polymers [61, 27, 28]. The generalisation of this idea to polygons is apparently not immediate, and is best explained by first considering a non-interacting model.

The *atmosphere* of a given lattice polygon can be defined by using Fig. 9.10. Consider first Fig. 9.10(I). On the left hand side one may replace one edge by three edges to create the conformation on the right hand side. This construction can be marked by placing a unit square, outlined in dotted lines on the left, on the edge. The total collection of unit squares which can be placed on a polygon where this construction can be performed, forms the *positive atmosphere* of the polygon. Choosing one such square from the positive atmosphere, and performing the construction, gives a polygon of length two longer than the original.

Next consider Fig. 9.10(II). On the left hand side three edges in a U-conformation can be replaced by a single edge to find a shorter polygon. The total collection of such U-conformations forms the *negative atmosphere* of the polygon. Choosing such a U-conformation and removing it gives a polygon of length two shorter than the original.

Consider next the sets of polygons of length $n$ and of length $n + 2$, illustrated schematically in Fig. 9.11. Polygons of length $n$ are mapped to polygons of length $n + 2$ by the construction in Fig. 9.10(I). Since a polygon has more than one positive atmospheric square, it is mapped to more than one polygon of length $n + 2$, depending on which atmospheric square is chosen to add the edges. This is illustrated by moving along the arrows in the figure.

Similarly, by removing two edges as in Fig. 9.10(II) to reduce the length of a polygon of length $n + 2$, a polygon of length $n$ is obtained. This is illustrated by moving against the arrows in Fig. 9.11.

Observe that each move on a positive atmospheric square can be reversed, since adding edges as in Fig. 9.10(I) creates a negative atmospheric U-conformation. Similarly, when removing a U-conformation which contributed to the negative atmosphere, a unit square contributing to the positive atmosphere is obtained.

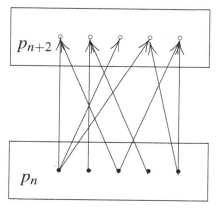

**Fig. 9.11** *Adding edges on the positive atmospheric squares of a polygon gives a polygon of length increased by two. Since any given polygon has a number of positive atmospheric squares, each atmospheric square may produce a different polygon. Each such construction results in the creation of a negative atmospheric U-shaped conformation. By reversing this, the original polygon is recovered. By counting the number of arrows in this diagram, and relating this number to $p_n$ and $p_{n+2}$, equation (9.69) is obtained.*

We proceed now by counting the number of arrows in figure 9.11. Let $p_n(a_+, a_-)$ be the number of polygons of length $n$, counted up to translations in the hypercubic lattice, and with positive atmosphere of size $a_+$ and negative atmosphere of size $a_-$. Then the number of arrows in figure 9.11 is given by

$$\text{Number of arrows} = \sum_{\substack{a_+ \geq 0 \\ a_- \geq 0}} a_+ p_n(a_+, a_-) = \sum_{\substack{a_+ \geq 0 \\ a_- \geq 0}} a_- p_{n+2}(a_+, a_-). \qquad (9.67)$$

Thus, if $\langle a_+ \rangle_n$ is the average size of the positive atmosphere for polygons of length $n$, and $\langle a_- \rangle_{n+2}$ is the average size of the negative atmosphere for polygons of length $n + 2$, then the last equation implies that

$$\langle a_+ \rangle_n = \frac{\sum_{\substack{a_+ \geq 0 \\ a_- \geq 0}} (a_+ p_n(a_+, a_-))}{p_n}$$

$$= \frac{\sum_{\substack{a_+ \geq 0 \\ a_- \geq 0}} (a_- p_{n+2}(a_+, a_-))}{p_{n+2}} \left( \frac{p_{n+2}}{p_n} \right)$$

$$= \langle a_- \rangle_{n+2} \left( \frac{p_{n+2}}{p_n} \right). \tag{9.68}$$

In other words, the ratio of the average sizes of the atmospheres gives

$$\frac{p_{n+2}}{p_n} = \frac{\langle a_+ \rangle_n}{\langle a_- \rangle_{n+2}}. \tag{9.69}$$

To implement this, assume that polygons of length $n$ have been sampled uniformly along a Markov chain by a Monte Carlo algorithm (for example, by the pivot algorithm). Assume similarly, that in a second simulation, polygons of length $n+2$ have been sampled. By computing estimates of the average atmospheres for length $n$ and $n+2$ in these simulations, independent estimates are obtained, and their ratio is an estimate of $p_{n+2}/p_n$, which should converge to $\mu^2$ with increasing $n$.

An alternative implementation would be to estimate atmospheres at a selected set of values of $n$, and then extrapolate by fitting as follows:

$$\langle a_+ \rangle_n = A_+ n + B_+ + C_+/n + \ldots,$$
$$\langle a_- \rangle_n = A_- n + B_- + C_-/n + \ldots \tag{9.70}$$

where we observed that the atmospheres are extensive quantities that grow proportional to $n$. In the $n \to \infty$ limit the ratio $\langle a_+ \rangle_n / \langle a_- \rangle_{n+2}$ approaches $A_+/A_-$, and this is an estimate of $\mu^2$. Evidently, the above implies that $\langle a_+ \rangle_n / \langle a_- \rangle_n$ should approach $\mu^2$ as $n \to \infty$. In addition, one may analyse other ratios, including $\langle a_-/a_+ \rangle_n$. In Fig. 9.12 the ratio $\langle a_+ \rangle_n / \langle a_- \rangle_n$ is plotted against $n$ on data collected by the pivot algorithm in two dimensions.

Implementing this procedure for interacting models is a straightforward generalisation of the above. Assume that a polygon $\omega$ has Boltzmann weight $e^{-\beta E(\omega)}$. If adding the edges on a positive atmospheric square on $\omega$ produces the polygon $\omega'$ of Boltzmann weight $e^{-\beta E(\omega')}$, then the atmospheric arrow is weighted by the ratio $e^{-\beta(E(\omega')-E(\omega))}$. Sampling canonically from the Boltzmann distribution and computing averages of the weighted atmospheres then shows that

$$\frac{Z_{n+2}(\beta)}{Z_n(\beta)} = \frac{\langle a_+(\beta) \rangle_n}{\langle a_-(\beta) \rangle_{n+2}}. \tag{9.71}$$

Assuming that

$$\lim_{n \to \infty} \left( \frac{Z_{n+2}(\beta)}{Z_n(\beta)} \right) = e^{2\mathscr{F}(\beta)} \tag{9.72}$$

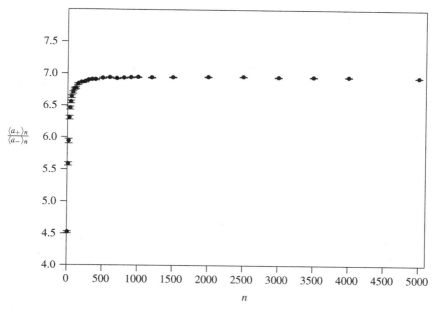

**Fig. 9.12** *The ratio* $\langle a_+ \rangle_n / \langle a_- \rangle_n$ *for two-dimensional lattice polygons computed by using the pivot algorithm. As* $n \to \infty$*, this ratio should approach* $\mu^2$*.*

then shows that one may obtain an estimate of the free energy. Of course, it is not known that the above limit exists in any given model, but the ratio $Z_{n+2}(\beta)/Z_n(\beta)$ is nevertheless expected to give a good approximation to $e^{2\mathscr{F}(\beta)}$, for large $n$, where $\mathscr{F}(\beta)$ is the limiting free energy in the model.

In Fig. 9.12 estimates of $p_{n+2}/p_n$ using atmospheric data by using the pivot algorithm is given. By extrapolating the data to $n \to \infty$ using least squares analysis and the four parameter model $\log \mu^2(n) = 2\log \mu + C_1 \log(1 + 2/n) + C_2/n + C_3/n^2$ gives acceptable fits at the 95% level if $n \geq n_{min} = 10$. If $n_{min} = 10$, then the regression gives the estimate $2.63735 \pm 0.00012$ and if $n_{min} = 16$, then the estimate is $2.63763 \pm 0.00011$, where the error bars are one standard deviation. A systematic error may be estimated by comparing these results. Taking the absolute difference suggests that the systematic error is $0.00028$. Taking our final error bar to be a 95% statistical confidence interval plus the systematic error then gives $\mu = 2.63763 \pm 0.00022 \pm 0.00028 = 2.63763 \pm 0.00050$. This compares well with the estimate from exact enumeration data $2.63815\ldots$ in reference [34]. While this estimate cannot be considered highly accurate compared to other known techniques for determining $\mu$, it can be improved with more computer resources, and this shows that the determination of the free energies using this technique should produce good data.

## 9.5 Conclusions

The two Monte Carlo algorithms discussed in this chapter are the main numerical techniques for sampling polygons. The BFACF algorithm, which is a grand canonical algorithm sampling polygons (usually) from a Boltzmann distribution over length at a given temperature, may suffer from long autocorrelation times which pose some difficulties in data analysis, but it can be used in three dimensions to sample lattice knots of fixed knot type. This algorithm has been used extensively for the sampling of knotted polygons, and in a few instances, also for the sampling of linked polygons [29].

The BFACF algorithm can be implemented in a number of ways. The simplest implementation is a Metropolis-style implementation involving the sampling of polygons along a Markov chain from a Boltzmann distribution, but this can be generalized to umbrella sampling by generalising the Boltzmann distribution to be an umbrella distribution spanning a range of temperatures or other parameters in the model. Finally, multiple Markov chain Monte Carlo implementation of the BFACF algorithm can also be done—the advantage with this implementation is that it typically shortens autocorrelation times while at the same time allowing the analysis of data over a range of values of the temperature and other parameters.

The pivot algorithm may also be implemented using an umbrella sampling technique, or as a multiple Markov chain Monte Carlo algorithm. The shorter autocorrelation times in the time series produced by this algorithm gives it an inherent advantage over the BFACF algorithm. However, the ergodicity properties of this algorithm for polygons of fixed knot types are not known, and since knotted conformations are rare in the state space of polygons of lengths up to millions of edges, it is not easily possible to sample knotted polygons of a given knot type using a pivot algorithm. However, the algorithm has been used effectively in sampling polygons in a variety of models, including models of ring polymer adsorption and collapse [22].

Some generalisation of the pivot algorithm is possible, but still unexplored. For example, an elementary move may be implemented which cuts a polygon into (say) $N$ sub-walks, then performing pivot moves on each, and reconnecting them in random order back into a polygon. Such moves will radically reorder the vertices in a given polygon with each successful attempt, and potential reductions in correlation times along the time series of sampled polygons may be achieved.

## References

1. C. Aragao de Carvalho, S. Caracciolo and J. Fröhlich: *Polymers and $g\phi^4$ Theory in Four Dimensions*. Nucl. Phys. B **215**, 209 (1983)
2. A. Beretti and A.D. Sokal: *New Monte Carlo Method for the Self-avoiding Walk*. J. Stat. Phys. **40**, 483 (1985)
3. B. Berg: *Markov Chain Monte Carlo Simulations and Their Statistical Analysis*. (World Scientific, Singapore 2004)

4. B. Berg and D. Foester: *Random Paths and Random Surfaces on a Digital Computer*. Phys. Lett. B **106**(4), 323 (1981)
5. S. Caracciolo, A.J. Guttmann, I. Jensen, A. Pelissetto, A.N. Rogers and A.D. Sokal: *Correction-to-Scaling Exponents for Two-Dimensional Self-Avoiding Walks*. J. Stat. Phys. **120**(5-6), 1037 (2005)
6. S. Caracciolo, A. Pellissetto and A.D. Sokal: *Nonlocal Monte Carlo Algorithm for Self-avoiding Walks with Fixed Endpoints*. J. Stat. Phys. **60**, 1 (1990)
7. J.L. Cardy: Conformal Invariance. In *Phase Transitions and Critical Phenomena* vol 11, ed. by C. Domb and J.L. Lebowitz (Academic Press, London 1984)
8. J.L. Cardy and A.J. Guttmann: *Universal Amplitude Combinations for Self-Avoiding Walks, Polygons and Trails*. J. Phys. A: Math. Gen. **26**, 2485 (1993)
9. N. Clisby: *An accurate estimate of $\nu$ for three-dimensional self-avoiding walks via the pivot algorithm*, in preparation. (2008)
10. N. Clisby, R. Liang and G. Slade: *Self-Avoiding Walk Enumeration via the Lace Expansion*. J. Phys. A: Math. Gen. **40**, 10973-11017 (2007)
11. P.G. de Gennes: *Scaling Concepts in Polymer Physics*. (Cornell University Press, Ithaca 1979)
12. M. Delbrück: *Knotting Problems in Biology*. Proc. Symp. Appl. Math. **14**, 55 (1962)
13. B. Duplantier and H. Saleur: *Exact Determination of the Percolation Hull Exponent in Two Dimensions*. Phys. Rev. Lett. **58**, 2325 (1987)
14. P.J. Flory: *Principles of Polymer Chemistry*. (Cornell University Press, Ithaca 1971)
15. H.L. Frisch and E. Wasserman: *Chemical Topology*. J. Amer. Chem. Soc. **83**, 3789 (1968)
16. C.J. Geyer: Markov Chain Monte Carlo Maximum Likelihood Computing Science and Statistics, In *Proceedings of the 23rd Symposium on the Interface*, ed. by E.M. Keramidas (Interface Foundation, Fairfax Station 1991)
17. C.J. Geyer and E.A. Thompson: *Annealing Markov Chain Monte Carlo with Applications to Ancestral Inference*. J. Amer. Stat. Assoc. **90**, 909 (1994)
18. J.M. Hammersley: *On the number of polygons on a lattice*. Math. Proc. Camb. Phil. Soc. **57**, 516 (1961)
19. T. Hara and G. Slade: *The Lace Expansion for Self-Avoiding Walk in Five or More Dimensions*. Rev. Math. Phys. **4**, 235 (1990)
20. E. Hille: *Functional Analysis and Semi-Groups*. AMS Colloq. Publ. **31**, (AMS, New York 1948)
21. E.J. Janse van Rensburg: *Ergodicity of the BFACF Algorithm in Three Dimensions*. J. Phys. A: Math. Gen. **25**, 1031 (1992)
22. E.J. Janse van Rensburg: *Collapsing and Adsorbing Polygons*. J. Phys. A: Math. Gen. **31**, 8295 (1998)
23. E.J. Janse van Rensburg: *The Statistical Mechanics of Interacting Walks, Polygons, Animals and Vesicles*. Oxford Lecture Series in Mathematics and its Applications **18**, (OUP: Oxford, 2000)
24. E.J. Janse van Rensburg: The Probability of Knotting in Lattice Polygons. In *Physical Knots: Knotting, Linking, and Folding Geometric Objects in $\mathscr{R}^3$*, ed. by J.A. Calvo, K.C. Millett, E.J. Rawdon. Contemporary Math. **304**, 125 (2002)
25. E.J. Janse van Rensburg: *Squeezing Knots*. J. Stat. Mech.: Theo. Exp. **03**, P03001 (2007)
26. E.J. Janse van Rensburg, S.G. Whittington and N. Madras: *The Pivot Algorithm and Polygons: Results on the FCC Lattice*. J. Phys. A: Math. Gen. **23**, 1589 (1990)
27. E.J. Janse van Rensburg and A. Rechnitzer: *High Precision Canonical Monte Carlo Determination of the Connective Constant of Self-Avoiding Trees*, Phys. Rev. E **67**, 036116-1 (2003)
28. E.J. Janse van Rensburg and A. Rechnitzer: *Multiple Markov Chain Monte Carlo Study of Adsorbing Self-Avoiding Walks in Two and in Three Dimensions*, J. Phys. A: Math. Gen. **37**, 6875 (2004)
29. E.J. Janse van Rensburg, D.W. Sumners and S.G. Whittington: *The Writhe of Knots and Links*. In: *Ideal Knots* ed. by A. Stasiak (Series on Knots and Everything, vol 19, World Scientific, Singapore 1999)
30. E.J. Janse van Rensburg and S.G. Whittington: *The Knot Probability of Lattice Polygons*. J. Phys. A: Math. Gen. **23**, 3573 (1990)

31. E.J. Janse van Rensburg and S.G. Whittington: *The BFACF Algorithm and Knotted Polygons.* J. Phys. A: Math. Gen. **24**, 5553 (1991)
32. E.J. Janse van Rensburg and S.G. Whittington: *The Dimensions of Knotted Polygons.* J. Phys. A: Math. Gen. **24**, 3935 (1991)
33. I. Jensen: *Size and Area of Square Lattice Polygons.* J. Phys. A: Math. Gen. **33**, 3533 (2000)
34. I. Jensen: *A Parallel Algorithm for the Enumeration of Self-Avoiding Polygons in the Square Lattice.* J. Phys. A: Math. Gen. **36**, 5731 (2003)
35. I. Jensen and A.J. Guttmann: *Self-avoiding Walks, Neighbour-avoiding Walks and Trials on Semi-regular Lattices.* J. Phys. A: Math. Gen. **31** 8137 (1998)
36. I. Jensen and A.J. Guttmann: *Self-avoiding Polygons on the Square Lattice.* J. Phys. A: Math. Gen. **32**, 4867 (1999)
37. T. Kennedy. *A faster implementation of the pivot algorithm for self-avoiding walks.* J. Stat. Phys, **106**, 407–429 (2002).
38. H. Kesten: *On the number of self-avoiding walks.* J. Math. Phys. **4**, 960 (1963)
39. H. Kesten: *On the number of self-avoiding walks II.* J. Math. Phys. **5**, 1128 (1964)
40. M. Lal: *Monte Carlo Computer Simulations of Chain Molecules I.* Mol. Phys. **17**, 57 (1969)
41. B. Li, N. Madras and A.D. Sokal: *Critical Exponents, Hyperscaling, and Universal Amplitude Ratios for Two and Three Dimensional Self-Avoiding Walks.* J. Stat. Phys. **80**, 661 (1995)
42. N. Madras: Unpublished (1986)
43. N. Madras: Bounds on the Critical Exponents of Self-Avoiding Polygons. In *Random Walks, Brownian Motion and Interacting Particle Systems* ed by R. Durrett and H. Kesten (Birkhauser: Boston 1991).
44. N. Madras: *A Rigorous Bound on the Critical Exponents for the Numbers of Lattice Trees, Animals and Polygons.* J. Stat. Phys. **78**, 681 (1995)
45. N. Madras, A. Orlitsky and L.A. Shepp: *Monte Carlo Generation of Self-Avoiding Walks with Fixed Endpoints and Fixed Length.* J. Stat. Phys. **58**, 159 (1990)
46. N. Madras and G. Slade: *The Self-Avoiding Walk.* (Birkhäuser, Boston 1993)
47. N. Madras and A.D. Sokal: *Nonergodicity of Local, Length-conserving Monte Carlo Algorithms for the Self-avoiding Walk.* J. Stat. Phys. **47**, 573 (1987)
48. N. Madras and A.D. Sokal: *The Pivot Algorithm: A Highly Efficient Monte Carlo Method for the Self-avoiding Walk.* J. Stat. Phys. **50**, 109 (1988)
49. N. Metropolis and S. Ulam: *The Monte Carlo method.* J. Amer. Stat. Ass. **44**(247), 335 (1949)
50. N. Metropolis, A.W. Rosenbluth, M.N. Rosenbluth, A.H. Teller and E. Teller: *Equation of State Calculations by Fast Computing Machines.* J. Chem. Phys. **21**, 1087 (1953)
51. M. Mezei: *Adaptive Umbrella Sampling: Self-consistent Determination of the Non-Boltzman Bias.* J. Comp. Phys. **68**, 237 (1987)
52. J.P.J. Michels and F.W. Wiegel: *The Probability of Knots in a Polymer Ring.* Phys. Lett. A **90**, 381 (1982)
53. R. Neal: *Probabilistic Inference using Markov Chain Monte Carlo Methods.* Technical Report CRG-TR-93-1. Department of Computer Science, University of Toronto, Toronto (1993)
54. B. Nienhuis: *Exact Critical Point and Critical Exponents of O(n) Models in Two Dimensions.* Phys. Rev. Lett. **49**, 1062 (1982)
55. B. Nienhuis: Coulomb Gas Formulation of Two-Dimensional Phase Transitions. In *Phase Transitions and Critical Phenomena*, vol 11, ed by C. Domb and J.L. Lebowitz. (Academic Press, London 1984), 1–53
56. E. Orlandini, M.C. Tesi, E.J. Janse van Rensburg and S.G. Whittington: *Entropic Exponents of Lattice Polygons with Specified Knot Type.* J. Phys. A: Math. Gen. **29**, L299 (1996)
57. E. Orlandini, M.C. Tesi, E.J. Janse van Rensburg and S.G. Whittington: *Asymptotics of Knotted Lattice Polygons.*, J. Phys. A: Math. Gen. **31**, 5953 (1998)
58. A.L. Owczarek and T. Prellberg: *Monte Carlo Investigation of Lattice Models of Polymer Collapse in Five Dimensions.* Int. J. Mod. Phys. C **14**(5), 621 (2003)
59. N. Pippenger: *Knots in Self-Avoiding Walks.* Disc. Appl. Math. **25**, 273 (1989)
60. D.C. Rapaport: *On Three-dimensional Self-Avoiding Walks.* J. Phys. A: Math. Gen. **18**, 113 (1985)

61. A. Rechnitzer and E.J. Janse van Rensburg: *Canonical Monte Carlo Determination of the Connective Constant of Self-Avoiding Walks.* J. Phys. A: Math. Gen. **35**, L605 (2002)
62. M.N. Rosenbluth and A.W. Rosenbluth: *Monte Carlo Calculation of the Average Extention of Molecular Chains.* J. Chem. Phys. **23**, 356 (1955)
63. R.Y. Rubinstein: *Simulation and the Monte Carlo Method.* (Wiley: New York 1981)
64. A.D. Sokal: Monte Carlo Methods for the Self-Avoiding Walk. In: *Monte Carlo and Molecular Dynamics Simulations on Polymer Science*, ed. by K. Binder (Oxford University Press: London 1995)
65. C.E. Soteros, D.W. Sumners and S.G. Whittington: *Entanglement Complexity of Graphs in* $\mathbb{Z}^3$. Math. Proc. Camb. Phil. Soc. **111**, 75 (1992)
66. D.W. Sumners and S.G. Whittington: *Knots in Self-Avoiding Walks.* J. Phys. A: Math. Gen. **21**, 1689 (1988)
67. M.C. Tesi, E.J. Janse van Rensburg, E. Orlandini, S.G. Whittington: *Interacting Self-Avoiding Walks and Polygons in Three Dimensions.* J. Phys. A: Math. Gen. **29**, 2451 (1996)
68. M.C. Tesi, E.J. Janse van Rensburg, E. Orlandini, S.G. Whittington: *Monte Carlo Study of the Interacting Self-Avoiding Walk Model in Three Dimensions.* J. Stat. Phys. **82**, 155 (1996)
69. G.M. Torrie and J.P. Valleau: *Monte Carlo Study of a Phase-separating Liquid Mixture by Umbrella Sampling.* J. Chem. Phys. **66**, 1402 (1977)
70. S.G. Whittington and E.J. Janse van Rensburg: Random Knots in Ring Polymers. In: *Proceedings of the Eighth International Conference on Mathematical and Computer Modelling. Mathematical Modelling and Scientific Computing*, vol 2, 741 (1992)

# Chapter 10
# Effect of Confinement: Polygons in Strips, Slabs and Rectangles

Anthony J Guttmann and Iwan Jensen

## 10.1 Introduction

In this chapter we will be considering the effect of confining polygons to lie in a bounded geometry. This has already been briefly discussed in Chapters 2 and 3, but here we give many more results. The simplest, non-trivial case is that of SAP on the two-dimensional square lattice $\mathbb{Z}^2$, confined between two parallel lines, say $x = 0$ and $x = w$. This problem is essentially 1-dimensional, and as such is in principle solvable. As we shall show, the solution becomes increasingly unwieldy as the distance $w$ between the parallel lines increases. Stepping up a dimension to the situation in which polygons in the simple-cubic lattice $\mathbb{Z}^3$ are confined between two parallel planes, that is essentially a two-dimensional problem, and as such is not amenable to exact solution.

Self-avoiding walks in slits were first treated theoretically by Daoud and de Gennes [4] in 1977, and numerically by Wall et al. [14] the same year. Wall et al. studied SAW on $\mathbb{Z}^2$, in particular the mean-square end-to-end distance. For a slit of width one they obtained exact results, and also obtained asymptotic results for a slit of width two. Around the same time, Wall and co-workers [13, 15] used Monte Carlo methods to study the width dependence of the growth constant for walks confined to strips of width $w$. In 1980 Klein [9] calculated the behaviour of SAW and SAP confined to strips in $\mathbb{Z}^2$ of width up to six, based on a transfer matrix formulation.

The interest in the problem arises from two separate aspects. Firstly, there is the intrinsic interest in the effect of geometrical constraints. Secondly, this confined geometry is appropriate to model polymeric properties, such as sensitised flocculation and steric stabilisation, again first discussed in this context by de Gennes [5] in 1979.

Anthony J Guttmann and Iwan Jensen
Department of Mathematics and Statistics, The University of Melbourne, Victoria, Australia, e-mail: tonyg@ms.unimelb.edu.au, e-mail: iwan@ms.unimelb.edu.au

The effect of confinement leads to a loss of configurational entropy, with the consequence that there is a repulsion exerted by the polygon on the confining walls. That this force is repulsive for all values of $w$ was proved by Hammersley and Whittington [7] in 1985, a result that was extended by Janse van Rensburg et al. [8] who showed that the force remains repulsive despite a certain level of interaction with the confining lines. If there is, in addition, an attractive interaction with the walls, there is then a competition between entropic repulsion and the attractive polymer adsorption.

In an earlier paper Di Marzio and Rubin [6] studied a random walk model of a polymer confined between two planes in $\mathbb{Z}^3$. The model included wall–walk interactions. In the absence of these interactions there is the expected loss of configurational entropy, and the walls exert an effective repulsion. If there is an attraction to only one of the two walls, this repulsion was found to persist. If however there is an equal attraction at both walls, then the more interesting situation in which the force is repulsive for weak wall–monomer interactions, but attractive for stronger wall–monomer interactions was found.

In the bulk, it is known (see Chapter 1) that self-avoiding walks and self-avoiding polygons have the same growth constant. However, this is not true for SAW and SAP confined to a slit, as proved by Soteros and Whittington [11]. In fact they proved that the growth constant for polygons in a slit of width $w$ is strictly less than the growth constant for SAW in a slit of the same (finite) width. This is a strictly two-dimensional phenomenon. It is not true in higher dimensions. That is to say SAW and SAP confined to lie between two parallel planes in $\mathbb{Z}^3$ have the same connective constant.

In a thorough and detailed study of SAW in strips, Ahlberg and Janson [1] in 1990 gave a transfer matrix formalism. Denoting the generating function for walks in a strip of width $L$, as usual, as $C_L(x) = \sum_n c_n^{(L)} x^n$, where $c_n^{(L)}$ is the number of translationally distinct SAW, they proved (a) that $c_n^{(L)} = \alpha \mu_L^n + o(\mu_L^n)$ as $n \to \infty$, and (b) $c_{n+1}^{(L+1)}/c_n^{(L)} \to \infty$, and (c) $c_n^{(L)}/\mu_L^n$ converges exponentially. They proved similar results for SAP in a slit, including the result of Soteros and Whittington that the growth constant for SAP in a slit of finite width is strictly less than the corresponding result for SAW. They obtained the growth constants for SAW in strips of width up to 10 steps, and, for SAW on a cylinder, in a cylinder up to 10 links in the circular direction. They also gave a detailed study of the properties of the transfer matrix, and obtained a Central Limit Theorem for the endpoint.

Very recently Alvarez et al. [2] studied SAW and SAP in a slit, with wall interactions. For SAP they found that, for any finite value of the wall–monomer interaction term, there is an infinite number of slit widths where a polygon will induce a repulsion between the confining lines.

In Chapter 3 we saw how SAP in a strip are completely encodable by the position of their horizontal edges. Indeed, it was shown that they could be encoded by a finite alphabet, and that alphabet was given for both a strip of width 2 and a strip of width 3. The number of states required to count polygons in a strip of width $w$ grows as $3^w$, which prevents this calculation from being pushed to very high values of $w$. More precisely, Klein [9] has shown that the number of states is given by

$$\frac{w(w+1)}{2} \, {}_2F_3\left(1, \frac{2-w}{2}, \frac{1-w}{2}; 2, 3; 4\right).$$

This gives a sequence $2, 5, 12, 30, 76, 196, 512, 1353, \ldots$ which is growing exponentially, proportional to $3^w$. As we saw in Chapter 3 this can be reduced by symmetry. By judicious use of the transfer matrix method, as discussed in Chapter 7, we have extended these encodings to strips of width $w = 17$ for square lattice polygons and for honeycomb lattice polygons (in both lattice directions[1]), and for triangular lattice polygons in strips of width up to 14.

The generating function in each case is rational, and the nature of the singularity at the radius of convergence, which we identify with the critical point, is just a simple pole. For square lattice polygons, we find the degree of the numerator, $N$ and denominator $D$ to be $(N, D) = (0, 1), (2, 4), (10, 14), (34, 40)$ for widths $w = 1, 2, 3, 4$ respectively. The value of the smallest real positive zero of the denominator polynomial gives the radius of convergence, and also the reciprocal of the growth constant for polygons, $\mu_w$, and this is a monotone increasing function of $w$. In this way we obtained the lower bounds $\mu(\text{square}) > 2.4537$, $\mu(\text{honey}) > 1.7759$ and $\mu(\text{tri}) > 3.7272$. (This isn't a particularly efficient way to obtain lower bounds, but is, rather, an additional outcome of the study.)

Daoud and de Gennes [4] developed the scaling theory that predicts how $\mu_w$ is expected to scale with width $w$. Their result was for SAW in a strip, but can be expected to hold mutatis mutandis for polygons in a strip. They find that

$$\log \mu - \log \mu_w \sim const. \times w^{-\phi},$$

where $\phi = 1/\nu = 4/3$. Recall that $\nu = 3/4$ is the mean-square end-to-end distance scaling exponent. For walks confined between planes in $\mathbb{Z}^3$, the same result is expected to hold, except now the value of $\nu$ is not known exactly, but to a good approximation is $\nu(3d) \approx 0.57\ldots$ Recall that we have very precise estimates of $\mu$ for all lattices (these are more precise in 2d than in 3d). Indeed, for the honeycomb lattice in 2d we believe the exact value to be $\mu(\text{honey}) = \sqrt{2 + \sqrt{2}}$.

In Table 10.1 we give the results for the growth constant for strips of width $d = w + 1$ sites, ($w$ is the width in bonds), for the square, triangular and honeycomb lattices. The monotone increasing values of $\mu_d$ can be readily seen. In Fig. 10.1 we plot $\log \mu - \log \mu_d$ against $\log d$ for the square lattice, and show the solid line of gradient $-4/3$. It can be seen that quite large values of $w$ are required before we reach the asymptotic regime, but that the scaling predictions are well supported by the data. That is to say, the later points do indeed seem to have a locus of the same gradient as the line drawn. The figures for the other lattices are qualitatively similar.

---

[1] We draw the honeycomb lattice as a brickwork lattice, so the lattice is not symmetrical in the two lattice directions.

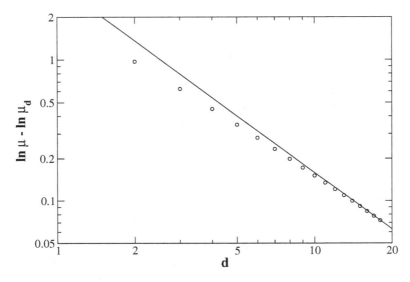

**Fig. 10.1** Plot of $\log \mu - \log \mu_d$ vs. $d$ on a logarithmic scale for the square lattice data. The straight line corresponds to the theoretical prediction of gradient $-4/3$.

## 10.2 Polygons in a Square

Consider now SAW confined to an $L \times L$ square. This problem has a long history, and a detailed discussion can be found in [3]. In that paper it was shown, among

**Table 10.1** $d = w + 1$ is the strip width, $\mu_d$ is the growth constant for square, triangular and honeycomb (direction 1 and 2) lattice polygons in the strip.

| $d$ | $\mu_d$(square) | $\mu_d$(tri) | $\mu_d$(honey1) | $\mu_d$(honey2) |
|---|---|---|---|---|
| 2 | 1.000000000 | 1.000000000 | 1.000000000 | 1.000000000 |
| 3 | 1.414213562 | 1.795088688 | 1.229990405 | 1.229990405 |
| 4 | 1.681759003 | 2.328493240 | 1.374333111 | 1.411814730 |
| 5 | 1.863069582 | 2.684771831 | 1.470190448 | 1.506504156 |
| 6 | 1.992445913 | 2.936411124 | 1.537116598 | 1.570592201 |
| 7 | 2.088633483 | 3.121963721 | 1.585935359 | 1.616386758 |
| 8 | 2.162502131 | 3.263502052 | 1.622845444 | 1.650489362 |
| 9 | 2.220732353 | 3.374453146 | 1.651575171 | 1.676722086 |
| 10 | 2.267631888 | 3.463397284 | 1.674476550 | 1.697434531 |
| 11 | 2.306090565 | 3.536045068 | 1.693096661 | 1.714142575 |
| 12 | 2.338112184 | 3.596328876 | 1.708490122 | 1.727863827 |
| 13 | 2.365125683 | 3.647036006 | 1.721398098 | 1.739304455 |
| 14 | 2.388174882 | 3.690191779 | 1.732355266 | 1.748968490 |
| 15 | 2.408038380 | 3.727299874 | 1.741756176 | 1.757224477 |
| 16 | 2.425307673 | | 1.749897782 | 1.764347596 |
| 17 | 2.440439441 | | 1.757007495 | 1.770547115 |
| 18 | 2.453791386 | | 1.763262171 | 1.775984775 |

other things, that the number of SAW starting at $(0,0)$ and ending at $(L,L)$ and never leaving the square, grows as $\lambda^{L^2}$. That is to say, if $C_L$ is the number of such walks, then $\lim_{L\to\infty} C_L^{1/L^2} = \lambda$. It was estimated that $\lambda = 1.744550 \pm 0.000005$. A related problem consists of estimating the number of *transverse walks,* defined as SAW that cross the square from any vertex on the left edge of the square (hence the $x$ co-ordinate is 0), to any vertex on the right edge (with $x$ co-ordinate $L$). In [3] it was proved that if $T_L$ denotes the number of such walks in an $L \times L$ square, then $\lim_{L\to\infty} T_L^{1/L^2} = \lambda$, with the same value of $\lambda$ as for $C_L$.

We now consider SAP that span the square. That is to say, one or more edges of the polygon must lie on each edge of the square. As far as we are aware, this problem has not previously been considered. Let $P_L$ denote the number of such polygons. A moment's reflection shows that $P_L < T_L$, and $P_{L+1} > C_L$, as one can readily construct a unique SAP occupying an $(L+1) \times (L+1)$ lattice from a SAW going from $(0,0)$ to $(L,L)$, by the addition of a step from $(L,L)$ to $(L,L+1)$, then another from $(L+1,L+1)$, then a ray from $(L+1,L+1)$, to $(L+1,-1)$, followed by a further ray from $(L+1,-1)$ to $(0-1)$, and then a final step to the origin. In this way we can prove that $\lim_{L\to\infty} P_L^{1/L^2} = \lambda$.

**Table 10.2** The number of self-avoiding polygons $P_L$ in a square of size $L \times L$.

| $L$ | $P_L$ |
|---|---|
| 1 | 1 |
| 2 | 5 |
| 3 | 106 |
| 4 | 6074 |
| 5 | 943340 |
| 6 | 419355340 |
| 7 | 554485727288 |
| 8 | 2208574156731474 |
| 9 | 26609978139626497670 |
| 10 | 973224195603423767343946 |
| 11 | 108342096917091380628767818812 |
| 12 | 367632110165285493100682241223368860 |
| 13 | 3804428704374943628459464430886149605492 |
| 14 | 120080993887856855992693253821542678777528272944 |
| 15 | 1155964922833172664443974642986506946314409762614495586 |
| 16 | 33934880416462899814285781006397200200998294954062388898965682 |

Several refinements or extensions of this problem remain to be considered. The number of steps (the perimeter) of a SAP in a square varies from a minimum of $4L$ to a maximum of $(L+1)^2$. It would be interesting to study the distribution of perimeters with $L$. A second aspect amenable to study would be to include an interaction between adjacent monomers of the polygon, and/or with the edges of the square. From the discussion in Chapter 12, and the section below, we have some understanding of what to expect in these cases, but it would still be of interest to see the details.

## 10.3 Polygons in a Strip Interacting with Walls

In the previous two sections we considered SAP in confined geometries, but apart from confinement, there was no additional constraint imposed by the walls. In this section we consider the situation where there is an interaction associated with edges of the polygon in, or adjacent to, a wall.

Very recently, Alvarez et al [2] have investigated the situation of SAP (and SAW) in strips of width $w$, interacting with the walls. It is necessary to consider not just the situation *at* the surfaces, but also immediately adjacent to the surfaces if a full range of behaviour is to be observed. This is because polygons are topologically circular. So that if they span a strip, the top edge can never reach the bottom of the strip (this phenomenon is particular to polygons in two dimensions). More precisely, if we define the *top edge* of the polygon as that part of the polygon between the first vertex lying in the top of the slit and the last vertex lying in the top of the slit, then no vertex in the top edge can lie in the bottom of the slit. This topological constraint has been overcome by considering interactions with the second row. This means the top (bottom) can interact with the second layer of the bottom (top). We shall adopt the notation of Alvarez et al. [2], and point out a six-dimensional vector $\mathbf{v} = (v_0, v_{0,1}, v_1, v_{w-1}, v_{w-1,w}, v_w)$ is required in order to keep track of the number of bonds $v_0$ lying in the slit edge at $y = 0$, $v_1$ lying in the row $y = 1$ (which is the row immediately above the bottom of the slit), $v_w$ lying in the slit edge at $y = w$, $v_{w-1}$ lying in the row $y = w - 1$ (which is the row immediately below the top of the slit). Finally $v_{0,1}$ is the number of *vertical* bonds between $y = 0$ and $y = 1$, and $v_{w-1,w}$ is the number of *vertical* bonds between $y = w - 1$ and $y = w$. We also introduce two corresponding vectors of Boltzmann factors, $\mathbf{a} = (a_0, a_{0,1}, a_1)$ and $\mathbf{b} = (b_{t-1}, b_{t-1,t}, b_t)$. Then if $p_n(\mathbf{v}, w)$ is the number of SAP in a slit of width $w$ with $n$ edges and restricted by having edges in various places as specified by $\mathbf{v}$, the partition function is

$$Z_n(\mathbf{a}, \mathbf{b}, w) = \sum_{\mathbf{v}} c_n(\mathbf{v}, w) a_0^{v_0} a_{0,1}^{v_{0,1}} a_1^{v_1} b_{t-1}^{v_{w-1}} b_{t-1,t}^{v_{w-1,w}} b_t^{v_w}. \tag{10.1}$$

The grand canonical partition function is then given by

$$H(\mathbf{a}, \mathbf{b}, w) = \sum_{n=0}^{\infty} Z_n(\mathbf{a}, \mathbf{b}, w) z^n, \tag{10.2}$$

and the corresponding free energy is

$$\kappa(\mathbf{a}, \mathbf{b}, w) = \lim_{n \to \infty} n^{-1} \log Z_n(\mathbf{a}, \mathbf{b}, w), \tag{10.3}$$

while the force exerted by the polygon on the confining walls is

$$f(\mathbf{a}, \mathbf{b}, w) := \frac{\partial}{\partial w} \kappa(\mathbf{a}, \mathbf{b}, w). \tag{10.4}$$

Alvarez *et al.* considered four special cases, which were:

(a) $a_{0,1} = a_1 = b_{t-1} = b_{t-1,t} = 1$, corresponding to a *single layer at both walls*. Here we have switched off interactions in the second layer, both at the top and at the bottom, and only interactions in the surfaces take place.

(b) $a_{0,1} = a_1 = b_{t-1,t} = 1$, and $b_{t-1} = b_t = b$ corresponding to a *double layer at the top wall*. Here the interactions occur with the top and next-to-top layer, and also with the bottom layer. All other interactions are switched off.

(c) $a_{0,1} = b_{t-1,t} = 1$, $a_0 = a_1 = a$ and $b_{t-1} = b_t = b$ corresponding to a *double layer at both walls*. Here the interactions occur with the top and next-to-top layer, and also with the bottom and next-to-bottom layer. All other interactions (between the two top and between the two bottom layers) are switched off.

(d) $a_0 = a_{0,1} = a_1 = a$, and $b_{t-1} = b_{t-1,t} = b_t = b$ corresponding to *fully interacting double layers*. Here all interactions are on. The only restriction is that the interactions are all equal at the top, and all equal at the bottom.

Series expansions by use of the transfer matrix method (see Chapter 7) were obtained for strips up to width 9. Two useful lemmas were also proved. They are:

**Lemma 1** For SAP in a slit in case (a), we have for any width $w > 0$ that the free energy difference produced by increasing the width by at least $w$ units is non-negative. That is to say

$$\kappa(\mathbf{a}, \mathbf{b}, w + i) - \kappa(\mathbf{a}, \mathbf{b}, w) \geq 0$$

for any integer $i > w$.

**Lemma 2** There are infinitely many values of $w$ for which the force for SAP in a slit of width $w$ in case (a) is always non-negative.

In Fig. 10.2 we show a plot of the force with $a = b$ line for SAP in slits of various widths in the single layer case (a). It is clear that the forces are always positive, corresponding to a purely repulsive regime. This observation is consistent with the above lemma. The result may indeed be true for all $w$, but this has not been proved. Also note that the force quickly drops off as $a$ increases.

The case of a double layer at the top wall, case (b) above, overcomes the shielding effect of case (a), in which topology prevents the top of the polygon reaching the bottom wall (and vice versa). SAPs now exhibit both an attractive and a repulsive regime. In Fig. 10.3 we show the zero-force curve for SAP in slits of various widths. The positive force regime (repulsive) is to the S-W of the curves, while the attractive regime is to the N-E. The minimum we observe means we have *re-entrant* behaviour. For wall interaction parameter $b = 3$ say, as we increase the value of $a$, the force changes from repulsive for small $a$, then to attractive for intermediate values of $a$, then back to repulsive for large values of $a$.

Figure 10.4 shows the force along the $a = b$ line for SAP in slits of various widths in case (b). Again we see that for small values of the wall interaction parameters, the force is repulsive, but as the interaction strength increases, it becomes attractive. This double layer model overcomes the screening that prevents the formation of an attractive regime in the single layer case.

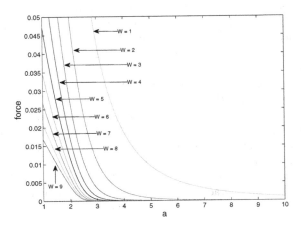

**Fig. 10.2** Force along the $a = b$ line for SAP in slits of various widths in the single layer case. Note the absence of any attractive regime (corresponding to a negative force).

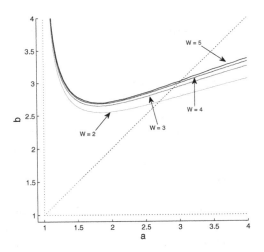

**Fig. 10.3** Zero-force curve for SAP in slits of various widths in the case of a double layer at the top wall and a single layer at the bottom wall. The two intersections with the line $b = 3$ signals *re-entrant* behaviour.

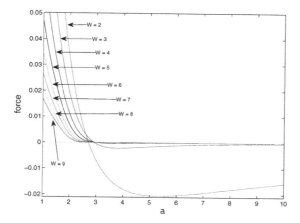

**Fig. 10.4**  Force along the $a = b$ line for SAP in slits of various widths in the case of a double layer at the top wall and a single layer at the bottom wall. Both attractive and repulsive regimes are evident.

The next situation considered is that of a double layer at both walls. In Fig. 10.5 we show the zero-force curve for SAP in slits of various widths. Unlike the previous case, we observe an expected symmetry about the line $a = b$. As in the previous case, the positive force regime (repulsive) is to the S-W of the curves, while the attractive regime is to the N-E. It is unclear whether the zero-force point along the line $a = b$ diverges as the strip width increases.

The next figure, Fig. 10.6, plots the force along the $a = b$ line for SAP in slits of various widths. Again, both attractive and repulsive regimes are seen. As the width increases, the attractive force is seen to become rather weak.

The final case considered has all interactions switched on, but with the restriction that all the interactions at the top wall are equal, as are all the interactions at the bottom wall. In Fig. 10.7 we plot the zero-force curve for SAP in slits of various widths. The picture looks qualitatively the same as Fig. 10.5, with re-entrant behaviour evident in some regions of the phase diagram.

In Fig. 10.8 the force along the $a = b$ line for SAP of various widths is shown. Again, we see qualitative similarity to the previous double layer case, and the remarks made about that situation apply here too.

In [2] Alvarez *et al.* give a similar analysis for SAW in strips. They observe some significant differences between SAW and SAP. These differences are beyond the scope of this chapter, and are also likely to be confined to the two-dimensional case, so we refer the interested reader to their article.

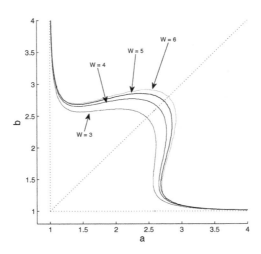

**Fig. 10.5** Zero-force curve for SAP in slits of various widths in the case of double layers at both walls. Both *re-entrant* behaviour and symmetry about the line $a = b$ are evident.

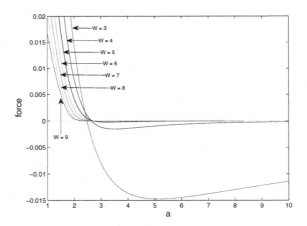

**Fig. 10.6** Force along the $a = b$ line for SAP in slits of various widths in the case of double layers at both walls. Both attractive and repulsive regimes are evident.

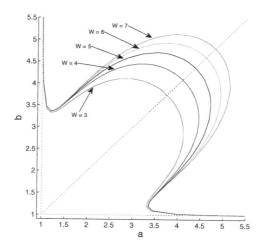

**Fig. 10.7**  Zero-force curve for SAP in slits of various widths in the case of fully interacting double layers at both walls. Both *re-entrant* behaviour and symmetry about the line $a = b$ are evident.

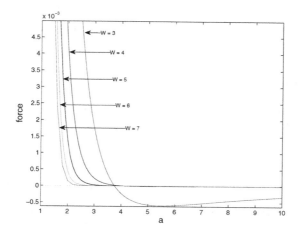

**Fig. 10.8**  Force along the $a = b$ line for SAP in slits of various widths in the case of fully interacting double layers at both walls. Both attractive and repulsive regimes are evident, though the attractive regime declines with increasing width.

## 10.4 Conclusion

The consequences of confining SAP (and SAW) to strips/slabs/prisms clearly produces a rich set of both combinatorial results and models of physical, and indeed biological interest. The fact that regimes can change from attractive to repulsive, and back, opens up the possibility of constructing simple models with quite complex behaviour. More significantly perhaps, it means that it is not necessary to postulate complex models to explain complex behaviour. The extension of the study of Alvarez *et al.* [2], discussed above, to three dimensions, would be extremely interesting, but is probably beyond current computational resources.

## References

1. Alm S E and Janson S, (1990) Random self-avoiding walks on one dimensional lattices Commun. Statist-Stochastic Models, **6** (2), 189-212
2. Alvarez J, Janse van Rensburg E J, Soteros E G and Whittington, S G (2008) Self avoiding polygons and walks in slits J. Phys A: Math. Gen, **41**
3. Bousquet-Mélou, M, Guttmann, A J and Jensen, I, (2005) Self-avoiding walks crossing a square J. Phys A: Math. Gen, **38** 9159-9181.
4. Daoud, M and de Gennes P G (1977) Statistics of macromolecular solutions trapped in small pores J. de Physique **38** 85-93
5. deGennes P G (1979) Scaling concepts in Polymer Physics (Ithaca: Cornell University Press)
6. Di Marzio E A and Rubin R J (1971) Adsorption of a chain polymer between two plates J. Chem. Phys, **55** 4318-36.
7. Hammersley J M and Whittington S G (1985) Self-avoiding walks in wedges J. Phys A: Math. Gen, **18** 101-11
8. Janse van Rensburg E J, Orlandini E and Whittington, S G (2006) Self-avoiding walks in a slab: rigorous results J. Phys A: Math. Gen, **39** 13869-13902.
9. Klein, D (1980) Asymptotic distributions for self-avoiding walks constrained to strips, cylinders and tubes J. Stat. Phys **23** 561-86.
10. Martin, R, Orlandini E, Owczarek A L, Rechnitzer A and Whittington, S G Exact enumerations and Monte Carlo results for self avoiding walks in a slab, (2007) J. Phys A: Math. Gen, **40** 7509-21.
11. Soteros C E and Whittington S G (1988) Polygons and stars in a slit geometry J. Phys A: Math. Gen, **21** L857-861.
12. Stilck J F and Machado K D Tension of polymers in a strip (1998) Eur. Phys. J. B **5** 899-904.
13. Wall, F T, Mandel, F and Chin J C (1976) Self-avoiding random walks subject to external spatial constraints J. Chem. Phys, **65** 2231-2234.
14. Wall, F T, Seitz W A and Chin J C (1977) Self-avoiding walks subject to boundary constraints J. Chem. Phys, **67** 434-8.
15. Wall, F T, Seitz W A Chin J C and deGennes P G (1978) Statistics of self-avoiding walks confined to strips and capillaries Proc. Nat. Acad. Sci **75** no 5 2069-70.

# Chapter 11
# Limit Distributions and Scaling Functions

Christoph Richard

## 11.1 Introduction

For a given combinatorial class of objects, such as polygons or polyhedra, the most basic question concerns the number of objects of a given size (always assumed to be finite), or an asymptotic estimate thereof. Informally stated, in this overview we will analyse the refined question:

What does a typical object look like?

In contrast to the combinatorial question about the number of objects of a given size, the latter question is of a probabilistic nature. For counting parameters in addition to object size, one asks for their (asymptotic) probability law. To give this question a meaning, an underlying ensemble has to be specified. The simplest choice is the uniform ensemble, where each object of a given size occurs with equal probability.

For self-avoiding polygons on the square lattice, size may be the number of edges of the polygon, and an additional counting parameter may be the area enclosed by the polygon. We will call this ensemble the *fixed perimeter ensemble*. For the *uniform* fixed perimeter ensemble, one assumes that, for a fixed number of edges, each polygon occurs with the same probability. Another ensemble, which we will call the *fixed area ensemble*, is obtained with size being the polygon area, and the number of edges being an additional counting parameter. For the *uniform* fixed area ensemble, one assumes that, for fixed area, each polygon occurs with the same probability.

To be specific, let $p_{m,n}$ denote the number of square lattice self-avoiding polygons of half-perimeter $m$ and area $n$. Discrete random variables $\widetilde{X}_m$ of area in the uniform fixed perimeter ensemble and of perimeter $\widetilde{Y}_n$ in the uniform fixed area ensemble are defined by

Christoph Richard
Fakultät für Mathematik, Universität Bielefeld, Postfach 10 01 31, 33501 Bielefeld, Germany, e-mail: richard@math.uni-bielefeld.de

$$\mathbb{P}(\widetilde{X}_m = n) = \frac{p_{m,n}}{\sum_n p_{m,n}}, \qquad \mathbb{P}(\widetilde{Y}_n = m) = \frac{p_{m,n}}{\sum_m p_{m,n}}.$$

We are interested in an asymptotic description of these probability laws, in the limit of infinite object size.

In statistical physics, certain non-uniform ensembles are important. For fixed object size, the probability of an object with value $n$ of the counting parameter (such as the area of a polygon) may be proportional to $a^n$, for some non-negative parameter $a = e^{-\beta E}$ of non-uniformity. Here $E$ is the energy of the object, and $\beta = 1/(k_B T)$, where $T$ is the temperature, and $k_B$ denotes Boltzmann's constant. A qualitative change in the behaviour of typical objects may then be reflected in a qualitative change in the probability law of the counting parameter w.r.t. $a$. Such a change is an indication of a phase transition, i.e., a non-analyticity in the free energy of the corresponding ensemble.

For self-avoiding polygons in the fixed perimeter ensemble, let $q$ denote the parameter of non-uniformity,

$$\mathbb{P}(\widetilde{X}_m(q) = n) = \frac{p_{m,n} q^n}{\sum_n p_{m,n} q^n}.$$

Polygons of large area are suppressed in probability for small values of $q$, such that one expects a typical self-avoiding polygon to closely resemble a branched polymer. Likewise, for large values of $q$, a typical polygon is expected to be inflated, closely resembling a ball (or square) shape. Let us define the *ball-shaped phase* by the condition that the mean area of a polygon grows quadratically with its perimeter. The ball-shaped phase occurs for $q > 1$ [31]. Linear growth of the mean area w.r.t. perimeter is expected to occur for all values $0 < q < 1$. This phase is called the *branched polymer phase*. Of particular interest is the point $q = 1$, at which a phase transition occurs [31]. This transition is called a *collapse transition*. Similar considerations apply for self-avoiding polygons in the fixed area ensemble,

$$\mathbb{P}(\widetilde{Y}_n(x) = m) = \frac{p_{m,n} x^m}{\sum_m p_{m,n} x^m},$$

with parameter of non-uniformity $x$, where $0 < x < \infty$.

For a given model, these effects may be studied using data from exact or Monte-Carlo enumeration and series extrapolation techniques. Sometimes, the underlying model is exactly solvable, i.e., it obeys a combinatorial decomposition, which leads to a recursion for the counting parameter. In that case, its (asymptotic) behaviour may be extracted from the recurrence.

A convenient tool is generating functions. The combinatorial information about the number of objects of a given size is coded in a one-variable (ordinary) generating function, typically of positive and finite radius of convergence. Given the generating function of the counting problem, the asymptotic behaviour of its coefficients can be inferred from the leading singular behaviour of the generating function. This is determined by the location and nature of the singularity of the generating function closest to the origin. There are elaborate techniques for studying this behaviour exactly [37] or numerically [43], see chapter 8.

The case of additional counting parameters leads to a multivariate generating function. For self-avoiding polygons, the half-perimeter and area generating function is

$$P(x,q) = \sum_{m,n} p_{m,n} x^m q^n.$$

For a fixed value of a non-uniformity parameter $q_0$, where $0 < q_0 \leq 1$, let $x_0$ be the radius of convergence of $P(x,q_0)$. The asymptotic law of the counting parameter is encoded in the singular behaviour of the generating function $P(x,q)$ about $(x_0,q_0)$. If locally about $(x_0,q_0)$ the nature of the singularity of $P(x,q)$ does not change, then distributions are expected to be concentrated, with a Gaussian limit law. This corresponds to the physical intuition that fluctuations of macroscopic quantities are asymptotically negligible away from phase transition points. If the nature of the singularity does change locally, we expect non-concentrated distributions, resulting in non-Gaussian limit laws. This is expected to be the case at phase transition points.

Qualitative information about the singularity structure is given by the *singularity diagram* (also called the *phase diagram*), compare chapter 2. It displays the region of convergence of the two-variable generating function, i.e., the set of points $(x,q)$ in the closed upper right quadrant of the plane, such that the generating function $P(x,q)$ converges. The set of boundary points with positive coordinates is a set of singular points of $P(x,q)$, called the *critical curve*. See Fig. 11.1 for a sketch of the singularity diagram of a typical polygon model such as self-avoiding polygons, counted by half-perimeter and area, with generating function $P(x,q)$ as above. There

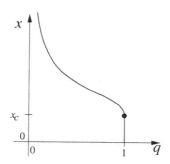

**Fig. 11.1** Singularity diagram of a typical polygon model counted by half-perimeter and area, with $x$ conjugate to half-perimeter and $q$ conjugate to area.

appear two lines of singularities, which intersect at the point $(x,q) = (x_c,1)$. Here $x_c$ is the radius of convergence of the half-perimeter generating function $P(x,1)$, also called the *critical point*. The nature of a singularity does not change along each of the two lines, and the intersection point $(x,q) = (x_c,1)$ of the two lines is a phase transition point. For $0 < q < 1$ fixed, denote by $x_c(q)$ the radius of convergence of $P(x,q)$. The branched polymer phase for the fixed perimeter ensemble $0 < q < 1$

(and also for the corresponding fixed area ensemble) is asymptotically described by the singularity of $P(x,q)$ about $(x_c(q),q)$. In the ball-shaped phase $q > 1$ of the fixed perimeter ensemble, the (ordinary) generating function does not seem the right object to study, since it has zero radius of convergence for fixed $q > 1$. The singularity of $P(x,q)$ about $(x,1)$ describes, for $0 < x < x_c$, a ball-shaped phase in the fixed area ensemble, with a *finite* average size of a ball.

For points $(x,q)$ within the region of convergence, both $x$ and $y$ positive, the generating function $P(x,q)$ is finite and positive. Thus, such points may be interpreted as parameters in a mixed infinite ensemble

$$\mathbb{P}(\widetilde{X}(x,q) = (m,n)) = \frac{p_{m,n}x^m q^n}{\sum_{m,n} p_{m,n}x^m q^n}.$$

The limiting law of the counting parameter in the fixed area or fixed perimeter ensemble can be extracted from the leading singular behaviour of the two-variable generating function. There are two different approaches to the problem. The first one consists in analysing, for fixed non-uniformity parameter $a$, the singular behaviour of the remaining one-parameter generating function and its derivatives w.r.t. $a$. This method is also called the *method of moments*. It can be successfully applied in the fixed perimeter ensemble at the phase transition point. Typically, this results in non-concentrated distributions.

The second approach derives an asymptotic approximation of the two-variable generating function. Away from a phase transition point, such an approximation can be obtained for some classes of models, typically resulting in concentrated distributions, with a Gaussian law for the centred and normalised random variable. However, it is usually difficult to extract such information at a phase transition point. The theory of tricritical scaling seeks to fill this gap, by suggesting and justifying a particular ansatz for an approximation using scaling functions. Knowledge of the approximation may imply knowledge of the quantities analysed in the first approach.

In the following, we give an overview of these two approaches. For the first approach, summarised by the title *limit distributions*, there are a number of rigorous results, which we will discuss. The second approach, summarised by the title *scaling functions*, is less developed. For that reason, our presentation will be more descriptive, stating important open questions. We will stress connections between the two approaches, thereby providing a probabilistic interpretation of scaling functions in terms of limit distributions.

## 11.2 Polygon Models and Generating Functions

Models of polygons, polyominoes or polyhedra have been studied intensively on the square and cubic lattices. It is believed that the leading asymptotic behaviour of such models, such as the type of limit distribution or critical exponents, is independent of the underlying lattice.

In two dimensions, a number of models of square lattice polygons have been enumerated according to perimeter and area and other parameters, see chapter 3 and [7] for a review of models with an exact solution. The majority of such models has an algebraic perimeter generating function. We mention prudent polygons [96, 22, 8] as a notable exception. Of particular importance for polygon models is the fixed perimeter ensemble, since it models two-dimensional vesicle collapse. Another important ensemble is the fixed area ensemble, which serves as a model of ring polymers. The fixed area ensemble may also describe percolation and cluster growth. For example, staircase polygons are models of *directed compact percolation* [26, 28, 29, 27, 12, 57]. This may be compared to the exactly solvable case of percolation on a tree [42]. The model of self-avoiding polygons is conjectured to describe the hull of critical percolation clusters [60].

In addition to perimeter, other counting parameters have been studied, such as width and height, generalisations of area [89], radius of gyration [53, 64], number of nearest-neighbour interactions [4], last column height [7], and site perimeter [20, 11]. Also, motivated by applications in chemistry, symmetry subclasses of polygon models have been analysed [63, 62, 40, 95]. Whereas this gives rise to a number of different ensembles, only a few of them have been asymptotically studied. Not all of them display phase transitions.

In three dimensions, models of polyhedra on the cubic lattice have been enumerated according to perimeter, surface area and volume, see [74, 102, 3] and the discussion in section 11.3.9. Various ensembles may be defined, such as the fixed surface area ensemble and the fixed volume ensemble. The fixed surface area ensemble serves as a model of three-dimensional vesicle collapse [104].

In this chapter, we will consider models of square lattice polygons, counted by half-perimeter and area. Let $p_{m,n}$ denote the (finite) number of such polygons of half-perimeter $m$ and area $n$. The numbers $p_{m,n}$ will always satisfy the following assumption.

**Assumption 1.** *For $m, n \in \mathbb{N}_0$, let non-negative integers $p_{m,n} \in \mathbb{N}_0$ be given. The numbers $p_{m,n}$ are assumed to satisfy the following properties.*

*i) There exist positive constants $A, B > 0$ such that $p_{m,n} = 0$ if $n \leq Am$ or if $n \geq Bm^2$.*
*ii) The sequence $\left( \sum_n p_{m,n} \right)_{m \in \mathbb{N}_0}$ has infinitely many positive elements and grows at most exponentially.*

**Remarks.** *i)* A sequence $(a_n)_{n \in \mathbb{N}_0}$ is said to *grow at most exponentially*, if there are positive constants $C$, $\mu$ such that $|a_n| \leq C\mu^n$ for all $n$.
*ii)* Condition *i)* reflects the geometric constraint that the area of a polygon grows at most quadratically and at least linearly with its perimeter. For self-avoiding polygons, we have $n \geq m - 1$. Since $p_{m,n} = 0$ if $m < 2$, we may choose $A = 1/3$. Since $n \leq m^2/4$ for self-avoiding polygons, we may choose $B = 1/3$. Condition *ii)* is a natural condition on the growth of the number of polygons of a given perimeter. For self-avoiding polygons, we may choose $C = 1$ and $\mu = 16$.
*iii)* For models with counting parameters different from area, or for models in higher dimensions, a modified assumption holds, with the growth condition *i)* being re-

placed by $n \leq Am^{k_0}$ and $n \geq Bm^{k_1}$, for appropriate values of $k_0$ and $k_1$. Counting parameters satisfying $p_{m,n} = 0$ for $n \geq Bm^k$ are called *rank k parameters* [25].

The above assumption imposes restrictions on the generating function of the numbers $p_{m,n}$. These explain the qualitative form of the singularity diagram Fig. 11.1.

**Proposition 1.** *For numbers $p_{m,n}$, let Assumption 1 be satisfied. Then, the generating function $P(x,q) = \sum_{m,n} p_{m,n} x^m q^n$ has the following properties.*

*i) The generating function $P(x,q)$ satisfies for $k \in \mathbb{N}$*

$$A^k \left( x \frac{\partial}{\partial x} \right)^k P(x,q) \ll \left( q \frac{\partial}{\partial q} \right)^k P(x,q) \ll B^k \left( x \frac{\partial}{\partial x} \right)^{2k} P(x,q),$$

*where $\ll$ denotes coefficient-wise domination.*
*ii) The evaluation $P(x,1)$ is a power series with radius of convergence $x_c$, where $0 < x_c \leq 1$.*
*iii) The generating function $P(x,q)$ diverges, if $x \neq 0$ and $|q| > 1$. It converges, if $|q| < 1$ and $|x| < x_c q^{-A}$. In particular, for $k \in \mathbb{N}_0$, the evaluations*

$$\frac{\partial^k}{\partial x^k} P(x,q) \bigg|_{x=x_c}$$

*are power series with radius of convergence 1.*
*iv) For $k \in \mathbb{N}_0$, the evaluations*

$$\frac{\partial^k}{\partial q^k} P(x,q) \bigg|_{q=1}$$

*are power series with radius of convergence $x_c$. They satisfy, for $|x| < x_c$,*

$$\frac{\partial^k}{\partial q^k} P(x,q) \bigg|_{q=1} = \lim_{\substack{q \to 1 \\ -1 < q < 1}} \frac{\partial^k}{\partial q^k} P(x,q).$$

*Proof (sketch).* The domination formula follows immediately from condition *i)*. The existence of the evaluations at $q = 1$ and $x = x_c$ as formal power series also follows from condition *i)*. Condition *ii)* ensures that $0 < x_c \leq 1$ for the radius of convergence of $P(x,1)$. Equality of the radii of convergence for the derivatives follows from condition *i)* by elementary estimates. The claimed analytic properties of $P(x,q)$ follow from conditions *i)* and *ii)* by elementary estimates. The claimed left-continuity of the derivatives in *iv)* is implied by Abel's continuity theorem for real power series.
□

**Remarks.** *i)* Proposition 1 implies that the critical curve $x_c(q)$ satisfies for $0 < q < 1$ the estimate $x_c(q) \geq x_c q^{-A}$. For self-avoiding polygons, the critical curve $x_c(q)$ is continuous for $0 < q < 1$. This follows from a certain supermultiplicative inequality for the numbers $p_{m,n}$ by convexity arguments [48].
*ii)* Of central importance in the sequel will be the power series

$$g_k(x) = \frac{1}{k!} \frac{\partial^k}{\partial q^k} P(x,q) \Big|_{q=1}. \tag{11.1}$$

They are called *factorial moment generating functions*, for reasons which will become clear later.

We continue studying analytic properties of the factorial moment generating functions. In the following, the notation $x \nearrow x_0$ denotes the limit $x \to x_0$ for sequences $(x_n)$ satisfying $|x_n| < x_0$. The notation $f(x) \sim g(x)$ as $x \nearrow x_0$ means that $g(x) \neq 0$ in a left neighbourhood of $x_0$ and that $\lim_{x \nearrow x_0} f(x)/g(x) = 1$. Likewise, $a_m \sim b_m$ as $m \to \infty$ for sequences $(a_m), (b_m)$ means that $b_m \neq 0$ for almost all $m$ and $\lim_{m \to \infty} a_m/b_m = 1$. The following lemma is a standard result.

**Lemma 1.** *Let $(a_m)_{m \in \mathbb{N}_0}$ be a sequence of real numbers, which asymptotically satisfy*

$$a_m \sim A x_c^{-m} m^{\gamma-1} \qquad (m \to \infty), \tag{11.2}$$

*for real numbers $A, x_c, \gamma$, where $A \neq 0$ and $x_c > 0$.*
*Then, the generating function $g(x) = \sum_{m=0}^{\infty} a_m x^m$ has radius of convergence $x_c$. If $\gamma \notin \{0, -1, -2, \ldots\}$, then there exists a power series $g^{(reg)}(x)$ with radius of convergence strictly larger than $x_c$, such that $g(x)$ satisfies*

$$\left( g(x) - g^{(reg)}(x) \right) \sim \frac{A\Gamma(\gamma)}{(1 - x/x_c)^{\gamma}} \qquad (x \nearrow x_c), \tag{11.3}$$

*where $\Gamma(z)$ denotes the Gamma function.*

**Remarks.** i) The above lemma can be proved using the analytic properties of the polylog function [32]. If $\gamma \in \{0, -1, -2, \ldots\}$, an asymptotic form similar to Eq. (11.3) is valid, which involves logarithms.
ii) The function $g^{(reg)}(x)$ in the above lemma is not unique. For example, if $\gamma > 0$, any polynomial in $x$ may be chosen. We demand $g^{(reg)}(x) \equiv 0$ in that case. If $\gamma < 0$ and $g^{(reg)}(x)$ is restricted to be a polynomial, it is uniquely defined. If $-1 < \gamma < 0$, we have $g^{(reg)}(x) \equiv g(x_c)$. In the general case, the polynomial has degree $\lfloor -\gamma \rfloor$, compare [32]. In the following, we will demand uniqueness by the above choice. The power series $g^{(sing)}(x) := \left( g(x) - g^{(reg)}(x) \right)$ is then called *the singular part* of $g(x)$.

Conversely, let a power series $g(x)$ with radius of convergence $x_c$ be given. In order to conclude from Eq. (11.3) the behaviour Eq. (11.2), certain additional analyticity assumptions on $g(x)$ have to be satisfied. To this end, a function $g(x)$ is called $\Delta(x_c, \eta, \phi)$-*regular* (or simply $\Delta$-*regular*) [30], if there is a positive real number $x_c > 0$, such that $g(x)$ is analytic in the *indented disc* $\Delta(x_c, \eta, \phi) := \{z \in \mathbb{C} : |z| \leq x_c + \eta, |\text{Arg}(z - x_c)| \geq \phi\}$, for some $\eta > 0$ and some $\phi$, where $0 < \phi < \pi/2$. Note that $x_c \notin \Delta$, where we adopt the convention $\text{Arg}(0) = 0$. The point $x = x_c$ is the only point for $|x| \leq x_c$, where $g(x)$ may possess a singularity.

**Lemma 2 ([35]).** *Let the function $g(x)$ be $\Delta$-regular and assume that*

$$g(x) \sim \frac{1}{(1 - x/x_c)^\gamma} \qquad (x \to x_c \text{ in } \Delta).$$

If $\gamma \notin \{0, -1, -2, \ldots\}$, we then have

$$[x^m]g(x) \sim \frac{1}{\Gamma(\gamma)} x_c^{-m} m^{\gamma-1} \qquad (m \to \infty),$$

where $[x^m]g(x)$ denotes the Taylor coefficient of $g(x)$ of order $m$ about $x = 0$.

**Remarks.** *i)* Note that the coefficients of the function $f(x) = (1 - x/x_c)^{-\gamma}$ with real exponent $\gamma \notin \{0, -1, -2, \ldots\}$ satisfy

$$[x^m]f(x) \sim \frac{1}{\Gamma(\gamma)} x_c^{-m} m^{\gamma-1} \qquad (m \to \infty). \tag{11.4}$$

This may be seen by an application of the binomial series and Stirling's formula. For functions $g(x) \sim f(x)$, the assumption of $\Delta$-regularity for $g(x)$ ensures that the same asymptotic estimate holds for the coefficients of $g(x)$.
*ii)* Theorems of the above type are called *transfer theorems* [35, 37]. The set of $\Delta$-regular functions with singularities of the above form is closed under addition, multiplication, differentiation, and integration [30].
*iii)* The case of a finite number of singularities on the circle of convergence can be treated by a straightforward extension of the above result [35, 37].

Lemma 1 implies a particular singular behaviour of the factorial moment generating functions, if the numbers $p_{m,n}$ satisfy certain typical asymptotic estimates. We write $(a)_k = a \cdot (a - 1) \cdot \ldots \cdot (a - k + 1)$ to denote the lower factorial.

**Proposition 2.** *For $m, n \in \mathbb{N}_0$, let real numbers $p_{m,n}$ be given. Assume that the numbers $p_{m,n}$ asymptotically satisfy, for $k \in \mathbb{N}_0$,*

$$\frac{1}{k!} \sum_n (n)_k p_{m,n} \sim A_k x_c^{-m} m^{\gamma_k - 1} \qquad (m \to \infty), \tag{11.5}$$

*for real numbers $A_k, x_c, \gamma_k$, where $A_k > 0$, $x_c > 0$, and $\gamma_k \notin \{0, -1, -2, \ldots\}$. Then, the factorial moment generating functions $g_k(x)$ satisfy*

$$g_k^{(sing)}(x) \sim \frac{f_k}{(1 - x/x_c)^{\gamma_k}} \qquad (x \nearrow x_c), \tag{11.6}$$

*where $f_k = A_k \Gamma(\gamma_k)$.*

**Remarks.** *i)* The above assumption on the growth of the coefficients in Eq. (11.5) is typical for polygon models, with $\gamma_k = (k - \theta)/\phi$, and $\phi > 0$.
*ii)* If the numbers $p_{m,n}$ satisfy, in addition to Eq.(11.5), Condition *i)* of Assumption 1, this implies for exponents of the form $\gamma_k = (k - \theta)/\phi$, where $\phi > 0$, the estimate $1/2 \le \phi \le 1$.

*iii)* The proposition implies that the singular part of the factorial moment generating function $g_k(x)$ is asymptotically equal to the singular part of the corresponding (ordinary) moment generating function,

$$\left(\frac{\partial^k}{\partial q^k}P(x,q)\Big|_{q=1}\right)^{(sing)} \sim \left(\left(q\frac{\partial}{\partial q}\right)^k P(x,q)\Big|_{q=1}\right)^{(sing)} \qquad (x \nearrow x_c).$$

We give a list of exponents and area limit distributions for a number of polygon models. An asterisk denotes that corresponding results rely on a numerical analysis. It appears that the value $(\theta, \phi) = (1/3, 2/3)$ arises for a large number of models. Furthermore, the exponent $\gamma_0$ seems to determine the area limit law. These two observations will be explained in the following section.

**Table 11.1** Exponents and area limit laws for prominent polygon models. An asterisk denotes a numerical analysis.

| Model | $\phi$ | $\theta$ | $\gamma_0$ | Area limit law |
|---|---|---|---|---|
| rectangles<br>convex polygons | $\frac{1}{2}$ | $-1$ | $2$ | $\beta_{1,1/2}$ |
| Ferrers diagrams<br>stacks | $\frac{1}{2}$ | $-\frac{1}{2}$ | $1$ | Gaussian |
| staircase polygons<br>bargraph polygons<br>column-convex polygons<br>directed column-convex polygons<br>diagonally convex directed polygons<br>rooted self-avoiding polygons* | $\frac{2}{3}$ | $\frac{1}{3}$ | $-\frac{1}{2}$ | Airy |
| directed convex polygons | $\frac{2}{3}$ | $-\frac{1}{3}$ | $\frac{1}{2}$ | meander |
| diagonally convex polygons* | | | $-\frac{1}{2}$ | |
| three-choice polygons | | | $0$ | |

## 11.3 Limit Distributions

In this section, we will concentrate on models of square lattice polygons in the fixed perimeter ensemble, and analyse their area law. The uniform ensemble is of particular interest, since non-Gaussian limit laws usually appear, due to expected phase transitions at $q = 1$. For non-uniform ensembles $q \neq 1$, Gaussian limit laws are expected, due to the absence of phase transitions.

There are effective techniques for the uniform ensemble, since the relevant generating functions are typically algebraic. This is different from the fixed area ensemble, where singularities are more difficult to analyse. It will turn out that the dominant singularity of the perimeter generating function determines the limiting area law of the model. We will first discuss several examples with different type of singularity. Then, we will describe a general result, by analysing classes of $q$-difference equations (see e.g. [103]), which exactly solvable polygon models obey. Whereas in the case $q \neq 1$ their theory is developed to some extent, the case $q = 1$ is more difficult to analyse. Motivated by the typical behaviour of polygon models, we assume that a $q$-difference equation reduces to an algebraic equation as $q$ approaches unity, and then analyse the behaviour of its solution about $q = 1$.

Useful background concerning a probabilistic analysis of counting parameters of combinatorial structures can be found in [37, Ch IX]. See [80, Ch 1] and [5, Ch 1] for background about asymptotic expansions. For properties of formal power series, see [39, Ch 1.1]. A useful reference on the Laplace transform, which will appear below, is [23].

## 11.3.1 An Illustrative Example: Rectangles

### 11.3.1.1 Limit Law of Area

Let $p_{m,n}$ denote the number of rectangles of half-perimeter $m$ and area $n$. Consider the uniform fixed perimeter ensemble, with a discrete random variable of area $\widetilde{X}_m$ defined by

$$\mathbb{P}(\widetilde{X}_m = n) = \frac{p_{m,n}}{\sum_n p_{m,n}}. \tag{11.7}$$

The $k$-th moments of $\widetilde{X}_m$ are given explicitly by

$$\mathbb{E}[\widetilde{X}_m^k] = \sum_{l=1}^{m-1} (l(m-l))^k \frac{1}{m-1}$$
$$\sim m^{2k} \int_0^1 (x(1-x))^k \mathrm{d}x = \frac{(k!)^2}{(2k+1)!} m^{2k} \qquad (m \to \infty),$$

where we approximated the Riemann sum by an integral, using the Euler-MacLaurin summation formula. Thus, the random variable $\widetilde{X}_m$ has mean $\mu_m \sim m^2/6$ and variance $\sigma_m^2 \sim m^4/180$. Since the sequence of random variables $(\widetilde{X}_m)$ does not satisfy the concentration property $\lim_{m \to \infty} \sigma_m/\mu_m = 0$, we expect a non-trivial limiting distribution. Consider the normalised random variable

$$X_m = \frac{2}{3} \frac{\widetilde{X}_m}{\mu_m} = 4 \frac{\widetilde{X}_m}{m^2}. \tag{11.8}$$

Since the moments of $X_m$ converge as $m \to \infty$, and the limit sequence $M_k := \lim_{m \to \infty} \mathbb{E}[X_m^k]$ satisfies the Carleman condition $\sum_k (M_{2k})^{-1/(2k)} = \infty$, they define [17, Ch 4.5] a unique random variable $X$ with moments $M_k$. Its moment generating function $M(t) = \mathbb{E}[e^{-tX}]$ is readily obtained as

$$M(t) = \sum_{k=0}^{\infty} \frac{\mathbb{E}[X^k]}{k!}(-t)^k = \frac{1}{2}\sqrt{\frac{\pi}{t}}\, e^t \,\mathrm{erf}\left(\sqrt{t}\right).$$

The corresponding probability distribution $p(x)$ is obtained by an inverse Laplace transform, and is given by

$$p(x) = \begin{cases} \frac{1}{2\sqrt{1-x}} & 0 \leq x \leq 1 \\ 0 & x > 1 \end{cases}. \tag{11.9}$$

This distribution is known as the beta distribution $\beta_{1,1/2}$. Together with [17, Thm 4.5.5], we arrive at the following result.

**Theorem 1.** *The area random variable $\widetilde{X}_m$ of rectangles Eq. (11.7) has mean $\mu_m \sim m^2/6$ and variance $\sigma_m^2 \sim m^4/180$. The normalised random variables $X_m$ Eq. (11.8) converge in distribution to a continuous random variable with limit law $\beta_{1,1/2}$ Eq. (11.9). We also have moment convergence.*

### 11.3.1.2 Limit Law via Generating Functions

We now extract the limit distribution using generating functions. Whereas the derivation is less direct than the previous approach, the method applies to a number of other cases, where a direct approach fails. Consider the half-perimeter and area generating function $P(x,q)$ for rectangles,

$$P(x,q) = \sum_{m,n} p_{m,n} x^m q^n.$$

The factorial moments of the area random variable $\widetilde{X}_m$ Eq. (11.7) are obtained from the generating function via

$$\mathbb{E}[(\widetilde{X}_m)_k] = \frac{\sum_n (n)_k p_{m,n}}{\sum_n p_{m,n}} = \frac{[x^m]\frac{\partial^k}{\partial q^k}P(x,q)\big|_{q=1}}{[x^m]P(x,1)},$$

where $(a)_k = a \cdot (a-1) \cdot \ldots \cdot (a-k+1)$ is the lower factorial. The generating function $P(x,q)$ satisfies [87, Eq. 5.1] the linear $q$-difference equation [103]

$$P(x,q) = x^2 q P(qx,q) + \frac{x^2 q(1+qx)}{1-qx}. \tag{11.10}$$

Due to the particular structure of the functional equation, the area moment generating functions

$$g_k(x) = \frac{1}{k!}\frac{\partial^k}{\partial q^k}P(x,q)\bigg|_{q=1}$$

are rational functions and can be computed recursively from the functional equation, by repeated differentiation w.r.t. $q$ and then setting $q=1$. (Such calculations are easily performed with a computer algebra system.) This gives, in particular,

$$g_0(x) = \frac{x^2}{(1-x)^2}, \qquad g_1(x) = \frac{x^2}{(1-x)^4},$$

$$g_2(x) = \frac{2x^3}{(1-x)^6}, \qquad g_3(x) = \frac{6x^4}{(1-x)^8},$$

$$g_4(x) = \frac{x^4(1+22x+x^2)}{(1-x)^{10}}, \qquad g_5(x) = \frac{12x^5(1+8x+x^2)}{(1-x)^{12}}.$$

Whereas the exact expressions get messy for increasing $k$, their asymptotic form about their singularity $x_c = 1$ is simply given by

$$g_k(x) \sim \frac{k!}{(1-x)^{2k+2}} \qquad (x \to 1). \tag{11.11}$$

The above result can be inferred from the functional equation, which induces a recursion for the functions $g_k(x)$, which in turn can be asymptotically analysed. This method is called *moment pumping* [36]. Below, we will extract the above asymptotic behaviour by the method of dominant balance.

The asymptotic behaviour of the moments of $\widetilde{X}_m$ can be obtained from singularity analysis of generating functions, as described in Lemma 2. Using the functional equation, it can be shown that all functions $g_k(x)$ are Laurent series about $x = 1$, with a finite number of terms. Hence the remark following Lemma 2 implies for the (factorial) moments of the random variable $X_m$ Eq. (11.8) the expression

$$\frac{\mathbb{E}[(X_m)^k]}{k!} \sim \frac{\mathbb{E}[(X_m)_k]}{k!} \sim \frac{k!}{\Gamma(2k+2)} = \frac{k!}{(2k+1)!} \qquad (m \to \infty),$$

in accordance with the previous derivation.

On the level of the moment generating function, an application of Watson's lemma [5, Sec 4.1] shows that the coefficients $k!$ in Eq. (11.11) appear in the asymptotic expansion of a certain Laplace transform of the (entire) moment generating function $\mathbb{E}[e^{-tX}]$,

$$\int_0^\infty e^{-st} \left( \sum_{k \geq 0} \frac{\mathbb{E}[X^k]}{k!} (-t^2)^k \right) t\,dt \sim \sum_{k \geq 0} (-1)^k k! s^{-(2k+2)} \qquad (s \to \infty).$$

Note that the r.h.s. is formally obtained by term-by-term integration of the l.h.s..

Using the arguments of [46, Ch 8.11], one concludes that there exists an $s_0 > 0$, such that there is a *unique* function $F(s)$ analytic for $\Re(s) \geq s_0$ with the above asymptotic expansion. It is given by

$$F(s) = \mathrm{Ei}(s^2) e^{s^2}, \tag{11.12}$$

where $\text{Ei}(z) = \int_1^\infty e^{-tz}/t\,dt$ is the exponential integral. The moment generating function $M(t) = \mathbb{E}[e^{-tX}]$ of the random variable $X$ is given by an inverse Laplace transform of $F(s)$,

$$\int_0^\infty e^{-st} M(t^2)t\,dt = F(s).$$

Since there are effective methods for computing inverse Laplace transforms [23], the question arises whether the function $F(s)$ can be easily obtained. It turns out that the functional equation Eq. (11.10) induces a differential equation for $F(s)$. This equation can be obtained in a mechanical way, using the method of dominant balance.

### 11.3.1.3 Dominant Balance

For a given functional equation, the method of dominant balance consists of a certain rescaling of the variables, such that the quantity of interest appears in the expansion of a rescaled variable to leading order. The method was originally used as a heuristic tool in order to extract the scaling function of a polygon model [84] (see the following section). In the present framework, it is a rigorous method.

Consider the half-perimeter and area generating function $P(x,q)$ as a formal power series. The substitution $q = 1 - \widetilde{\varepsilon}$ is valid, since the coefficients of the power series $P(x,q)$ in $x$ are polynomials in $q$. We get the power series in $\widetilde{\varepsilon}$,

$$H(x,\widetilde{\varepsilon}) = \sum_{k\geq 0}(-1)^k g_k(x)\widetilde{\varepsilon}^k.$$

whose coefficients $(-1)^k g_k(x)$ are power series in $x$. The functional equation Eq. (11.10) induces an equation for $H(x,\widetilde{\varepsilon})$, from which the factorial area moment generating functions $g_k(x)$ may be computed recursively.

Now replace $g_k(x)$ by its expansion about $x = 1$,

$$g_k(x) = \sum_{l\geq 0}\frac{f_{k,l}}{(1-x)^{2k+2-l}}.$$

Introducing $\widetilde{s} = 1 - x$, this leads to a power series $E(\widetilde{s},\widetilde{\varepsilon})$ in $\widetilde{\varepsilon}$,

$$E(\widetilde{s},\widetilde{\varepsilon}) = \sum_{k\geq 0}(-1)^k \left(\sum_{l\geq 0}\frac{f_{k,l}}{\widetilde{s}^{2k+2-l}}\right)\widetilde{\varepsilon}^k,$$

whose coefficients are Laurent series in $\widetilde{s}$. As above, the functional equation induces an equation for the power series $E(\widetilde{s},\widetilde{\varepsilon})$ in $\widetilde{\varepsilon}$, from which the expansion coefficients may be computed recursively.

We infer from the previous equation that

$$E(s\varepsilon, \varepsilon^2) = \frac{1}{\varepsilon^2} \sum_{l \geq 0} \left( \sum_{k \geq 0} (-1)^k \frac{f_{k,l}}{s^{2k+2-l}} \right) \varepsilon^l = \frac{1}{\varepsilon^2} F(s, \varepsilon). \tag{11.13}$$

Write $F(s, \varepsilon) = \sum_{l \geq 0} F_l(s)\varepsilon^l$. By construction, the (formal) series $F_0(s) = F(s, 0)$ coincides with the asymptotic expansion of the desired function $F(s)$ Eq. (11.12) about infinity.

The above example suggests a technique for computing $F_0(s)$. The functional equation Eq. (11.10) for $P(x, q)$ induces, after reparametrisation, differential equations for the functions $F_l(s)$, from which $F_0(s)$ may be obtained explicitly. These may be computed by first writing

$$P(x, q) = \frac{1}{1-q} F\left( \frac{1-x}{(1-q)^{1/2}}, (1-q)^{1/2} \right), \tag{11.14}$$

and then introducing variables $s$ and $\varepsilon$, by setting $x = 1 - s\varepsilon$ and $q = 1 - \varepsilon^2$. Expand the equation to leading order in $\varepsilon$. This yields, to order $\varepsilon^0$, the first order differential equation

$$sF_0'(s) + 2 - 2s^2 F_0(s) = 0.$$

The above equation translates into a recursion for the coefficients $f_{k,0}$, from which $f_{k,0} = k!$ can be deduced. In addition, the equation has a unique solution with the prescribed asymptotic behaviour Eq. (11.13), which is given by $F_0(s) = \text{Ei}(s^2) e^{s^2}$.

As we will argue in the next section, Eq. (11.14) is sometimes referred to as a *scaling Ansatz*, the function $F(s, 0)$ appears as a *scaling function*, the functions $F_l(s)$, for $l \geq 1$, appear as *correction-to-scaling functions*. In our formal framework, where the series $F_l(s)$ are rescaled generating functions for the coefficients $f_{k,l}$, their derivation is rigorous.

### 11.3.2 A General Method

In the preceding two subsections, we described a method for obtaining limit laws of counting parameters, via a generating function approach. Since this method will be important in the remainder of this section, we summarise it here. Its first ingredient is based on the so-called method of moments [17, Thm 4.5.5].

**Proposition 3.** *For $m, n \in \mathbb{N}_0$, let real numbers $p_{m,n}$ be given. Assume that the numbers $p_{m,n}$ asymptotically satisfy, for $k \in \mathbb{N}_0$,*

$$\frac{1}{k!} \sum_n (n)_k p_{m,n} \sim A_k x_c^{-m} m^{\gamma_k - 1} \qquad (m \to \infty), \tag{11.15}$$

*where $A_k$ are positive numbers, and $\gamma_k = (k - \theta)/\phi$, with real constants $\theta$ and $\phi > 0$. Assume that the numbers $M_k := A_k/A_0$ satisfy the Carleman condition*

$$\sum_{k=1}^{\infty} (M_{2k})^{-1/(2k)} = +\infty.$$ (11.16)

*Then the following conclusions hold.*

*i) For almost all m, the random variables $\widetilde{X}_m$*

$$\mathbb{P}(\widetilde{X}_m = n) = \frac{p_{m,n}}{\sum_n p_{m,n}}$$ (11.17)

*are well defined. We have*

$$X_m := \frac{\widetilde{X}_m}{m^{1/\phi}} \xrightarrow{d} X,$$ (11.18)

*for a unique random variable X with moments $M_k$, where $d$ denotes convergence in distribution. We also have moment convergence.*

*ii) If the numbers $M_k$ satisfy for all $t \in \mathbb{R}$ the estimate*

$$\lim_{k \to \infty} \frac{M_k t^k}{k!} = 0,$$ (11.19)

*then the moment generating function $M(t) = \mathbb{E}[e^{-tX}]$ of X is an entire function. The coefficients $A_k \Gamma(\gamma_k)$ are related to $M(t)$ by a Laplace transform which has, for $\theta > 0$, the asymptotic expansion*

$$\int_0^{\infty} e^{-st} \left( \sum_{k \geq 0} \frac{\mathbb{E}[X^k]}{k!} (-t^{1/\phi})^k \right) \frac{1}{t^{1-\gamma_0}} \, dt$$
$$\sim \frac{1}{A_0} \sum_{k \geq 0} (-1)^k A_k \Gamma(\gamma_k) s^{-\gamma_k} \qquad (s \to \infty).$$ (11.20)

*Proof (sketch).* A straightforward calculation using Eq. (11.15) leads to

$$\frac{\mathbb{E}[(\widetilde{X}_m)_k]}{k!} \sim \frac{A_k}{A_0} m^{k/\phi} \qquad (m \to \infty).$$

This implies that the same asymptotic form holds for the (ordinary) moments $\mathbb{E}[(\widetilde{X}_m)^k]$. Due to the growth condition Eq. (11.16), the sequence $(M_k)$ defines a unique random variable X with moments $M_k$. Also, moment convergence of the sequence $(X_m)$ to X implies convergence in distribution, see [17, Thm 4.5.5]. Due to the growth condition Eq. (11.19), the function $M(t)$ is entire. Hence the conditions of Watson's Lemma [5, Sec 4.1] are satisfied, and we obtain Eq. (11.20).  □

**Remarks.** *i)* The growth condition Eq. (11.19) implies the Carleman condition Eq. (11.16). All examples below have entire moment generating functions $M(t)$.
*ii)* If $\gamma_0 < 0$, a modified version of Eq. (11.20) can be given, see for example staircase polygons below.

Proposition 2 states that assumption Eq. (11.15) translates, at the level of the half-perimeter and area generating function $P(x,q) = \sum_{m,n} p_{m,n} x^m q^n$, to a certain asymptotic expression for the factorial moment generating functions

$$g_k(x) = \frac{1}{k!} \frac{\partial^k}{\partial q^k} P(x,q) \bigg|_{q=1}.$$

Their asymptotic behaviour follows from Eq. (11.15), and is

$$g_k^{(sing)}(x) \sim \frac{f_k}{(1-x/x_c)^{\gamma_k}} \qquad (x \nearrow x_c),$$

where $f_k = A_k \Gamma(\gamma_k)$. Adopting the generating function viewpoint, the amplitudes $f_k$ determine the numbers $A_k$, hence the moments $M_k = A_k/A_0$ of the limit distribution. The series $F(s) = \sum_{k \geq 0} (-1)^k f_k s^{-\gamma_k}$ will be of central importance in the sequel.

**Definition 1 (Area amplitude series).** Let Assumption 1 be satisfied. Assume that the generating function $P(x,q) = \sum_{m,n} p_{m,n} x^m q^n$ satisfies asymptotically

$$\left( \frac{1}{k!} \frac{\partial^k}{\partial q^k} P(x,q) \bigg|_{q=1} \right)^{(sing)} \sim \frac{f_k}{(1-x/x_c)^{\gamma_k}} \qquad (x \nearrow x_c),$$

with exponents $\gamma_k \notin \{0, -1, -2, \ldots\}$. Then, the formal series

$$F(s) = \sum_{k \geq 0} (-1)^k \frac{f_k}{s^{\gamma_k}}$$

is called the *area amplitude series*.

**Remarks.** *i)* Proposition 3 states that the area amplitude series appears in the asymptotic expansion about infinity of a Laplace transform of the moment generating function of the area limit distribution. The probability distribution of the limiting area distribution is related to $F(s)$ by a double Laplace transform.
*ii)* For typical polygon models, all derivatives of $P(x,q)$ w.r.t. $q$, evaluated at $q = 1$, exist and have the same radius of convergence, see Proposition 1. Typical polygon models do have factorial moment generating functions of the above form, see the examples below.

The second ingredient of the method consists in applying the method of dominant balance. As described above, this may result in a differential equation (or in a difference equation [90]) for the function $F(s)$. Its applicability has to be tested for each given type of functional equation. Typically, it can be applied if the factorial area moment generating functions $g_k(x)$ Eq. (11.1) have, for values $x < x_c$, a local expansion about $x = x_c$ of the form

$$g_k^{(sing)}(x) = \sum_{l \geq 0} \frac{f_{k,l}}{(1-x/x_c)^{\gamma_{k,l}}},$$

where $\gamma_{k,l} = (k - \theta_l)/\phi$ and $\theta_{l+1} > \theta_l$. If a transfer theorem such as Lemma 2 applies, then the differential equation for $F(s)$ induces a recurrence for the moments of the limit distribution. If the differential equation can be solved in closed form, inverse Laplace transform techniques may be applied in order to obtain explicit expressions for the moment generating function and the probability density. Also, higher order corrections to the limiting behaviour may be analysed, by studying the functions $F_l(s)$, for $l \geq 1$. See [87] for examples.

### 11.3.3 Further Examples

Using the general method as described above, area limit laws for the other exactly solved polygon models can be derived. A model with the same area limit law as rectangles is convex polygons, compare [87]. We will discuss some classes of polygon models with different area limit laws.

#### 11.3.3.1 Ferrers Diagrams

In contrast to the previous example, the limit distribution of area of Ferrers diagrams is concentrated.

**Proposition 4.** *The area random variable $\widetilde{X}_m$ of Ferrers diagrams has mean $\mu_m \sim m^2/8$. The normalised random variables $X_m$ Eq. (11.18) converge in distribution to a random variable with density $p(x) = \delta(x - 1/8)$.*

**Remark.** It should be noted that the above convergence statement already follows from the concentration property $\lim_{m \to \infty} \sigma_m/\mu_m = 0$, with $\sigma_m^2 \sim m^3/48$ the variance of $X_m$, by an explicit analysis of the first three factorial moment generating functions. (By Chebyshev's inequality, the concentration property implies convergence in probability, which in turn implies convergence in distribution.) We will give a proof via the moment method in the following proof. This will serve as a transparent example for the methodology introduced above. Moreover, it explains the occurrence of the particular "scaling function" $F(s)$ in Eq. (11.21) below.

*Proof.* Ferrers diagrams, counted by half-perimeter and area, satisfy the linear $q$-difference equation [87, Eq (5.4)]

$$P(x,q) = \frac{qx^2}{(1-qx)^2}P(qx,q) + \frac{qx^2}{(1-qx)^2}.$$

The perimeter generating function $g_0(x) = x^2/(1-2x)$ is obtained by setting $q = 1$ in the above equation. Hence $x_c = 1/2$. Using the functional equation, it can be shown by induction on $k$ that all area moment generating functions $g_k(x)$ are rational in $g_0(x)$ and its derivatives. Hence all $g_k(x)$ are rational functions. Since the area of a polygon grows at most quadratically with the perimeter, we have a bound on the exponent, $\gamma_k \le 2k + 1$, of the leading singular part of $g_k(x)$. Given this bound, the method of dominant balance can be applied. We set

$$P(x,q) = \frac{1}{(1-q)^{\frac{1}{2}}} F\left( \frac{1-2x}{(1-q)^{\frac{1}{2}}}, (1-q)^{\frac{1}{2}} \right),$$

and introduce new variables $s$ and $\varepsilon$ by $q = 1 - \varepsilon^2$ and $2x = 1 - s\varepsilon$. Then an expansion of the functional equation yields, to order $\varepsilon^0$, the ODE of first order $F'(s) = 4sF(s) - 1$, whose unique solution with the prescribed asymptotic behaviour is

$$F(s) = \sqrt{\frac{\pi}{8}} \operatorname{erfc}\left( \sqrt{2}s \right) e^{2s^2}. \tag{11.21}$$

It can be inferred from the differential equation that all coefficients in the asymptotic expansion of $F(s)$ at infinity are nonzero. Hence, the above exponent bound is tight. It can be inferred from the functional equation by induction on $k$ that each $g_k(x)$ is a Laurent polynomial about $x_c = 1/2$. Thus, Lemma 2 applies, and we obtain the moment generating function of the corresponding random variable Eq. (11.18) as $M(s) = \exp(-s/8)$. This is readily recognised as the moment generating function of a probability distribution concentrated at $x = 1/8$. □

A sequence of random variables, which satisfies the concentration property, often leads to a Gaussian limit law, after centring and suitable normalisation. This is also the case for Ferrers diagrams.

**Theorem 2 ([97]).** *The area random variable $\widetilde{X}_m$ of Ferrers diagrams has mean $\mu_m \sim m^2/8$ and variance $\sigma_m^2 \sim m^3/48$. The centred and normalised random variables*

$$X_m = \frac{\widetilde{X}_m - \mu_m}{\sigma_m} \tag{11.22}$$

*converge in distribution to a Gaussian random variable.*

**Remarks.** *i)* It is possible to prove this result by the moment method, combined with the method of dominant balance. The idea of proof consists in studying the functional equation of the generating function for the "centred coefficients" $p_{m,n} - \mu_m$. *ii)* The above arguments can also be applied to stack polygons to yield the concentration property and a central limit theorem.

### 11.3.3.2 Staircase Polygons

The limit law of area of staircase polygons is the Airy distribution. This distribution (see [34] and the survey [52]) is conveniently defined via its moments.

**Definition 2 (Airy distribution [34]).** The random variable $Y$ is said to be Airy distributed if

$$\frac{\mathbb{E}[Y^k]}{k!} = \frac{\Gamma(\gamma_0)}{\Gamma(\gamma_k)} \frac{\phi_k}{\phi_0},$$

where $\gamma_k = 3k/2 - 1/2$, and the numbers $\phi_k$ satisfy, for $k \geq 1$, the quadratic recurrence

$$\gamma_{k-1}\phi_{k-1} + \frac{1}{2} \sum_{l=0}^{k} \phi_l \phi_{k-l} = 0,$$

with initial condition $\phi_0 = -1$.

**Remarks ([34, 58]).** *i)* The first moment is $\mathbb{E}[Y] = \sqrt{\pi}$. The sequence of moments can be shown to satisfy the Carleman condition. Hence the distribution is uniquely determined by its moments.

*ii)* The numbers $\phi_k$ appear in the asymptotic expansion of the logarithmic derivative of the Airy function at infinity,

$$\frac{d}{ds} \log \mathrm{Ai}(s) \sim \sum_{k \geq 0} (-1)^k \frac{\phi_k}{2^k} s^{-\gamma_k} \qquad (s \to \infty),$$

where $\mathrm{Ai}(x) = \frac{1}{\pi} \int_0^\infty \cos(t^3/3 + tx)\, dt$ is the Airy function.

*iii)* Explicit expressions for the numbers $\phi_k$ are known [58]. They are, for $k \geq 1$, given by

$$\phi_k = 2^{k+1} \frac{3}{4\pi^2} \int_0^\infty \frac{x^{3(k-1)/2}}{\mathrm{Ai}(x)^2 + \mathrm{Bi}(x)^2}\, dx,$$

where $\mathrm{Bi}(z)$ is the second standard solution of the Airy differential equation $f''(z) - zf(z) = 0$.

*iv)* The Airy distribution appears in a variety of contexts [34]. In particular, the random variable $Y/\sqrt{8}$ describes the law of the area of a Brownian excursion. See also [76] for an overview from a physical perspective.

Explicit expressions have been derived for the moment generating function of the Airy distribution and for its density.

**Fact 1 ([19, 66, 99, 34]).** *The moment generating function $M(t) = \mathbb{E}[e^{-tY}]$ of the Airy distribution satisfies the modified Laplace transform*

$$\frac{1}{\sqrt{2\pi}} \int_0^\infty (e^{-st} - 1) M(2^{-3/2} t^{3/2}) \frac{1}{t^{3/2}}\, dt = 2^{1/3} \left( \frac{\mathrm{Ai}'(2^{1/3}s)}{\mathrm{Ai}(2^{1/3}s)} - \frac{\mathrm{Ai}'(0)}{\mathrm{Ai}(0)} \right). \qquad (11.23)$$

*The moment generating function $M(t)$ is given explicitly by*

$$M(2^{-3/2}t) = \sqrt{2\pi} t \sum_{k=1}^{\infty} \exp\left(-\beta_k t^{2/3} 2^{-1/3}\right),$$

*where the numbers $-\beta_k$ are the zeros of the Airy function. Its density $p(x)$ is given explicitly by*

$$2^{3/2}p(2^{3/2}x) = \frac{2\sqrt{6}}{x^2}\sum_{k=1}^{\infty}e^{-v_k}v_k^{2/3}U\left(-\frac{5}{6},\frac{4}{3};v_k\right),$$

*where $v_k = 2\beta_k^3/(27x^2)$ and $U(a,b,z)$ is the confluent hypergeometric function.*

**Remarks.** *i)* The confluent hypergeometric function $U(a,b;z)$ is defined as [1]

$$U(a,b;z) = \frac{\pi}{\sin\pi b}\left(\frac{{}_1F_1[a,b;z]}{\Gamma(1+a-b)\Gamma(b)} - \frac{z^{1-b}{}_1F_1[1+a-b,2-b;z]}{\Gamma(a)\Gamma(2-b)}\right),$$

where ${}_1F_1[a;b;z]$ is the hypergeometric function

$$_1F_1[a;b;z] = 1 + \frac{a}{b}\frac{z}{1!} + \frac{a(a+1)}{b(b+1)}\frac{z^2}{2!} + \cdots$$

*ii)* The moment generating function and its density are obtained by two consecutive inverse Laplace transforms of Eq. (11.23), see [67, 68] and [99, 54].
*iii)* In the proof of the following theorem, we will derive Eq. (11.23) using the model of staircase polygons. This shows, in particular, that the coefficients $\phi_k$ appear in the asymptotic expansion of the Airy function.

**Theorem 3.** *The normalised area random variables $X_m$ of staircase polygons Eq. (11.18) satisfy*

$$\frac{X_m}{\sqrt{\pi}/4} \xrightarrow{d} \frac{Y}{\sqrt{\pi}} \quad (m \to \infty),$$

*where $Y$ is Airy distributed according to Definition 2. We also have moment convergence.*

**Remark.** Given the functional equation of the half-perimeter and area generating function of staircase polygons,

$$P(x,q) = \frac{x^2q}{1 - 2xq - P(qx,q)} \tag{11.24}$$

(see [88] for a recent derivation), this result is a special case of Theorem 4 below, which is stated in [25].

*Proof.* We use the method of dominant balance. From the functional equation Eq. (11.24), we infer $g_0(x) = 1/4 + \sqrt{1-4x}/2 + (1-4x)/4$. Hence $x_c = 1/4$. The structure of the functional equation implies that all functions $g_k(x)$ can be written as Laurent series in $s = \sqrt{1-4x}$, see also Proposition 7 below. Explicitly, we get $g_1(x) = x^2/(1-4x)$. This suggests $\gamma_k = (3k-1)/2$. An upper bound of this form on the exponent $\gamma_k$ can be derived without too much effort from the functional equation, by an application of Faà di Bruno's formula, see also [89, Prop (4.4)]. Thus, the method of dominant balance can be applied. We set

$$P(x,q) = \frac{1}{4} + (1-q)^{1/3}F\left(\frac{1-4x}{(1-q)^{2/3}},(1-q)^{1/3}\right)$$

and introduce variables $s, \varepsilon$ by $4x = 1 - s\varepsilon^2$ and $q = 1 - \varepsilon^3$. In the above equation, we excluded the constant $1/4 =: P^{(reg)}(x,q)$, since it does not contribute to the moment asymptotics. Expanding the functional equation to order $\varepsilon^2$ gives the Riccati equation

$$F'(s) + 4F(s)^2 - s = 0. \tag{11.25}$$

It follows that the coefficients $f_k$ of $F(s)$ satisfy, for $k \geq 1$, the quadratic recursion

$$\gamma_{k-1} f_{k-1} + 4 \sum_{l=0}^{k} f_l f_{k-l} = 0,$$

with initial condition $f_0 = -1/2$. A comparison with the definition of the Airy distribution shows that $\phi_k = 2^{2k+1} f_k$. Using the closure properties of $\Delta$-regular functions, it can be inferred from the functional equation that (the analytic continuation of) each factorial moment generating function $g_k(x)$ is $\Delta$-regular, with $x_c = 1/4$, see also Proposition 7 below. Hence the transfer theorem Lemma 2 can be applied. We obtain $4X_m \xrightarrow{d} Y$ in distribution and for moments, where $Y$ is Airy distributed.   $\square$

**Remarks.** *i)* The unique solution $F(s)$ of the differential equation in the above proof Eq. (11.25), satisfying the prescribed asymptotic behaviour, is given by

$$F(s) = \frac{1}{4} \frac{d}{ds} \log Ai(4^{1/3} s). \tag{11.26}$$

The moment generating function $M(t)$ of the limiting random variable $X = \lim_{m \to \infty} X_m$ is related to the function $F(s)$ via the modified Laplace transform

$$\int_0^\infty (e^{-st} - 1) M(t^{3/2}) \frac{1}{t^{3/2}} \, dt = 4\sqrt{\pi}(F(s) - F(0)),$$

where the modification has been introduced in order to ensure a finite integral about the origin. This result relates the above proof to Fact 1.

*ii)* The method of dominant balance can be used to obtain corrections $F_l(s)$ to the limiting behaviour [87].

The fact that the area law of staircase polygons is, up to normalisation, the same as that of the area under a Brownian excursion, suggests that there might be a combinatorial explanation. Indeed, as is well known, there is a bijection [21, 98] between staircase polygons and Dyck paths, a discrete version of Brownian excursions [2], see Fig. 11.2 [88]. Within this bijection, the polygon area corresponds to the sum of peak heights of the Dyck path, but not to the area below the Dyck path. For more about this connection, see the remark at the end of the following subsection.

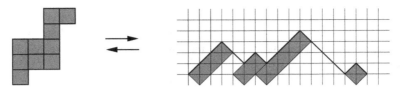

**Fig. 11.2** [88] A combinatorial bijection between staircase polygons and Dyck paths [21, 98]. Column heights of a polygon correspond to peak heights of a path.

### 11.3.4 *q-Difference Equations*

All polygon models discussed above have an algebraic perimeter generating function. Moreover, their half-perimeter and area generating function satisfies a functional equation of the form

$$P(x,q) = G(x,q,P(x,q),P(qx,q)),$$

for a real polynomial $G(x,q,y_0,y_1)$. Since, under mild assumptions on $G$, the equation reduces to an algebraic equation for $P(x,1)$ in the limit $q \to 1$, it may be viewed as a "deformation" of an algebraic equation. In this subsection, we will analyse equations of this type at the special point $(x,q) = (x_c,1)$, where $x_c$ is the radius of convergence of $P(x,1)$. It will appear that the methods used in the above examples also can be applied to this more general case.

The above equation falls into the class of *q-difference equations* [103]. While particular examples appear in combinatorics in a number of places, see e.g. [37], the asymptotic behaviour of equations of the above form seems to have been systematically studied initially in [25, 87]. The study can be done in some generality, e.g., also for non-polynomial power series $G$, for replacements more general than $x \mapsto qx$, and for multivariate generalisations, see [89] and [25]. For simplicity, we will concentrate on polynomial $G$, and then briefly discuss generalisations. Our exposition closely follows [89, 87].

#### 11.3.4.1 Algebraic *q*-Difference Equations

**Definition 3 (Algebraic *q*-difference equation [25, 87]).** An algebraic *q*-difference equation is an equation of the form

$$P(x,q) = G(x,q,P(x,q),P(qx,q),\ldots,P(q^N x,q)), \qquad (11.27)$$

where $G(x,q,y_0,y_1,\ldots,y_N)$ is a complex polynomial. We require that

$$G(0,q,0,0,\ldots,0) \equiv 0, \qquad \frac{\partial G}{\partial y_k}(0,q,0,0,\ldots,0) \equiv 0 \qquad (k = 0,1,\ldots,N).$$

**Remarks.** *i)* See [103] for an overview of the theory of $q$-difference equations. As $q$ approaches unity, the above equation reduces to an algebraic equation.
*ii)* Asymptotics for solutions of algebraic $q$-difference equations have been considered in [25]. The above definition is a special case of [89, Def 2.4], where a multivariate extension is considered, and where $G$ may be non-polynomial. Also, replacements more general than $x \mapsto f(q)x$ are allowed. Such equations are called $q$-functional equations in [89]. The results presented below apply mutatis mutandis also to $q$-functional equations.

The algebraic $q$-difference equation in Definition 3 uniquely defines a (formal) power series $P(x,q)$ satisfying $P(0,q) \equiv 0$. This is shown by analysing the implied recurrence for the coefficients $p_m(q)$ of $P(x,q) = \sum_{m>0} p_m(q)x^m$, see also [89, Prop 2.5]. In fact, $p_m(q)$ is a polynomial in $q$. The growth of its degree in $m$ is not larger than $cm^2$ for some positive constant $c$, hence the counting parameters are rank 2 parameters [25]. In our situation, such a bound holds, since the area of a polygon grows at most quadratically with its perimeter.

From the preceding discussion, it follows that the factorial moment generating functions

$$g_k(x) = \frac{1}{k!} \frac{\partial^k}{\partial q^k} P(x,q)\Big|_{q=1}$$

are well-defined as formal power series. In fact, they can be recursively determined from the $q$-difference equation by implicit differentiation, as a consequence of the following proposition.

**Proposition 5 ([87, 89]).** *Consider the derivative of order $k > 0$ of an algebraic $q$-difference equation Eq. (11.27) w.r.t. $q$, evaluated at $q = 1$. It is linear in $g_k(x)$, and its r.h.s. is a complex polynomial in the power series $g_l(x)$ and its derivatives up to order $k - l$, where $l = 0, \ldots, k$.*

**Remarks.** *i)* This statement can be shown by analysing the $k$-th derivative of the $q$-difference equation, using Faà di Bruno's formula [18].
*ii)* It follows that every function $g_k(x)$ is rational in $g_l(x)$ and its derivatives up to order $k - l$, where $0 \le l < k$. Since $G$ is a polynomial, $g_k(x)$ is *algebraic*, by the closure properties of algebraic functions.

We discuss analytic properties of the (analytic continuations of the) factorial moment generating functions $g_k(x)$. These are determined by the analytic properties of $g_0(x) = P(x,1)$. We discuss the case of a square-root singularity of $P(x,1)$, which often occurs for combinatorial structures, and which is well studied, see e.g. [79, Thm 10.6] or [37, Ch VII.4]. Other cases may be treated similarly. We make the following assumption:

**Assumption 2.** *The $q$-difference equation in Definition 3 has the following properties:*

*i) All coefficients of the polynomial $G(x,q,y_0,y_1,\ldots,y_N)$ are non-negative.*

*ii) The polynomial $Q(x,y) := G(x,1,y,y,\ldots,y)$ satisfies $Q(x,0) \not\equiv 0$ and has degree at least two in $y$.*

*iii) $P(x,1) = \sum_{m\geq 1} p_m x^m$ is aperiodic, i.e., there exist indices $1 \leq i < j < k$ such that $p_i p_j p_k \neq 0$, while $\gcd(j-i,k-i) = 1$.*

**Remarks.** *i)* The positivity assumption is natural for combinatorial constructions. There are, however, $q$-difference equations with negative coefficients, which arise from systems of $q$-difference equations with non-negative coefficients by reduction. Examples are convex polygons [87, Sec 5.4] and directed convex polygons, see below.

*ii)* Assumptions *i)* and *ii)* result in a square-root singularity as the dominant singularity of $P(x,1)$.

*iii)* Assumption *iii)* implies that there is only one singularity of $P(x,1)$ on its circle of convergence. Since $P(x,1)$ has non-negative coefficients only, it occurs on the positive real half-line. The periodic case can be treated by a straightforward extension [37].

An application of the (complex) implicit function theorem ensures that $P(x,1)$ is analytic at the origin. It can be analytically continued, as long as the defining algebraic equation remains invertible. Together with the positivity assumption, one can conclude that there is a number $0 < x_c < \infty$, such that the analytic continuation of $P(x,1)$ satisfies $y_c = \lim_{x \nearrow x_c} P(x,1) < \infty$, with

$$Q(x_c, y_c) = y_c, \quad \frac{\partial}{\partial y} Q(x_c, y) \bigg|_{y=y_c} = 1.$$

With the positivity assumption on the coefficients, it follows that

$$B := \frac{1}{2} \frac{\partial^2}{\partial y^2} Q(x_c, y) \bigg|_{y=y_c} > 0, \quad C := \frac{\partial}{\partial x} Q(x, y_c) \bigg|_{x=x_c} > 0. \tag{11.28}$$

These conditions characterise the singularity of $P(x,1)$ at $x = x_c$ as a square-root. It can be shown that there exists a locally convergent expansion of $P(x,1)$ about $x = x_c$, and that $P(x,1)$ is analytic for $|x| < x_c$. We have the following result. Recall that a function $f(z)$ is $\Delta$-regular if it is analytic in the indented disc $\Delta = \{z : |z| \leq x_c + \eta, |\mathrm{Arg}(z - x_c)| \geq \phi\}$ for some $\eta > 0$ and some $\phi$, where $0 < \phi < \pi/2$.

**Proposition 6 ([79, 37, 89]).** *Given Assumption 2, the power series $P(x,1)$ is analytic at $x = 0$, with radius of convergence $x_c$. Its analytic continuation is $\Delta$-regular, with a square-root singularity at $x = x_c$ and a local Puiseux expansion*

$$P(x,1) = y_c + \sum_{l=0}^{\infty} f_{0,l}(1 - x/x_c)^{1/2+l/2},$$

*where $y_c = \lim_{x \nearrow x_c} P(x,1) < \infty$ and $f_{0,0} = -\sqrt{x_c C/B}$, for constants $B > 0$ and $C > 0$ as in Eq. (11.28). The numbers $f_{0,l}$ can be recursively determined from the $q$-difference equation.*

The asymptotic behaviour of $P(x,1) = g_0(x)$ carries over to the factorial moment generating functions $g_k(x)$.

**Proposition 7 ([89]).** *Given Assumption 2, all factorial moment generating functions $g_k(x)$ are, for $k \geq 1$, analytic at $x = 0$, with radius of convergence $x_c$. Their analytic continuations are $\Delta$-regular, with local Puiseux expansions*

$$g_k(x) = \sum_{l=0}^{\infty} f_{k,l}(1 - x/x_c)^{-\gamma_k + l/2},$$

*where $\gamma_k = 3k/2 - 1/2$. The numbers $f_{k,0} = f_k$ are, for $k \geq 2$, characterised by the recursion*

$$\gamma_{k-1} f_{k-1} + \frac{1}{4f_1} \sum_{l=0}^{k} f_l f_{k-l} = 0,$$

*and the numbers $f_0 < 0$ and $f_1 > 0$ are given by*

$$f_0 = -\sqrt{\frac{Cx_c}{B}}, \qquad 4f_1 = \frac{\sum_{k=1}^{N} k \frac{\partial G}{\partial y_k}(x_c, 1, y_c, y_c, \dots, y_c)}{B}, \tag{11.29}$$

*for constants $B > 0$ and $C > 0$ as in Eq. (11.28).*

**Remarks.** *i)* This result can be obtained by a direct analysis of the $q$-difference equation, applying Faa di Bruno's formula, see also [87, Sec 2.2].
*ii)* Alternatively, it can be obtained by applying the method of dominant balance to the $q$-difference equation. To this end, one notes that all functions $g_k(x)$ are Laurent series in $\sqrt{1 - x/x_c}$, and that their leading exponents are bounded from above by $\gamma_k$. (An upper bound on an exponent is usually easier to obtain than its exact value, since cancellations can be ignored.) With these two ingredients, the method of dominant balance, as described above, can be applied. The differential equation of the function $F(s)$ then translates, via a transfer theorem, into the above recursion for the coefficients. See [89, Sec 5].

The above result can be used to infer the limit distribution of area, along the lines of Section 11.3.2.

**Theorem 4 ([25, 89]).** *Let Assumption 2 be satisfied. For the solution of an algebraic $q$-difference equation $P(x,q) = \sum_{m,n} p_{m,n} x^m q^n$, let $\widetilde{X}_m$ denote the random variable*

$$\mathbb{P}(\widetilde{X}_m = n) = \frac{p_{m,n}}{\sum_n p_{m,n}}$$

*(which is well-defined for almost all m). The mean of $\widetilde{X}_m$ is given by*

$$\mathbb{E}[\widetilde{X}_m] \sim 2\sqrt{\pi} \frac{f_1}{|f_0|} m^{3/2} \qquad (m \to \infty),$$

*where the numbers $f_0$ and $f_1$ are given in Eq. (11.29). The sequence of normalised random variables $X_m$ converges in distribution,*

$$X_m = \frac{\widetilde{X}_m}{\mathbb{E}[\widetilde{X}_m]} \xrightarrow{d} \frac{Y}{\sqrt{\pi}} \qquad (m \to \infty),$$

*where $Y$ is Airy distributed according to Definition 2. We also have moment convergence.*

**Remarks.** *i)* An explicit calculation shows that $\phi_k = |f_0|^{-1} \left(\frac{|f_0|}{2f_1}\right)^k f_k$. Together with Proposition 7, the claim of the proof follows by standard reasoning, as in the examples above.

*ii)* The above theorem appears in [25, Thm 3.1], together with an indication of the arguments of a proof. [There is a misprint in the definition of $\gamma$ in [25, Thm 3.1]. In our notation $\gamma = 4Bf_1$.] Within the more general setup of $q$-functional equations, the theorem is a special case of [89, Thm 1.5].

*iii)* The above theorem is a kind of central limit theorem for combinatorial constructions, since the Airy distribution arises under natural assumptions for a large class of combinatorial constructions. For a connection to certain Brownian motion functionals, see below.

### 11.3.4.2 $q$-Functional Equations and Other Extensions

We discuss extensions of the above result. Generically, the dominant singularity of $P(x, 1)$ is a square-root. The case of a simple pole as dominant singularity, which generalises the example of Ferrers diagrams, has been discussed in [87]. Under weak assumptions, the resulting limit distribution of area is concentrated. Other singularities can also be analysed, as shown in the examples of rectangles above and of directed convex polygons in the following subsection. Compare also [90].

The case of non-polynomial $G$ can be discussed along the same lines, with certain assumptions on the analyticity properties of the series $G$. In the undeformed case $q = 1$, it is a classical result [37, Ch VII.3] that the generating function has a square-root as dominant singularity, as in the polynomial case. One can then argue along the above lines that an Airy distribution emerges as the limit law of the deformation variable [89, Thm 1.5]. Such an extension is relevant, since prominent combinatorial models, such as the Cayley tree generating function, fall into that class. See also the discussion of self-avoiding polygons below.

The above statements also remain valid for more general classes of replacements $x \mapsto qx$, e.g., for replacements $x \mapsto f(q)x$, where $f(q)$ is analytic for $0 \le q \le 1$, with non-negative series coefficients about $q = 0$. More interestingly, the idea of introducing a $q$-deformation may be iterated [25], leading to equations such as

$$P(x, q_1, \ldots, q_M) = G(x, P(xq_1 \cdot \ldots \cdot q_M, q_1q_2 \cdot \ldots \cdot q_M, q_2q_3 \cdot \ldots \cdot q_M, \ldots, q_M)).$$
$$(11.30)$$

The counting parameters corresponding to $q_k$ are rank $k + 1$ parameters, and limit distributions for such quantities have been derived for some types of singularities [77, 78, 88]. There is a central limit result for the generic case of a square-root singularity [89]. This generalisation applies to counting parameters, which decompose linearly under a combinatorial construction. These results can also be obtained by an alternative method, which generalises to non-linear parameters, see [51].

The case where the limit $q$ to unity in a $q$-difference equation is not algebraic, has not been discussed. For example, if $G(x, q, P(x, q), P(qx, q)) = 0$ for some polynomial $G$, the limit $q$ to unity might lead to an algebraic differential equation for $P(x, 1)$. This may be seen by noting that

$$\lim_{q \to 1} \frac{f(x) - f(qx)}{(1 - q)} = xf'(x),$$

for $f(x)$ differentiable at $x$. Such equations are possibly related to polygon models such as three-choice polygons [44] or punctured staircase polygons [45]. Their perimeter generating function is not algebraic, hence the models do not satisfy an algebraic $q$-difference equation as in Definition 3.

### 11.3.4.3 A Stochastic Connection

Lastly, we indicate a link to Brownian motion, which appears in [99, 100] and was further developed in [77, 78, 89, 88]. As we saw in Section 11.3.2, limit distributions can, under certain conditions, be characterised by a certain Laplace transform of their moment generating functions. This approach, which arises naturally from the viewpoint of generating functions, can be applied to discrete versions of Brownian motion, excursions, bridges or meanders. Asymptotic results are results for the corresponding stochastic objects. In fact, distributions of some functionals of Brownian motion have apparently first been obtained using this approach [99, 100].

Interestingly, a similar characterisation appears in stochastics for functionals of Brownian motion, via the Feynman-Kac formula. For example, Louchard's formula [66] relates the logarithmic derivate of the Airy function to a certain Laplace transform of the moment generating function of the law of the Brownian excursion area. Distributions of functionals of Brownian motion can also be obtained by a path integral approach, see [75] for a recent overview.

The discrete approach provides an alternative method for obtaining information about distributions of certain functionals of Brownian motion. For such functionals, it provides an alternative proof of Louchard's formula [77, 78]. It leads, via the method of dominant balance, quite directly to moment recurrences for the underlying distribution. These have been studied in the case of rank $k$ parameters for discrete models of Brownian motion. In particular, they characterise the distributions of integrals over $(k - 1)$-th powers of the corresponding stochastic objects [77, 78, 89, 88]. Such results have apparently not been previously derived using stochastic methods. The generating function approach can also be applied to classes of $q$-functional equations with singularities different from those connected to Brownian motion. For a related generalisation, see [10].

Vice versa, results and techniques from stochastics can be (and have been) analysed in order to study asymptotic properties of polygons. An example is the contour process of simply generated trees [38], which asymptotically describes the area of a staircase polygon. See also [69, 70, 71, 59].

### 11.3.5 Directed Convex Polygons

We show that the limit law of area of directed convex polygons in the uniform fixed perimeter ensemble is that of the area of the Brownian meander.

**Fact 2 ([100, Thm 2]).** *The random variable $Z$ of area of the Brownian meander is characterised by*

$$\frac{\mathbb{E}[Z^k]}{k!} = \frac{\Gamma(\alpha_0)}{\Gamma(\alpha_k)} \frac{\omega_k}{\omega_0} \frac{1}{2^{k/2}},$$

*where $\alpha_k = 3k/2 + 1/2$. The numbers $\omega_k$ satisfy for $k \geq 1$ the quadratic recurrence*

$$\alpha_{k-1}\omega_{k-1} + \sum_{l=0}^{k} \phi_l 2^{-l}\omega_{k-l} = 0,$$

*with initial condition $\omega_0 = 1$, where the numbers $\phi_k$ appear in the Airy distribution as in Definition 2.*

**Remarks.** *i)* This result has been derived using a discrete meander, whose length and area generating function is described by a system of two algebraic $q$-difference equations, see [77, Prop 1].
*ii)* We have $\mathbb{E}[Z] = 3\sqrt{2\pi}/8$ for the mean of $Z$. The random variable $Z$ is uniquely determined by its moments. The numbers $\omega_k$ appear in the asymptotic expansion [100, Thm 3]

$$\Omega(s) = \frac{1 - 3\int_0^s \mathrm{Ai}(t)\,dt}{3\,\mathrm{Ai}(s)} \sim \sum_{k\geq 0} (-1)^k \omega_k s^{-\alpha_k} \qquad (s \to \infty),$$

where $\mathrm{Ai}(x) = \frac{1}{\pi}\int_0^\infty \cos(t^3/3 + tx)\,dt$ is the Airy function.

Explicit expressions have been derived for the moment generating function and for the distribution function of $Z$.

**Fact 3 ([100, Thm 5]).** *The moment generating function $M(t) = \mathbb{E}[e^{-tZ}]$ of $Z$ satisfies the Laplace transform*

$$\int_0^\infty e^{-st} M(\sqrt{2}t^{3/2})\frac{1}{t^{1/2}}\,dt = \sqrt{\pi}\,\Omega(s). \qquad (11.31)$$

*It is explicitly given by*

$$M(t) = 2^{-1/6} t^{1/3} \sum_{k=1}^{\infty} R_k \exp(-\beta_k t^{2/3} 2^{-1/3})$$

for $\Re(t) > 0$, where the numbers $-\beta_k$ are the zeroes of the Airy function, and where

$$R_k = \frac{\beta_k (1 + 3 \int_0^{\beta_k} \mathrm{Ai}(-t)\,dt)}{3\,\mathrm{Ai}'(-\beta_k)}.$$

The random variable $Z$ has a continuous density $p(y)$, with distribution function $R(x) = \int_0^x p(y)\,dy$ given by

$$R(x) = \frac{\sqrt{\pi}}{(18)^{1/6} x} \sum_{k=1}^{\infty} R_k\, e^{-v_k}\, v_k^{-1/3}\, \mathrm{Ai}((3 v_k/2)^{2/3}),$$

where $v_k = (\beta_k)^3/(27 x^2)$.

**Remark.** The moment generating function and the distribution function are obtained by two consecutive inverse Laplace transforms of Eq. (11.31).

**Theorem 5.** *The normalised area random variables $X_m$ of directed convex polygons Eq. (11.18) satisfy*

$$X_m \xrightarrow{d} \frac{1}{2} Z \qquad (m \to \infty),$$

*where $Z$ is the area random variable of the Brownian meander as in Fact 2. We also have moment convergence.*

*Proof.* A system of $q$-difference equations for the generating function $Q(x,y,q)$ of directed convex polygons, counted by width, height and area, has been given in [9, Lemma 1.1]. It can be reduced to a single equation,

$$
\begin{aligned}
&q(qx-1)Q(x,y,q) + ((1+q)(P(x,y,q)+y))Q(qx,y,q)+\\
&\left(xyq - y^2 + P(x,y,q)(qx-y-1)\right)Q(q^2 x,y,q)\\
&- q^2 xy\,(y + P(x,y,q) - 1) = 0,
\end{aligned}
\tag{11.32}
$$

where $P(x,y,q)$ is the width, height and area generating function of staircase polygons. Setting $q=1$ and $x=y$ yields the half-perimeter generating function

$$g_0(x) = \frac{x^2}{\sqrt{1-4x}}.$$

Hence $x_c = 1/4$ for the radius of convergence of $Q(x,x,1)$.

It is possible to derive from Eq. (11.32) a $q$-difference equation for the (isotropic) half-perimeter and area generating function $Q(x,q) = Q(x,x,q)$ of directed convex polygons. This is due to the symmetry $Q(x,y,q) = Q(y,x,q)$, which results from invariance of the set of directed convex polygons under reflection along the negative diagonal $y = -x$. Since this equation is quite long, we do not give it here. By arguments analogous to those of the previous subsection, it can be deduced from this equation that all area moment generating functions $g_k(x)$ of $Q(x,1)$ are Laurent series in $s = \sqrt{1-4x}$, see also [89, Prop (4.3)]. The leading singular exponent of $g_k(x)$, defined by $g_k(x) \sim h_k(1-x/x_c)^{-\alpha_k}$ as $x \nearrow x_c$, can be bounded from above by $\alpha_k \leq 3k/2+1/2$, see also [89, Prop (4.4)] for the argument. We apply the method of dominant balance, in order to prove that $\alpha_k = 3k/2 + 1/2$ and to yield recurrences for the coefficients $h_k$. We define

$$P(x,q) = \frac{1}{4} + (1-q)^{1/3}F\left(\frac{1-4x}{(1-q)^{2/3}}, (1-q)^{1/3}\right),$$

$$Q(x,q) = (1-q)^{-1/3}H\left(\frac{1-4x}{(1-q)^{2/3}}, (1-q)^{1/3}\right),$$

where $F(s) = F(s,0)$ has already been determined in Eq. (11.26). We set $4x = 1 - s\varepsilon^2$, $q = 1 - \varepsilon^3$, and expand the $q$-difference equation to leading order in $\varepsilon$. We get for $H(s) := H(s,0)$ the inhomogeneous linear differential equation of first order

$$H'(s) + 4H(s)F(s) + \frac{1}{8} = 0.$$

This implies for the coefficients $h_k$ of $H(s) = \sum_{k\geq 0} h_k s^{-\alpha_k}$ and $f_k$ of $F(s) = \sum_{k\geq 0} f_k s^{-\gamma_k}$ for $k \geq 1$ the quadratic recursion

$$\alpha_{k-1}h_{k-1} + 4\sum_{l=0}^{k} f_l h_{k-l} = 0,$$

where $h_0 = 1/16$. Using $f_k = 2^{-2k-1}\phi_k$, we obtain the meander recursion in Fact 2 by setting $h_k = 2^{-k-4}\omega_k$. It can be inferred from the functional equation that (the analytic continuations of) all factorial moment generating functions are $\Delta$-regular, with $x_c = 1/4$. Thus Lemma 2 applies, and we conclude $X_m \xrightarrow{d} Z/2$.                       □

**Remarks.** *i)* The above theorem states that the limit distribution of area of directed convex polygons coincides, up to normalisation, with the area distribution of the Brownian meander [100]. This suggests that there might exist a combinatorial bijection to discrete meanders, in analogy to that between staircase polygons and Dyck paths. Up to now, a "nice" bijection has not been found, see however [6, 72] for combinatorial bijections to discrete bridges.
*ii)* The above proof relies on a $q$-difference equation for the isotropic generating function $Q(x,x,q)$. Up to normalisation, the meander distribution also appears for the anisotropic model $Q(x,y,q)$, where $0 < y < 1/2$ is fixed, as can be shown by a considerably simpler calculation. The normalisation constant coincides with that of the isotropic model for $y = 1/2$. The latter statement is also a consequence of the fact that the height random variable of directed polygons is asymptotically Gaussian, after centring and normalisation. Analogous considerations apply to the relation between isotropic and anisotropic versions of the other polygon classes.

## 11.3.6 Limit Laws Away From $(x_c, 1)$

As indicated in the introduction, limit laws in the fixed perimeter ensemble for $q \neq 1$ are expected to be Gaussian. The same remark holds for the fixed area ensemble for $x \neq x_c$. There are partial results for the model of staircase polygons. The fixed area ensemble can, for $x < x_c$ and $q$ near unity, be analysed using Fact 7 of the following section. For staircase polygons in the uniform fixed area ensemble $x = 1$, the following result holds.

**Fact 4 ([37, Prop IX.11]).** *Consider the perimeter random variable of staircase polygons in the uniform fixed area ensemble,*

$$\mathbb{P}(\widetilde{Y}_n = m) = \frac{p_{m,n}}{\sum_m p_{m,n}}.$$

*The variable $\widetilde{Y}_n$ has mean $\mu_n \sim \mu \cdot n$ and standard deviation $\sigma_n \sim \sigma \sqrt{n}$, where the numbers $\mu$ and $\sigma$ satisfy*

$$\mu = 0.8417620156\ldots, \qquad \sigma = 0.4242065326\ldots$$

*The centred and normalised random variables*

$$Y_n = \frac{\widetilde{Y}_n - \mu_n}{\sigma_n},$$

*converge in distribution to a Gaussian random variable.*

**Remark.** The above result is proved using an explicit expression for the half-perimeter and area generating function, as a ratio of two $q$-Bessel functions. It can be shown that this expression is meromorphic about $(x, q) = (1, q_c)$ with a simple pole, where $q_c$ is the radius of convergence of the generating function $P(1, q)$. The explicit form of the singularity about $(1, q_c)$ yields a Gaussian limit law.

There are a number of results for classes of column-convex polygons in the uniform fixed area ensemble, typically leading to Gaussian limit laws. The upper and lower shape of a polygon can be described by Brownian motions. See [69, 70, 71] for details. It would be interesting to prove convergence to a Gaussian limit law within a more general framework, such as $q$-difference equations. Analogous questions for other functional equations, describing counting parameters such as horizontal width, have been studied in [24].

## 11.3.7 Self-Avoiding Polygons

A numerical analysis of self-avoiding polygons, using data from exact enumeration [91, 92], supports the conjecture that the limit law of area is, up to normalisation, the Airy distribution.

Let $p_{m,n}$ denote the number of square lattice self-avoiding polygons of half-perimeter $m$ and area $n$. Exact enumeration techniques have been applied to obtain the numbers $p_{m,n}$ for all values of $n$ for given $m \leq 50$. Numerical extrapolation techniques yield very accurate estimates of the asymptotic behaviour of the coefficients of the factorial moment generating functions. To leading order, these are given by

$$[x^m]g_k(x) = \frac{1}{k!}\sum_n (n)_k p_{m,n} \sim A_k x_c^{-m} m^{3k/2-3/2-1} \qquad (m \to \infty), \qquad (11.33)$$

for positive amplitudes $A_k$. The above form has been numerically checked [91, 92] for values $k \leq 10$ and is conjectured to hold for arbitrary $k$. The value $x_c$ is the radius of convergence of the half-perimeter generating function of self-avoiding polygons. The amplitudes $A_k$ have been extrapolated to at least five significant digits. In particular, we have

$$x_c = 0.14368062927(2), \qquad A_0 = 0.09940174(4), \qquad A_1 = 0.0397886(1),$$

where the numbers in brackets denote the uncertainty in the last digit. An exact value of the amplitude $A_1 = 1/(8\pi)$ has been predicted [15] using field-theoretic arguments.

The particular form of the exponent implies that the model of *rooted* self-avoiding polygons $\widetilde{p}_{m,n} = m p_{m,n}$ has the same exponents $\phi = 2/3$ and $\theta = 1/3$ as staircase polygons. In particular, it implies a square-root as dominant singularity of the half-perimeter generating function. Together with the above result for $q$-functional equations, this suggests that (rooted) self-avoiding polygons might obey the Airy distribution as a limit law of area.

A natural method to test this conjecture consists in analysing ratios of moments, such that a normalisation constant is eliminated. Such ratios are also called *universal amplitude ratios*. If the conjecture were true, we would have asymptotically

$$\frac{\mathbb{E}[\widetilde{X}_m^k]}{\mathbb{E}[\widetilde{X}_m]^k} \sim k! \frac{\Gamma(\gamma_1)^k}{\Gamma(\gamma_k)\Gamma(\gamma_0)^{k-1}} \frac{\phi_k \phi_0^{k-1}}{\phi_1^k} \qquad (m \to \infty),$$

for the area random variables $\widetilde{X}_m$ as in Eq. (11.17). The numbers $\phi_k$ and exponents $\gamma_k$ are those of the Airy distribution as in Definition 2. The above form was numerically confirmed for values $k \leq 10$ to a high level of numerical accuracy. The normalisation constant is obtained by noting that $\mathbb{E}[Y] = \sqrt{\pi}$.

**Conjecture 1 (cf [91, 92]).** *Let $p_{m,n}$ denote the number of square lattice self-avoiding polygons of half-perimeter $m$ and area $n$. Let $\widetilde{X}_m$ denote the random variable of area in the uniform fixed perimeter ensemble,*

$$\mathbb{P}(\widetilde{X}_m = n) = \frac{p_{m,n}}{\sum_n p_{m,n}}.$$

*We conjecture that*

$$\frac{\widetilde{X}_m}{\mathbb{E}[\widetilde{X}_m]} \xrightarrow{d} \frac{Y}{\sqrt{\pi}},$$

*where Y is Airy distributed according to Definition 2.*

**Remarks.** *i)* Field theoretic arguments [15] yield $A_1 = 1/(8\pi)$.

*ii)* References [91, 92] contain conjectures for the scaling function of self-avoiding polygons and rooted self-avoiding polygons, see the following section. In fact, the numerical analysis in [91, 92] mainly concerns the area amplitudes $A_k$, which determine the limit distribution of area.

*iii)* The area law of self-avoiding polygons has also been studied [91, 92] on the triangular and hexagonal lattices. As for the square lattice, the area limit law appears to be the Airy distribution, up to normalisation.

*iv)* It is an open question whether there are non-trivial counting parameters other than the area, whose limit law (in the fixed perimeter ensembles) coincides between self-avoiding polygons and staircase polygons. See [88] for a negative example. This indicates that underlying stochastic processes must be quite different.

*v)* A proof of the above conjecture is an outstanding open problem. It would be interesting to analyse the emergence of the Airy distribution using stochastic Loewner evolution [60]. Self-avoiding polygons at criticality are conjectured to describe the hull of critical percolation clusters and the outer boundary of two-dimensional Brownian motion [60].

A numerical analysis of the fixed area ensemble along the above lines again shows behaviour similar to that of staircase polygons. This supports the following conjecture.

**Conjecture 2.** *Consider the perimeter random variable of self-avoiding polygons in the uniform fixed area ensemble,*

$$\mathbb{P}(\widetilde{Y}_n = m) = \frac{p_{m,n}}{\sum_m p_{m,n}}.$$

*The random variable $\widetilde{Y}_n$ is conjectured to have mean $\mu_n \sim \mu \cdot n$ and standard deviation $\sigma_n \sim \sigma\sqrt{n}$, where the numbers $\mu$ and $\sigma$ satisfy*

$$\mu = 1.855217(1), \qquad \sigma^2 = 0.3259(1),$$

*where the number in brackets denotes the uncertainty in the last digit. The centred and normalised random variables*

$$Y_n = \frac{\widetilde{Y}_n - \mu_n}{\sigma_n},$$

*are conjectured to converge in distribution to a Gaussian random variable.*

The above conjectures, together with the results of the previous subsection, also raise the question whether rooted square-lattice self-avoiding polygons, counted by half-perimeter and area, might satisfy a $q$-functional equation. In particular, it would be interesting to consider whether rooted self-avoiding polygons might satisfy

$$P(x) = G(x, P(x)), \tag{11.34}$$

for some power series $G(x, y)$ in $x, y$. If the perimeter generating function $P(x)$ is not algebraic, this excludes polynomials $G(x, y)$ in $x$ and $y$. Note that the anisotropic perimeter generating function of self-avoiding polygons is not $D$-finite [86]. It is thus unlikely that the isotropic perimeter generating function is $D$-finite and, in particular, algebraic. On the other hand, solutions of Eq. (11.34) need not be algebraic or $D$-finite. An example is the Cayley tree generating function $T(x)$ satisfying $T(x) = x\exp(T(x))$, see [33].

### 11.3.8 Punctured Polygons

Punctured polygons are self-avoiding polygons with internal holes, which are also self-avoiding polygons. The polygons are also mutually avoiding. The perimeter of a punctured polygon is the sum of the lengths of its boundary curves, the area of a punctured polygon is the area of the outer polygon minus the area of the holes. Apart from intrinsic combinatorial interest, models of punctured polygons may be viewed as arising from two-dimensional sections of three-dimensional self-avoiding vesicles. Counted by area, they may serve as an approximation to the polyomino model.

We consider, for a given subclass of self-avoiding polygons, punctured polygons with holes from the same subclass. The case of a bounded number of punctures of bounded size can be analysed in some generality. The case of a bounded number of punctures of unbounded size leads to simple results if the critical perimeter generating function of the model without punctures is finite.

For a given subclass of self-avoiding polygons, the number $p_{m,n}$ denotes the number of polygons with half-perimeter $m$ and area $n$. Let $p_{m,n}^{(r,s)}$ denote the number of polygons with $r \geq 1$ punctures whose half-perimeter sum equals $s$. Let $p_{m,n}^{(r)}$ denote the number of polygons with $r \geq 1$ punctures of arbitrary size.

**Theorem 6 ([94, Thms 1,2]).** *Assume that, for a class of self-avoiding polygons without punctures, the area moment coefficients $p_m^{(k)} = \sum_{n \geq 0} n^k p_{m,n}$ have, for $k \in \mathbb{N}_0$, the asymptotic form*

$$p_m^{(k)} \sim A_k x_c^{-m} m^{\gamma_k - 1} \qquad (m \to \infty),$$

*for numbers $A_k > 0$, for $0 < x_c \leq 1$ and for $\gamma_k = (k - \theta)/\phi$, where $0 < \phi < 1$. Let $g_0(x) = \sum_{m \geq 0} p_m^{(0)} x^m$ denote the half-perimeter generating function.*

*Then, the area moment coefficient $p_m^{(r,k,s)} = \sum_n n^k p_{m,n}^{(r,s)}$ of the polygon class with $r \geq 1$ punctures whose half-perimeter sum equals $s$ is, for $k \in \mathbb{N}_0$, asymptotically* *given by*

$$p_m^{(r,k,s)} \sim A_k^{(r,s)} x_c^{-m} m^{\gamma_k+r-1} \qquad (m \to \infty),$$

*where* $A_k^{(r,s)} = \frac{A_{k+r}}{r!} x_c^s [x^s](g_0(x))^r.$

If $\theta > 0$, *the area moment coefficient* $p_m^{(r,k)} = \sum_n n^k p_{m,n}^{(r)}$ *of the polygon class with* $r \geq 1$ *punctures of arbitrary size satisfies, for* $k \in \mathbb{N}_0$, *asymptotically*

$$p_m^{(r,k)} \sim A_k^{(r)} x_c^{-m} m^{\gamma_k+r-1} \qquad (m \to \infty),$$

*where the amplitudes* $A_k^{(r)}$ *are given by*

$$A_k^{(r)} = \frac{A_{k+r}(g_0(x_c))^r}{r!}.$$

**Remarks.** *i)* The basic argument in the proof of the preceding result involves an estimate of interactions of hole polygons with one another or with the boundary of the external polygon, which are shown to be asymptotically irrelevant. This argument also applies in higher dimensions, as long as the exponent $\phi$ satisfies $0 < \phi < 1$.
*ii)* In the case of an infinite critical perimeter generating function, such as for subclasses of convex polygons, boundary effects are asymptotically relevant, if punctures of unbounded size are considered. The case of an unbounded number of punctures, which approximates the polyomino problem, is unsolved.
*iii)* The above result leads to new area limit distributions. For rectangles with $r$ punctures of bounded size, we get $\beta_{r+1,1/2}$ as the limit distribution of area. For staircase polygons with punctures, we obtain generalisations of the Airy distribution, which are discussed in [94]. In contrast, for Ferrers diagrams with punctures of bounded size, the limit distribution of area stays concentrated.
*iv)* The theorem also applies to models of punctured polygons, which do not satisfy an algebraic $q$-difference equation. An example is given by staircase polygons with a staircase hole of unbounded size, whose perimeter generating function is not algebraic [45].

## *11.3.9 Models in Three Dimensions*

There are very few results for models in higher dimensions, notably for models on the cubic lattice. There are a number of natural counting parameters for such objects. We restrict consideration to area and volume, which is the three-dimensional analogue of perimeter and area of two-dimensional models.

One prominent model is self-avoiding surfaces on the cubic lattice, also studied as a model of three-dimensional vesicle collapse. We follow the review in [102] (see also the references therein) and consider closed orientable surfaces of genus zero, i.e., surfaces homeomorphic to a sphere. Numerical studies indicate that the surface generating function displays a square-root $\gamma = -1/2$ as the dominant singularity.

Consider the fixed surface area ensemble with weights proportional to $q^n$, with $n$ the volume of the surface. One expects a deflated phase (branched polymer phase) for small values of $q$ and an inflated phase (spherical phase) for large values of $q$. In the deflated phase, the mean volume of a surface should grow proportionally to the area $m$ of the surface, in the inflated phase the mean volume should grow like $m^{3/2}$ with the surface. Numerical simulations suggest a phase transition at $q = 1$ with exponent $\phi = 1$. This indicates that a typical surface resembles a branched polymer, and a concentrated distribution of volume is expected. Note that this behaviour differs from that of the two-dimensional model of self-avoiding polygons.

Even relatively simple subclasses of self-avoiding surfaces such as rectangular boxes [73] and plane partition vesicles [50], generalising the two-dimensional models of rectangles and Ferrers diagrams, display complicated behaviour. Let $p_{m,n}$ denote the number of surfaces of area $m$ and volume $n$ and consider the generating function $P(x,q) = \sum_{m,n} p_{m,n} x^m q^n$. For rectangular box vesicles, we apparently have $P(x,1) \sim A |\log(1-x)|/(1-x)^{3/2}$ as $x \to 1^-$, some constant $A > 0$, see [73, Eq (35)]. In the fixed surface area ensemble, a linear polymer phase $0 < q < 1$ is separated from a cubic phase $q > 1$. At $q = 1$, we have $\phi = 2/3$, such that typical rectangular boxes are expected to attain a cubic shape. We expect a limit distribution which is concentrated. For plane partition vesicles, it is conjectured on the basis of numerical simulations [50, Sec 4.1.1] that $P(x,1) \sim A \exp(\alpha/(x_c - x)^{1/3})/(x_c - x)^\gamma$, where $\gamma \approx 1.7$ at $x_c = 0.8467(3)$, for non-vanishing constants $A$ and $\alpha$. It is expected that $\phi = 1/2$.

As in the previous subsection, three-dimensional models of punctured vesicles may be considered. The above arguments hold, if the exponent $\phi$ satisfies $0 < \phi < 1$. A corresponding result for punctures of unbounded size can be stated if the critical surface area generating function is finite.

## *11.3.10 Summary*

In this section, we described methods to extract asymptotic area laws for polygon models on the square lattice, and we applied these to various classes of polygons. Some of the laws were found to coincide with those of the (absolute) area under a Brownian excursion and a Brownian meander. A combinatorial explanation for the latter result has not been given. Is there a simple polygon model with the same area limit law as the area under a Brownian bridge? The connection to stochastics deserves further investigation. In particular, it would be interesting to identify underlying stochastic processes. For an approach to a number of different random combinatorial structures starting from a probabilistic viewpoint, see [82].

Area laws of polygon models in the uniform fixed perimeter ensemble $q = 1$ have been understood in some generality, by an analysis of the singular behaviour of $q$-functional equations about the point $(x,q) = (x_c, 1)$. Essentially, the type of singularity of the half-perimeter generating function determines the limit law. A refined analysis can be done, leading to local limit laws and providing convergence rates.

Also, limit distributions describing corrections to the asymptotic behaviour can be derived. They seem to coincide with distributions arising in models of punctured polygons, see [94].

For non-uniform ensembles, concentrated distributions are expected, but general results, e.g. for $q$-functional equations, are lacking. These may be obtained by multivariate singularity analysis, see also [24, 65].

The underlying structure of $q$-functional equations appears in a number of other combinatorial models, such as models of two-dimensional directed walks, counted by length and area between the walk and the $x$-axis, models of simply generated trees, counted by the number of nodes and path length, and models which appear in the average case analysis of algorithms, see [34, 37]. Thus, the above methods and results can be applied to such models. In statistical physics, this mainly concerns models of (interacting) directed walks, see [48] for a review. There is also an approach to the behaviour of such walks from a stochastic viewpoint, see e.g. the review [101].

There are exactly solvable polygon models, which do not satisfy an algebraic $q$-difference equation, such as three-choice polygons [44], punctured staircase polygons [45], prudent polygon subclasses [96], and possibly diagonally convex polygons. For a rigorous analysis of the above models, it may be necessary to understand $q$-difference equations with more general holonomic solutions, as $q$ approaches unity.

Focussing on self-avoiding polygons, it might be interesting to analyse whether the perimeter generating function of rooted self-avoiding polygons might satisfy an implicit equation Eq. (11.34). Asymptotic properties of the area can possibly be studied using stochastic Loewner evolution [60]. Another open question concerns the area limit law for $q \neq 1$ or the perimeter limit law for $x \neq x_c$, where Gaussian behaviour is expected. At present, even the simpler question of analyticity of the critical curve $x_c(q)$ for $0 < q < 1$ is open.

Most results of this section concerned area limit laws of polygon models. Similarly, one can ask for perimeter laws in the fixed area ensemble. Results have been given for the uniform ensemble. Generally, Gaussian limit laws are expected away from criticality, i.e., away from $x = x_c$. Perimeter laws are more difficult to extract from a $q$-functional equation than area laws. We will however see in the following section that, surprisingly, under certain conditions, knowledge of the area limit law can be used to infer the perimeter limit law at criticality.

## 11.4 Scaling Functions

From a technical perspective, the focus in the previous section was on the singular behaviour of the single-variable factorial moment generating function $g_k(x)$ Eq. (11.1), and on the associated asymptotic behaviour of their coefficients. This yielded the limiting area distribution of some polygon models.

In this section, we discuss the more general problem of the singular behaviour of the two-variable perimeter and area generating function of a polygon model. Near the special point $(x,q) = (x_c,1)$, the perimeter and area generating function $P(x,q) = \sum_{m \geq 0} p_m(q)x^m = \sum_{n \geq 0} a_n(x)q^n$ is expected to be approximated by a scaling function, and the corresponding coefficient functions $p_m(q)$ and $a_n(x)$ are expected to be approximated by finite size scaling functions. As we will see, scaling functions encapsulate information about the limit distributions discussed in the previous section, and thus have a probabilistic interpretation.

We will give a focussed review, guided by exactly solvable examples, since singularity analysis of multivariate generating functions is, in contrast to the one-variable case, not very well developed, see [81] for a recent overview. Methods of particular interest to polygon models concern asymptotic expansions about multi-critical points, which are discussed for special examples in [80, 5]. Conjectures for the behaviour of polygon models about multicritical points arise from the physical theory of tricritical scaling [41], see the review [61], which has been adapted to polygon models [14, 13]. There are few rigorous results about scaling behaviour of polygon models, which we will discuss. This will complement the exposition in [47]. See also [42, Ch 9] for the related subject of scaling in percolation.

### *11.4.1 Scaling and Finite Size Scaling*

The half-perimeter and area generating function of a polygon model $P(x,q)$ about $(x,q) = (x_c,1)$ is expected to be approximated by a scaling function. This is motivated by the following heuristic argument. Assume that the factorial area moment generating functions $g_k(x)$ Eq. (11.1) have, for values $x < x_c$, a local expansion about $x = x_c$ of the form

$$g_k(x) = \sum_{l \geq 0} \frac{f_{k,l}}{(1 - x/x_c)^{\gamma_{k,l}}},$$

where $\gamma_{k,l} = (k - \theta_l)/\phi$ and $\theta_{l+1} > \theta_l$. Disregarding questions of analyticity, we argue

$$P(x,q) \approx \sum_{k \geq 0} (-1)^k \left( \sum_{l \geq 0} \frac{f_{k,l}}{(1 - x/x_c)^{\gamma_{k,l}}} \right) (1-q)^k$$

$$\approx \sum_{l \geq 0} (1-q)^{\theta_l} \left( \sum_{k \geq 0} (-1)^k f_{k,l} \left( \frac{1 - x/x_c}{(1-q)^\phi} \right)^{-\gamma_{k,l}} \right).$$

In the above calculation, we replaced $P(x,q)$ by its Taylor series about $q = 1$, and then replaced the Taylor coefficients by their expansion about $x = x_c$. The preceding heuristic calculation has, for some polygon models and on a formal level, a rigorous counterpart, see the previous section. In the above expression, the r.h.s. depends on series $\mathscr{F}_l(s) = \sum_{k \geq 0} (-1)^k f_{k,l} s^{-\gamma_{k,l}}$ of a single variable of combined argument

$s = (1 - x/x_c)/(1 - q)^\phi$. Restricting to the leading term $l = 0$, this motivates the following definition. For $\phi > 0$ and $x_c > 0$, we define for numbers $s_-, s_+ \in [-\infty, +\infty]$ the domain

$$D(s_-, s_+) = \{(x,q) \in (0,\infty) \times (0,1) : s_- < (1 - x/x_c)/(1-q)^\phi < s_+)\}.$$

**Definition 4 (Scaling function).** For numbers $p_{m,n}$ with generating function $P(x,q) = \sum_{m,n} p_{m,n} x^m q^n$, let Assumption 1 be satisfied. Let $0 < x_c \leq 1$ be the radius of convergence of $P(x,1)$. Assume that there exist constants $s_-, s_+ \in [-\infty, +\infty]$ satisfying $s_- < s_+$ and a function $\mathscr{F} : (s_-, s_+) \to \mathbb{R}$, such that $P(x,q)$ satisfies, for real constants $\theta$ and $\phi > 0$,

$$P^{(sing)}(x,q) \sim (1-q)^\theta \mathscr{F}\left(\frac{1 - x/x_c}{(1-q)^\phi}\right) \qquad (x,q) \to (x_c, 1) \text{ in } D(s_-, s_+). \quad (11.35)$$

Then, the function $\mathscr{F}(s)$ is called an *(area) scaling function*, and $\theta$ and $\phi$ are called *critical exponents*.

**Remarks.** *i)* In analogy to the one-variable case, the above asymptotic equality means that there exists a power series $P^{(reg)}(x,q)$ convergent for $|x| < x_1$ and $|q| < q_1$, where $x_1 > x_c$ and $q_1 > 1$, such that the function $P^{(sing)}(x,q) := P(x,q) - P^{(reg)}(x,q)$ is asymptotically equal to the r.h.s..
*ii)* Due to the region $D(s_-, s_+)$ where the limit $(x,q) \to (x_c, 1)$ is taken, admissible values $(x,q)$ satisfy $0 < q < 1$ and $0 < x < x_0(q)$, where $x_0(q) = x_c(1 - s_-(1-q)^\phi)$, if $s_- \neq -\infty$. Thus, in this case, the critical curve $x_c(q)$ satisfies $x_c(q) \geq x_0(q)$ as $q$ approaches unity. Note that equality need not hold in general.
*iii)* The method of dominant balance was originally applied in order to obtain a defining equation for a scaling function $\mathscr{F}(s)$ from a given functional equation of a polygon model. This assumes the existence of a scaling function, together with additional analyticity properties. See [84, 91, 87].
*iv)* For particular examples, an analytic scaling function $\mathscr{F}(s)$ exists, with an asymptotic expansion about infinity, and the area amplitude series $F(s)$ agrees with the asymptotic series, see below.
*v)* There is an alternative definition of a scaling function [31] by demanding

$$P^{(sing)}(x,q) \sim \frac{1}{(1 - x/x_c)^{-\theta/\phi}} \mathscr{H}\left(\frac{1-q}{(1 - x/x_c)^{1/\phi}}\right) \qquad (x,q) \to (x_c, 1) \quad (11.36)$$

in a suited domain, for a function $\mathscr{H}(t)$ of argument $t = (1-q)/(1 - x/x_c)^{1/\phi}$. Such a scaling form is also motivated by the above argument. One may then call such a function $\mathscr{H}(t)$ a *perimeter scaling function*. If $\mathscr{F}(s)$ is a scaling function, then a function $\mathscr{H}(t)$, satisfying Eq. (11.36) in a suited domain, is given by

$$\mathscr{H}(t) = t^\theta \mathscr{F}(t^{-\phi}).$$

If $s_- \leq 0$ and $s_+ = \infty$, the particular scaling form Eq. (11.35) implies a certain asymptotic behaviour of the critical area generating function and of the half-perimeter generating function. The following lemma is a consequence of Definition 4.

**Lemma 3.** *Let the assumptions of Definition 4 be satisfied.*

i) *If $s_+ = \infty$ and if the scaling function $\mathscr{F}(s)$ has the asymptotic behaviour*

$$\mathscr{F}(s) \sim f_0 s^{-\gamma_0} \qquad (s \to \infty),$$

*then $\gamma_0 = -\frac{\theta}{\phi}$, and the half-perimeter generating function $P(x,1)$ satisfies*

$$P^{(sing)}(x,1) \sim f_0 (1 - x/x_c)^{\theta/\phi} \qquad (x \nearrow x_c).$$

ii) *If $s_- \leq 0$ and if the scaling function $\mathscr{F}(s)$ has the asymptotic behaviour*

$$\mathscr{F}(s) \sim h_0 s^{\alpha_0} \qquad (s \searrow 0),$$

*then $\alpha_0 = 0$, and the critical area generating function $P(x_c, q)$ satisfies*

$$P^{(sing)}(x_c, q) \sim h_0 (1 - q)^{\theta} \qquad (q \nearrow 1).$$

A sufficient condition for equality of the area amplitude series and the scaling function is stated in the following lemma, which is an extension of Lemma 3.

**Lemma 4.** *Let the assumptions of Definition 4 be satisfied.*

i) *Assume that the relation Eq. (11.35) remains valid under arbitrary differentiation w.r.t. q. If $s_+ = \infty$, if the scaling function $\mathscr{F}(s)$ has an asymptotic expansion*

$$\mathscr{F}(s) \sim \sum_{k \geq 0} (-1)^k f_k s^{-\gamma_k} \qquad (s \to \infty),$$

*and if an according asymptotic expansion is true for arbitrary derivatives, then the following statements hold.*

a) *The exponent $\gamma_k$ is, for $k \in \mathbb{N}_0$, given by*

$$\gamma_k = \frac{k - \theta}{\phi}.$$

b) *The scaling function $\mathscr{F}(s)$ determines the asymptotic behaviour of the factorial area moment generating functions via*

$$\left( \frac{1}{k!} \frac{\partial^k}{\partial q^k} P(x,q) \Big|_{q=1} \right)^{(sing)} \sim \frac{f_k}{(1 - x/x_c)^{\gamma_k}} \qquad (x \nearrow x_c).$$

ii) *Assume that the relation Eq. (11.35) remains valid under arbitrary differentiation w.r. to x. If $s_- \leq 0$, and if the scaling function $\mathscr{F}(s)$ has an asymptotic expansion*

$$\mathscr{F}(s) \sim \sum_{k \geq 0} (-1)^k h_k s^{\alpha_k} \qquad (s \searrow 0),$$

and if an according asymptotic expansion is true for arbitrary derivatives, then the following statements hold.

a) The exponent $\alpha_k$ is, for $k \in \mathbb{N}_0$, given by $\alpha_k = k$.
b) The scaling function determines the asymptotic behaviour of the factorial perimeter moment generating functions at $x = x_c$ via

$$\left( \frac{1}{k!} \frac{\partial^k}{\partial x^k} P(x,q) \Big|_{x=x_c} \right)^{(\text{sing})} \sim \frac{x_c^{-k} h_k}{(1-q)^{\beta_k}} \qquad (q \nearrow 1),$$

where $\beta_k = k\phi - \theta$.

**Remarks.** Lemma 4 states conditions under which the area amplitude series coincides with the scaling function. Given these conditions, the scaling function also determines the perimeter law of the polygon model at criticality.

In the one-variable case, the singular behaviour of a generating function translates, under suitable assumptions, to the asymptotic behaviour of its coefficients. We sketch the analogous situation for the asymptotic behaviour of a generating function involving a scaling function.

**Definition 5 (Finite size scaling function).** For numbers $p_{m,n}$ with generating function $P(x,q) = \sum_{m,n} p_{m,n} x^m q^n$, let Assumption 1 be satisfied. Let $0 < x_c \leq 1$ be the radius of convergence of the generating function $P(x,1)$.

i) Assume that there exist a number $t_+ \in (0,\infty]$ and a function $f : [0,t_+] \to \mathbb{R}$, such that the perimeter coefficient function asymptotically satisfies, for real constants $\gamma_0$ and $\phi > 0$,

$$[x^m]P(x,q) \sim x_c^{-m} m^{\gamma_0-1} f(m^{1/\phi}(1-q)) \qquad (q,m) \to (1,\infty),$$

where the limit is taken for $m$ a positive integer and for real $q$, such that $m^{1/\phi}(1-q) \in [0,t_+]$. Then, the function $f(t)$ is called a *finite size (perimeter) scaling function*.

ii) Assume that there exist constants $t_- \in [-\infty,0)$, $t_+ \in (0,\infty]$, and a function $h : [t_-,t_+] \to \mathbb{R}$, such that the area coefficient function asymptotically satisfies, for real constants $\beta_0$ and $\phi > 0$,

$$[q^n]P(x,q) \sim n^{\beta_0-1} h(n^\phi(1-x/x_c)) \qquad (x,n) \to (x_c,\infty),$$

where the limit is taken for $n$ a positive integer and real $x$, such that $n^\phi(1-x/x_c) \in [t_-,t_+]$. Then, the function $h(t)$ is called a *finite size (area) scaling function*.

**Remarks.** i) The following heuristic calculation motivates the expectation that a finite size scaling function approximates the coefficient function. For the perimeter coefficient function, assume that the exponents $\gamma_k$ of the factorial area moment generating functions are of the special form $\gamma_k = (k-\theta)/\phi$. We argue

$$[x^m]P(x,q) \approx [x^m] \sum_{k=0}^{\infty} (-1)^k \frac{f_k}{(1-x/x_c)^{\gamma_k}} (1-q)^k$$

$$\approx x_c^{-m} m^{\gamma_0-1} \sum_{k=0}^{\infty} (-1)^k \frac{f_k}{\Gamma(\gamma_k)} \left( m^{1/\phi}(1-q) \right)^k.$$

In the above expression, the r.h.s. depends on a function $f(t)$ of a single variable of combined argument $t = m^{1/\phi}(1-q)$.

For the area coefficient function, we assume that $\beta_k = k\phi - \theta$ and argue as above,

$$[q^n]P(x,q) \approx [q^n] \sum_{k=0}^{\infty} (-1)^k \frac{h_k}{(1-q)^{\beta_k}} (1-x/x_c)^k$$

$$\approx n^{\beta_0-1} \sum_{k=0}^{\infty} (-1)^k \frac{h_k}{\Gamma(\beta_k)} \left( n^{\phi}(1-x/x_c) \right)^k.$$

In the above expression, the r.h.s. depends on a function $h(t)$ of a single variable of combined argument $t = n^{\phi}(1 - x/x_c)$.

*ii)* The above argument suggests that a scaling function and a finite size scaling function may be related by a Laplace transformation. A comparison with Eq.(11.20) leads one to expect that finite size scaling functions are moment generating functions of the limit laws of area and perimeter.

*iii)* Sufficient conditions under which knowledge of a scaling function implies the existence of a finite size scaling function have been given for the finite size area scaling function [13] using Darboux's theorem.

A scaling function describes the leading singular behaviour of the generating function $P(x,q)$ in some region about $(x,q) = (x_c,1)$. A particular form of subsequent correction terms has been argued for at the beginning of the section.

**Definition 6 (Correction-to-scaling functions).** For numbers $p_{m,n}$ with generating function $P(x,q) = \sum_{m,n} p_{m,n} x^m q^n$, let Assumption 1 be satisfied. Let $0 < x_c \leq 1$ be the radius of convergence of the generating function $P(x,1)$. Assume that there exist constants $s_-, s_+ \in [-\infty, +\infty]$ satisfying $s_- < s_+$, and functions $\mathscr{F}_l : (s_-, s_+) \to \mathbb{R}$ for $l \in \mathbb{N}_0$, such that the generating function $P(x,q)$ satisfies, for real constants $\phi > 0$ and $\theta_l$, where $\theta_{l+1} > \theta_l$,

$$P^{(sing)}(x,q) \sim \sum_{l \geq 0} (1-q)^{\theta_l} \mathscr{F}_l \left( \frac{1-x/x_c}{(1-q)^{\phi}} \right) \qquad (x,q) \to (x_c,1) \text{ in } D(s_-,s_+).$$

Then, the function $\mathscr{F}_0(s)$ is a scaling function, and for $l \leq 1$, the functions $\mathscr{F}_l(s)$ are called *correction-to-scaling functions*.

**Remarks.** *i)* In the above context, the symbol $\sim$ denotes a (generalised) asymptotic expansion (see also [80, Ch 1]): Let $(G_k(\boldsymbol{x}))_{k \in \mathbb{N}_0}$ be a sequence of (multivariate) functions satisfying for all $k$ the estimate $G_{k+1}(\boldsymbol{x}) = o(G_k(\boldsymbol{x}))$ as $\boldsymbol{x} \to \boldsymbol{x}_c$ in some prescribed region. For a function $G(x)$, we then write $G(x) \sim \sum_{k=0}^{\infty} G_k(x)$ as $x \to x_c$, if for all $n$ we have $G(x) = \sum_{k=0}^{n-1} G_k(x) + \mathcal{O}(G_n(x))$ as $x \to x_c$.

*ii)* The previous section yielded effective methods for obtaining area amplitude functions. These are candidates for correction-to-scaling functions, see also [87].

## 11.4.2 Squares and Rectangles

We consider the models of squares and rectangles, whose scaling behaviour can be explicitly computed. Their half-perimeter and area generating function can be written as a single sum, to which the Euler-MacLaurin summation formula [80, Ch 8] can be applied. We first discuss squares.

**Fact 5 (cf [49, Thm 2.4]).** *For $0 < x, q < 1$, the generating function $P(x,q) = \sum_{m=0}^{\infty} x^m q^{m^2/4}$ of squares, counted by half-perimeter and area, is given by*

$$P(x,q) = \frac{1}{\sqrt{|\log q|}} \mathcal{F}\left(\frac{|\log x|}{\sqrt{|\log q|}}\right) + \frac{1}{2} + R(x,q),$$

*with $\mathcal{F}(s) = \sqrt{\pi} e^{s^2} \operatorname{erfc}(s)$, where the remainder term $R(x,q)$ is bounded by*

$$|R(x,q)| \le \frac{1}{6} |\log x|.$$

**Remarks.** *i)* The remainder term differs from that in [49, Thm 2.4], where it was estimated by an integral with lower bound one instead of zero [49, Eq. (46)].

*ii)* With $x_c = 1$, $s_- = 0$ and $s_+ = \infty$, the function $\mathcal{F}(s)$ is a scaling function according to the above definition. The remainder term is uniformly bounded in any rectangle $[x_0, 1) \times [q_0, 1)$ for $0 < x_0, q_0 < 1$, and so the approximation is uniform in this rectangle.

*iii)* The generating function $P(x,q)$ satisfies the quadratic $q$-difference equation $P(x,q) = 1 + x q^{1/4} P(q^{1/2}x, q)$. Using the methods of the previous section, the area amplitude series of the model can be derived. It coincides with the above scaling function $\mathcal{F}(s)$. This particular form is expected, since the distribution of area is concentrated, $p(x) = \delta(x - 1/4)$, compare also with Ferrers diagrams.

*iv)* It has not been studied whether the scaling region can be extended to values $x > 1$ near $(x,q) = (1,1)$. It can be checked that the scaling function $\mathcal{F}(s)$ also determines the asymptotic behaviour of the perimeter moment generating functions, via its expansion about the origin. As expected, they indicate a concentrated distribution.

The half-perimeter and area generating function of rectangles is given by

$$P(x,q) = \sum_{r=1}^{\infty} \sum_{s=1}^{\infty} x^{r+s} q^{rs} = \sum_{r=1}^{\infty} \frac{x(qx)^r}{1 - q^r x}.$$

We have $P(x,1) = x^2/(1-x)^2$, and it can be shown that $P(1,q) \sim -\frac{\log(1-q)}{1-q}$ as $q \nearrow 1$, see [85, 49]. The latter result implies that a scaling form as in Definition 4, with $s_- \leq 0$, does not exist for rectangles. We have the following result.

**Fact 6 ([49, Thm 3.4]).** *For $0 < q < 1$ and $0 < qx < 1$, the generating function $P(x,q)$ of rectangles satisfies*

$$P(x,q) = \frac{x}{|\log q|} \left( \frac{|\log q|}{|\log x|} - \mathrm{LerchPhi}\left( qx, 1, \frac{|\log x|}{|\log q|} \right) \right) + R(x,q),$$

*with the Lerch Phi-function* $\mathrm{LerchPhi}(z,a,v) = \sum_{n=0}^{\infty} \frac{z^n}{(v+n)^a}$, *where the remainder term $R(x,q)$ is bounded by*

$$|R(x,q)| \leq \frac{x^2 q}{1 - qx} \left( \frac{1}{2} + \frac{|\log x|}{6} \right) + \frac{x^2 q}{(1-qx)^2} \frac{|\log q|}{6}.$$

**Remarks.** *i)* The theorem implies that, for every $q_0 \in (0,1)$, the function $(1 - qx)^2 P(x,q)$ is uniformly approximated for points $(x,q)$ satisfying $q_0 < q < 1$ and $0 < x < x_c(q)$, where $x_c(q) = 1/q$ is the critical curve.
*ii)* Rectangles cannot have a scaling function $\mathscr{F}(s)$ as in Definition 4 with $s_- \leq 0$, since the area generating function diverges with a logarithmic singularity. This is reflected in the above approximation.
*iii)* It has not been studied whether the area moments or the perimeter moments at criticality can be extracted from the above approximation.
*iv)* The relation of the above approximation to the area amplitude series of rectangles of the previous section, $F(s) = \mathrm{Ei}(s^2) e^{s^2}$, is not understood. Interestingly, the expansion of $F(s)$ about $s = 0$ resembles a logarithmic divergence. It is not clear whether its expansion at the origin is related to the asymptotic behaviour of the perimeter moment generating functions.

### 11.4.3 Ferrers Diagrams

The singularity diagram of Ferrers diagrams is special, since the value $x_c(1) := \lim_{q \nearrow 1} x_c(q)$ does not coincide with the radius of convergence $x_c$ of the half-perimeter generating function $P(x,1)$. (The function $q \mapsto x_c(q)$ is continuous on $(0,1]$, as may be inferred from the exact solution.) Thus, there are two special points in the singularity diagram, namely $(x,q) = (x_c, 1)$ and $(x,q) = (x_c(1), 1)$. Scaling behaviour about the latter point has apparently not been studied, see also [85].

About the former point $(x,q) = (x_c, 1)$, scaling behaviour is expected. The area amplitude series $F(s)$ of Ferrers diagrams is given by the *entire* function

$$F(s) = \sqrt{\frac{\pi}{8}} \, \mathrm{erfc}\left( \sqrt{2}s \right) e^{2s^2}.$$

A numerical analysis indicates that its Taylor coefficients about $s = 0$ coincide with the perimeter moment amplitudes at criticality, which characterise a concentrated distribution. There is no singularity of $F(s)$ on the negative real axis at any finite value of $s$, in accordance with the fact that the critical line at $q = 1$ extends above $x = x_c$.

It is not known whether a scaling function exists for Ferrers diagrams, or whether it would coincide with the amplitude generating function, see also the recent discussion [50, Sec 2.3]. A rigorous study may be possible, by first rewriting the half-perimeter and area generating function as a contour integral. A further analysis then reveals a saddle point coalescing with the integration boundary at criticality. For such phenomena, uniform asymptotic expansions can be obtained by Bleistein's method [80, Ch 9.9]. The approach proposed above is similar to that for the staircase model [83] in the following subsection.

### 11.4.4 Staircase Polygons

For staircase polygons, counted by width, height, and area with associated variables $x, y, q$, the existence of an area scaling function has been proved. The derivation starts from an exact expression for the generating function, which has then been written as a complex contour integral. About the point $(x, q) = (x_c, 1)$, this led to a saddle-point evaluation with the effect of two coalescing saddles.

**Fact 7 (cf [83, Thm 5.3]).** *Consider $0 < x, y, q < 1$ such that the generating function $P(x, y, q)$ of staircase polygons, counted by width, height and area, is convergent. Set $q = e^{-\varepsilon}$ for $\varepsilon > 0$. Then, as $\varepsilon \searrow 0$, we have*

$$P(x,y,q) = \left( \frac{1-x-y}{2} + \right.$$
$$\left. + \alpha^{-1/2}\varepsilon^{1/3} \frac{\text{Ai}'(\alpha\varepsilon^{-2/3})}{\text{Ai}(\alpha\varepsilon^{-2/3})} \sqrt{\left(\frac{1-x-y}{2}\right)^2 - xy} \right) (1+\mathcal{O}(\varepsilon))$$

*uniformly in $x, y$, where $\alpha = \alpha(x, y)$ satisfies the implicit equation*

$$\frac{4}{3}\alpha^{3/2} = \log(x)\frac{\log(z_m - \sqrt{d})}{\log(z_m + \sqrt{d})} + 2\text{Li}_2(z_m - \sqrt{d}) - 2\text{Li}_2(z_m + \sqrt{d}),$$

*where $z_m = (1+y-x)/2$ and $d = z_m^2 - y$, and $\text{Li}_2(t) = -\int_0^t \frac{\log(1-u)}{u}du$ is the Euler dilogarithm.*

**Remarks.** *i)* The characterisation of $\alpha^{3/2}$ given in [83, Eq (4.21)] has been used.
*ii)* The above approximation defines an area scaling function. For $x = y$ and $x_c = 1/4$, we obtain the approximation [83, Eq (1.14)]

$$P(x,q) \sim \frac{1}{4} + 4^{-2/3}\varepsilon^{1/3}\frac{\mathrm{Ai}'(4^{4/3}(1/4-x)\varepsilon^{-2/3})}{\mathrm{Ai}(4^{4/3}(1/4-x)\varepsilon^{-2/3})}$$

as $(x,q) \to (x_c,1)$ within the region of convergence of $P(x,q)$. It follows by comparison that the area amplitude series coincides with the area scaling function.

*iii)* An area amplitude series for the anisotropic model has been given in [56], by a suitable refinement of the method of dominant balance.

*iv)* It is expected that the perimeter law at $x = x_c$ may be inferred from the Taylor expansion of the scaling function $\mathscr{F}(s)$ at $s = 0$. A closed form for the moment generating function or the probability density has not been given. The right tail of the distribution has been analysed via the asymptotic behaviour of the moments [57, 55]. See also the next subsection.

*v)* The above expression gives the singular behaviour of $P(x,q)$ as $q$ approaches unity, *uniformly* in $x,y$. Restricting to $x = y$, it describes the singular behaviour along the line $q = 1$ for $0 < x < x_c$. In the compact percolation picture, this line describes compact percolation below criticality. Perimeter limit laws away from criticality may be inferred along the above lines. (Asymptotic expansions which are uniform in an additional parameter appear also for solutions of differential equations near singular points [80].)

*vi)* By analytic continuation, it follows that the critical curve $x_c(q)$ for $P(x,x,q)$ coincides near $q = 1$ with the upper boundary curve $x_0(q) = (1 - s_-(1-q)^{2/3})/4$ of the scaling domain, where the value $s_-$ is determined by the singularity of smallest modulus of the scaling function on the negative real axis, hence by the first zero of the Airy function. This leads to a simple pole singularity in the generating function, which describes the branched polymer phase close to $q = 1$.

## *11.4.5 Self-Avoiding Polygons*

In the previous section, a conjecture for the limit distribution of area for self-avoiding polygons and rooted self-avoiding polygons was stated. We further explain the underlying numerical analysis, following [91, 92, 93]. The numerically established form Eq. (11.33) implies for the area moment generating functions for $k \neq 1$ singular behaviour of the form

$$g_k^{(sing)}(x) \sim \frac{f_k}{(1 - x/x_c)^{\gamma_k}} \qquad (x \nearrow x_c),$$

with critical point $x_c = 0.14368062927(2)$ and $\gamma_k = 3k/2 - 3/2$, where the numbers $f_k$ are related to the amplitudes $A_k$ in Eq. (11.33) by

$$A_k = \frac{f_k}{\Gamma(\gamma_k)}.$$

For $k = 1$, we have $\gamma_k = 0$, and a logarithmic singularity is expected, $g_1(x) \sim f_1 \log(1 - x/x_c)$, with $f_1 = A_1$. Similar to Conjecture 1, this leads to a corresponding conjecture for the area amplitude series of self-avoiding polygons. If the area amplitude series was a scaling function, we would expect that it also describes the limit law of perimeter at criticality $x = x_c$, via its expansion about the origin. (Interestingly, these moments are related to the moments of the Airy distribution of negative order, see [93, 34].) This prediction was confirmed in [93], up to numerical accuracy, for the first ten perimeter moments. Also, the crossover behaviour to the branched polymer phase has been found to be consistent with the corresponding scaling function prediction. As was argued in the previous subsection, the critical curve $x_c(q)$ close to unity should coincide with the upper boundary curve $x_0(q) = x_c(1 - s_-(1 - q)^{2/3})$, where the point $s_-$ is related to the first zero of the Airy function on the negative real axis, $s_- = -0.2608637(5)$. The latter two observations support the following conjecture.

**Conjecture 3 ([87, 93]).** *Let* $p_{m,n}$ *denote the number of self-avoiding polygons of half-perimeter $m$ and area $n$, with generating function $P(x,q) = \sum_{m,n} p_{m,n} x^m q^n$. Let $x_c = 0.14368062927(2)$ be the radius of convergence of the half-perimeter generating function $P(x,1)$. Assume that*

$$\sum_n p_{m,n} \sim A_0 x_c^{-m} m^{-5/2} \qquad (m \to \infty),$$

*where $A_0$ is estimated by $A_0 = 0.09940174(4)$. Let the number $s_-$ be such that $(4A_0)^{\frac{2}{3}} \pi s_-$ coincides with the zero of the Airy function on the negative real axis of smallest modulus. We have $s_- = -0.2608637(5)$.*

i) *For rooted self-avoiding polygons with half-perimeter and area generating function $P^{(r)}(x,q) = x\frac{d}{dx}P(x,q)$, the conjectured form of a scaling function $\mathscr{F}^{(r)}(s) : (s_-, \infty) \to \mathbb{R}$ as in Definition 4 is*

$$\mathscr{F}^{(r)}(s) = \frac{x_c}{2\pi} \frac{d}{ds} \log \mathrm{Ai}\left((4A_0)^{\frac{2}{3}} \pi s\right),$$

*with critical exponents $\theta = 1/3$ and $\phi = 2/3$.*
ii) *The conjectured scaling behaviour of (unrooted) self-avoiding polygons is*

$$\left(P^{(sing)}(x,q) - \frac{1}{12\pi}(1-q)\log(1-q)\right) \sim (1-q)^\theta \mathscr{F}\left(\frac{1-x/x_c}{(1-q)^\phi}\right) \tag{11.37}$$

$$(x,q) \to (x_c, 1) \text{ in } D(s_-, \infty),$$

*with scaling function $\mathscr{F}(s) : (s_-, \infty) \to \mathbb{R}$ obtained by integration,*

$$\mathscr{F}(s) = -\frac{1}{2\pi} \log \mathrm{Ai}\left((4A_0)^{\frac{2}{3}} \pi s\right),$$

*and with critical exponents $\theta = 1$ and $\phi = 2/3$.*

**Remarks.** *i)* The above conjecture is essentially based on the conjecture of the previous section that both staircase polygons and rooted self-avoiding polygons have, up to normalisation constants, the same limiting distribution of area in the uniform ensemble $q = 1$. For a numerical investigation of the implications of the scaling function conjecture, see the preceding discussion.

*ii)* A field-theoretical justification of the above conjecture has been proposed [16]. Also, the values of $A_1 = 1/(8\pi)$ and the prefactor $1/(12\pi)$ in Eq. (11.37) have been predicted using field-theoretic methods [15], see also the discussion in [93].

## *11.4.6 Models in Higher Dimensions*

Only very few models of vesicles have been studied in three dimensions. For the simple model of cubes, the scaling behaviour in the perimeter-area ensemble is the same as for squares [49, Thm 2.4]. The scaling form in the area-volume ensemble has been given [49, Thm 2.8]. The asymptotic behaviour of rectangular box vesicles has been studied to some extent [73]. Explicit expressions for scaling functions have not been derived.

## *11.4.7 Open Questions*

The mathematical problem of this section concerns the local behaviour of multivariate generating functions about non-isolated singularities. If such behaviour is known, it may, under appropriate conditions, be used to infer asymptotic properties such as limit distributions. Along lines of the same singular behaviour in the singularity diagram, expressions uniform in the parameters are expected. This may lead to Gaussian limit laws [37]. Parts of the theory of such asymptotic expansions have been developed using methods of several complex variables [81]. The case of several coalescing lines of different singularities is more difficult. Non-Gaussian limit laws are expected, and this case is subject to recent mathematical research [81].

Our approach is motivated by certain models of statistical physics. It relies on the observation that the singular behaviour of their generating function is described by a scaling function. There are major open questions concerning scaling functions. On a conceptual level, the transfer problem [35] should be studied in more detail, i.e., conditions under which the existence of a scaling function implies the existence of the finite-size scaling function. Also, conditions have to be derived such that limit laws can be extracted from scaling functions. This is related to the question when can an asymptotic relation be differentiated. Real analytic methods, in conjunction with monotonicity properties of the generating function, might prove useful [80].

For particular examples, such as models satisfying a linear $q$-difference equation or directed convex polygons, scaling functions may be extracted explicitly. It would be interesting to prove scaling behaviour for classes of polygon models from their defining functional equation. Furthermore, the staircase polygon result indicates that some generating functions may have in fact asymptotic expansions for $q \nearrow 1$, which are valid uniformly in the perimeter variable (i.e., not only in the limit $x \nearrow x_c$). Such expansions would yield scaling functions and correction-to-scaling functions, thereby extending the formal results of the previous section. This might be worked out for specific models, at least in the relevant example of staircase polygons.

## Acknowledgements

The author would like to thank Tony Guttmann and Iwan Jensen for comments on the manuscript, and Nadine Eisner, Thomas Prellberg and Uwe Schwerdtfeger for helpful discussions.

## References

1. M. Abramowitz and I.A. Stegun. *Handbook of Mathematical Functions with Formulas, Graphs, and Mathematical Tables*, volume 18. National Bureau of Standards Applied Mathematics Series, 1964. Reprint Dover 1973.
2. D.J. Aldous. The continuum random tree II: An overview. In M.T. Barlow and N.H. Bingham, editors, *Stochastic Analysis*, pages 23–70. Cambridge University Press, Cambridge, 1991.
3. G. Aleksandrowicz and G. Barequet. Counting $d$-dimensional polycubes and nonrectangular planar polyominoes. In *Proc. 12th Ann. Int. Computing and Combinatorics Conf. (COCOON), Taipei, Taiwan*, volume 4112 of *Springer Lecture Notes in Computer Science*, pages 418–427. Springer, 2006.
4. D. Bennett-Wood, I.G. Enting, D.S. Gaunt, A.J. Guttmann, J.L. Leask, A.L. Owczarek, and S.G. Whittington. Exact enumeration study of free energies of interacting polygons and walks in two dimensions. *J. Phys. A: Math. Gen*, 31:4725–4741, 1998.
5. N. Bleistein and R.A. Handelsman. *Asymptotic Expansions of Integrals*. Holt, Rinehart and Winston, New York, 1975.
6. M. Bousquet-Mélou. Une bijection entre les polyominos convexes dirigés et les mots de Dyck bilatéres. *RAIRO Inform. Théor. Appl.*, 26:205–219, 1992.
7. M. Bousquet-Mélou. A method for the enumeration of various classes of column-convex polygons. *Discrete Math.*, 154:1–25, 1996.
8. M. Bousquet-Mélou. Families of prudent self-avoiding walks. Preprint arXiv:0804.4843, 2008.
9. M. Bousquet-Mélou and J.-M. Fédou. The generating function of convex polyominoes: the resolution of a $q$-differential system. *Discr. Math.*, 137:53–75, 1995.
10. M. Bousquet-Mélou and S. Janson. The density of the ISE and local limit laws for embedded trees. *Ann. Appl. Probab.*, 16:1597–1632, 2006.
11. M. Bousquet-Mélou and A. Rechnitzer. The site-perimeter of bargraphs. *Adv. in Appl. Math.*, 31:86–112, 2003.
12. R. Brak and J.W. Essam. Directed compact percolation near a wall. III. Exact results for the mean length and number of contacts. *J. Phys. A: Math. Gen.*, 32:355–367, 1999.
13. R. Brak and A.L. Owczarek. On the analyticity properties of scaling functions in models of polymer collapse. *J. Phys. A: Math. Gen.*, 28:4709–4725, 1995.
14. R. Brak, A.L. Owczarek, and T. Prellberg. A scaling theory of the collapse transition in geometric cluster models of polymers and vesicles. *J. Phys. A: Math. Gen.*, 26:4565–5479, 1993.

15. J. Cardy. Mean area of self-avoiding loops. *Phys. Rev. Lett.*, 72:1580–1583, 1994.
16. J. Cardy. Exact scaling functions for self-avoiding loops and branched polymers. *J. Phys. A: Math. Gen.*, 34:L665–L672, 2001.
17. K.L. Chung. *A Course in Probability Theory.* Academic Press, New York, 2nd edition, 1974.
18. G.M. Constantine and T.H. Savits. A multivariate Faa di Bruno formula with applications. *Trans. Amer. Math. Soc.*, 348:503–520, 1996.
19. D.A. Darling. On the supremum of a certain Gaussian process. *Ann. Probab.*, 11:803–806, 1983.
20. M.-P. Delest, D. Gouyou-Beauchamps, and B. Vauquelin. Enumeration of parallelogram polyominoes with given bond and site perimeter. *Graphs Combin.*, 3:325–339, 1987.
21. M.-P. Delest and X.G. Viennot. Algebraic languages and polyominoes enumeration. *Theor. Comput. Sci.*, 34:169–206, 1984.
22. J.C. Dethridge, T.M. Garoni, A.J. Guttmann, and I. Jensen. Prudent walks and polygons. Preprint arXiv:0810:3137, 2008.
23. G. Doetsch. *Introduction to the Theory and Application of the Laplace Transform.* Springer, New York, 1974.
24. M. Drmota. Systems of functional equations. *Random Structures Algorithms*, 10:103–124, 1997.
25. P. Duchon. $q$-grammars and wall polyominoes. *Ann. Comb.*, 3:311–321, 1999.
26. J.W. Essam. Directed compact percolation: Cluster size and hyperscaling. *J. Phys. A: Math. Gen.*, 22:4927–4937, 1989.
27. J.W. Essam and A.J. Guttmann. Directed compact percolation near a wall. II. Cluster length and size. *J. Phys. A: Math. Gen.*, 28:3591–3598, 1995.
28. J.W. Essam and D. Tanlakishani. Directed compact percolation. II. Nodal points, mass distribution, and scaling. In *Disorder in physical systems*, volume 67, pages 67–86. Oxford Univ. Press, New York, 1990.
29. J.W. Essam and D. Tanlakishani. Directed compact percolation near a wall. I. Biased growth. *J. Phys. A: Math. Gen.*, 27:3743–3750, 1994.
30. J.A. Fill, P. Flajolet, and N. Kapur. Singularity analysis, Hadamard products, and tree recurrences. *J. Comput. Appl. Math.*, 174:271–313, 2005.
31. M.E. Fisher, A.J. Guttmann, and S.G. Whittington. Two-dimensional lattice vesicles and polygons. *J. Phys. A: Math. Gen.*, 24:3095–3106, 1991.
32. P. Flajolet. Singularity analysis and asymptotics of Bernoulli sums. *Theoret. Comput. Sci.*, 215:371–381, 1999.
33. P. Flajolet, S. Gerhold, and B. Salvy. On the non-holonomic character of logarithms, powers, and the $n$-th prime function. *Electronic Journal of Combinatorics*, 11:A2:1–16, 2005.
34. P. Flajolet and G. Louchard. Analytic variations on the Airy distribution. *Algorithmica*, 31:361–377, 2001.
35. P. Flajolet and A. Odlyzko. Singularity analysis of generating functions. *SIAM J. Discr. Math.*, 3:216–240, 1990.
36. P. Flajolet, P. Poblete, and A. Viola. On the analysis of linear probing hashing. Average-case analysis of algorithms. *Algorithmica*, 22:37–71, 1998.
37. P. Flajolet and R. Sedgewick. *Analytic Combinatorics.* Book in preparation, 2008.
38. B. Gittenberger. On the contour of random trees. *SIAM J. Discr. Math.*, 12:434–458, 1999.
39. I.P. Goulden and D.M. Jackson. *Combinatorial enumeration.* John Wiley & Sons, New York, 1983.
40. D. Gouyou-Beauchamps and P. Leroux. Enumeration of symmetry classes of convex polyominoes on the honeycomb lattice. *Theoret. Comput. Sci.*, 346:307–334, 2005.
41. R.B. Griffiths. Proposal for notation at tricritical points. *Phys. Rev. B*, 7:545–551, 1973.
42. G. Grimmett. *Percolation.* Springer, Berlin, 1999. 2nd ed.
43. A.J. Guttmann. Asymptotic analysis of power-series expansions. In C. Domb and J.L. Lebowitz, editors, *Phase Transitions and Critical Phenomena*, volume 13, pages 1–234. Academic, New York, 1989.
44. A.J. Guttmann and I. Jensen. Fuchsian differential equation for the perimeter generating function of three-choice polygons. *Séminaire Lotharingien de Combinatoire*, 54:B54c, 2006.

45. A.J. Guttmann and I. Jensen. The perimeter generating function of punctured staircase polygons. *J. Phys. A: Math. Gen.*, 39:3871–3882, 2006.
46. G.H. Hardy. *Divergent Series*. Clarendon Press, Oxford, 1949.
47. E.J. Janse van Rensburg. *The Statistical Mechanics of Interacting Walks, Polygons, Animals and Vesicles*, volume 18 of *Oxford Lecture Series in Mathematics and its Applications*. Oxford University Press, Oxford, 2000.
48. E.J. Janse van Rensburg. Statistical mechanics of directed models of polymers in the square lattice. *J. Phys. A: Math. Gen.*, 36:R11–R61, 2003.
49. E.J. Janse van Rensburg. Inflating square and rectangular lattice vesicles. *J. Phys. A: Math. Gen.*, 37:3903–3932, 2004.
50. E.J. Janse van Rensburg and J. Ma. Plane partition vesicles. *J. Phys. A: Math. Gen.*, 39:11171–11192, 2006.
51. S. Janson. The Wiener index of simply generated random trees. *Random Structures Algorithms*, 22:337–358, 2003.
52. S. Janson. Brownian excursion area, Wright's constants in graph enumeration, and other Brownian areas. *Probab. Surv.*, 4:80–145, 2007.
53. I. Jensen. Perimeter generating function for the mean-squared radius of gyration of convex polygons. *J. Phys. A: Math. Gen.*, 38:L769–775, 2005.
54. B.McK. Johnson and T. Killeen. An explicit formula for the c.d.f. of the $l_1$ norm of the Brownian bridge. *Ann. Prob.*, 11:807–808, 1983.
55. J.M. Kearney. On a random area variable arising in discrete-time queues and compact directed percolation. *J. Phys. A: Math. Gen.*, 37:8421–8431, 2004.
56. M.J. Kearney. Staircase polygons, scaling functions and asymmetric compact percolation. *J. Phys. A: Math. Gen.*, 35:L731–L735, 2002.
57. M.J. Kearney. On the finite-size scaling of clusters in compact directed percolation. *J. Phys. A: Math. Gen.*, 36:6629–6633, 2003.
58. M.J. Kearney, S.N. Majumdar, and R.J. Martin. The first-passage area for drifted Brownian motion and the moments of the Airy distribution. *J. Phys. A: Math. Theor.*, 40:F863–F869, 2007.
59. J.-M. Labarbe and J.-F. Marckert. Asymptotics of Bernoulli random walks, bridges, excursions and meanders with a given number of peaks. *Electronic J. Probab.*, 12:229–261, 2007.
60. G.F. Lawler, O. Schramm, and W. Werner. On the scaling limit of planar self-avoiding walk. In *Fractal Geometry and Applications: A Jubilee of Benoît Mandelbrot, Part 2*, volume 72 of *Proceedings of Symposia in Pure Mathematics*, pages 339–364. Amer. Math. Soc., Providence, RI, 2004.
61. I.D. Lawrie and S. Sarbach. Theory of tricritical points. In C. Domb and J.L. Lebowitz, editors, *Phase Transitions and Critical Phenomena*, volume 9, pages 1–161. Academic Press, London, 1984.
62. P. Leroux and É. Rassart. Enumeration of symmetry classes of parallelogram polyominoes. *Ann. Sci. Math. Québec*, 25:71–90, 2001.
63. P. Leroux, É. Rassart, and A. Robitaille. Enumeration of symmetry classes of convex polyominoes in the square lattice. *Adv. in Appl. Math*, 21:343–380, 1998.
64. K.Y. Lin. Rigorous derivation of the perimeter generating functions for the mean-squared radius of gyration of rectangular, Ferrers and pyramid polygons. *J. Phys. A: Math. Gen.*, 39:8741–8745, 2006.
65. M. Lladser. *Asymptotic enumeration via singularity analysis*. PhD thesis, Ohio State University, 2003. Doctoral dissertation.
66. G. Louchard. Kac's formula, Lévy's local time and Brownian excursion. *J. Appl. Probab.*, 21:479–499, 1984.
67. G. Louchard. The Brownian excursion area: A numerical analysis. *Comput. Math. Appl.*, 10:413–417, 1985.
68. G. Louchard. Erratum: "The Brownian excursion area: A numerical analysis". *Comput. Math. Appl.*, 12:375, 1986.
69. G. Louchard. Probabilistic analysis of some (un)directed animals. *Theoret. Comput. Sci.*, 159:65–79, 1996.

70. G. Louchard. Probabilistic analysis of column-convex and directed diagonally-convex animals. *Random Structures Algorithms*, 11:151–178, 1997.
71. G. Louchard. Probabilistic analysis of column-convex and directed diagonally-convex animals. II. Trajectories and shapes. *Random Structures Algorithms*, 15:1–23, 1999.
72. A. Del Lungo, M. Mirolli, R. Pinzani, and S. Rinaldi. A bijection for directed-convex polyominoes. *Discr. Math. Theo. Comput. Sci.*, AA (DM-CCG):133–144, 2001.
73. J. Ma and E.J. Janse van Rensburg. Rectangular vesicles in three dimensions. *J. Phys. A: Math. Gen.*, 38:4115–4147, 2005.
74. N. Madras and G. Slade. *The Self-Avoiding Walk*. Birkhäuser Boston, Boston, MA, 1993.
75. S.N. Majumdar. Brownian functionals in physics and computer science. *Current Sci.*, 89:2076–2092, 2005.
76. S.N. Majumdar and A. Comtet. Airy distribution function: From the area under a Brownian excursion to the maximal height of fluctuating interfaces. *J. Stat. Phys.*, 119:777–826, 2005.
77. M. Nguyễn Thế. Area of Brownian motion with generatingfunctionology. In C. Banderier and C. Krattenthaler, editors, *Discrete Random Walks, DRW'03*, Discrete Mathematics and Theoretical Computer Science Proceedings, AC, pages 229–242. Assoc. Discrete Math. Theor. Comput. Sci., Nancy, 2003.
78. M. Nguyễn Thế. Area and inertial moment of Dyck paths. *Combin. Probab. Comput.*, 13:697–716, 2004.
79. A.M. Odlyzko. Asymptotic enumeration methods. In R.L. Graham, M. Grötschel, and L. Lovász, editors, *Handbook of Combinatorics*, volume 2, pages 1063–1229. Elsevier, Amsterdam, 1995.
80. F.W.J. Olver. *Asymptotics and Special Functions*. Academic Press, New York, 1974.
81. R. Pemantle and M. Wilson. Twenty combinatorial examples of asymptotics derived from multivariate generating functions. *SIAM Rev.*, 50:199–272, 2008.
82. J. Pitman. *Combinatorial Stochastic Processes*, volume 1875 of *Lecture Notes in Mathematics*. Springer-Verlag, Berlin, 2006.
83. T. Prellberg. Uniform *q*-series asymptotics for staircase polygons. *J. Phys. A: Math. Gen.*, 28:1289–1304, 1995.
84. T. Prellberg and R. Brak. Critical exponents from nonlinear functional equations for partially directed cluster models. *J. Stat. Phys.*, 78:701–730, 1995.
85. T. Prellberg and A.L. Owczarek. Stacking models of vesicles and compact clusters. *J. Stat. Phys.*, 80:755–779, 1995.
86. A. Rechnitzer. Haruspicy 2: The anisotropic generating function of self-avoiding polygons is not *D*-finite. *J. Combin. Theory Ser. A*, 113:520–546, 2006.
87. C. Richard. Scaling behaviour of two-dimensional polygon models. *J. Stat. Phys.*, 108:459–493, 2002.
88. C. Richard. Staircase polygons: Moments of diagonal lengths and column heights. *J. Phys.: Conf. Ser.*, 42:239–257, 2006.
89. C. Richard. On *q*-functional equations and excursion moments. *Discr. Math., in press*, 2008. math.CO/0503198.
90. C. Richard and A.J. Guttmann. *q*-linear approximants: Scaling functions for polygon models. *J. Phys. A: Math. Gen.*, 34:4783–4796, 2001.
91. C. Richard, A.J. Guttmann, and I. Jensen. Scaling function and universal amplitude combinations for self-avoiding polygons. *J. Phys. A: Math. Gen.*, 34:L495–L501, 2001.
92. C. Richard, I. Jensen, and A.J. Guttmann. Scaling function for self-avoiding polygons. In D. Iagolnitzer, V. Rivasseau, and J. Zinn-Justin, editors, *Proceedings of the International Congress on Theoretical Physics TH2002 (Paris), Supplement*, pages 267–277. Birkhäuser, Basel, 2003.
93. C. Richard, I. Jensen, and A.J. Guttmann. Scaling function for self-avoiding polygons revisited. *J. Stat. Mech.: Th. Exp.*, page P08007, 2004.
94. C. Richard, I. Jensen, and A.J. Guttmann. Area distribution and scaling function for punctured polygons. *Electronic Journal of Combinatorics*, 15:#R53, 2008.
95. C. Richard, U. Schwerdtfeger, and B. Thatte. Area limit laws for symmetry classes of staircase polygons. Preprint arXiv:0710:4041, 2007.

96. U. Schwerdtfeger. Exact solution of two classes of prudent polygons. Preprint arXiv:0809:5232, 2008.

97. U. Schwerdtfeger. Volume laws for boxed plane partitions and area laws for Ferrers diagrams. In *Fifth Colloquium on Mathematics and Computer Science*, Discrete Mathematics and Theoretical Computer Science Proceedings, AG, pages 535–544. Assoc. Discrete Math. Theor. Comput. Sci., Nancy, 2008.

98. R.P. Stanley. *Enumerative Combinatorics*, volume 2. Cambridge University Press, Cambridge, Cambridge.

99. L. Takács. On a probability problem connected with railway traffic. *J. Appl. Math. Stochastic Anal.*, 4:1–27, 1991.

100. L. Takács. Limit distributions for the Bernoulli meander. *J. Appl. Prob.*, 32:375–395, 1995.

101. R. van der Hofstad and W. König. A survey of one-dimensional random polymers. *J. Statist. Phys.*, 103:915–944, 2001.

102. C. Vanderzande. *Lattice Models of Polymers*, volume 11 of *Cambridge Lecture Notes in Physics*. Cambridge University Press, Cambridge, 1998.

103. L. Di Vizio, J.-P. Ramis, J. Sauloy, and C. Zhang. Équations aux $q$-différences. *Gaz. Math.*, 96:20–49, 2003.

104. S.G. Whittington. Statistical mechanics of three-dimensional vesicles. *J. Math. Chem.*, 14:103–110, 1993.

# Chapter 12
# Interacting Lattice Polygons

Aleks L Owczarek and Stuart G Whittington

## 12.1 Introduction

A polymer is a long chain molecule of repeated chemical units, monomers. A *ring polymer* is simply a polymer whose ends have been joined so that topologically the molecule forms a circle. Lattice polygons are useful models of the configurational properties of flexible ring polymers in dilute solution in so-called "good" solvents. Good solvents are those where any attractive interactions between parts of the polymer have been effectively screened by the solvent molecules, leaving only entropic repulsion. The model of ring polymers as lattice polygons then can be modified by adding interactions to mimic phenomena such as ring polymer *adsorption* and *collapse* . Lattice polygons play the same role for ring polymers as self-avoiding walks do for linear polymers.

Polymers in dilute solution interacting with an impenetrable surface to which the monomers are attracted can undergo a phase transition, known as the adsorption transition. At high temperatures the polymer is repelled entropically from the wall and has very few monomers in contact with the wall: this is known as the desorbed phase. There is a phase transition at some particular temperature and at low temperatures the system behaves differently: in this adsorbed phase there is a positive density of monomers in contact with the wall. See Fig. 12.1. The lattice polygon model of ring polymer adsorption will be discussed in Section 12.2.

When a polymer is in dilute solution in a good solvent the polymer forms an open random coil and its root-mean-square radius of gyration scales like $n^v$ where $n$ is the degree of polymerization and $v$ is about 0.588 (in three dimensions). In so-called "poor" solvent conditions the monomer–solvent contacts are energetically

Aleks Owczarek
Department of Mathematics and Statistics, The University of Melbourne, Victoria, Australia, e-mail: A.Owczarek@ms.unimelb.edu.au

Stuart Whittington
Department of Chemistry, University of Toronto, Toronto, Ontario, Canada, e-mail: swhittin@chem.utoronto.ca

Desorbed Phase                              Absorbed Phase

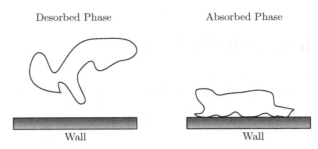

Wall                                              Wall

**Fig. 12.1** A schematic picture of a ring polymer in the desorbed (left-hand side) and adsorbed phases (right-hand side).

unfavourable and the polymer collapses to a compact ball to favour monomer–monomer contacts rather than monomer–solvent contacts. See Fig. 12.2. This collapse phenomenon has been observed for linear polymers by light scattering measurements. A useful model of the collapse transition for ring polymers is a lattice polygon with an additional vertex-vertex interaction which can be varied to favour or disfavour the collapse. In a sense this vertex-vertex interaction can be thought of as a potential of mean force which takes account of solvent–monomer interactions. The model will be discussed in Section 12.3.

'Good' solvents                              'Poor' solvents

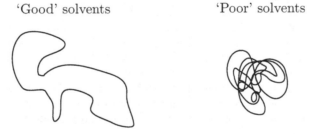

**Fig. 12.2** A schematic picture of a ring polymer in a good solvent as an open coil (left-hand side) and in a poor solvent as a compact globule (right-hand side).

In principle the phenomena of adsorption and collapse can both occur and the situation then results in a rich phase diagram which is discussed in Section 12.4.

An interesting question which occurs in each of these three cases is whether the free energies of the walk and polygon models are identical in the infinite size limit. This will be a particular focus of this chapter and we shall say something both about what is known rigorously and about some numerical studies of this question.

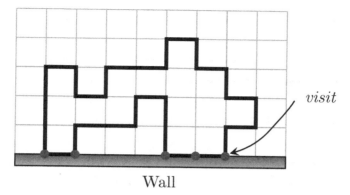

**Fig. 12.3** A polygon on the square lattice with $v = 5$ *visits* to the surface (wall).

## 12.2 The Adsorption Transition

Polymer molecules in dilute solution in a good solvent can adsorb at an impenetrable surface. This phenomenon plays an important role in such phenomena as steric stabilisation of dispersions [11]. At high temperatures the polymer will be desorbed and will have only a vanishingly small fraction of its monomers near the wall while at low temperatures it will adsorb and have a positive density of its monomers near the wall.

A natural model of this phenomenon is a self-avoiding walk (for linear polymers) or a lattice polygon (for ring polymers) with an interaction with the wall. We shall describe both models. Consider the simple cubic lattice $\mathbb{Z}^3$, though most things go through to $\mathbb{Z}^d$, $d > 3$, with no real difficulty. The $d = 2$ case turns out to have some special features and we consider this case separately. Attach a coordinate system $(x, y, z)$ to the vertices so that $x$, $y$ and $z$ are integers. Suppose that the impenetrable adsorbing surface is the plane $z = 0$ and that the solvent corresponds to the half-space $z > 0$. Suppose that $c_n^+(v)$ is the number of self-avoiding walks with $n$ edges, starting at the origin, having no vertices with negative $z$-coordinate and having $v + 1$ vertices in the plane $z = 0$. We say that such a walk *visits* the plane $z = 0$ $v$ times or that it has $v$ *visits*. Similarly let $p_n(v)$ be the number of lattice polygons with $n$ edges, having no vertices with negative $z$-coordinate and having $v \geq 2$ vertices in $z = 0$. See Fig. 12.3. We shall be interested in the partition functions

$$C_n(\alpha) = \sum_{v=0}^{n} c_n(v) e^{\alpha v} \tag{12.1}$$

and

$$P_n(\alpha) = \sum_{v=2}^{n} p_n(v) e^{\alpha v}. \tag{12.2}$$

We can define corresponding (reduced, intensive) free energies

$$\kappa_n(\alpha) = n^{-1} \log C_n(\alpha) \tag{12.3}$$

and

$$\kappa_n^0(\alpha) = n^{-1} \log P_n(\alpha) \tag{12.4}$$

and we shall be interested in quantities such as

$$\rho_n(\alpha) = \frac{\partial \kappa_n(\alpha)}{\partial \alpha} = \frac{1}{n} \frac{\sum_v v c_n(v) e^{\alpha v}}{\sum_v c_n(v) e^{\alpha v}} \tag{12.5}$$

and

$$\rho_n^0(\alpha) = \frac{\partial \kappa_n^0(\alpha)}{\partial \alpha} = \frac{1}{n} \frac{\sum_v v p_n(v) e^{\alpha v}}{\sum_v p_n(v) e^{\alpha v}}. \tag{12.6}$$

These are the mean fractions of visits for the two models and we expect that these will be small (in fact zero in the infinite $n$ limit) when the polymer is desorbed. For the infinite $n$ case we expect behaviour similar to that sketched in Fig. 12.4.

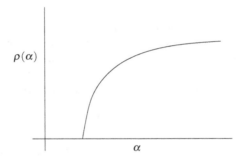

**Fig. 12.4** The expected dependence of the mean fraction of visits on $\alpha$.

Rigorous results are available [5, 15] about the existence of the limits in (12.3) and (12.4). It is known rigorously that these two free energies are equal for $d \geq 3$ for all $\alpha$ [15] and that they are non-analytic functions of $\alpha$. This means that the two models have a phase transition (corresponding to adsorption) and that this transition is at the same place for the two models. Bounds are available on the location of the transition [5, 7] but these are weak and one must turn to numerical methods for estimates of the location of the transition [3, 6, 8]. Similarly the order of the transition is not known rigorously but numerical results are available, though there is some disagreement [3, 6, 8] about the value of the crossover exponent $\phi$, defined below, see eqn. (12.14).

Other interesting properties include the $n$-dependence of the mean span of the polygon or walk in the $z$-direction and in the $x$- (or $y$-) direction as a function of $\alpha$. Equivalently one could look at the various components of the radius of gyration. All of these can act as signals that the walk or polygon is desorbed or adsorbed. These quantities can be estimated by Monte Carlo methods [6, 8] or by exact enumeration techniques.

## 12.2.1 Rigorous Results

We focus on the results for polygons though the theorems were originally proved for self-avoiding walks [5]. At first we confine ourselves to $d = 3$ though the results go over in a straightforward way to $d > 3$. Concatenation arguments can be used to prove the existence of the limit

$$\lim_{n\to\infty} \kappa_n^0(\alpha) \equiv \kappa^0(\alpha) \tag{12.7}$$

for all $\alpha < \infty$. Similarly it is possible to prove that $\kappa^0(\alpha)$ is a convex function of $\alpha$ and is therefore continuous. Moreover $\kappa^0(\alpha)$ is differentiable almost everywhere.

For $\alpha \geq 0$ it is easy to see that $P_n(\alpha) \leq p_n e^{\alpha n}$ where $p_n = \sum_v p_n(v)$ is the number of $n$-edge polygons on $\mathbb{Z}^3$. Writing $\lim_{n\to\infty} n^{-1} \log p_n = \kappa_3$ we have

$$\kappa^0(\alpha) \leq \kappa_3 + \alpha, \quad \alpha \geq 0. \tag{12.8}$$

By picking out a particular term we have

$$P_n(\alpha) \geq p_n(n) e^{\alpha n} \tag{12.9}$$

and, by monotonicity in $\alpha$, $P_n(\alpha) \geq P_n(0) = p_n$. Note that $p_n(n)$ counts polygons on $\mathbb{Z}^2$. If we write $\lim_{n\to\infty} n^{-1} \log p_n(n) = \kappa_2$ we then obtain

$$\kappa^0(\alpha) \geq \max[\kappa_3, \kappa_2 + \alpha], \quad \alpha \geq 0. \tag{12.10}$$

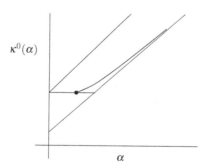

**Fig. 12.5** The dependence of the reduced limiting free energy, $\kappa^0(\alpha)$ on $\alpha$. The straight lines are bounds and the filled circle shows the location of the adsorption transition.

For $\alpha \leq 0$ we have $P_n(\alpha) \leq P_n(0) = p_n$ by monotonicity and hence $\kappa^0(\alpha) \leq \kappa_3$. Again we can pick out a particular term to give

$$P_n(\alpha) \geq p_n(2) e^{2\alpha} \tag{12.11}$$

and, since $p_n(2) \geq p_{n-2}$, we have $\kappa^0(\alpha) \geq \kappa_3$. This means that the free energy is equal to $\kappa_3$, independent of $\alpha$, for $\alpha \leq 0$ but strictly greater than $\kappa_3$ for $\alpha > \kappa_3 - \kappa_2$. Hence $\kappa^0(\alpha)$ has a singular point at $\alpha = \alpha_0$ with $0 \leq \alpha_0 \leq \kappa_3 - \kappa_2$. With a little more work these two inequalities can be made strict [5, 7]. In addition it can be shown that $\kappa^0(\alpha)$ is asymptotic to the line $\kappa_2 + \alpha$ as $\alpha \to \infty$. The behaviour is sketched in Fig. 12.5.

Soteros [15] showed that $\kappa(\alpha) = \kappa^0(\alpha)$ so the two limiting free energies are equal and, in particular, the adsorption points are the same for the two models.

If we are interested in the mean fraction of visits in the infinite $n$ limit we need to look at

$$\rho^0(\alpha) = \lim_{n \to \infty} \rho_n^0(\alpha) = \lim_{n \to \infty} \frac{\partial \kappa_n^0(\alpha)}{\partial \alpha}. \tag{12.12}$$

$\kappa_n^0(\alpha)$ is a convex function of $\alpha$ so the order of the limit and derivative can be interchanged so that

$$\rho^0(\alpha) = \frac{\partial \kappa^0(\alpha)}{\partial \alpha} \tag{12.13}$$

and hence $\rho^0(\alpha)$ is zero for $\alpha < \alpha_0$. This is the desorbed phase where the mean fraction of visits vanishes in the infinite $n$ limit. When $\alpha > \alpha_0$ $\rho^0(\alpha) > 0$ and we are in the adsorbed phase. It is not known rigorously whether or not $\rho^0(\alpha)$ is continuous at $\alpha = \alpha_0$.

The behaviour of the free energy near $\alpha = \alpha_0$ is governed by the crossover exponent, $\phi$. Formally this is defined as

$$\frac{1}{\phi} = \lim_{\alpha \to \alpha_0+} \frac{\kappa^0(\alpha) - \kappa^0(\alpha_0)}{\alpha - \alpha_0}. \tag{12.14}$$

It is not known rigorously that the limit exists but, if it does, the above result of Soteros [15] means that the crossover exponent has the same value for the walk and polygon models, for $d \geq 3$.

When we turn to examine the situation for $d = 2$ there are some important differences. One can prove that the limiting free energies $\kappa^0(\alpha)$ and $\kappa(\alpha)$ exist and both functions are convex in $\alpha$. For the walk problem we have

$$\max[\kappa_2, \alpha] \leq \kappa(\alpha) \leq \kappa_2 + \alpha, \quad \alpha \geq 0 \tag{12.15}$$

and

$$\kappa(\alpha) = \kappa_2, \quad \alpha \leq 0 \tag{12.16}$$

which is analogous to the results for $d \geq 3$ for walks. For polygons in $\mathbb{Z}^2$ we have $\kappa^0(\alpha) = \kappa_2$ for $\alpha \leq 0$. However, for $\alpha \geq 0$ the maximum number of vertices of the polygon which can be in the line $y = 0$ is $n/2$. Hence

$$\max[\kappa_2, \alpha/2] \leq \kappa^0(\alpha) \leq \kappa_2 + \alpha/2, \quad \alpha \geq 0. \tag{12.17}$$

This means that $\kappa^0(\alpha)$ has an asymptote with slope $1/2$ as $\alpha \to \infty$ (cf walks which have an asymptote with unit slope). Therefore $\kappa(\alpha) > \kappa^0(\alpha)$ for large enough values of $\alpha$ and the two free energies are not identical. Both $\kappa(\alpha)$ and $\kappa^0(\alpha)$ have singular points (at $\alpha_w$ and $\alpha_0$ respectively) and $\alpha_0 \geq \alpha_w$. Whether these are distinct is not known.

## 12.2.2 Numerical Results

The two primary numerical methods which have been used to investigate this problem are exact enumeration coupled with series analysis and Monte Carlo methods. In addition there are some transfer matrix calculations in two dimensions. The main quantities of interest are:

1. The temperature dependence of the free energy.
2. The location of the phase transition.
3. The shape of the free energy curve close to the phase transition, in the low temperature (adsorbed) phase, characterized by the crossover exponent $\phi$, and hence the value of $\phi$.
4. Various metric quantities such as the mean-square radius of gyration or the mean distance of a vertex from the surface, as a function of temperature.

It isn't too difficult to use exact enumeration methods to calculate the values of $\kappa_n(\alpha)$ and $\kappa_n^0(\alpha)$, defined by (12.3) and (12.4), for modest values of $n$. Series analysis techniques, discussed in Chapter 8, such as ratio methods, can then be used to estimate $\kappa(\alpha)$ and $\kappa^0(\alpha)$, which we know to be identical in three and higher dimensions. The difficulty is to extract a reliable estimate of the critical value of $\alpha$ since one is asking where a function stops being constant. It is easy enough to get a reasonable upper bound on $\alpha_0$ (or on $\alpha_w$ in two dimensions where $\alpha_0$ and $\alpha_w$ may be different) but it is extraordinarily difficult to estimate a reliable lower bound. One might hope to examine the corresponding fluctuation quantity $\partial^2 \kappa_n(\alpha)/\partial \alpha^2$ but this doesn't behave well for small $n$ and is difficult to extrapolate. As a result the most reliable estimates of $\alpha_0$ (and $\alpha_w$) come from Monte Carlo calculations.

We are not aware of any Monte Carlo calculations for adsorption of ring polymers (i.e. polygons) but there are many studies of the adsorption of linear polymers (i.e. self-avoiding walks) both in two and in three dimensions. In three dimensions we know, vide supra, that polygons and walks have the same limiting free energies so calculations for walks give useful information for polygons, in the thermodynamic limit. We shall not attempt to give a systematic survey of the literature on Monte Carlo studies for walks but we content ourselves with mentioning some recent papers and pointing to an interesting open question.

Hegger and Grassberger [6] carried out a very thorough Monte Carlo study of adsorption of self-avoiding walks in three dimensions (on the simple cubic lattice), obtaining data for values of $n$ up to about 2000. At the time of their work the value of $\alpha_0$ was not known very precisely but the balance of evidence suggested a value around 0.285. Hegger and Grassberger examined a variety of different properties and concluded that $\alpha_0$ was probably between 0.2857 and 0.2861, with a preferred value of about 0.2859. They also noticed strong corrections to scaling which make it difficult to extrapolate from data for short walks. There were several previous attempts to estimate the value of $\phi$ in three dimensions and they estimated

$$\phi = 0.496 \pm 0.004 \qquad (12.18)$$

which was somewhat smaller than previous estimates.

Janse van Rensburg and Rechnitzer [8] attacked the problem somewhat differently by producing very high quality Monte Carlo data on shorter walks (with $n$ values up to 120) and then doing a careful statistical analysis in which they attempted to incorporate correction to scaling terms. They estimated $\alpha_0 = 0.288 \pm 0.020$ for the simple cubic lattice and

$$\phi = 0.5005 \pm 0.0036. \qquad (12.19)$$

The two approaches give values of $\alpha_0$ which are in reasonable agreement and both studies are consistent with a value of $\phi = 1/2$. Since there is good evidence that $\phi = 1/2$ also in two dimensions this would imply a super-universality for $\phi$ in dimensions 2 to 4 [6]. However, Grassberger [3] returned to the problem using a somewhat different algorithm and obtained data for walks with $n$ up to 8000. He gave the very precise estimate

$$\alpha_0 = 0.28567 \pm 0.00008 \qquad (12.20)$$

but gave a lower estimate for $\phi$, namely

$$\phi = 0.484 \pm 0.002. \qquad (12.21)$$

If this value is correct then $\phi$ is not super-universal.

In two dimensions we know that walks and polygons do not have the same limiting free energy and we do not know of any Monte Carlo calculations for polygons in two dimensions.

## 12.3 The Collapse Transition

In dilute solution in a good solvent, polymers are typically expanded coils. In these conditions the monomer–solvent contacts are favourable and monomers tend to be surrounded by solvent. In a poor solvent monomer–solvent contacts become less favourable and the polymer collapses to a compact ball producing monomer–monomer contacts at the expense of monomer–solvent contacts. Typically the solvent becomes worse as the temperature decreases and there is a temperature, the $\theta$-temperature, at which the polymer collapses.

This situation for a ring polymer can be modelled by considering polygons with $n$ edges where we keep track of the number of pairs of vertices which are unit distance apart but are not joined by an edge of the polygon. We call these *contacts*. See Fig. 12.6 for an example on the square lattice. We can weight polygons according to the number of contacts. Suppose that $p_n(k)$ is the number of $n$-edge polygons with $k$ contacts. For example, on the square lattice $p_8(2) = 6$ and $p_8(0) = 1$.

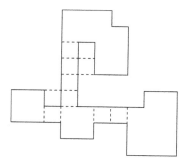

**Fig. 12.6** A polygon with 13 contacts. The contacts are indicated by dashed lines.

Define the partition function

$$Z_n^0(\beta) = \sum_k p_n(k) e^{\beta k} \tag{12.22}$$

and the corresponding free energy

$$F_n^0(\beta) = n^{-1} \log Z_n^0(\beta). \tag{12.23}$$

We expect that the limit $F^0(\beta) = \lim_{n \to \infty} F_n^0(\beta)$ will exist and that $F^0(\beta)$ will have a singularity at $\beta = \beta_c^0$ corresponding to the $\theta$-point.

One can define a similar model for self-avoiding walks. Let $c_n(k)$ be the number of $n$-edge self-avoiding walks with $k$ contacts. Define the partition function

$$Z_n(\beta) = \sum_k c_n(k) e^{\beta k} \tag{12.24}$$

and the corresponding free energy

$$F_n(\beta) = n^{-1} \log Z_n(\beta). \tag{12.25}$$

Again we expect that the limit $F(\beta) = \lim_{n \to \infty} F_n(\beta)$ will exist and that $F(\beta)$ will be singular at some $\beta = \beta_c$. Some natural questions which arise are whether $F^0(\beta) = F(\beta)$. If not is $\beta_c^0$ different from $\beta_c$?

If we write $S_n(\beta)$ and $S_n^0(\beta)$ for the mean-square radius of gyration of $n$-edge walks and polygons at parameter $\beta$, it is natural to expect that $S_n(\beta) \sim An^\nu$ for $\beta < \beta_c$, where $\nu$ is the exponent corresponding to a good solvent. For $\beta > \beta_c$ (i.e. at low temperatures) $S_n(\beta) \sim Bn^{1/d}$, with similar expressions for the polygon problem. That is, the transition is associated with a change in the radius of gyration exponent.

## 12.3.1 Rigorous Results

Remarkably little is known rigorously about this problem, either for the polygon model or for the self-avoiding walk model. Concatenation arguments [16] can be used to prove the existence of the limit

$$F^0(\beta) = \lim_{n\to\infty} n^{-1} \log Z_n^0(\beta).$$ (12.26)

In addition it can be shown that $F^0(\beta)$ is a convex function of $\beta$ and hence is continuous. Tesi et al. [16] also showed that, if $\beta \le 0$, the limiting free energy $F(\beta) = \lim_{n\to\infty} F_n(\beta)$ exists and that $F(\beta) = F^0(\beta)$. So the polygon and walk models have the same free energy for $\beta \le 0$, i.e. for repulsive interactions. To get a bound in one direction one deletes an edge from a polygon giving

$$c_{n-1}(k+1) \ge p_n(k)$$ (12.27)

since the edge deletion creates a contact. This immediately gives

$$\liminf_{n\to\infty} F_n(\beta) \ge F^0(\beta)$$ (12.28)

and this bound works for all $\beta$. The idea behind the bound in the other direction is to relate polygons and walks to *unfolded walks* and to use a theorem about unfolded walks due to Hammersley and Welsh [4]. This does not work for $\beta > 0$ because unfolding can destroy contacts. The existence of the thermodynamic limit for the walk model is an open question.

Tesi et al. proved another potentially useful lemma [16]. Suppose that $K_n(\beta)$ is the mean number of contacts for $n$-edge self-avoiding walks at parameter $\beta$, i.e.

$$K_n(\beta) = \frac{\sum_k k c_n(k) e^{\beta k}}{\sum_k c_n(k) e^{\beta k}}$$ (12.29)

and that $K_n^0(\beta)$ is the corresponding quantity for the polygon model. Tesi et al. showed that if $K_n^0(\beta) \ge K_n(\beta)$ for all sufficiently large even $n$ and for all $\beta > 0$ then $\lim_{n\to\infty} F_n(\beta)$ exists and $F(\beta) = F^0(\beta)$ for all $\beta$.

## 12.3.2 Numerical Results

The question of whether or not the limiting free energies of walks and polygons are equal for $\beta > 0$ has been addressed numerically [2, 16]. Tesi et al. [16] used Monte Carlo methods (in fact umbrella sampling and multiple Markov chain methods, as described in Chapter 9) to study self-interacting polygons on the simple cubic lattice. They estimated the heat capacity $\partial^2 F_n^0(\beta)/\partial \beta^2$ as a function of $n$ and $\beta$ and observed peaks in the heat capacity (as a function of $\beta$) which increase in height as $n$ increases, consistent with a second order phase transition at a critical value of $\beta$ which they estimated to be $0.2782 \pm 0.0070$. The corresponding estimate for the walk problem is $0.2779 \pm 0.0041$, so the results are consistent with walks and polygons collapsing at the same temperature.

Tesi et al. [16] also estimated the free energy difference

$$\Delta F_n(\beta) = [F_n^0(\beta) - F_n^0(0)] - [F_n(\beta) - F_n(0)] \tag{12.30}$$

(recall that $F(0) = F^0(0)$). They observed that $\Delta F_n(\beta)$ is positive for $\beta > 0$ but decreases as $n$ increases, again consistent with the free energies being equal in the infinite $n$ limit.

To test this further they estimated the ratio $K_n^0(\beta)/K_n(\beta)$ as a function of $n$ and $\beta$. The ratio is greater than unity for the range of $\beta$ and $n$ values studied, going through a maximum at fixed $n$ and decreasing towards unity as $\beta$ increases. The height of the maximum decreases as $n$ increases. Using the result discussed in Section 12.3.1, this is strong evidence that the two models have the same limiting free energy for all values of $\beta$. Incidentally, this implies that they have the same value for the crossover exponent, $\phi$.

For the square lattice in two dimensions the critical value of $\beta$ has been estimated for the walk and polygon problems and the results are consistent with a common value $\beta_c = 0.663 \pm 0.016$. The evidence is reviewed briefly in [2].

Bennett-Wood et al. [2] derived exact enumeration data for the square lattice for $n \leq 29$ for the walk model and for $n \leq 42$ for the polygon model, enabling them to calculate $Z_n(\beta)$ and $Z_n^0(\beta)$ for these values of $n$. They computed $K_n(\beta)$ and $K_n^0(\beta)$. For $\beta < 0.6$ (i.e. at high temperature) they observed that $K_n^0(\beta) > K_n(\beta)$ at the largest values of $n$ considered, and gave evidence that $K_n^0(\beta) - K_n(\beta) \to 0$ as $n \to \infty$, for $\beta < 0.663$. For larger values of $\beta$ (beyond the collapse transition) $K_n(\beta) > K_n^0(\beta)$ for the values of $n$ considered so the Lemma of Tesi et al. [16] does not apply.

To investigate the situation at low temperatures ($\beta > 0.663$) Bennett-Wood et al. [2] defined

$$Q_n = \frac{\sqrt{Z_{n+1}^0 Z_{n-1}^0}}{Z_n} \tag{12.31}$$

for $n$ odd and $Q_n = Z_n^0/Z_n$ for $n$ even. They used series analysis techniques to estimate $\lim_{n \to \infty} Q_n^{1/n}$ for various values of $\beta$. For $0 < \beta < 1.5$ (so well into the collapsed phase) they estimated that $\lim_{n \to \infty} Q_n^{1/n}$ is unity within the estimated error bars. This is consistent with the equality of the limiting free energies for the walk and polygon models for all values of $\beta$ which were considered.

## 12.4 Adsorption and Collapse

One can also consider the situation where a polymer can adsorb at a surface and collapse into a compact ball. This involves having two different energy terms, one corresponding to the attraction of a monomer to the surface at which adsorption can occur, and another corresponding to the monomer–monomer attraction which can lead to collapse.

Consider the simple cubic lattice and the half-space $z \geq 0$. Suppose that $p_n(v,k)$ is the number of $n$-edge polygons with $v$ vertices in the plane $z = 0$ ($v \geq 2$) and with $k$ contacts. The appropriate partition function is now

$$Z_n^0(\alpha,\beta) = \sum_{v,k} p_n(v,k) e^{\alpha v + \beta k} \tag{12.32}$$

and the corresponding (intensive) free energy is

$$F_n^0(\alpha,\beta) = n^{-1} \log Z_n^0(\alpha,\beta). \tag{12.33}$$

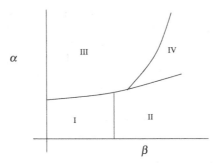

**Fig. 12.7** The phase diagram for polygons which can both adsorb at a surface and collapse. The four phases are I: desorbed and expanded, II: desorbed and collapsed, III: adsorbed and expanded, and IV: adsorbed and collapsed.

One might expect four different phases in the $(\alpha,\beta)$-plane. At small $\alpha$ and small $\beta$ the polymer should be desorbed and expanded. As $\alpha$ increases at small fixed $\beta$ the polymer should adsorb to give a phase where the polymer is adsorbed and expanded. Increasing $\beta$ at fixed small $\alpha$ should give a desorbed and collapsed phase. From the interior of the adsorbed and expanded phase increasing $\beta$ should lead to collapse in the adsorbed phase, to give a collapsed phase at the surface. See Fig. 12.7 for a sketch of a possible phase diagram.

### 12.4.1 Rigorous Results

A few rigorous results for this model have been obtained by Vrbovà and Whittington [18]. They used concatenation arguments to prove the existence of the limit

$$\lim_{n \to \infty} F_n^0(\alpha, \beta) \equiv F^0(\alpha, \beta) \tag{12.34}$$

for all $\alpha < \infty$ and $\beta < \infty$. They also showed that $F^0(\alpha, \beta)$ is doubly convex (i.e. convex as a surface, which is a stronger statement than being convex in both arguments), and hence a continuous function of $\alpha$ and $\beta$.

The arguments of Section 12.2.1 can be extended to show that polygons exhibit an adsorption transition at some critical value of $\alpha = \alpha_c(\beta)$ for all $\beta < \infty$. This establishes the existence of the phase boundary between the desorbed phases (I and II) and the adsorbed phases (III and IV) in Figure 12.7. It is not known if the phase boundary $\alpha = \alpha_c(\beta)$ is a continuous function of $\beta$. It is also not known rigorously that there is a collapse transition at some $\beta = \beta_o$ when $\alpha = 0$. However, if these two conditions are met then the phase boundary between the desorbed and expanded phase (I) and the desorbed and collapsed phase (II) is a straight line and the critical value of $\beta$ is independent of $\alpha$.

### 12.4.2 Numerical Results

The polygon problem does not seem to have been studied numerically but one would expect the same phase diagram for walks and for polygons so we discuss some numerical results for the corresponding walk problem. Vrbovà and Whittington [19] used Monte Carlo methods to investigate this problem for a self-avoiding walk model on the simple cubic lattice. They found clear evidence for four phases: desorbed-expanded, desorbed-collapsed, adsorbed-expanded and adsorbed-collapsed, with a phase diagram qualitatively similar to that shown in Figure 12.7. In particular they found evidence for two triple points and a phase boundary between the adsorbed-expanded and desorbed-collapsed phases. Vrbovà and Procházka [17] found evidence from Monte Carlo data that the phase boundary between the desorbed-expanded and adsorbed-expanded phases is a horizontal line in Fig. 12.7. That is, the adsorption critical point doesn't depend on $\beta$ until the collapse point is reached.

Singh and coworkers [12, 14] studied the same problem in three dimensions, using exact enumeration techniques. Although there was some initial disagreement as to whether the phase diagram had two triple points or a quadruple point (where four phases coexist) there now seems to be general agreement that there are two triple points, as sketched in Fig. 12.7. Vrbovà and Whittington [20] also studied a similar model with adsorption at a penetrable surface and found a similar phase diagram.

Krawczyk et al. [9] have investigated the problem in three dimensions also with a Monte Carlo technique known as flatPERM, which allowed the consideration of a large part of the phase space at once. Walks attached to the surface up to length 256 were considered. While unusual features appear at low temperatures for finite lengths they find a phase diagram in agreement with that in Fig. 12.7 from Vrbovà and Whittington [19].

Singh et al. [13] (see also [10]) suggested the existence of an additional *surface attached globule phase* though it seems [9] that this doesn't correspond to a bulk phase transition. That is, its phase boundary does not correspond to singularities in the limiting free energy defined in (12.34) but to singularities in a suitably defined surface free energy.

Bachmann and Janke [1] have simulated a variant of the problem where the walk is not attached to the attractive surface. As a consequence they need to place a second wall some distance from the first. Walks of length 100 were simulated and the pseudo phase diagram suggests extra phases. It will be interesting for future work to delineate the exact difference between the tethered and non-tethered cases.

# References

1. M. Bachmann and W. Janke: *Conformational transitions of nongrafted polymers near an adsorbing substrate* Phys. Rev. Lett. **95**, 058102 (2005)
2. D. Bennett-Wood, I. G. Enting, D. S. Gaunt, A. J. Guttmann, J. L. Leask, A. L. Owczarek and S. G. Whittington: *Exact enumeration study of free energies of interacting polygons and walks in two dimensions* J. Phys. A: Math. Gen. **31**, 4725–4741 (1998)
3. P. Grassberger: *Simulations of grafted polymers in a good solvent* J. Phys. A: Math. Gen. **38**, 323–331 (2005)
4. J. M. Hammersley and D. J. A. Welsh: *Further results on the rate of convergence to the connective constant of the hypercubic lattice* Quart. J. Math. Oxford **18**, 108–110 (1962)
5. J. M. Hammersley, G. M. Torrie and S. G. Whittington: *Self-avoiding walks interacting with a surface* J. Phys. A: Math. Gen. **15**, 539–571 (1982)
6. R. Hegger and P. Grassberger: *Chain polymers near an adsorbing surface* J. Phys. A: Math. Gen. **27**, 4069–4081 (1994)
7. E. J. Janse van Rensburg: *Collapsing and adsorbing polygons* J. Phys. A: Math. Gen, **31**, 8295–8306 (1998)
8. E. J. Janse van Rensburg and A. Rechnitzer: *Multiple Markov chain Monte Carlo study of adsorbing self-avoiding walks in two and three dimensions* J. Phys. A: Math. Gen. **37**, 6875–6898 (2004)
9. J. Krawczyk, A. L. Owczarek, T. Prellberg and A. Rechnitzer: *Layering transitions for adsorbing polymers in poor solvents* Europhysics Lett., **70**, 726–732 (2005)
10. P. Mishra, D. Giri, S. Kumar and Y. Singh: *Does a surface attached globule phase exists?* Physica A **318**, 171–178 (2003)
11. D. Napper, *Polymeric Stabilisation of Colloidal Dispersions*, Academic Press, London, 1983.
12. R. Rajesh, D. Dhar, D. Giri, S. Kumar and Y. Singh: *Adsorption and collapse transitions in a linear polymer chain near an attractive wall* Phys. Rev. E **65**, 056124 (2002)
13. Y. Singh, D. Giri and S. Kumar: *Crossover of a polymer chain from bulk to surface states* J. Phys. A: Math. Gen. **34**, L67–L74 (2001)
14. Y. Singh, S. Kumar and D. Giri: *Surface adsorption and collapse transition of a linear polymer chain in three dimensions* J. Phys. A: Math. Gen. **32**, L407–L411 (1999)

15. C. E. Soteros: *Adsorption of lattice animals with specified topology* J. Phys. A: Math. Gen. **25**, 3153–3173 (1992)
16. M. C. Tesi, E. J. Janse van Rensburg, E. Orlandini and S. G. Whittington: *Interacting self-avoiding walks and polygons in three dimensions* J. Phys. A: Math. Gen. **29**, 2451–2463 (1996)
17. T. Vrbová and K. Procházka: *Adsorption of self-avoiding walks at an impenetrable plane in the expanded phase: a Monte Carlo study* J. Phys. A: Math. Gen. **32**, 5469–5476 (1999)
18. T. Vrbová and S. G. Whittington: *Adsorption and collapse of self-avoiding walks and polygons in three dimensions* J. Phys. A: Math. Gen. **29**, 6253–6264 (1996)
19. T. Vrbová and S. G. Whittington: *Adsorption and collapse of self-avoiding walks in three dimensions: A Monte Carlo study* J. Phys. A: Math. Gen. **31**, 3989–3998 (1998)
20. T. Vrbová and S. G. Whittington: *Adsorption and collapse of self-avoiding walks at a defect plane* J. Phys. A: Math. Gen. **31**, 7031–7041 (1998)

# Chapter 13
# Fully Packed Loop Models on Finite Geometries

Jan de Gier

## 13.1 Fully Packed Loop Models on the Square Lattice

A fully packed loop (FPL) model on the square lattice is the statistical ensemble of all loop configurations, where loops are drawn on the bonds of the lattice, and each loop visits every site once [4, 18]. On finite geometries, loops either connect external terminals on the boundary, or form closed circuits, see for example Fig. 13.1. In this chapter we shall be mainly concerned with FPL models on squares and rectangles with an alternating boundary condition where every other boundary terminal is covered by a loop segment, see Fig. 13.1.

An FPL model thus describes the statistics of closely packed polygons on a finite geometry. Polygons may be nested, corresponding to punctures studied in Chapter 8. FPL models can be generalised to include weights. In particular we will study FPL models where a weight $\tau$ is given to each straight local loop segment. The partition function of an FPL model on various geometries can be computed exactly using its relation to the solvable six-vertex lattice model. It is well known that the model undergoes a bulk phase transition at $\tau = 2$.

We furthermore study nests of polygons connected to the boundary. In the case of FPL models with mirror or rotational symmetry, the probability distribution function of such nests is known analytically, albeit conjecturally. FPL models undergo another phase transition as a function of the boundary nest fugacity. At criticality, we derive a scaling form for the nest distribution function which displays an unusual non-Gaussian cubic exponential behaviour.

The purpose of this chapter is to collect and discuss known results for FPL models which may be relevant to polygon models. For that reason we have not put an emphasis on derivations, many of which are well-documented in the existing literature, but rather on interpretations of results.

Jan de Gier

Department of Mathematics and Statistics, The University of Melbourne, Victoria, Australia, e-mail: degier@ms.unimelb.edu.au

**Fig. 13.1** Fully packed loops inside a square with alternating boundary condition.

### 13.1.1 Bijection with the Six-Vertex Model, Alternating-Sign Matrices and Height Configurations

There is a well-known one-to-one correspondence between FPL, six-vertex and alternating-sign configurations [69, 25]. In the six-vertex model, to each bond of the square lattice is associated an arrow, such that at each vertex there are two in- and two out-pointing arrows, see e.g. [7]. There are six local vertex configurations which are given in the top row of Fig. 13.2. The six-vertex and FPL configurations are related in the following way. The square lattice is divided into two sublattices, even (*A*) and odd (*B*). For each arrow configuration we draw only those bonds on which the arrow points to the even sublattice. If we choose the vertex in the upper

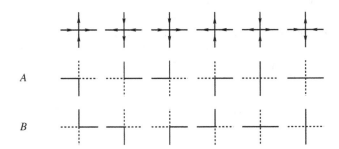

**Fig. 13.2** Bijection between six-vertex and FPL vertices. The correspondence is different on the two sublattices *A* and *B*.

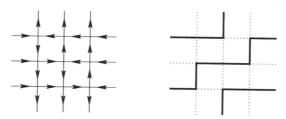

**Fig. 13.3** An equivalent six-vertex and fully packed loop configuration.

left corner to belong to the even sublattice, the six-vertex and FPL configuration in Fig. 13.3 are equivalent, as can be seen from the correspondence in Fig. 13.2.

Alternating sign matrices (ASMs) were introduced by Mills, Robbins and Rumsey [51, 52] and are matrices with entries in $\{-1, 0, 1\}$ such that the entries in each column and each row add up to 1 and the non-zero entries alternate in sign. A well-known subclass of ASMs are the permutation matrices. Let us also introduce the height interpretation of an ASM. Let $A = (a_{ij})_{i,j=1}^n$ be an ASM, then define the heights $h_{ij}$ by

$$h_{ij} = n - i - j + 2 \sum_{i' \leq i,\, j' \leq j} a_{i'j'}. \tag{13.1}$$

This rule ensures that neighbouring heights differ by one. The correspondence between the three objects is given in Fig. 13.4. An example of a six vertex and its corresponding height configuration is given in Fig. 13.5 for the $3 \times 3$ identity matrix.

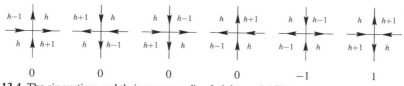

**Fig. 13.4** The six vertices and their corresponding heights and ASM entries.

## 13.1.2 Structure

As each external terminal, or outgoing bond, is connected to another terminal, FPL diagrams can be naturally labeled by link patterns, or equivalently, two-row Young tableaux or Dyck paths. For example, the diagram in Fig. 13.6 has link pattern $((()(()))(()))$ which is short hand for saying that 1 is connected to 12, 2 is connected to 11, 3 to 4 etc. The information about connectivities can also be coded in two-row standard Young tableaux. The entries of the first row of the Young tableau correspond to the positions of opening parentheses '(' in a link pattern, and the entries of

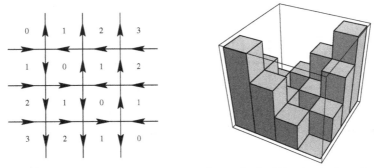

**Fig. 13.5** Vertex and height interpretation corresponding to the $3 \times 3$ identity matrix.

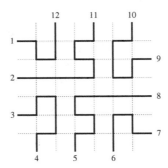

**Fig. 13.6** An FPL diagram with link pattern $((\,()\,(())\,()\,))$.

the second row to the positions of the closing parentheses ')'. The FPL diagram of
Fig. 13.6 carries as a label the standard Young tableau given in Fig. 13.7.

| 1 | 2 | 3 | 5 | 6 | 9 |
| 4 | 7 | 8 | 10 | 11 | 12 |

**Fig. 13.7** Standard Young tableau corresponding to the FPL diagram in Fig. 13.6.

Yet another way of coding the same information uses Dyck paths. Each entry in
the first row of the standard Young tableau represents an up step, while those in the
second row represent down steps. The Dyck path corresponding to Fig. 13.7 is given
in Fig. 13.8.

In this section we collect some structural results regarding local update moves of
FPL models. Following Wieland [78], we define operators $G_{ij}$ that act on the height
configurations as follows. They act as the identity on each square except on the
square at $(i, j)$ where they either increase or lower the height by 2 if it is allowed.
A change of height is allowed if neighbouring heights still differ by one after the
change. If it is not allowed, $G_{ij}$ acts as the identity. For future convenience we also

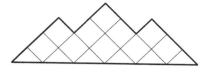

**Fig. 13.8** Dyck path corresponding to the FPL diagram in Fig. 13.6 and the standard Young tableau in Fig. 13.7.

define the operators

$$G_0 = \prod_{(i,j)\in S_0} G_{ij}, \quad G_1 = \prod_{(i,j)\in S_1} G_{ij}. \tag{13.2}$$

where $G_0$ and $G_1$ denote the even and odd sublattice of the square lattice respectively.

Starting from an initial height configuration, such as the one in Fig. 13.5, the operators $G_{ij}$ generate all height configurations. Put in other words, if we denote the height configuration corresponding to the unit matrix by $Z_1$, all other allowed height configurations correspond to a word in the operators $G_{ij}$ acting on $Z_1$.

On a plaquette of an FPL configuration, the involution $G$ acts as

$$G: \quad \boxed{|\ |} \quad \leftrightarrow \quad \boxed{\phantom{|}} \tag{13.3}$$

while on other types of plaquettes $G$ acts as the identity. Wieland [78] observed that the operator $G_0 \circ G_1$ "gyrates" a link pattern and that the number of FPL configurations is an invariant under gyration.

We define two other operations on the FPL diagrams, $U_{ij}$ and $O_{ij}$, that leave the link pattern invariant but that generate all diagrams belonging to a fixed link pattern. The operator $U$ acts on two plaquettes, either horizontally or vertically. Where it acts non-trivially it is given by,

$$U: \tag{13.4}$$

The operator $O$ acts on three plaquettes, either horizontally or vertically. Where it acts non-trivially, it is given by,

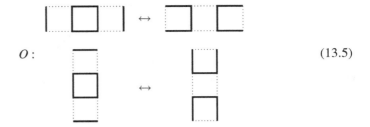

$$O: \qquad\qquad\qquad\qquad\qquad\qquad\qquad\qquad\qquad\qquad (13.5)$$

It is easy to see that both $U$ and $O$ leave the link pattern external to the plaquettes on which they act invariant. It is also not difficult to see that on a horizontal strip of arbitrary length, such that only the leftmost and rightmost edge are connected to the outside world, the operators $U$ and $O$ generate all possible FPL diagrams leaving the link pattern invariant. A similar argument holds for vertical strips. This proves that acting with $U$ and $O$ on an FPL diagram with given link pattern, one generates all FPL diagrams corresponding to that link pattern, and no more.

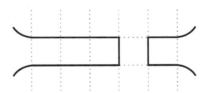

**Fig. 13.9** An isolated row inside an FPL configuration: only the leftmost and rightmost edge are connected to the rest of the FPL configuration. The operators $U$ and $O$ generate all possible configurations within the row.

## 13.2 Partition Function

To each local FPL vertex we assign a weight $w_i$ and define the statistical mechanical partition function $Z_n$ as the sum over all FPL configurations of the product of the vertex weights,

$$Z_n = \sum_{\text{configurations}} \prod_{i=1}^{6} w_i^{k_i}, \qquad\qquad (13.6)$$

where $k_i$ is the number of vertices of type $i$. We will consider only the case where the weights on the two sublattices are the same, i.e. in the six-vertex representation the weights are invariant under arrow reversal. Using standard six-vertex notation we write $w_1 = w_2 = a$, $w_3 = w_4 = b$ and $w_5 = w_6 = c$, see Fig. 13.10.

It is convenient to parametrise $a$, $b$ and $c$ in the following way,

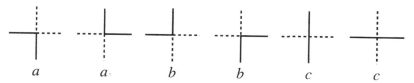

**Fig. 13.10** Weights of the six local FPL vertices.

$$a = \sin(\gamma - u), \qquad b = \sin(\gamma + u), \qquad c = \sin(2\gamma), \qquad (13.7)$$

and to introduce the $u$-independent quantity $\tau$ by

$$\tau^2 = \frac{c^2 - (a-b)^2}{ab} = 2(1-\Delta) = 4\cos^2\gamma, \qquad (13.8)$$

where $\Delta$ is the standard notation for the anisotropy parameter of the six-vertex model defined by

$$\Delta = \frac{a^2 + b^2 - c^2}{2ab} = -\cos(2\gamma). \qquad (13.9)$$

When $a = b$, $\tau = c/a$ gives a weight to straight loop segments. It is therefore expected that for some critical value of $\tau$ there is an ordering transition in the FPL model from a disorder phase to a phase where the vertex with weight $c$ dominates and the polygons are elongated. We will see below that this transition takes place at $\tau = 2$. For $a > b + c$ or $b > a + c$ there is another ordering transition at $\tau = 0$ where the vertices with weight $a$ or $b$, respectively, dominate.

The partition function $Z_n$ can be computed exactly for finite $n$ applying methods of solvable lattice models to the six-vertex model with domain wall boundary conditions. This was first done by Korepin and Izergin [39, 36, 37] who derived the following determinant expression for $Z_n$,

$$Z_n = \frac{(\sin(\gamma + u)\sin(\gamma - u))^{n^2}}{\left(\prod_{k=0}^{n-1} k!\right)^2} \sigma_n, \qquad (13.10)$$

where $\sigma_n$ is the Hankel determinant

$$\sigma_n = \det\left(\frac{d^{i+k-2}\phi}{du^{i+k-2}}\right)_{1 \le i,k \le n}, \qquad (13.11)$$

and

$$\phi(u) = \frac{\sin(2\gamma)}{\sin(\gamma + u)\sin(\gamma - u)}. \qquad (13.12)$$

Using the height representation (13.1) it is possible to introduce elliptic weights, rather than the trigonometric weights (13.7). The partition function in that case has been computed by Rosengren [70].

## 13.2.1 Another Form of the Partition Function

Independent of Izergin and Korepin, in the case $a = b$ (i.e. $u = 0$), another form of $Z_n$ was discovered conjecturally by Robbins in the context of alternating-sign matrices (ASMs) and symmetry classes thereof, see [68]. As can be easily seen from Figures 13.2 and 13.4, a $\tau$ weighted FPL configuration, where each straight loop segment is assigned a weight $\tau$, is equal to the generating of weighted ASMs where each nonzero entry is assigned a weight $\tau$. Up to a simple factor, this is also the generating function $A_n(\tau^2)$ of $\tau^2$-weighted ASMs of size $n \times n$ where each $-1$ is assigned a weight $\tau^2$ (each additional $-1$ in an ASM also introduces an additional $+1$). The latter was conjectured by Robbins [68] to equal

$$A_{2n}(\tau^2) = 2T_n(\tau^2)R_{n-1}(\tau^2), \qquad A_{2n+1}(\tau^2) = T_n(\tau^2)R_n(\tau^2). \tag{13.13}$$

where

$$T_n(\tau^2) = \det_{1 \le i,j \le n} \left( \sum_{r=0}^{2n} \binom{i-1}{r-i} \binom{j}{2j-r} \tau^{2(2j-r)} \right), \tag{13.14}$$

and

$$R_n(\tau^2) = \det_{0 \le i,j \le n-1} \left( \sum_{r=0}^{2n-1} Y_{i,r,\mu} Y_{j,r,0} \, \tau^{2(2j+1-r)} \right), \tag{13.15}$$

where

$$Y_{i,r,\mu} = \binom{i+\mu}{2i+1+\mu-r} + \binom{i+1+\mu}{2i+1+\mu-r}. \tag{13.16}$$

The precise correspondence using the notation of the previous section is

$$Z_{2n} = (\sin\gamma)^{2n(2n-1)} A_{2n}(4\cos^2\gamma), \tag{13.17a}$$

$$Z_{2n+1} = 2\cos\gamma \, (\sin\gamma)^{2n(2n+1)} A_{2n+1}(4\cos^2\gamma). \tag{13.17b}$$

In fact, Robbins' conjecture was slightly more general and gave a generating function for refined ASMs. The generating functions $R$ and $T$ appear naturally in weighted enumerations of cyclically symmetric plane partitions [53].

The equivalence of the homogeneous limit of Izergin's determinant and Robbins' conjecture, i.e. equation (13.17), is only proved for $\tau = 1$ [81, 42]. Kuperberg and Robbins [43, 68] noticed several other such equivalences between homogeneous Izergin or Tsuchiya[1] type determinants and generating functions of the form (13.14) or (13.15). Some of these were recently proved in [34] using a technique which seems immediately applicable to all the cases considered by Kuperberg and Robbins.

---

[1] The Tsuchiya determinant is the generating function of horizontally or vertically symmetric FPL diagrams [77]

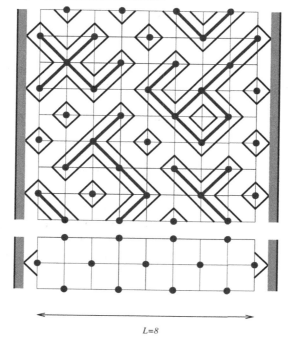

$L=8$

**Fig. 13.11** Bond percolation clusters and O(1) cluster boundaries on a semi-infinite strip. Configurations are generated by repeated concatenation of double rows using the double-row transfer matrix. The particular boundary conditions chosen here are called *closed* or *reflecting*.

## 13.3 Bond Percolation, the O($n = 1$) Model and the Razumov-Stroganov Conjecture

In this section we mention a (partially conjectural) relation between FPL diagrams and the O(1) loop model. We will use this relation to generate FPL statistics in a relatively easy way, without having to explicitly enumerate FPL diagrams.

Imagine that each site of the square lattice is a reservoir of water. With probability $p$, water percolates between reservoirs along a bond of the square lattice. At $p = 1/2$, the model is critical, and equivalent to the dense O($n = 1$) loop model [12] on a square lattice. The loops of the O(1) model describe the boundaries of the percolation clusters, see Fig. 13.11. Many asymptotic properties such as critical exponents of correlation functions can be computed for the O($n$) model using Coulomb gas techniques and conformal field theory, see Chapter 14 for an exhaustive overview. More recently, geometric properties of conformally invariant loops have been analysed using the stochastic Loewner evolution (SLE), see Chapter 15 of this book.

Configurations of the O(1) loop model can be generated using a transfer matrix, see Fig. 13.11 for the particular case of *closed* or *reflecting* boundary conditions. Schematically, the local blocks of the O(1) transfer matrix are given by

$$\square = \frac{1}{2} \diagup + \frac{1}{2} \diagdown .$$  (13.18)

The closed loops of the O(1) loop model have weight $n = 1$. Loops ending on the boundary of the strip define a link pattern. For example, the link pattern corresponding to the bottom side of Fig. 13.11 has link pattern $()()()()$. The transfer matrix $T$ of the O(1) loop model therefore acts on states indexed by a link pattern.

### 13.3.1 The Razumov-Stroganov Conjecture

The largest eigenvalue of the transfer matrix of the O(1) has eigenvalue 1. It was found in [5, 64, 65, 32, 59] that the corresponding groundstate eigenvector surprisingly is related to the statistics of FPL models. Denoting a link pattern by $\alpha$ and forming a vector space with basis elements $|\alpha\rangle$, the groundstate eigenvector satisfies

$$T|\psi\rangle = |\psi\rangle, \qquad |\Psi\rangle = \sum_\alpha \psi_\alpha |\alpha\rangle.$$  (13.19)

In the case of periodic boundary conditions, Razumov and Stroganov formulated the following important conjecture:

> The coefficient $\psi_\alpha$ equals the number of FPL diagrams with link pattern $\alpha$.

The RS conjecture generalises to other boundary conditions, in which case the eigenvector coefficient $\psi_\alpha$ of the corresponding transfer matrix enumerates symmetry classes of FPL diagrams, to be discussed below. This is explained in detail in [30]. The case that will be treated in most detail here is the O(1) model on a strip, as in Fig. 13.11, for which $\psi_\alpha$ conjecturally enumerates horizontally symmetric FPL diagrams.

Assuming the RS conjecture, we will use the O(1) loop model to generate FPL statistics by solving (13.19), and variants thereof for other boundary conditions. The particular boundary conditions we will use are *periodic*, *cylindrical* and *closed*. See e.g. [54, 30, 24, 62, 84, 85, 76] for examples and other types of boundary conditions not considered here.

Let us define the norm $\mathcal{N}_L$ of $|\Psi\rangle$ by

$$\mathcal{N}_L = \sum_\alpha \psi_\alpha,$$  (13.20)

and denote the largest element of $|\Psi\rangle$ by $\psi_{\max}$. The result of solving (13.19) for various boundary conditions is shown in Table 13.1, where the numbers $A$, $A_{HT}$ and $A_V$ are defined by:

- The number of $n \times n$ ASMs,

$$A(n) = \prod_{k=0}^{n-1} \frac{(3k+1)!}{(n+k)!} = 1, 2, 7, 42, \ldots \tag{13.21}$$

- The number of $n \times n$ half turn symmetric ASMs,

$$A_{\mathrm{HT}}(2n) = A(n)^2 \prod_{k=0}^{n-1} \frac{3k+2}{3k+1} = 2, 10, 140, 5544, \ldots$$

$$A_{\mathrm{HT}}(2n-1) = \prod_{k=1}^{n-1} \frac{4}{3} \left( \frac{(3k)!(k!)}{(2k)!^2} \right)^2 = 1, 3, 25, 588, \ldots \tag{13.22}$$

- The number of $(2n-1) \times (2n-1)$ horizontally (or vertically) symmetric ASMs,

$$A_{\mathrm{V}}(2n-1) = \prod_{k=1}^{n-1} (3k-1) \frac{(6k-3)!(2k-1)!}{(4k-2)!(4k-1)!} = 1, 1, 3, 26, 646, \ldots \tag{13.23}$$

and its related version for even sizes (also denoted by $N_8$ in [13]),

$$A_{\mathrm{V}}(2n) = \prod_{k=1}^{n-1} (3k+1) \frac{(6k)!(2k)!}{(4k)!(4k+1)!} = 1, 2, 11, 170, \ldots \tag{13.24}$$

**Table 13.1** The norm and largest eigenvalue of transfer matrices of various types.

| Type | $\mathcal{N}_L$ | $\psi_{\max}$ |
|---|---|---|
| Periodic, $L$ even | $A(L/2)$ | $A(L/2-1)$ |
| Cylindrical, $L$ even | $A_{\mathrm{HT}}(L)$ | $A_{\mathrm{HT}}(L-1)$ |
| Cylindrical, $L$ odd | $A_{\mathrm{HT}}(L)$ | $A((L-1)/2)^2$ |
| Closed, $L$ even | $A_{\mathrm{V}}(L+1)$ | $A_{\mathrm{V}}(L)$ |
| Closed, $L$ odd | $A_{\mathrm{V}}(L+1)$ | $A_{\mathrm{V}}(L)$ |

Mills et al. conjectured the number of ASMs to be $A(n)$, which was proved more than a decade later by Zeilberger [81] and in an entirely different way by Kuperberg [42]. Kuperberg made essential use of the connection to the six-vertex model and its integrability. Conjectured enumerations of symmetry classes were given by Robbins [68], many of which were subsequently proved by Kuperberg [43]. The properties and history of ASMs are reviewed in the book by Bressoud [13], as well as by Robbins [67] and Propp [61].

### 13.3.2 Proofs and Other Developments

The sum rules listed in Table 13.1, relating the norms (13.20) of $|\Psi\rangle$ for different boundary conditions to symmetry classes of alternating-sign matrices, were originally obtained conjecturally. These sum rules have been proved algebraically using an inhomogeneous extension of the transfer matrix, a method initiated and developed by Di Francesco and Zinn-Justin [21, 19]. This has led to further interesting directions, not pursued here, such as the connections between weighted FPL diagrams (or ASMs), plane partitions and the $q$-deformed Knizhnik-Zamolodchikov equation [57, 19, 22, 20, 23, 34].

In an alternative interpretation, the O(1) model is equivalent to a stochastic model defined on link patterns, the so called raise and peel model [33]. It is an open question how to define a stochastic model directly on FPL diagrams, by say the Wieland involutions $G$ describe in Section 13.1.2, such that it has an equipartite stationary state and reduces to the raise and peel model when the action of the operators $O$ and $U$ of Section 13.1.2 is divided out. Such a process would result in a direct proof of the Razumov-Stroganov conjecture.

## 13.4 Symmetry Classes of FPL Diagrams

We will now focus on FPL models defined on rectangular grids, corresponding to certain symmetry classes of square FPL diagrams. The two main reasons are that for such FPL models there is a natural boundary giving rise to additional structure, and that at the time of writing, for these models more results are known which are relevant to polygon models.

### 13.4.1 Horizontally Symmetric FPL Diagrams

For horizontally symmetric FPL diagrams (HSFPLs) one only has to consider the lower half of an FPL diagram. As explained in [30], due to geometric constraints one can further reduce the size of such half diagrams. Therefore, for $L$ even, the reduced lower half of a horizontally symmetric FPL diagram of size $(L+1) \times (L+1)$ is an FPL diagram of size $(L-1) \times L/2$. The total number $Z_{\text{HSFPL}}(2n)$ of horizontally (or vertically) symmetric FPL diagrams of size $(2n-1) \times n$ is known, and can be computed from the Tsuchiya determinant [77, 43],

$$Z_{\text{HSFPL}}(2n) = A_{\text{V}}(2n+1) = \prod_{k=1}^{n}(3k-1)\frac{(6k-3)!(2k-1)!}{(4k-2)!(4k-1)!}. \tag{13.25}$$

As can be seen from Table 13.1, this number is equal to the norm $\mathcal{N}_{2n}$ for the O(1) model with closed boundary conditions and $L = 2n$. For odd system sizes,

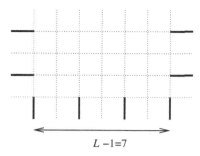

**Fig. 13.12** Boundary conditions for an HSFPL diagram of size $(2n - 1) \times n = 7 \times 4$. The number of external terminals equals $2n = 8$, hence the statistics of this diagram is generated from the $O(1)$ model with $L = 8$.

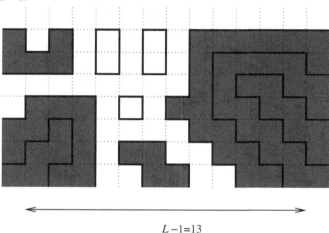

$$L - 1 = 13$$

**Fig. 13.13** An FPL diagram of size $(L - 1) \times L/2 = 13 \times 7$ with four nests.

$L = 2n + 1$, the norm $\mathcal{N}_L$ equals the number of FPL diagrams of size $L \times (L - 1)/2$, which we will denote by $Z_{\text{HSFPL}}(2n + 1)$.

There are two interesting and natural statistics on HSFPLs which we will explain now. As noted above, to each FPL diagram is associated a link pattern. Each link pattern factorises in sets of completed links where, in terms of the parenthesis notation, the number of closing parentheses equals the number of opening parentheses. For example,

$$(()())((((())())() = (()()) \cdot ((((())()) \cdot ().$$

Such completed links are called nests , and they provide a statistic for HSFPLs. An example of an HSFPL diagram of size $13 \times 7$ with four nests is given in Fig. 13.13.

Another natural statistic is the number $d^*$ of loops connecting the leftmost loop terminals with the rightmost ones, i.e. loops connecting terminal $i$ with $2\lfloor L/2 \rfloor - i + 1$ for $i = 1, \ldots, d^*$. It will be convenient to define $d$ by

$$d = \left\lfloor \frac{L-1}{2} \right\rfloor - d^*, \tag{13.26}$$

where $d$ is called the depth of an HSFPL diagram. An example of an HSFPL diagram of size $(L-1) \times L/2 = 13 \times 7$ with three nests and depth $d = 4$ ($d^* = 2$) is given in Fig. 13.14.

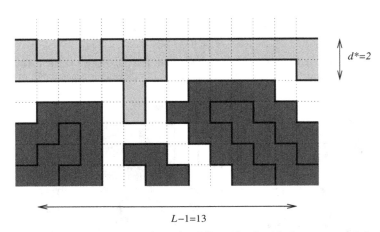

**Fig. 13.14** An FPL diagram of size $(L-1) \times L/2 = 13 \times 7$ with three nests and depth $d = 4$ ($d^* = 2$).

### 13.4.2 Depth-Nest Enumeration of HSFPLs

In this section we will say that an FPL diagram is of size $L$, if it is of size $(L-1) \times L/2$ if $L$ is even, or of size $L \times (L-1)/2$ if $L$ is odd. Let $P(L,d,m)$ be the number of such FPL diagrams of size $L$, depth $d$ and having $m+1$ nests. The nest generating function for diagrams of size $L$ and depth $d$ is defined by

$$\mathscr{P}(L,d;z) = \sum_{m=0}^{d} P(L,d,m)z^m. \tag{13.27}$$

Let $S(L,d)$ be the total number of HSFPL diagrams at a given size $L$ and depth $d$. Obviously we have

$$S(L,d) = \mathscr{P}(L,d;1) = P(L,d+1,0), \tag{13.28}$$

and

$$Z_{\mathrm{HSFPL}}(L) = S(L,\lfloor \tfrac{L-1}{2} \rfloor) = \mathscr{P}(L,\lfloor \tfrac{L-1}{2} \rfloor;1). \tag{13.29}$$

Based on the RS conjecture, Mitra et al. and Pyatov have conjectured the exact form of $S(L,d)$ [54, 62]. Here we give this conjecture in the following form:

**Conjecture 1** The total number of HSFPL diagrams at a given size $L$ and depth $d$ is given by

$$S(L,d) = \prod_{k=0}^{d} \frac{\Gamma(L-k+1)}{2^k(1/2)_k\Gamma(L-2k+1)} \frac{\Gamma(\frac{2L+2k+3}{6})\Gamma(\frac{L-2k+3}{3})}{\Gamma(\frac{2L-k+3}{6})\Gamma(\frac{2L-k+6}{6})}. \tag{13.30}$$

Assuming the RS conjecture, the formula for $S(L,d)$ has recently been proved [34].

Pyatov also found an exact formula for $P(L,d,m)$ [62] which fits exact data for small system sizes ($L \leq 18$). He conjectured that this formula holds for all $L$, $d$ and $m$. In terms of the nest generating function this conjecture can be stated as follows.

**Conjecture 2** The nest generating function is given by

$$\mathscr{P}(L,d;z) = S(L,d-1)\,{}_3F_2\left(\begin{array}{c} -d,L-2d,L-d+\frac{1}{2} \\ -2d,2L-2d+1 \end{array};4z\right). \tag{13.31}$$

Note that Conjecture 1 follows from Conjecture 2 due to the evaluation

$${}_3F_2\left(\begin{array}{c} -d,L-2d,L-d+\frac{1}{2} \\ -2d,2L-2d+1 \end{array};4\right) = \frac{S(L,d)}{S(L,d-1)}, \tag{13.32}$$

which is a consequence of one of the strange evaluations of Gessel and Stanton [28]. For $d = \lfloor(L-1)/2\rfloor$, the formulas in Conjecture 1 and Conjecture 2 were given in [30].

By convention, $P(L,d,m)$ have the following boundary values:

$$P(L,d,m=-1) = P(L,d,m=-2) = P(L,d,m=d+1) = 0, \tag{13.33}$$

and we also note the boundary condition

$$P(L,d,m=1) = (L-2d)S(L,d-1). \tag{13.34}$$

It was found in [1] that the function $P(L,d,m)$ is completely determined by these boundary conditions and the following interesting bilinear relation called the *split hexagon relation*,

$$P(L+1,d+1,m)S(L-1,d-1)$$
$$= P(L-1,d,m)S(L+1,d)+P(L,d-1,m-2)S(L,d+1). \tag{13.35}$$

Summing up over $m = 0,1,\ldots d+1$ in (13.35) reproduces the hexagon relation, or discrete Boussinesq equation, for $S(L,d)$, see [62].

### 13.4.2.1 Cyclically Symmetric Transpose Complement Plane Partitions

Somewhat outside the scope of this book, we note the following interesting fact observed in [34]. The total number of nests at a given depth, $S(L,d)$, is equal to the number of punctured cyclically symmetric transpose complement plane partitions [14], see Fig. 13.15. This can be seen by enumerating the number of non-intersecting lattice paths in the South-East fundamental domain of the plane partition. Using the Gessel-Viennot-Lindström method [29, 49] one obtains a determinant of the type (13.14) with $\tau = 1$, which can be evaluated in factorised form [14]. This form equals the expression in (13.30).

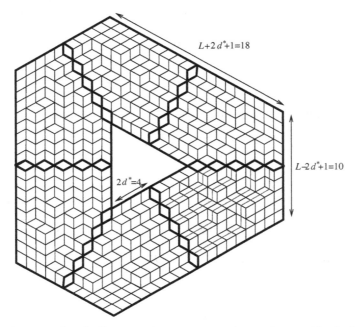

**Fig. 13.15** A punctured cyclically symmetric transpose complement plane partition for $L = 13$ and $d^* = 2$.

## 13.4.3 Average Number of Nests in HSFPL Diagrams

The average number of nests in HSFPL diagrams at depth $d$ and size $L$, denoted by, $\langle 1 + m \rangle_{d^*}$, is defined as

$$\langle 1+m\rangle_{d^*} = \frac{1}{Z_{\text{HSFPL}}(L)} \sum_{m=0}^{d} (1+m)P(L,d,m)$$

$$\equiv \frac{S(L,d)}{Z_{\text{HSFPL}}(L)} \left(1 + \langle m\rangle_{d^*}^{\text{c}}\right). \tag{13.36}$$

For notational clarity we will suppress the dependence of $\langle 1+m\rangle_{d^*}$ on $L$ and recall that

$$d = \left\lfloor \frac{L-1}{2} \right\rfloor - d^*.$$

With the data $P(L,d,m)$ we can calculate $\langle m\rangle_{d^*}^{\text{c}}$:

$$\langle m\rangle_{d^*}^{\text{c}} = \frac{1}{S(L,d)} \sum_{m=1}^{d} mP(L,d,m)$$

$$= \frac{\text{d}}{\text{d}z}\bigg|_{z=1} \log \mathscr{P}(L,d;z). \tag{13.37}$$

The `Mathematica` implementation of the Gosper-Zeilberger algorithm [35, 79, 80] by Paule and Schorn [58], is able to recognise $\langle m\rangle_{d^*}$ in an almost factorised form. Let $L = 2n$, then define $\mu_n(d^*)$ by

$$\mu_n(d^*) = \sum_{m=0}^{n-1-d^*} \left(3m + 4(d^*+1)\right) \frac{P(2n, n-1-d^*, m)}{P(2n, n-1-d^*, 0)}$$

$$= \left(3\langle m\rangle_{d^*}^{\text{c}} + 4(d^*+1)\right) \frac{S(2n, n-1-d^*)}{S(2n, n-2-d^*)}. \tag{13.38}$$

The expression $\mu_n(d^*)$ turns out to be summable in factorised form, giving rise to

$$\langle m\rangle_{d^*}^{\text{c}} = -\frac{2}{3}(L-2d) + 2^{2/3} \frac{\Gamma\left(\frac{2L+2d+5}{6}\right)\Gamma\left(\frac{2L-d+3}{3}\right)\Gamma\left(\frac{L-2d+1}{3}\right)}{\Gamma\left(\frac{2L+2d+3}{6}\right)\Gamma\left(\frac{2L-d+2}{3}\right)\Gamma\left(\frac{L-2d}{3}\right)}. \tag{13.39}$$

This formula also holds for odd values of $L$.

### 13.4.4 Half-Turn Symmetric FPL Diagrams

In the case of half turn symmetric FPL diagrams (HTSFPLs) it also suffices to consider only the lower half of an FPL diagram, but the boundary conditions on the top row of the half diagram are different from HSFPLs, see Fig. 13.16. The total number of HTSFPL diagrams is given by [43]

$$Z_{\text{HT}}(2n) = A_{\text{HT}}(2n) = 2 \prod_{k=1}^{n-1} \frac{3(3k+2)!(3k-1)!k!(k-1)!}{4(2k+1)!^2(2k-1)!^2}. \tag{13.40}$$

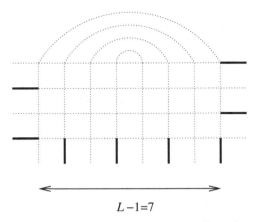

$$L-1=7$$

**Fig. 13.16** Boundary conditions for an HTSFPL diagram of size $(2n-1) \times n = 7 \times 4$. The arcs at the top are additional edges which may contain loop segments. The number of external terminals equals $2n = 8$, hence the statistics of this diagram is generated from the periodic $O(1)$ model with $L = 8$.

Care has to be taken when defining link patterns and nests for HTSFPL diagrams. External terminals can be connected in two distinct ways depending on whether the corresponding loop runs over an odd or even number of the arcs on the top of the diagram. In the case of an odd number of arcs, we exchange the parentheses denoting the connection of a pair of sites. For example, the connectivity of the HTSFPL diagram in Fig. 13.17 is denoted by

$$) \cdot () \cdot () \cdot (() ,$$

where the dots again denote the factorisation of link pattern into nests. Figure 13.17 thus denotes an HTSFPL diagram with three nests.

As in the case of horizontal symmetry, there exists a conjecture for the nest distribution function [30], but in this case only for $L = 2n$ and $d^* = 0$. Let $P(L, m)$ denote the number of half-turn symmetric FPL diagrams with $m + 1$ nests, and define the nest generating function by

$$\mathscr{P}(L; z) = \sum_{m=0}^{n-1} P(L, m) z^m. \tag{13.41}$$

**Conjecture 3** The nest generating function for half-turn symmetric FPL diagrams is given by

$$\mathscr{P}(2n; z) = Z_{\text{HTSFPL}}(2n) \frac{3n}{4n^2 - 1} \, {}_3F_2 \left( \begin{array}{c} 3/2, 1-n, 1+n \\ 2-2n, 2+2n \end{array} ; 4z \right) .$$

The average number of nests in HTSFPL diagrams of size $L = 2n$ having $1 + m$ nests, denoted by $\langle 1 + m \rangle$, is defined as

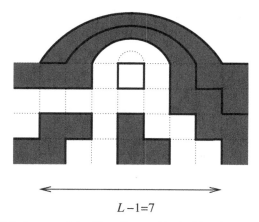

$$L-1=7$$

**Fig. 13.17** An HTSFPL diagram of size $(2n-1) \times n = 7 \times 4$ with link pattern $) \cdot () \cdot () \cdot (()$, having three nests.

$$\langle 1+m \rangle = \frac{1}{Z_{\mathrm{HTSFPL}}(L)} \sum_{m=0}^{n-1} (1+m)P(L,d,m) = 1 + z\frac{\mathrm{d}}{\mathrm{d}z} \log \mathscr{P}(L;z) \qquad (13.42)$$

Knowing the nest generating function we may compute $\langle 1+m \rangle$, which turns out to be summable [30].

**Conjecture 4** The average number of nests in HTSFPL diagrams of size $L$, is given by

$$\langle 1+m \rangle = n \prod_{j=1}^{n-1} \frac{3j+1}{3j+2}.$$

## 13.5 Phase Transitions

### 13.5.1 Bulk Asymptotics and Phase Diagram

The phase diagram of the FPL model can be derived from the asymptotics of the partition function $Z_n$ defined in (13.10). The leading asymptotics of $Z_n$ for general values of $\tau$ has been computed by Korepin and Zinn-Justin [41] using the Toda equation satisfied by $\sigma_n$ [74],

$$\sigma_n \frac{\mathrm{d}^2 \sigma_n}{\mathrm{d}u^2} - \left( \frac{\mathrm{d}\sigma_n}{\mathrm{d}u} \right)^2 = \sigma_{n+1}\sigma_{n-1}. \qquad (13.43)$$

Writing $\sigma_n$ as a matrix model integral [83], further subleading asymptotics were computed by Bleher and Fokin [10] and Bleher and Liechty [11] using orthogonal polynomials. For special values of $\gamma$ this method was first employed by Colomo and Pronko [17]. The final result for $0 < \tau^2 = 4\cos^2\gamma < 4$ ($1 > \Delta > -1$) is that for some $\varepsilon > 0$,

$$Z_n = Cn^\kappa \exp\left[fn^2(1 + \mathcal{O}(n^{-\varepsilon}))\right], \tag{13.44}$$

where $C$ is a constant and

$$f = \frac{\pi \sin(\gamma+u)\sin(\gamma-u)}{2\gamma\cos(\pi u/2\gamma)}, \tag{13.45}$$

$$\kappa = \frac{1}{12} - \frac{2\gamma^2}{3\pi(\pi - 2\gamma)}. \tag{13.46}$$

The result (13.44) is valid in the so-called disordered (D) phase $0 < \tau^2 < 4$. There is a phase transition to an ordered phase at $\tau^2 = 4$ where the vertices with weight $c$ are favoured and the perimeters of the polygons in the FPL model consist of elongated straight lines. In terms of the six-vertex model this is called the antiferromagnetic (AF) phase. At $\tau = 0$, i.e. $a = b + c$ or $b = a + c$, there is another phase transition to a so-called ferromagnetic phase, where, respectively, the $a$- or $b$-type vertices dominate. The complete phase diagram is given in Fig. 13.18.

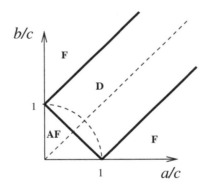

**Fig. 13.18** Bulk phase diagram of the FPL model. The phases are traditionally called disordered (D), ferro-electric (F) and anti-ferro-electric (AF), cf. the six-vertex model. The arc corresponds to the free-fermion condition $\Delta = 0$ ($\tau^2 = 2$), and the line $a = b$ corresponds to $\tau^2$-enumerations of ASMs. On this line $\tau^2 = c^2/a^2 = 2(1 - \Delta)$, and the D-AF phase transition takes place at $\tau^2 = 4$.

The phase diagram and the Bethe-Ansatz solution of the six-vertex model for *periodic* and *anti-periodic* boundary conditions are thoroughly discussed in the works of Lieb [45, 46, 47], Lieb and Wu [48], Sutherland [73], Baxter [7], and Batchelor et al. [3].

## 13.5.2 Asymptotics for Symmetry Classes at $\tau = 1$

In this section we determine the asymptotics of equally weighted horizontally and half-turn symmetric FPL diagrams for $\tau = 1$ or $\gamma = 2\pi/3$, corresponding to the numbers given in Table 13.1. The leading asymptotic form of these numbers, which are all products over factorials, can be computed using the Euler-Maclaurin approximation. Full asymptotics can easily be derived using Barnes' $G$-function [2], which satisfies

$$G(z+1) = \Gamma(z)G(z), \qquad G(1) = 1, \tag{13.47}$$

and whose leading asymptotic behaviour is given by (see e.g. [56]),

$$\log(G(z+1)) = z^2 \left( \frac{1}{2}\log z - \frac{3}{4} \right) + \frac{1}{2}z\log 2\pi - \frac{1}{12}\log z + \mathscr{O}(1). \tag{13.48}$$

In the case of $A(n)$ and $A_{HT}(n)$, a detailed asymptotic analysis including the lower order terms was carried out by Mitra and Nienhuis [55]. Here we list only the leading asymptotics of the FPL numbers relevant to the current context. The generic asymptotic form of the numbers is

$$\log Z_L = s_0 \, \text{Area} + f_0 \, \text{Surface} + x\log(\text{Length}) + \mathscr{O}(1), \tag{13.49}$$

where the bulk and boundary entropies are given by

$$s_0 = \log\left( \frac{3\sqrt{3}}{4} \right), \qquad f_0 = \log\left( \frac{3\sqrt{3}}{4\sqrt{2}} \right). \tag{13.50}$$

The critical exponent $x$ is a universal quantity. In detail, the cases relevant for this chapter are

- FPL diagrams, $L$ even
  The number $A(L/2)$ counts FPL configurations on an $L/2 \times L/2$ square grid, the area of which is $\frac{1}{4}L^2$. We thus find,

$$\log Z(L) = \log A(L/2) = \frac{1}{4}s_0 L^2 - \frac{5}{36}\log L + O(1). \tag{13.51}$$

- Half turn symmetric FPL diagrams, $L$ even
  $A_{HT}(L)$ counts the number of FPL configurations on half an $L \times L$ square grid, the area of which is $\frac{1}{2}L^2$. We thus find for $L$ even,

$$\log Z_{HT}(L) = \log A_{HT}(L) = \frac{1}{2}s_0 L^2 + \frac{1}{18}\log L + O(1). \tag{13.52}$$

- Half turn symmetric FPL diagrams, $L$ odd
  $A_{HT}(L)$ counts the number of FPL configurations on a square grid of dimension $L \times (L-1)/2$, the area of which is $\frac{1}{2}L(L-1)$. We thus find for $L$ odd,

$$\log Z_{\mathrm{HT}}(L) = \log A_{\mathrm{HT}}(L) = \frac{1}{2}s_0 L(L-1) + \frac{1}{36}\log L^2 + O(1). \tag{13.53}$$

- Horizontally symmetric FPL diagrams, $L$ even
  $A_{\mathrm{V}}(L+1)$ counts the number of FPL configurations on an $(L-1) \times L/2$ rectangular grid. We find,

$$\log Z_{\mathrm{HSFPL}}(L) = \log A_{\mathrm{V}}(L+1) = \frac{1}{2}s_0 L(L-1) + f_0 L - \frac{5}{72}\log L + O(1). \tag{13.54}$$

- Horizontally symmetric FPL diagrams, $L$ odd
  For $L$ odd, $A_{\mathrm{V}}(L+1)$ counts the number of FPL configurations on a $L \times (L-1)/2$ rectangular grid. We find,

$$\log Z_{\mathrm{HSFPL}}(L) = \log A_{\mathrm{V}}(L+1) = \frac{1}{2}s_0 L(L-1) + f_0 L + \frac{7}{72}\log L + O(1). \tag{13.55}$$

Note that because the upper boundary for FPL diagrams corresponding to HSFPLs is not fixed, see e.g. Fig. 13.12, there is a nonzero boundary entropy in $\log Z_{\mathrm{HSFPL}}(L)$.

### 13.5.3 Nest Phase Transitions

From Section 13.4.3 we recall that the average number of nests is given by $\frac{S(L,d)}{Z_{\mathrm{HSFPL}}(L)}(1 + \langle m \rangle^{\mathrm{c}}_{d^*})$ where

$$\langle m \rangle^{\mathrm{c}}_{d^*} = z\frac{\mathrm{d}}{\mathrm{d}z}\log \mathscr{P}(L,d;z), \tag{13.56}$$

with $\mathscr{P}(L,d;z)$ given in Conjecture 2. The asymptotics for $\langle m \rangle^{\mathrm{c}}_{d^*}$ as $L \to \infty$ can be derived from the hypergeometric equation satisfied by $\mathscr{P}(L,d;z)$. Taking $L = 2n$ this gives

$$\theta(\theta + 1 + 2d^* - 2n)(\theta + 2 + 2d^* + 2n)\mathscr{P}(2n,d;z) =$$
$$4z(\theta + 2 + d^*)(\theta + 1 + d^* - n)(\theta + 3/2 + n)\mathscr{P}(2n,d;z), \tag{13.57}$$

where $\theta = z\mathrm{d}/\mathrm{d}z$ and

$$d = \left\lfloor \frac{L-1}{2} \right\rfloor - d^* = n - 1 - d^*. \tag{13.58}$$

Assuming that $d^* = \mathscr{O}(1)$, we discriminate the cases $z < 1$, $z = 1$ and $z > 1$.

- $z < 1$
  In this case, up to an overall constant factor, the leading asymptotics of $\mathscr{P}(2n,d;z)$ will be polynomial in $n$. Neglecting lower order terms, equation (13.57) reduces to,

$$\theta\mathscr{P}(2n,d;z) = z(\theta + 2 + d^*)\mathscr{P}(2n,d;z). \tag{13.59}$$

We thus we find $(1-z)\mathscr{P}'(2n,d;z) = (2+d^*)\mathscr{P}(2n,d;z)$ and

$$\langle m \rangle^c_{d^*} = (2+d^*)\frac{z}{1-z} \qquad (n \to \infty). \tag{13.60}$$

- $z = 1$

In (13.39) an exact expression was given for $\langle m \rangle^c_{d^*}$ at $z = 1$. Asymptotically we find that for $L - 2d = \mathscr{O}(1)$,

$$\langle m \rangle^c_{d^*} \approx \frac{\Gamma(\frac{L-2d+1}{3})}{\Gamma(\frac{L-2d}{3})}L^{2/3} + \mathscr{O}(1), \tag{13.61}$$

which for $L = 2n$ can be written as

$$\langle m \rangle^c_{d^*} \approx \frac{\Gamma(\frac{2d^*+3}{3})}{\Gamma(\frac{2d^*+2}{3})}(2n)^{2/3} + \mathscr{O}(1), \tag{13.62}$$

- $z > 1$

In this case, and when $d$ is of order $n$, the leading asymptotics of $\mathscr{P}(2n,d;z)$ will be of the form $p(n)z^n$, where $p(n)$ is a polynomial in $n$. This means that $\theta\mathscr{P}(2n,d;z)$ is of the same order as $n\mathscr{P}(2n,d;z)$ and (13.57) reduces in leading order to

$$(\theta^3 - 4n^2\theta)\mathscr{P}(2n,d;z) = 4z(\theta^3 - n^2\theta)\mathscr{P}(2n,d;z). \tag{13.63}$$

Using (13.56) one can derive the following equation for $\langle m \rangle^c_{d^*}$,

$$4n^2(z-1)\langle m \rangle^c_{d^*} = (4z-1)\left(\theta^2\langle m \rangle_{d^*} + 3\langle m \rangle^c_{d^*}\theta\langle m \rangle^c_{d^*} + (\langle m \rangle^c_{d^*})^3\right), \tag{13.64}$$

which in leading order when $\langle m \rangle^c_{d^*} \sim n$ reduces to $(4z-1)(\langle m \rangle^c_{d^*})^2 = 4n^2(z-1)$ and thus

$$\langle m \rangle^c_{d^*} \approx \sqrt{\frac{z-1}{4z-1}}L + \mathscr{O}(1). \tag{13.65}$$

The scaling behaviour near the phase transition at $z = 1$ is governed by a single exponent, the cross-over exponent $\phi$ [9]. On general grounds one expects,

$$\langle m \rangle^c_{d^*} \sim \begin{cases} L(z-1)^{1/\phi-1} & (z > 1) \\ L^\phi & (z = 1) \\ (1-z)^{-1} & (z < 1) \end{cases}. \tag{13.66}$$

Indeed, we find such scaling behaviour for $\langle m \rangle^c_{d^*}$ with $\phi = 2/3$.

### 13.5.3.1 Scaling Function

In [33] an analysis has been carried out to obtain the nest scaling function for $L = 2n$ and $d = n - 1$, i.e. $d^* = 0$. Following Polyakov [60], we expect the following scaling form of the nest distribution function at the critical point,

$$\frac{P(2n, n-1, m)}{S(2n, n-1)} \sim \frac{1}{\langle 1+m \rangle_0} f\left(\frac{1+m}{\langle 1+m \rangle_0}\right) \qquad (n \to \infty), \qquad (13.67)$$

where $\langle 1+m \rangle_0 = 1 + \langle m \rangle_0^c$. The large $x$ behaviour of $f(x)$ is related to the exponent $\phi$ [15],

$$\lim_{x \to \infty} f(x) \sim x^s e^{-ax^\delta}, \qquad \delta = \frac{1}{1-\phi}, \qquad (13.68)$$

where $a$ and $s$ are constants. The behaviour of $f(x)$ for small $x$ is related to the large $n$ behaviour of the probability $P(2n, n-1, m)/S(2n, n-1)$,

$$\lim_{x \to 0} f(x) = bx^\vartheta \quad \Rightarrow \quad b = \lim_{m \to 0} \lim_{n \to \infty} (1 + \langle m \rangle_0)^{1+\vartheta} \frac{P(2n, n-1, m)}{S(2n, n-1)}, \qquad (13.69)$$

from which we find

$$\vartheta = 1, \qquad b = \frac{3}{\Gamma(2/3)^3}. \qquad (13.70)$$

Assuming that the *full* scaling function is of the form $x^\vartheta e^{-ax^\delta}$, for all values of $x$, and using the normalisation condition

$$\int_0^\infty f(x)\,dx = 1, \qquad (13.71)$$

we find that

$$f(x) = bx\, e^{-bx^3/3}. \qquad (13.72)$$

In Fig. 13.19 we compare the scaling function (13.72) with a numerical evaluation of (13.67) for $n = 300$. When $\delta^* > 0$, the value of $\phi$, and hence that of $\delta$, is not changed, but it follows from (13.69) that the value of the exponent $\vartheta$ changes to

$$\vartheta = 1 + 2d^*. \qquad (13.73)$$

The full scaling function is not known in this case.

## 13.5.4 Half Turn Symmetry

The following analysis closely follows that of the previous section. We are interested in the asymptotics as $n \to \infty$ of the average number of nests defined in (13.56). This can be inferred from the hypergeometric equation for $\mathscr{P}(2n; z)$,

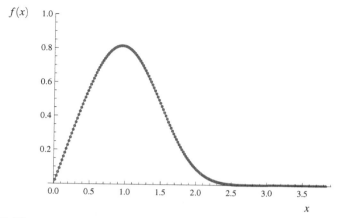

**Fig. 13.19** The scaling function $f(x)$ defined in (13.72) compared to a numerical evaluation (dots) of (13.67) for $L = 2n = 600$. It can be seen that these are indistinguishable.

$$\theta(\theta + 1 + 2n)(\theta + 1 - 2n)\mathscr{P}(2n;z) = 4z(\theta + 3/2)(\theta + 1 - n)(\theta + 1 + n)\mathscr{P}(2n;z).$$
(13.74)

Again we discriminate the cases $z < 1$, $z = 1$ and $z > 1$ and remind the reader that $L = 2n$.

- $z < 1$

  In this case, up to an overall constant prefactor, $\mathscr{P}(2n;z)$ will grow as a polynomial in $n$ and, neglecting lower order terms, (13.74) reduces to,

  $$\theta\mathscr{P}(2n;z) = z(\theta + 3/2)\mathscr{P}(2n;z),$$
  (13.75)

  so that we find $(1 - z)\mathscr{P}'(2n;z) = \frac{3}{2}\mathscr{P}(2n;z)$ and thus

  $$\langle 1 + m \rangle \approx \frac{2 + z}{2(1 - z)} + \mathscr{O}(1).$$
  (13.76)

- $z = 1$

  For this case, and exact expression was given for $\langle m + 1 \rangle$ in Conjecture 4. Asymptotically we find

  $$\langle 1 + m \rangle = n\prod_{j=1}^{n-1}\frac{3j+1}{3j+2} \approx \frac{\Gamma(5/6)}{\sqrt{\pi}}L^{2/3} + \mathscr{O}(1).$$
  (13.77)

- $z > 1$

  Here, up to an overall constant prefactor, $\mathscr{P}(2n;z)$ will grow as $p(n)z^n$ where $p(n)$ is a polynomial in $n$. This means that $\theta\mathscr{P}(2n;z)$ will be of the same order as $n\mathscr{P}(2n;z)$ and (13.74) reduces in leading order to

$$(\theta^3 - 4n^2\theta)\mathscr{P}(2n;z) = 4z(\theta^3 - n^2\theta)\mathscr{P}(2n;z), \qquad (13.78)$$

which is the same as (13.63). We thus find that

$$\langle 1+m \rangle \approx \sqrt{\frac{z-1}{4z-1}}\, L + \mathscr{O}(1). \qquad (13.79)$$

For half-turn symmetric FPL diagrams we find the same cross-over exponent $\phi = 2/3$ as for horizontally symmetric FPL diagrams.

## 13.6 Conclusion

We have described a model of tightly packed, nested polygons on the square lattice. We hope that the study of such tightly packed polygons is relevant to other polygon models described in this book. The advantage of the model described in this chapter is that many exact results can be obtained, even on finite geometries, due to its relation with the exactly solvable six-vertex and $O(n = 1)$ lattice models. In particular, the statistical mechanical partition function can be obtained rigorously on finite square patches of the square lattice. The free energy can then be obtained analytically and in the thermodynamic limit. The fully packed loop model undergoes a well-known bulk order–disorder phase transition as a function of an anisotropy parameter associated to the straight segments of the polygon boundary.

We have furthermore shown that it is possible to obtain closed-form expressions for partition functions of important subsets of fully packed loop configurations. Two examples of such subsets are horizontally symmetric fully packed loop models of depth $d$, and half-turn symmetric fully packed loop models of depth $d$. These closed-form expressions have been obtained experimentally, and remain conjectures at the time of this writing. In addition, using hypergeometric summation identities, we were able to compute the average number of polygon nests at the boundary in closed form. Asymptotic analyses allowed us to study a boundary phase transition as a function of the nest fugacity, and we obtained a crossover exponent $\phi = 2/3$. At criticality, we derive a scaling form for the nest distribution function which displays an unusual non-Gaussian cubic exponential behaviour.

To conclude, it should be said that while some of the exact results presented in this chapter are more than what one would hope for from a physicist's perspective, where numerical techniques are often all what is available, they are just the starting point for a mathematician. Although great progress has been made in recent years in understanding fully packed loop models, proving conjectures such as the nest distribution function and many related combinatorial results, remains a fascinating and completely open problem. It is for reasons such as these that polygon models in all shapes and sizes will continue to inspire future research.

# References

1. F.C. Alcaraz, P. Pyatov and V. Rittenberg, *Density profiles in the raise and peel model with and without a wall. Physics and combinatorics*, J. Stat. Mech. (2008), P01006; arXiv:0709.4575.

2. E. W. Barnes, *Genesis of the double gamma function*, Proc. London Math. Soc. **31** (1899), 358–381; *The theory of the G-function*, Quart. J. Math. **31** (1900), 264–314; *The theory of the double gamma function*, Philos. Trans. R. Soc. London Ser. A **196** (1901), 265–388.

3. M. T. Batchelor, R. J. Baxter, M. J. O'Rourke and C. M. Yung, *Exact solution and interfacial tension of the six-vertex model with anti-periodic boundary conditions*, J. Phys. A **28** (1995), 2759–2770.

4. M. T. Batchelor, H. W. J. Blöte, B. Nienhuis and C. M. Yung, *Critical behaviour of the fully packed loop model on the square lattice*, J. Phys. A, 29 (1996), L399–L404.

5. M. T. Batchelor, J. de Gier and B. Nienhuis, *The quantum symmetric XXZ chain at $\Delta = -\frac{1}{2}$, alternating-sign matrices and plane partitions*, J. Phys. A, 34 (2001), L265–L270; arXiv:cond-mat/0101385.

6. M. T. Batchelor, J. de Gier and B. Nienhuis, *The rotor model and combinatorics*, Int. J. Mod. Phys. B, 16 (2002), 1883–1889.

7. R. J. Baxter, *Exactly solved models in statistical mechanics*, Academic Press, San Diego (1982).

8. R. J. Baxter, *Solving models in statistical mechanics*, Adv. Stud. Pure Math., 19 (1989), 95–116.

9. K. Binder, in *Phase Transitions and Critical Phenomena* **8**, eds. C. Domb and J.L. Lebowitz, Academic Press Inc., London (1983), 2.

10. P.M. Bleher and V.V. Fokin, *Exact Solution of the Six-Vertex Model with Domain Wall Boundary Conditions. Disordered Phase*, Commun. Math. Phys. **268** (2006), 223–284; arXiv:math-ph/0510033.

11. P.M. Bleher and K. Liechty, *Exact Solution of the Six-Vertex Model with Domain Wall Boundary Conditions. Ferroelectric phase*, arXiv:0712.4091.

12. H. W. J. Blöte and B. Nienhuis, *Critical behaviour and conformal anomaly of the O(n) model on the square lattice*, J. Phys. A **9** (1989), 1415–1438.

13. D. M. Bressoud, *Proofs and Confirmations: The story of the Alternating Sign Matrix Conjecture*, Cambridge University Press, Cambridge (1999).

14. M. Ciucu and C. Krattenthaler, *Plane partitions II: $5\frac{1}{2}$ symmetry classes*, in: Combinatorial Methods in Representation Theory, M. Kashiwara, K. Koike, S. Okada, I. Terada, H. Yamada, eds., Advanced Studies in Pure Mathematics, vol. **28**, RIMS, Kyoto, 2000, 83–103; arXiv:math/9808018.

15. J. des Cloizeaux and G. Jannink, *Polymers in Solution*, Clarendon Press, Oxford (1990).

16. M. Ciucu and C. Krattenthaler, *Enumeration of lozenge tilings of hexagons with cut off corners*, J. Combin. Theory Ser. A **100** (2002), 201–231.

17. F. Colomo and A.G. Pronko, *Square ice, alternating sign matrices, and classical orthogonal polynomials*, J.Stat.Mech. 0501 (2005), P005; arXiv:math-ph/0411076.

18. D. Dei Cont and B. Nienhuis, *The packing of two species of polygons on the square lattice*; J. Phys. A **37** (2004), 3085–3100; arXiv:cond-mat/0311244.

19. P. Di Francesco, *Boundary qKZ equation and generalized Razumov–Stroganov sum rules for open IRF models*, J. Stat. Mech. (2005), P11003; arXiv:math-ph/0509011.

20. P. Di Francesco, *Open boundary Quantum Knizhnik–Zamolodchikov equation and the weighted enumeration of symmetric plane partitions*, J. Stat. Mech. (2007), P01024; arXiv:math-ph/0611012.

21. P. Di Francesco and P. Zinn-Justin, *Around the Razumov-Stroganov conjecture: proof of a multi-parameter sum rule*, 2005 Electron. J. Combin. **12**, R6; arXiv:math-ph/0410061.

22. P. Di Francesco and P. Zinn-Justin, *Quantum Knizhnik-Zamolodchikov equation, generalized Razumov–Stroganov sum rules and extended Joseph polynomials*, J. Phys. A **38** (2005), L815–L822; arXiv:math-ph/0508059.

23. P. Di Francesco and P. Zinn-Justin, *Quantum Knizhnik–Zamolodchikov equation: reflecting boundary conditions and combinatorics*, J. Stat. Mech. (2007), P12009,; arXiv:0709.3410.

24. Ph. Duchon, *On the link pattern distribution of quarter-turn symmetric FPL configurations*, arXiv:0711.2871.

25. N. Elkies, G. Kuperberg, M. Larsen, and J. Propp, *Alternating-Sign Matrices and Domino Tilings*, J. Algebraic Combin., 1 (1992), pp. 111–132 and 219–234.

26. V. Fridkin, Yu. G. Stroganov and D. Zagier, *Groundstate of the quantum symmetric finite-size XXZ spin chain with anisotropy parameter* $\Delta = \frac{1}{2}$, J. Phys. A **33** (2000), L121–L125; arXiv:hep-th/9912252

27. V. Fridkin, Yu. G. Stroganov and D. Zagier, *Finite-size XXZ spin chain with anisotropy parameter* $\Delta = \frac{1}{2}$, J. Stat. Phys. **102** (2001), 781–794; arXiv:nlin.SI/0010021.

28. I. Gessel and D. Stanton, *Strange evaluations of hypergeometric series*, SIAM J. Math. Anal. **13** (1982), 295–308.

29. I. Gessel and X. Viennot, *Binomial determinants, paths, and hook length formulae*, Adv. Math. **58** (1985), 300–321.

30. J. de Gier, *Loops, matchings and alternating-sign matrices*, Discr. Math. **298** (2005), 365–388, arXiv:math.CO/0211285.

31. J. de Gier, M.T. Batchelor, B. Nienhuis and S. Mitra, *The XXZ chain at* $\Delta = -1/2$: *Bethe roots, symmetric functions and determinants*, J. Math. Phys. **43** (2002), 4135–4146; texttarXiv:math-ph/0110011

32. J. de Gier, B. Nienhuis, P. A. Pearce and V. Rittenberg, *A new universality class for dynamical processes*, Phys. Rev. E **67** (2002), 016101–016104; arXiv:cond-mat/0205467, arXiv:cond-mat/0108051.

33. J. de Gier, B. Nienhuis, P. A. Pearce and V. Rittenberg, *The raise and peel model of a fluctuating interface*, J. Stat. Phys. **114** (2004), 1–35; arXiv:cond-mat/0301430

34. J. de Gier, P. Pyatov and P. Zinn-Justin, *Punctured plane partitions and the q-deformed Knizhnik-Zamolodchikov and Hirota equations*, arXiv:0712.3584.

35. R.W. Gosper, *Decision procedure for indefinite hypergeometric summation*, Proc. Natl. Acad. Sci. USA **75** (1978), 40–42.

36. A. G. Izergin, *Partition function of the six-vertex model in a finite volume*, Dokl. Akad. Nauk SSSR **297** (1987), 331–333 (Sov. Phys. Dokl. **32** (1987), 878–879).

37. A. G. Izergin, D. A. Coker and V. E. Korepin, *Determinant formula for the six-vertex model*, J. Phys. A **25** (1992), 4315–4334.

38. N. Kitanine, J. M. Maillet, N. A. Slavnov and V. Terras, *Emptiness formation probability of the XXZ spin-1/2 Heisenberg chain at* $\Delta = 1/2$, J. Phys. A **35** (2002), L385–L388; arXiv:hep-th/0201134.

39. V. E. Korepin, *Calculation of norms of Bethe wave functions*, Commun. Math. Phys. **86** (1982), 391–418.

40. V. E. Korepin, N. M. Bogoliubov and A. G. Izergin, *Quantum inverse scattering method and correlation functions*, Cambridge University Press, Cambridge (1993).

41. V. Korepin and P. Zinn-Justin, *Thermodynamic limit of the six-vertexmodel with domain wall boundary conditions*, J. Phys. A **33** (2000), 7053–7066; arXiv:cond-mat/0004250.

42. G. Kuperberg, *Another proof of the alternating sign matrix conjecture*, Invent. Math. Res. Notes, (1996), 139–150; arXiv:math.CO/9712207.

43. G. Kuperberg, *Symmetry classes of alternating-sign matrices under one roof*, Ann. Math. **156** (2002), 835–866; arXiv:math.CO/0008184.

44. D. Levy, *Algebraic structure of translation invariant spin-1/2 XXZ and q-Potts quantum chains*, Phys. Rev. Lett. **67** (1991), 1971-1974.

45. E.H. Lieb, *Exact solution of the problem of the entropy of two-dimensional ice*, Phys. Rev. Lett. **18** (1967), 692.

46. E.H. Lieb, *Exact solution of the two-dimensional Slater KDP model of an antiferroelectric*, Phys. Rev. Lett. **18** (1967), 1046–1048.

47. E. H. Lieb, *Exact solution of the two-dimensional Slater KDP model of a ferroelectric*, Phys. Rev. Lett. **19** (1967), 108–110.

48. E. H. Lieb and F. Y. Wu, *Two dimensional ferroelectric models*, in Phase transitions and critical phenomena **1** (Academic Press, 1972), C. Domb and M. Green eds., 331–490.

49. B. Lindström, *On the vector representations of induced matroids*, Bull. London Math. Soc. **5** (1973), 85–90.

50. P. P. Martin *Potts models and related problems in statistical mechanics*, World Scientific, Singapore (1991).

51. W. H. Mills, D. P. Robbins and H. Rumsey, *Proof of the MacDonald conjecture*, Invent. Math., 66 (1982), 73–87.

52. W. H. Mills, D. P. Robbins and H. Rumsey, *Alternating-sign matrices and descending plane partitions*, J. Combin. Theory Ser. A **34** (1983), 340–359.

53. W. H. Mills, D. P. Robbins and H. Rumsey, *Enumeration of a symmetry class of plane partitions*, Discr. Math. **67** (1987), 43–55.

54. S. Mitra, B. Nienhuis, J. de Gier and M. T. Batchelor, JSTAT (2004), P09010; arXiv:cond-mat/0401245.

55. S. Mitra and B. Nienhuis, *Exact conjectured expressions for correlations in the dense O(1) loop model on cylinders*, JSTAT (2004), P10006; arXiv:cond-mat/0407578.

56. S. Mitra, *Exact asymptotics of the characteristic polynomial of the symmetric Pascal matrix*, J. Combin. Theory Ser. A, in press; arXiv:0708.1763

57. V. Pasquier, *Quantum incrompressibility and Razumov Stroganov type conjectures*, Ann. Henri Poincare' **7** (2006) 397–421; arXiv:cond-mat/0506075.

58. P. Paule and M. Schorn, *A Mathematica version of Zeilberger's algorithm for proving binomial coefficient identities*, J. Symb. Comp. **20** (1995), 673–698.

59. P. A. Pearce, V. Rittenberg, J. de Gier and B. Nienhuis, *Temperley-Lieb stochastic processes* J. Phys. A **35** (2002), L661–L668; arXiv:math-ph/0209017.

60. V. Polyakov, *Zh. Eksp. Theor. Fiz.* **59** (1970), 542.

61. J. Propp, *The many faces of the alternating-sign matrices*, Discrete Mathematics and Theoretical Computer Science Proceedings AA (2001), 43–58.

62. P. Pyatov, *Raise and Peel Models of fluctuating interfaces and combinatorics of Pascal's hexagon*, J. Stat. Mech. (2004), P09003; arXiv:math-ph/0406025.

63. A. V. Razumov and Yu. G. Stroganov *Spin chains and combinatorics*, J. Phys. A **34** (2001), 3185–3190; arXiv:cond-mat/0012141.

64. A. V. Razumov and Yu. G. Stroganov, *Combinatorial nature of ground state vector of O(1) loop model*, Theor. Math. Phys. **138** (2004), 333–337; Teor. Mat. Fiz. **138** (2004), 395–400; arXiv:math.CO/0104216.

65. A. V. Razumov and Yu. G. Stroganov, *O(1) loop model with different boundary conditions and symmetry classes of alternating-sign matrices*, Theor. Math. Phys. **142** (2005) 237–243; Teor. Mat. Fiz. **142** (2005), 284–292; arXiv:math.CO/0108103.

66. A. V. Razumov and Yu. G. Stroganov, *Spin chains and combinatorics: twisted boundary conditions*, J. Phys. A **34** (2001), 5335–5340; arXiv:cond-mat/0102247.

67. D.P. Robbins, *The story of* 1, 2, 7, 42, 429, 7436, ..., Math. Intelligencer **13** (1991), 12–19.

68. D.P. Robbins, *Symmetry classes of alternating sign matrices*, (2000); arXiv:math.CO/0008045.

69. D.P. Robbins and H. Rumsey, *Determinants and alternating-sign matrices*, Adv. Math. **62** (1986), 169–184.

70. H. Rosengren, *An Izergin-Korepin-type identity for the 8VSOS model, with applications to alternating sign matrices*, arXiv:0801.1229.

71. Yu. G. Stroganov, *The importance of being odd*, J. Phys. A **34** (2001), L179-L185; arXiv:cond-mat/0012035.

72. Yu. G. Stroganov, *A new way to deal with Izergin-Korepin determinant at root of unity* (2002); arXiv:math-ph/0204042.

73. B. Sutherland, *Exact solution of a two-dimensional model for hydrogen-bonded crystals*, Phys. Rev. Lett. **19** (1967), 103–104.

74. K. Szogo, *Toda molecule equation and quotient-difference method*, J. Phys. Soc. Japan **62** (1993), 1081–1084.

75. H. N. V. Temperley and E. H. Lieb, *Relations between the 'percolation' and 'colouring' problem and other graph-theoretical problems associated with regular planar lattices: some exact results for the 'percolation' problem*, Proc. R. Soc. London A **322** (1971), 251–280.

76. J. Thapper, *Refined counting of fully packed loop configurations*, Sem. Lothar. Combin. **56** (2006/07), Art. B56e, 27 pp.

77. A. Tsuchiya, *Determinant formula for the six-vertex model with reflecting end*, J. Math. Phys. **39** (1998), 5946–5951; arXiv:solv-int/9804010.

78. B. Wieland, *A large dihedral symmetry of the set of alternating sign matrices*, Electron. J. Combin. **7** (2000), R37; arXiv:math/0006234.

79. D. Zeilberger, *A fast algorithm for proving terminating hypergeometric identities*, Discr. Math. **80** (1990), 207–211.

80. D. Zeilberger, *The method of creative telescoping*, J. Symb. Comp. **11** (1991), 195–204.

81. D. Zeilberger, *Proof of the alternating sign matrix conjecture*, Electr. J. Combin. **3** (1996), R13.

82. D. Zeilberger, *Proof of the refined alternating sign matrix conjecture*, New York J. Math. **2** (1996), 59–68.

83. P. Zinn-Justin, *Six-vertex model with domain wall boundary conditions and one-matrix model*, Phys. Rev. E **62** (2000), 3411–3418; arXiv:math-ph/0005008.

84. P. Zinn-Justin, *Loop model with mixed boundary conditions, qKZ equation and Alternating Sign Matrices*, J. Stat. Mech. (2007), P01007; arXiv:math-ph/0610067.

85. J.-B. Zuber, *On the counting of fully packed loop configurations; some new conjectures*; Electr. J. Combin. **11** (2004), R13; arxiv:math-ph/0309057

# Chapter 14
# Conformal Field Theory Applied to Loop Models

Jesper Lykke Jacobsen

## 14.1 Introduction

The application of methods of quantum field theory to problems of statistical mechanics can in some sense be traced back to Onsager's 1944 solution [1] of the two-dimensional Ising model. It does however appear fair to state that the 1970's witnessed a real gain of momentum for this approach, when Wilson's ideas on scale invariance [2] were applied to study critical phenomena, in the form of the celebrated renormalisation group [3]. In particular, the so-called $\varepsilon$ expansion permitted the systematic calculation of critical exponents [4], as formal power series in the space dimensionality $d$, below the upper critical dimension $d_c$. An important lesson of these efforts was that critical exponents often do not depend on the precise details of the microscopic interactions, leading to the notion of a restricted number of distinct universality classes.

Meanwhile, further exact knowledge on two-dimensional models had appeared with Lieb's 1967 solution [5] of the six-vertex model and Baxter's subsequent 1971 generalisation [6] to the eight-vertex model. These solutions challenged the notion of universality class, since they provided examples of situations where the critical exponents depend continuously on the parameters of the underlying lattice model. On the other hand, the techniques of integrability used relied crucially on certain exact microscopic conservation laws, thus placing important restrictions on the models which could be thus solved.

An important break-through occurred in 1984 when Belavin, Polyakov and Zamolodchikov [7] applied ideas of conformal invariance to classify the possible types of critical behaviour in two dimensions. These ideas had emerged earlier in string theory and mathematics, and in fact go back to earlier (1970) work of Polyakov [8] in which global conformal invariance is used to constrain the form of correlation functions in $d$-dimensional theories. It is however only by imposing

Jesper Lykke Jacobsen
Laboratoire de Physique Théorique, École Normale Supérieure (LPTENS), 24 rue Lhomond, 75231 Paris Cedex 05, France, e-mail: jesper.jacobsen@ens.fr

*local* conformal invariance in $d = 2$ that this approach becomes really powerful. In particular, it immediately permitted a full classification of an infinite family of conformally invariant theories (the so-called "minimal models") having a finite number of fundamental ("primary") fields, and the exact computation of the corresponding critical exponents. In the aftermath of these developments, conformal field theory (CFT) became for some years one of the most hectic research fields of theoretical physics, and indeed has remained a very active area up to this date.

Despite the amazing classification powers of CFT, it remains a tricky question to make the link between a given critical lattice model and the corresponding CFT. This is particularly true for geometrically defined models, such as percolation and self-avoiding polygons (SAP) and walks (SAW), since then typically the usual assumptions of minimality and unitarity (roughly speaking, positive definite Boltzmann weights) fail. Such models are however well treated by the so-called Coulomb gas (CG) approach, in which the geometric degrees of freedom are directly identified with the level lines of one or more free bosonic "height" fields. This approach preceded [9, 10, 11] the break-through of CFT, but shortly thereafter it was used (in a more formal and less geometrically inspired form) for the extensive computation of correlation functions in minimal models by Dotsenko and Fateev [12, 13]. A large number of applications in polymer physics was developed by Duplantier and Saleur [14].

The goal of this chapter is to present the application of CFT—with special emphasis on the CG approach—to two-dimensional models of self-avoiding loops, each loop occurring with a fugacity $n$. Self-avoiding polygons and walks then appear as special cases in the limit $n \to 0$. In section 14.2 we outline the key concepts of CFT. The aim is to make the presentation self-contained while remaining rather brief; the reader interested in more details should turn to the comprehensive textbook [15] or the Les Houches volume [16]. The geometric CG approach is introduced in section 14.3, and is shown to lead to a CFT of the Liouville type. The presence of screening charges is linked to a marginality requirement [17, 18] that ensures the exact solvability of the model.

The CG approach is subsequently applied in section 14.4 to the computation of bulk critical properties of SAP's and SAW's. It should be emphasised that most of the material is presented within the general framework of loop models, taking the SAP limit ($n \to 0$) only at the end of the computations. SAW's are then obtained by inserting appropriate defects before taking the limit. There are good reasons for this manner of presentation: first, more general results are obtained at no extra expense; second, a number of general concepts emerge more clearly; and third, the example of the $\Theta$-point collapse transition [19] shows that also $n \neq 0$ is of relevance to polymer problems. In fact, loop models furnish a nice illustration of most of the key concepts of two-dimensional CFT.

While section 14.3 focuses on loop models with a scalar height field, equivalent to the standard O($n$) and $Q$-state Potts models (with $n = \sqrt{Q}$), a new class of loop models with vectorial target spaces is introduced in section 14.5. In these models, first solved by Kondev and collaborators [20, 17, 21], the loops are fully packed on the lattice, and become Hamiltonian circuits or walks in the $n \to 0$ limit.

In sections 14.6–14.7 we illustrate the importance of the topology of the space in which the loops are embedded. The preceding discussion in fact pertained to the geometry of the (punctured) plane. In contrast, section 14.6 is devoted to the half-plane geometry, in which the loops undergo specific interactions with the surface. The appropriate theoretical setup is that of boundary CFT, a subject pioneered by Cardy [22]. Finally, in section 14.7 the loops are embedded in a torus and the fundamental requirement of modular invariance [23] is exploited to write down modular invariant partition functions in the continuum limit. Using a similar approach, exact continuum limit partition functions are written down in the annulus geometry as well.

## 14.2 Basic Concepts of CFT

### 14.2.1 Global Conformal Invariance

A conformal transformation in $d$ dimensions is an invertible mapping $\mathbf{x} \to \mathbf{x}'$ which multiplies the metric tensor $g_{\mu\nu}(\mathbf{x})$ by a space-dependent scale factor:

$$g'_{\mu\nu}(\mathbf{x}') = \Lambda(\mathbf{x})g_{\mu\nu}(\mathbf{x}). \qquad (14.1)$$

Note that such a mapping preserves angles. Therefore, just as Wilson [2] suggested using *global* scale invariance as the starting point for investigating a system at its critical point, Polyakov [8] proposed imposing the *local* scale invariance (14.1) as the fundamental requirement for studying a critical system in which the microscopic interactions are short ranged. A priori, a geometrical model of self-avoiding objects such as SAP's and SAW's does not seem to be governed by short-range interactions; that this is nevertheless true will be shown in section 14.3.2 where we shall make explicit the locality of such models.

The group of conformal transformations is easily shown to be generated by translations, dilations, rotations, and the so-called special conformal transformations (which are just the composition of an inversion $x^\mu \to x^\mu/\mathbf{x}^2$, a translation, and another inversion). Writing down the commutation rules of the generators, one establishes that the conformal group is isomorphic to the pseudo-orthogonal group $SO(d+1,1)$ with $\frac{1}{2}(d+1)(d+2)$ real parameters.

The connection between a statistical mechanics model and quantum field theory is made as usual by writing the partition function and correlation functions of the former as functional integrals in the latter:

$$Z = \int \mathscr{D}\Phi\, e^{-S[\Phi]}$$

$$\langle \phi_1(\mathbf{x}_1)\ldots\phi_k(\mathbf{x}_k)\rangle = Z^{-1}\int \mathscr{D}\Phi\, \phi_1(\mathbf{x}_1)\ldots\phi_k(\mathbf{x}_k)e^{-S[\Phi]} \qquad (14.2)$$

Here $S[\Phi]$ is the euclidean action, $\Phi$ the collection of fields, and $\phi_i \in \Phi$. In other words, $Z^{-1}e^{-S[\Phi]}\mathcal{D}\Phi$ is the Gibbs measure in the continuum limit. Paradoxically, in many cases the hypothesis of conformal invariance may permit one to classify and precisely characterise the possible continuum theories without ever having to write down explicitly the action $S[\Phi]$.

A field $\phi(\mathbf{x})$, here supposed spinless for simplicity, is called *quasi-primary* provided it transforms covariantly under the conformal transformation (14.1):

$$\phi(\mathbf{x}) \to \phi'(\mathbf{x}') = \left|\frac{\partial \mathbf{x}'}{\partial \mathbf{x}}\right|^{-\Delta/d} \phi(\mathbf{x}). \tag{14.3}$$

The number $\Delta = \Delta_\phi$ is a property of the field and is called its *scaling dimension*. Using this, conformal invariance completely fixes [8] the form of the two- and three-point correlation functions:

$$\langle \phi_1(\mathbf{x}_1)\phi_2(\mathbf{x}_2)\rangle = \frac{\delta_{\Delta_1,\Delta_2}}{x_{12}^{2\Delta_1}}, \tag{14.4}$$

$$\langle \phi_1(\mathbf{x}_1)\phi_2(\mathbf{x}_2)\phi_3(\mathbf{x}_3)\rangle = \frac{C_{123}}{x_{12}^{\Delta_1+\Delta_2-\Delta_3} x_{23}^{\Delta_2+\Delta_3-\Delta_1} x_{31}^{\Delta_3+\Delta_1-\Delta_2}} \tag{14.5}$$

where we have introduced $x_{ij} = |\mathbf{x}_i - \mathbf{x}_j|$. The fields have here been normalised so that the coefficient in (14.4) is unity. The structure constants $C_{123}$ appearing in (14.5) are then fundamental dynamical quantities characterising the theory at hand. With four or more points, the correlation functions are no longer completely fixed, due to the existence of conformally invariant functions of four points, $\eta = x_{12}x_{34}/x_{13}x_{24}$, the so-called *anharmonic ratios*.

## 14.2.2 Two Dimensions and Local Conformal Invariance

Conformal invariance is especially powerful in two dimensions for reasons that we shall expose presently. For the moment, we work in the geometry of the Riemann sphere, i.e., the plane with a point at infinity, and we shall write the coordinates as $\mathbf{x} = (x^1, x^2)$. Under a general coordinate transformation $x^\mu \to x'^\mu = w^\mu(x^1, x^2)$ application of (14.1) implies the Cauchy-Riemann equations, $\partial w^2/\partial x^1 = \pm \partial w^1/\partial x^2$ and $\partial w^1/\partial x^1 = \mp \partial w^2/\partial x^2$, i.e., $\mathbf{w}(\mathbf{x})$ is either a holomorphic or an antiholomorphic function. Important simplifications will therefore result upon introducing the complex coordinates $z \equiv x^1 + ix^2$ and $\bar{z} \equiv x^1 - ix^2$. A conformal mapping then reads simply $z \to z' = w(z)$.

The identification of two-dimensional conformal transformations with analytic maps $w(z)$ could have been anticipated from the well-known fact that the latter are angle-preserving. It should be noted that an analytic map is defined (via its Laurent series) by an *infinite* number of parameters. This does not contradict the result of section 14.2.1 that the set of global conformal transformations is defined by only

$\frac{1}{2}(d+1)(d+2) = 6$ real parameters, since analytic maps are not necessarily invertible and defined in the whole complex plane. Global conformal transformations in $d = 2$ take the form of the projective transformations

$$w(z) = \frac{a_{11}z + a_{12}}{a_{21}z + a_{22}} \tag{14.6}$$

with $a_{ij} \in \mathbb{C}$ and the constraint $\det a_{ij} = 1$, i.e., they form the group $SL(2, \mathbb{C}) \simeq SO(3, 1)$.

In complex coordinates, the transformation law (14.3) becomes

$$\phi'(w, \bar{w}) = \left(\frac{\mathrm{d}w}{\mathrm{d}z}\right)^{-h} \left(\frac{\mathrm{d}\bar{w}}{\mathrm{d}\bar{z}}\right)^{-\bar{h}} \phi(z, \bar{z}) \tag{14.7}$$

where the *real* parameters $(h, \bar{h})$ are called the *conformal weights*. The combinations $\Delta = h + \bar{h}$ and $s = h - \bar{h}$ are called respectively the scaling dimension and the spin of $\phi$. A field $\phi$ satisfying (14.7) for any projective transformation (resp. any analytic map) $w(z)$ is called quasi-primary (resp. primary). An example of a quasi-primary field which is not primary is furnished by the stress tensor (see below).

The expressions (14.4)–(14.5) for the two- and three-point correlation functions still hold true with the obvious modification that the dependence in $z_{ij} \equiv z_i - z_j$ (resp. in $\bar{z}_{ij}$) goes with the conformal weights $h$ (resp. $\bar{h}$).

### 14.2.3 Stress Tensor and Ward Identities

The stress tensor $T^{\mu\nu}$ is the conserved Noether current associated with the conformal symmetry. It can be defined[1] as the response of the partition function to a local change in the metric:

$$T^{\mu\nu}(\mathbf{x}) = -\frac{1}{2\pi} \frac{\delta \log Z}{\delta g_{\mu\nu}(\mathbf{x})} \tag{14.8}$$

Translational and rotational invariances imply the conservation law $\partial_\mu T^{\mu\nu} = 0$ as well as the symmetry $T^{\mu\nu} = T^{\nu\mu}$. Scale invariance further implies the tracelessness $T^\mu_\mu = 0$; in general the trace would be proportional to the beta function, which vanishes at a renormalisation group fixed point.

Rewriting this in complex coordinates, one finds that $T_{z\bar{z}} = T_{\bar{z}z} = 0$, while the conservation law takes the form $\partial_{\bar{z}} T(z) = \partial_z \bar{T}(\bar{z})$, where we have defined $T(z) \equiv T_{zz}$ and $\bar{T}(\bar{z}) \equiv T_{\bar{z}\bar{z}}$. So $T(z)$ is analytic, while $\bar{T}(\bar{z})$ is antianalytic. This is a very important element in the solvability of two-dimensional CFT. Following Fateev and Zamolodchikov [24] it is even possible to go (much) further: CFT's in which the conformal symmetry is enhanced with other, so-called extended, symmetries (superconformal, parafermionic, W algebra,...) can be constructed by requiring more

---

[1] Note the analogy with the theory of integrable systems, where the conserved charges are obtained as derivatives of the transfer matrix with respect to the anisotropy (spectral parameter).

analytic currents and making them coexist with $T(z)$ by imposing certain associativity requirements.

Consider now the change in the metric induced by an infinitesimal conformal transformation $z' = z + \varepsilon(z)$. Its effects on an arbitrary product of primary fields $X = \prod_j \phi_j(z_j, \bar{z}_j)$ can be written in terms of $T(z)$ as

$$\oint_C \langle T(z)X \rangle \varepsilon(z)\, dz = \sum_j \left( h_j \varepsilon'(z_j) + \varepsilon(z_j)\partial_{z_j} \right) \langle X \rangle, \tag{14.9}$$

where $C$ is any counterclockwise contour encircling $\{z_j\}$. This is called the *conformal Ward identity*. By the Cauchy theorem, this is equivalent to

$$T(z)\phi_j(z_j, \bar{z}_j) = \frac{h_j}{(z - z_j)^2} \phi_j(z_j, \bar{z}_j) + \frac{1}{z - z_j} \partial_{z_j}\phi(z_j, \bar{z}_j) + \mathcal{O}(1). \tag{14.10}$$

This is our first example of an *operator product expansion* (OPE), i.e., a formal power series in the coordinate difference that expresses the effect of bringing close together two operators. Several remarks are in order. First, it is tacitly understood that OPE's only have a sense when placed between the brackets $\langle \cdots \rangle$ of a correlation function. Second, we generically expect singularities to arise when approaching two local operators in a quantum field theory; in particular the average of a field over some small volume will have a variance that diverges when that volume is taken to zero. Third, an OPE should be considered an exact identity rather than an approximation, provided the formal expansion is written out to arbitrarily high order. In our example, (14.9) only determines the first two terms in the OPE (14.10). Fourth, contracting any field $\phi$ with $T(z)$ and comparing with (14.10) is actually a useful practical means of determining its primarity and its conformal dimension $h_\phi$.

It is not difficult to see from (14.8) that on dimensional grounds $T$ itself is a quasi-primary field of conformal dimension $h = 2$. However, the average $\langle T(z_1)T(z_2) \rangle \sim (z_1 - z_2)^{-4}$ has no reason to vanish, and so the OPE of $T$ with itself takes the form

$$T(z_1)T(z_2) = \frac{c/2}{(z_1 - z_2)^4} + \frac{2T(z_2)}{(z_1 - z_2)^2} + \frac{\partial T(z_2)}{z_1 - z_2} + \mathcal{O}(1). \tag{14.11}$$

In particular, $T$ is *not* primary. The constant $c$ appearing in (14.11) is called the *central charge*. Considering two non-interacting CFT's as a whole, one has from (14.8) that their stress tensors, and hence their central charges, add up, and so $c$ can be considered as a measure of the number of quantum degrees of liberty in the CFT. It is straightforward to establish that $c = 1/2$ for a free fermion and $c = 1$ for a free boson. We shall see later that standard SAP's have $c = 0$.

As $T$ is not primary, it cannot transform like (14.7) under a finite conformal transformation $z \to w(z)$. We can always write the modified transformation law as

$$T'(w) = \left( \frac{dw}{dz} \right)^{-2} \left[ T(z) - \frac{c}{12}\{w; z\} \right]. \tag{14.12}$$

To determine what $\{w; z\}$ represents, we use the constraint due to two successive applications of (14.12) and the fact that $\{w; z\} = 0$ for projective conformal transformations, since $T$ is quasi-primary. The result is that $\{w; z\}$ is the Schwarzian derivative

$$\{w; z\} = \frac{d^3 w/dz^3}{dw/dz} - \frac{3}{2} \left( \frac{d^2 w/dz^2}{dw/dz} \right)^2 . \tag{14.13}$$

### 14.2.4 Finite-Size Scaling on a Cylinder

The central charge $c$ is ubiquitous in situations where the CFT is placed in a finite geometry, i.e., interacts with some boundary condition. An important example is furnished by conformally mapping the plane to a cylinder of circumference $L$ by means of the transformation

$$w(z) = \frac{L}{2\pi} \log z . \tag{14.14}$$

This transformation can be visualised by viewing the cylinder in perspective, with one rim contracting to the origin and the other expanding to form the point at infinity. Taking the expectation value of (14.12), and using the fact that $\langle T(z) \rangle = 0$ in the plane on symmetry grounds, one finds that $\langle T(w) \rangle = -\pi^2 c/6L^2$ on the cylinder. Applying (14.8) then implies that the free energy per unit area $f_0(L)$ satisfies [25]

$$f_0(L) = f_0(\infty) - \frac{\pi c}{6L^2} + o(L^{-2}) . \tag{14.15}$$

This is a very useful result for obtaining $c$ for a concrete statistical model, since $f(L)$ can usually be determined from the corresponding transfer matrix, either numerically for small $L$ by using exact diagonalisation techniques, or analytically in the Bethe Ansatz context by using the Euler-McLauren formula.

It is also of interest to study such finite-size effects on the level of the two-point correlation function of a primary field $\phi$. Again using the mapping (14.14), the covariance property (14.7) and the form (14.4) of the correlator in the plane can be used to deduce its form on the cylinder. Assuming for simplicity $h = \bar{h} = \Delta/2$, and writing the coordinates on the cylinder as $w = t + ix$, with $t \in \mathbb{R}$ and $x \in [0, L)$, one arrives at

$$\langle \phi(t_1, x_1) \phi(t_2, x_2) \rangle = \left( \frac{2\pi}{L} \right)^{2\Delta} \left[ 2 \cosh \left( \frac{2\pi t_{12}}{L} \right) - 2 \cos \left( \frac{2\pi x_{12}}{L} \right) \right]^{-\Delta} , \tag{14.16}$$

where $t_{12} = t_1 - t_2$ and $x_{12} = x_1 - x_2$. In the limit of a large separation of the fields, $t_{12} \to \infty$, this decays like $e^{-t_{12}/\xi}$ with correlation length $\xi = L/2\pi\Delta$. But this decay can also be written $(\Lambda_\phi/\Lambda_0)^{-t_{12}}$, where $\Lambda_0$ is the largest eigenvalue of the transfer matrix, and $\Lambda_\phi$ is the largest eigenvalue compatible with the constraint that an operator $\phi$ has been inserted at each extremity $t = \pm\infty$ of the cylinder. Denoting the corresponding free energies per unit area $f(L) = -L^{-1} \log \Lambda$, we conclude that [26]

$$f_\phi(L) - f_0(L) = \frac{2\pi\Delta}{L^2} + o(L^{-2}).$$ (14.17)

This is as useful as (14.15) in (numerical or analytical) transfer matrix studies, since the constraint imposed by $\phi$ can usually be related explicitly to properties of the transfer matrix spectrum.

### 14.2.5 Virasoro Algebra and Its Representation Theory

Up to this point, we have worked in a setup where the fields were seen as functionals of the complex coordinates $z$, $\bar{z}$. To obtain an operator formalism, one must impose a quantisation scheme, i.e., single out a time and a space direction. The transfer matrix then propagates the system from one time slice to the following and is written as the exponential of the Hamiltonian $\mathscr{H}$, i.e., the energy operator on a fixed-time surface. In the continuum limit, one may freely choose the time direction. In CFT this is most conveniently done by giving full honours to the scale invariance of the theory, viz., by using for $\mathscr{H}$ the dilatation operator (to be precise, $\mathscr{H} = (2\pi/L)(L_0 + \bar{L}_0 - c/12)$)

$$\mathscr{D} = \frac{1}{2\pi i} \oint_C z T(z)\,\mathrm{d}z - \frac{1}{2\pi i} \oint_C \bar{z}\bar{T}(\bar{z})\,\mathrm{d}\bar{z} = L_0 + \bar{L}_0,$$ (14.18)

where $C$ is a counterclockwise contour enclosing the origin. This is called the *radial quantisation* scheme: the constant-time surfaces are concentric circles around the origin. Under the map (14.14) the time becomes simply the coordinate along the cylinder axis. The usual time ordering of operators then becomes a prescription of radial ordering.

In (14.18) we have anticipated the definition of the mode operators

$$L_n = \frac{1}{2\pi i} \oint_C z^{n+1} T(z)\,\mathrm{d}z, \qquad \bar{L}_n = \frac{1}{2\pi i} \oint_C \bar{z}^{n+1}\bar{T}(\bar{z})\,\mathrm{d}\bar{z}.$$ (14.19)

Using the radial ordering, the OPE (14.11) becomes, after a deformation of contours, the commutation relations

$$[L_n, L_m] = (n - m)L_{n+m} + \frac{c}{12}n(n^2 - 1)\delta_{n+m,0}$$ (14.20)

with a similar expression for $[\bar{L}_n, \bar{L}_m]$, whereas $[L_n, \bar{L}_m] = 0$. The algebra defined by (14.20) is called the *Virasoro algebra*. Importantly, the decoupling into two isomorphic Virasoro algebras, one for $L_n$ and another for $\bar{L}_n$, means that in the geometry chosen we can focus exclusively on $L_n$. It should be stressed that in the geometry of a torus, the two algebras couple non-trivially, in a way that is revealed by imposing modular invariance (see section 14.2.7 below).

We now describe the structure of the Hilbert space in radial quantisation. The vacuum state $|0\rangle$ must be invariant under projective transformations, whence $L_{\pm 1}|0\rangle = 0$, and we fix the ground state energy by $L_0|0\rangle = 0$. Non-trivial eigenstates of $\mathscr{H}$ are created by action with a primary field, $|h, \bar{h}\rangle = \phi(0, 0)|0\rangle$. Translating (14.10) into operator language implies then in particular $L_0|h, \bar{h}\rangle = h|h, \bar{h}\rangle$. We must also impose the highest-weight condition $L_n|h, \bar{h}\rangle = \bar{L}_n|h, \bar{h}\rangle = 0$ for $n > 0$. Excited states with respect to the primary $\phi$ then read

$$\phi^{\{n, \bar{n}\}} \equiv L_{-n_1} L_{-n_2} \cdots L_{-n_k} \bar{L}_{-\bar{n}_1} \bar{L}_{-\bar{n}_2} \cdots \bar{L}_{-\bar{n}_{\bar{k}}} |h, \bar{h}\rangle \qquad (14.21)$$

with $1 \leq n_1 \leq n_2 \leq \cdots \leq n_k$ and similarly for $\{\bar{n}\}$. These states are called the *descendents* of $\phi$ at level $\{N, \bar{N}\}$, where $N = \sum_{i=1}^{k} n_i$. A primary state and its descendents form a highest weight representation (or Verma module) of the Virasoro algebra.

Correlation functions of descendent fields can be obtained by acting with appropriate differential operators on the correlation functions of the corresponding primary fields. To see this, consider first for $n \geq 1$ the descendent $(L_{-n}\phi)(w)$ of the primary field $\phi(w)$, and let $X = \prod_j \phi_j(w_j)$ be an arbitrary product of other primaries as in the conformal Ward identity (14.9). Using (14.19) and (14.10) we have then

$$\langle (L_{-n}\phi)(w)X \rangle = \frac{1}{2\pi i} \oint_z dz\, (z - w)^{1-n} \langle T(z)\phi(w)X \rangle \qquad (14.22)$$

$$= -\frac{1}{2\pi i} \oint_{\{w_j\}} dz\, (z - w)^{1-n} \sum_j \left\{ \frac{\partial_{w_j}}{z - w_j} + \frac{h_j}{(z - w_j)^2} \right\} \langle \phi(w)X \rangle$$

where the minus sign comes from turning the integration contour inside out, so that it surrounds all the points $\{w_j\}$. In other words, a descendent in a correlation function may be replaced by the corresponding primary

$$\langle (L_{-n}\phi)(w)X \rangle = \mathscr{L}_{-n} \langle \phi(w)X \rangle \qquad (14.23)$$

provided that we act instead on the correlator with the linear differential operator

$$\mathscr{L}_{-n} \equiv \sum_j \left\{ \frac{(n - 1)h_j}{(w_j - w)^n} - \frac{\partial_{w_j}}{(w_j - w)^{n-1}} \right\} \qquad (14.24)$$

It is readily seen that a general descendent (14.21) is similarly dealt with by replacing each factor $L_{-n_i}$ by the corresponding factor of $\mathscr{L}_{-n_i}$ in (14.23).

We can now write the general form of the OPE of two primary fields $\phi_1$ and $\phi_2$. It reads

$$\phi_1(z, \bar{z})\phi_2(0, 0) = \sum_p C_{12p} \sum_{\{n, \bar{n}\} \cup \{0, 0\}} C_{12p}^{\{n, \bar{n}\}} z^{h_p - h_1 - h_2 + N} \bar{z}^{\bar{h}_p - \bar{h}_1 - \bar{h}_2 + \bar{N}} \phi_p^{\{n, \bar{n}\}}(0, 0),$$

$$(14.25)$$

where the summation is over a certain set of primaries $\phi_p \equiv \phi_p^{\{0, 0\}}$ as well as their descendents. The coefficients $C_{12p}^{\{n, \bar{n}\}}$ (we have set $C_{12p}^{\{0, 0\}} = 1$) can be determined by acting with all combinations of positive-index mode operators on both sides of (14.25) and solving the resulting set of linear equations. In contradistinction, the coefficients $C_{12p}$ are fundamental quantities, easily shown to coincide with those appearing in the three-point functions (14.5). They can be computed by the so-called conformal bootstrap method, i.e., by assuming crossing symmetry of the four-point functions.

## 14.2.6 Minimal Models

Denote by $\mathcal{V}(c,h)$ the highest weight representation (Verma module) generated by the mode operators $\{L_n\}$ acting on a highest weight state $|h\rangle$ in a CFT of central charge $c$. The Hilbert space of the CFT can then be written

$$\bigoplus_{h,\bar{h}} n_{h,\bar{h}} \mathcal{V}(c,h) \otimes \mathcal{V}(c,\bar{h}) \tag{14.26}$$

where the multiplicities $n_{h,\bar{h}}$ indicate the number of distinct primaries of conformal weights $(h,\bar{h})$ that are present in the theory. A *minimal model* is a CFT for which the sum in (14.26) is finite.

The Hermitian conjugate of a mode operator is defined by $L_n^\dagger = L_{-n}$; this induces an inner product on the Verma module. The *character* $\chi_{(c,h)}$ of the module $\mathcal{V}(c,h)$ can then be defined as

$$\chi_{(c,h)}(\tau) = \mathrm{Tr}\, q^{L_0 - c/24}, \tag{14.27}$$

where $\tau \in \mathbb{C}$ is the so-called *modular parameter* (see section 14.2.7 below) and $q = e^{2\pi i \tau}$. Since the number of descendents of $|h\rangle$ at level $N$ is just the number $p(N)$ of integer partitions of $N$, cf. (14.21), we have simply

$$\chi_{(c,h)}(\tau) = \frac{q^{h-c/24}}{P(q)}, \tag{14.28}$$

where

$$\frac{1}{P(q)} \equiv \prod_{n=1}^{\infty} \frac{1}{1-q^n} = \sum_{n=0}^{\infty} p(n) q^n \tag{14.29}$$

is the generating function of partition numbers; this is also often expressed in terms of the Dedekind function

$$\eta(\tau) = q^{1/24} P(q). \tag{14.30}$$

However, the generic Verma module is not necessarily irreducible, so further work is needed.

For certain values of $h$, it may happen that a specific linear combination $|\chi\rangle$ of the descendents of $|h\rangle$ at level $N$ is itself primary, i.e., $L_n|\chi\rangle = 0$ for $n > 0$. In other words, $|\chi\rangle$ is primary and descendent at the same time, and it generates its own Verma module $\mathcal{V}_\chi(c,h) \subset \mathcal{V}(c,h)$. One easily shows that the states in $\mathcal{V}_\chi(c,h)$ are orthogonal to those in $\mathcal{V}(c,h)$, and so in particular they have zero norm. A Verma module $\mathcal{V}(c,h)$ containing one or more such *null fields* $|\chi\rangle$ is called reducible, and can be turned into an irreducible Verma module $\mathcal{M}(c,h)$ by quotienting out the null fields, i.e., by setting $|\chi\rangle = 0$. The Hilbert space is then given by (14.26) with $\mathcal{V}$

replaced by $\mathcal{M}$; since it contains fewer states the corresponding characters (14.27) are *not* given by the simple result (14.28).

The concept of null states is instrumental in constructing *unitary* representations of the Virasoro algebra (14.20), i.e., representations in which no state of *negative* norm occurs. An important first step is the calculation of the Kac determinant $\det M^{(N)}$ of inner products between descendents at level $N$. Its roots can be expressed through the following parameterisation:

$$c(m) = 1 - \frac{6}{m(m+1)}$$

$$h(m) = h_{r,s}(m) \equiv \frac{[(m+1)r - ms]^2 - 1}{4m(m+1)} \tag{14.31}$$

where $r, s \geq 1$ are integers with $rs \leq N$. The condition for unitarity of models with $c < 1$, first found by Friedan, Qiu and Shenker [27] reads: $m, r, s \in \mathbb{Z}$ with $m \geq 2$, and $(r,s)$ must satisfy $1 \leq r < m$ and $1 \leq s \leq m$.

According to (14.23) the presence of a descendent field in a correlation function can be replaced by the action of a differential operator (14.24). Now let

$$\chi(w) = \sum_{Y, |Y|=N} \alpha_Y L_{-Y} \phi(w) \tag{14.32}$$

be an arbitrary null state. Here, $\alpha_Y$ are some coefficients, and we have introduced the abbreviations

$$Y = \{r_1, r_2, \ldots, r_k\}$$
$$|Y| = r_1 + r_2 + \ldots + r_k \tag{14.33}$$
$$L_{-Y} = L_{-r_1} L_{-r_2} \cdots L_{-r_k}$$

with $1 \leq r_1 \leq r_2 \leq \cdots \leq r_k$. A correlation function involving $\chi$ must vanish (since we have in fact set $\chi = 0$), and so

$$\langle \chi(w) X \rangle = \sum_{Y, |Y|=N} \alpha_Y \mathcal{L}_{-Y}(w) \langle \phi(w) X \rangle = 0 \tag{14.34}$$

Solving this $N$th order linear differential equation is a very useful practical means of computing the four-point correlation functions of a given CFT, provided that the level of degeneracy $N$ is not too large. Indeed, since the coordinate dependence is through a single anharmonic ratio $\eta$, one has simply an ordinary linear differential equation.

Moreover, requiring consistency with (14.25) places restrictions on the primaries that can occur on the right-hand side of the OPE. One can then study the conditions under which this so-called *fusion algebra* closes over a finite number of primaries. The end result is that the minimal models are given by

$$c = 1 - \frac{6(m - m')^2}{mm'}$$

$$h_{r,s} = \frac{(mr - m's)^2 - (m - m')^2}{4mm'} \tag{14.35}$$

with $m, m', r, s \in \mathbb{Z}$, and the allowed values of $(r, s)$ are restricted by $1 \le r < m'$ and $1 \le s < m$. The corresponding $h_{r,s}$ are referred to as the *Kac table* of conformal weights. The corresponding fusion algebra reads (for clarity we omit scaling factors, structure constants, and descendents):

$$\phi_{(r_1, s_1)} \phi_{(r_2, s_2)} = \sum_{r,s} \phi_{(r,s)} \tag{14.36}$$

where $r$ runs from $1 + |r_1 - r_2|$ to $\min(r_1 + r_2 - 1, 2m' - 1 - r_1 - r_2)$ in steps of 2, and $s$ runs from $1 + |s_1 - s_2|$ to $\min(s_1 + s_2 - 1, 2m - 1 - s_1 - s_2)$ in steps of 2.

The Kac table (14.35) is the starting point for elucidating the structure of the reducible Verma modules $\mathcal{V}_{r,s}$ for minimal models, and for constructing the proper irreducible modules $\mathcal{M}_{r,s}$. The fundamental observation is that

$$h_{r,s} + rs = h_{r,-s}. \tag{14.37}$$

Using the symmetry property $h_{r,s} = h_{m'-r,m-s}$ and the periodicity property $h_{r,s} = h_{r+m',s+m}$ it is seen that $h_{r,s} + rs = h_{m'+r,m-s}$ and that $h_{r,s} + (m'-r)(m-s) = h_{r,2m-s}$. This means that $\mathcal{V}_{r,s}$ contains two submodules, $\mathcal{V}_{m'+r,m-s}$ and $\mathcal{V}_{r,2m-s}$, at levels $rs$ and $(m'-r)(m-s)$ respectively, and these must correspond to null vectors. To construct the irreducible module $\mathcal{M}_{r,s}$ one might at first think that it suffices to quotient out these two submodules. However, iterating the above observations, the two submodules are seen to share two sub-submodules, and so on. So $\mathcal{M}_{r,s}$ is constructed from $\mathcal{V}_{r,s}$ by an infinite series of inclusions-exclusions of pairs of submodules. This allows us in particular to compute the irreducible characters of minimal models as

$$\chi_{(r,s)}(\tau) = K_{r,s}^{(m,m')}(q) - K_{r,-s}^{(m,m')}(q), \tag{14.38}$$

where the infinite addition-subtraction scheme has been tucked away in the functions

$$K_{r,s}^{(m,m')}(q) = \frac{q^{-1/24}}{P(q)} \sum_{n \in \mathbb{Z}} q^{(2mm'n + mr - m's)^2 / 4mm'}. \tag{14.39}$$

This should be compared with the generic character (14.28). Note also the similarity between (14.37) and (14.38) on the level of the indices.

It is truly remarkable that the above classification of minimal models has been achieved without ever writing down the action $S$ appearing in (14.2). In fact, an effective Landau-Ginzburg Lagrangian description for the unitary minimal models ($m' = m + 1$) has been suggested a posteriori by Zamolodchikov [28]. It suggests that the minimal models can be interpreted physically as an infinite series of multicritical versions of the Ising model. Indeed, the Ising model can be identified with the first non-trivial member in the series, $m = 3$, and the following, $m = 4$, with the tricritical Ising model.

To finish this section, we comment on the relation with SAP's. In section 14.3 we shall see that these (to be precise, the dilute $O(n \to 0)$ model) can be identified with the minimal model $m = 2$, $m' = 3$. Note that this is *not* a unitary theory. The central charge is $c = 0$, and the only field in the Kac table—modulo the symmetry property given after (14.37)—is the identity operator with conformal weight $h_{1,1} = 0$. Seemingly we have learnt nothing more than the trivial statement $Z = 1$. However, the operators of interest are of a *non-local* nature, and it is a pleasant surprise to find that their dimensions fit perfectly well into the Kac formula, although they are situated *outside* the "allowed" range of $(r,s)$ values, and sometimes require the indices $r,s$ to be half-integer. So the Kac formula, and the surrounding theoretical framework, is still a most useful tool for investigating these types of models.

### 14.2.7 Modular Invariance

In section 14.2.3 we have seen that conformal symmetry makes the stress tensor decouple into its holomorphic and antiholomorphic components, $T(z)$ and $\bar{T}(\bar{z})$, implying in particular that the corresponding mode operators, $L_n$ and $\bar{L}_n$, form two non-interacting Virasoro algebras (14.20). As a consequence, the key results of section 14.2.6 could be derived by considering only the holomorphic sector of the CFT. There are however constraints on the ways in which the two sectors may ultimately couple, the diagonal coupling (14.26) being just the simplest example in the context of minimal models. As first pointed out by Cardy [23], a powerful tool for examining which couplings are allowed—and for placing constraints on the operator content and the conformal weights—is obtained by defining the CFT on a torus and imposing the constraint of *modular invariance*.

In this section we expose the principles of modular invariance and apply them to a CFT known as the *compactified boson*, which is going to play a central role in the Coulomb gas approach of section 14.3. Many other applications, including a detailed study of the minimal models, can be found in Ref. [15].

Let $\omega_1, \omega_2 \in \mathbb{C} \setminus \{0\}$ such that $\tau \equiv \omega_2/\omega_1 \notin \mathbb{R}$. A torus is then defined as $\mathbb{C}/(\omega_1 \mathbb{Z} + \omega_2 \mathbb{Z})$, i.e., by identifying points in the complex plane that differ by an element in the lattice spanned by $\omega_1, \omega_2$. The numbers $\omega_1, \omega_2$ are called the *periods* of the lattice, and $\tau$ the *modular parameter*. Without loss of generality we can assume $\omega_1 \in \mathbb{R}$ and $\Im \tau > 0$.

Instead of using the radial quantisation scheme of section 14.2.5 we now define the time (resp. space) direction to be the imaginary (resp. real) axis in $\mathbb{C}$. The partition function on the torus may then be written $Z(\tau) = \mathrm{Tr} \exp[-(\Im \omega_2)\mathcal{H} - (\Re \omega_2)\mathcal{P}]$, where $\mathcal{H} = (2\pi/\omega_1)(L_0 + \bar{L}_0 - c/12)$ is the Hamiltonian and $\mathcal{P} = (2\pi/i\omega_1)(L_0 - \bar{L}_0 - c/12)$ the momentum operator. This gives

$$Z(\tau) = \mathrm{Tr}\left(q^{L_0 - c/24} \bar{q}^{\bar{L}_0 - c/24}\right), \tag{14.40}$$

where we have defined $q = \exp(2\pi i \tau)$. Comparing with (14.26)–(14.27) we have also

$$Z(\tau) = \sum_{h,\bar{h}} n_{h,\bar{h}} \chi_{(c,h)}(\tau) \bar{\chi}_{(c,\bar{h})}(\tau). \tag{14.41}$$

An explicit computation of $Z(\tau)$ will therefore give information on the coupling $n_{h,\bar{h}}$ between the holomorphic and antiholomorphic sectors. In many cases, but not all, the coupling turns out to be simply diagonal, $n_{h,\bar{h}} = \delta_{h,\bar{h}}$.

The fundamental remark is now that $Z(\tau)$ is invariant upon making a different choice $\omega_1', \omega_2'$ of the periods, inasmuch as they span the same lattice as $\omega_1, \omega_2$. Any two set of equivalent periods must therefore be related by $\omega_i' = \sum_j a_{ij} \omega_j$, where $\{a_{ij}\} \in \mathrm{Mat}(2,\mathbb{Z})$ with $\det a_{ij} = 1$. Moreover, an overall sign change, $a_{ij} \to -a_{ij}$ is immaterial, so the relevant symmetry group is the so-called *modular group* $\mathrm{SL}(2,\mathbb{Z})/\mathbb{Z}_2 \simeq \mathrm{PSL}(2,\mathbb{Z})$.

The remainder of this section is concerned with the construction of modular invariant partition functions for certain bosonic systems on the torus. As a warmup we consider the free boson, defined by the action

$$S[\phi] = \frac{g}{2} \int d^2\mathbf{x} \, (\nabla \phi)^2 \tag{14.42}$$

and $\phi(\mathbf{x}) \in \mathbb{R}$. Comparing (14.40) with (14.27)–(14.29), and bearing in mind that $c = 1$, we would expect the corresponding partition function to be of the form $Z_0(\tau) \propto 1/|\eta(\tau)|^2$. Fixing the proportionality constant is somewhat tricky [29]. In a first step, $\phi$ is decomposed on the normalised eigenfunctions of the Laplacian, and $Z_0(\tau)$ is expressed as a product over the eigenvalues. This product however diverges, due to the presence of a zero-mode, and must be regularised. A sensible result is obtained by a shrewd analytic continuation, the so-called $\zeta$-function regularisation technique [29]:

$$Z_0(\tau) = \frac{\sqrt{4\pi g}}{\sqrt{\Im \tau} \, |\eta(\tau)|^2} \tag{14.43}$$

The CFT which is of main interest for the CG technique is the so-called *compactified boson* in which $\phi(\mathbf{x}) \in \mathbb{R}/(2\pi a R \mathbb{Z})$. In other words, the field lives on a circle of radius $aR$ (the reason for the appearance of *two* parameters, $a$ and $R$, will become clear shortly). In this context, suitable periodic boundary conditions are specified by a pair of numbers, $m, m' \in a\mathbb{Z}$, so that for any $k, k' \in \mathbb{Z}$

$$\phi(z + k\omega_1 + k'\omega_2) = \phi(z) + 2\pi R(km + k'm') \tag{14.44}$$

It is convenient to decompose $\phi = \phi_{m,m'} + \phi_0$, where

$$\phi_{m,m'} = \frac{2\pi R}{\bar{\tau} - \tau} \left[ \frac{z}{\omega_1}(m\bar{\tau} - m') - \frac{\bar{z}}{\bar{\omega}_1}(m\tau - m') \right] \tag{14.45}$$

is the classical solution satisfying the topological constraint, and $\phi_0$ represents the quantum fluctuations, i.e., is a standard free boson satisfying standard periodic boundary conditions.

Integrating over $\phi_0$ as before, and keeping $m, m'$ fixed, gives the partition function

$$Z_{m,m'}(\tau) = Z_0(\tau) \exp\left(-2\pi^2 g R^2 \frac{|m\tau - m'|^2}{\Im\tau}\right). \qquad (14.46)$$

It is easy to see that this is not modular invariant. A modular invariant is however obtained by summing over all possible values of $m, m'$:

$$Z(\tau) \equiv \frac{R}{\sqrt{2}} Z_0(\tau) \sum_{m,m' \in a\mathbb{Z}} \exp\left(-2\pi^2 g R^2 \frac{|m\tau - m'|^2}{\Im\tau}\right) \qquad (14.47)$$

The prefactor $R/\sqrt{2}$ is again a subtle effect of the zero-mode integration. It is actually most easily justified a posteriori by requiring the correct normalisation of the identity operator in (14.48) below.

A more useful, and more physically revealing, form of (14.47) is obtained by using the Poisson resummation formula to replace the sum over $m' \in a\mathbb{Z}$ by a sum over the dual variable $e \in \mathbb{Z}/a$. The result is

$$Z(\tau) = \frac{1}{|\eta(\tau)|^2} \sum_{e \in \mathbb{Z}/a, \, m \in a\mathbb{Z}} q^{h_{e,m}} \bar{q}^{\bar{h}_{e,m}}, \qquad (14.48)$$

with

$$h_{e,m} = \frac{1}{2}\left(\frac{e}{R\sqrt{4\pi g}} + \frac{mR}{2}\sqrt{4\pi g}\right)^2, \qquad \bar{h}_{e,m} = \frac{1}{2}\left(\frac{e}{R\sqrt{4\pi g}} - \frac{mR}{2}\sqrt{4\pi g}\right)^2. \qquad (14.49)$$

Comparing now with (14.40) and (14.27)–(14.29) we see that (14.49) is nothing else than the conformal weights of the CFT at hand.

The requirement of modular invariance has therefore completely specified the operator content of the compactified boson system. An operator is characterised by two numbers, $e \in \mathbb{Z}/a$ and $m \in a\mathbb{Z}$, living on mutually dual lattices. A physical interpretation will be furnished by the CG formalism of section 14.3: $e$ is the "electric" charge of a vertex operator (spin wave), and $m$ is the "magnetic" charge of a topological defect (screw dislocation in the field $\phi$). Let us write for later reference the corresponding scaling dimension and spin:

$$\Delta_{e,m} = \frac{e^2}{4\pi g R^2} + m^2 \pi g R^2, \qquad s_{e,m} = em \qquad (14.50)$$

Observe in particular that the spin is integer, as expected for a bosonic system.

The reader will notice that the three constants $R$, $a$ and $g$ are related by the fact that they always appear in the dimensionless combination $R^2 a^2 g$. Field-theoretic literature often makes the choice $a = 1$ and $g = 1/4\pi$ in order to simplify formulae such as (14.49). In the CG approach—the subject of section 14.3—one starts from a geometrical construction (mapping to a height model) in which a convention for $a$ must be chosen. The compactification radius $aR$ then follows from a "geometrical" computation (identification of the ideal state lattice), and the correct coupling constant $g$ is only fixed in the end by a field-theoretic argument (marginality requirement of the Liouville potential). Needless to say, the results, such as (14.50) for the dimensions of physical operators, need (and will) be independent of the initial choice made for $a$.

To conclude, note that the roles of $e$ and $m$ in (14.49) are interchanged under the transformation $Ra\sqrt{2\pi g} \to (Ra\sqrt{2\pi g})^{-1}$, which leaves (14.48) invariant. This is another manifestation of the electro-magnetic duality. Ultimately, the distinction between $e$ and $m$ comes down to the choice of transfer direction. In the geometry of the torus this choice is immaterial, of course. In sections 14.3.4–14.3.5 we shall compare the geometries of the cylinder and the annulus; these are related by interchanging the space and time directions, and accordingly the electric and magnetic charges switch role when going from one to the other.

## 14.2.8 Boundary CFT

The aspects of CFT exposed to this point pertain to unbounded geometries, either that of the infinite plane (Riemann sphere) or, in section 14.2.7, that of the torus (which is really a finite geometry made unbounded through the periodic boundary conditions). In contrast, boundary conformal field theory (BCFT) describes surface critical behaviour, i.e., a critical system confined to a bounded geometry. The simplest such geometry, and probably the most relevant from the point of view of polymer physics, is that of the upper half plane $\{z \,|\, \Im z \geq 0\}$, where the real axis $\mathbb{R}$ acts as the boundary (one-dimensional "surface").

The foundations of BCFT were set by Cardy [22] who also initiated many of the subsequent developments and applications (see [15, 30] for reviews). A useful review of the status of boundary critical phenomena before the advent of CFT was given by Binder [31].

To convey an idea of which phase transitions may result from the interplay between bulk and boundary degrees of freedom, and what may be the corresponding boundary conditions, we begin by a qualitative discussion of a simple magnetic spin system. We denote the local order parameter (magnetisation) by $\phi$. When the boundary spins enjoy free boundary conditions, they interact more weekly than the bulk spins, since microscopically they are coupled to fewer neighbouring spins. Upon lowering the temperature, the bulk will therefore order before the surface: this is the so-called *ordinary transition*. Now consider placing the system slightly below the bulk critical temperature. Then $\phi$ is non-zero deep inside the bulk, and will decrease upon approaching the boundary. One can argue that in the continuum limit $\phi$ will vanish exactly on the boundary. Thus, the Dirichlet boundary condition $\phi|_{\mathbb{R}} = 0$ is the appropriate choice for describing the ordinary transition.

Let us now introduce a coupling $J_s$ between nearest-neighbour spins on the boundary which may be different from the usual bulk coupling constant $J$. Taking $J_s > J$ one may "help" the boundary to order more easily.[2] When $J_s$ takes a certain critical value we are at the *special transition*, at which the bulk and the boundary order simultaneously. Finally, when $J_s \to \infty$ the boundary spins are always completely ordered[3], a fact which changes the nature of the ordering transition of the bulk, now referred to as the *extraordinary transition*. This corresponds to the Dirichlet boundary condition $\phi|_\mathbb{R} = \infty$ in the continuum limit. Note that in the application of boundary CFT to loop models (see section 14.6) the meaning of $J_s$ is to give a specific fugacity to monomers on the boundary.

The control parameter $J_s$ can be thought of in a renormalisation group sense, and is readily seen to be irrelevant at the ordinary and extraordinary transitions. Accordingly we expect a boundary RG flow to go from the special to either of the two other transitions. (In the case of the Ising model, the special and extraordinary transitions actually coincide.)

In our subsequent application to loop models (see section 14.6) we rather think of $\phi$ as a height field which is dual to the system of oriented loops (this is the so-called Coulomb gas approach, see section 14.3). In other words, the loops are level lines of $\phi$. Dirichlet boundary conditions then describe a situation in which loops are reflected off the boundary, and adjoining two different Dirichlet conditions forces one or more "loop ends" to emanate from the boundary. One may also impose Neumann boundary conditions, $\partial\phi/\partial y|_\mathbb{R} = 0$, meaning that the "loops" coming close to the boundary must in fact terminate perpendicular to it. Clearly the non-local aspects of these situations call for a more detailed discussion, which will be postponed to section 14.6.

The allowed conformal mappings in BCFT must keep invariant both the boundary itself and the boundary conditions imposed along it. For the global conformal transformations (14.6) the invariance of the real axis forces $a_{ij} \in \mathbb{R}$, i.e., they form the group $SL(2, \mathbb{R})$ and the number of parameters is halved from 6 to 3. For an infinitesimal local conformal transformation $z \to w(z) = z + \varepsilon(z)$ the requirement reads $\varepsilon(\bar{z}) = \bar{\varepsilon}(z)$. This property can be used to eliminate the $\bar{\varepsilon}(z)$ part altogether, since it is just the analytic continuation of $\varepsilon(\bar{z})$ into the lower half plane. It follows that $\bar{L}_n = L_{-n}$, and so one half of the conformal generators has been eliminated.

At the level of the stress tensor, the requirement is $T(\bar{z}) = \bar{T}(z)$. In Cartesian coordinates this reads $T_{xy} = 0$ on the real axis, the so-called *conformal boundary conditions*. Its physical meaning is that there is no energy-momentum flow across $\mathbb{R}$. This has important consequences on the conformal Ward identity (14.9) where $T(z)$ is applied to a product of primary fields $X = \prod_j \phi_j(z_j, \bar{z}_j)$ situated in the upper half plane. The contour $C$ surrounding all $z_j$ can then be taken as a large semicircle with the diameter parallel to the real axis. However, writing the same identity for

---

[2] A similar effect could be obtained by adding a surface magnetic field, but here we do not wish to break the symmetry of the model [typically $O(n)$ in applications to loop models].

[3] This should not (as is sometimes seen in the literature) be confused with imposing *fixed* boundary conditions, which would rather correspond to an infinite symmetry-breaking field applied on the boundary.

$\bar{T}(\bar{z})$ yields another Ward identity involving the conjugate semicircle contour $\bar{C}$, and since $\bar{T} = T$ when $z \in \mathbb{R}$, the two contours can be fused into a complete circle surrounding both $z_j$ and $\bar{z}_j$. The end result, cf. (14.10), is thus

$$T(z)X = \sum_j \left( \frac{h_j}{(z - z_j)^2} + \frac{\partial_{z_j}}{z - z_j} + \frac{\bar{h}_j}{(\bar{z} - \bar{z}_j)^2} + \frac{\partial_{\bar{z}_j}}{\bar{z} - \bar{z}_j} \right) X . \tag{14.51}$$

In conclusion, everything happens as if each primary field in the upper half plane were accompanied by a mirror field in the lower half plane. This means that computations in the BCFT can be done using a *method of images* similar to that used in electrostatics when solving the Laplace equation with boundary conditions. Correlation functions are computed as if the theory were defined on the whole complex plane, and governed by a single Virasoro algebra (14.20): the physical fields are then situated in the upper half plane, and their unphysical mirror images in the lower half plane. The simplification of getting rid of $\bar{L}_n$ has thus been achieved at the price of doubling the number of points in correlation functions. In practice, the former simplification largely outweighs the latter complication.

In particular, the $n$-point boundary correlation functions satisfy the very same differential equations (14.34) as $2n$-point bulk correlation functions, but with different boundary conditions. The most interesting cases are $n = 1$ and $n = 2$, both tractable in the bulk picture in several situations of practical importance. As examples of the physical information which can be extracted from these cases we should mention, for $n = 1$, the probability profile of finding a monomer of a loop at a certain distance from the boundary, and for $n = 2$, the probability that a polymer comes close to the boundary at two prescribed points [32]. A particularly celebrated application of the $n = 2$ case is Cardy's computation [33] of the *crossing probability* that a percolation cluster traverses a large rectangle, as a function of the aspect ration of the latter.

The radial quantisation scheme of section 14.2.5 still makes sense in BCFT. The associated conformal mapping

$$w(z) = \frac{L}{\pi} \log z \tag{14.52}$$

transforms the upper half plane into a semi-infinite strip of width $L$ with non-periodic transverse boundary conditions. The two rims of the strip are then the images of the positive and the negative real axis, and the time (resp. space) direction is parallel (resp. perpendicular) to the axis of the strip. The dilatation operator reads $\mathscr{D} = L_0$ and the Hamiltonian $\mathscr{H} = (\pi/L)(L_0 - c/24)$. Non-trivial eigenstates of $\mathscr{H}$ are formed by a *boundary operator* $\phi_j(0)$ acting on the vacuum state, $|h\rangle = \phi_j(0)|0\rangle$.

In general, we expect boundary operators to have different scaling dimensions than bulk operators. This can be understood from the method of images: when a primary field approaches the boundary it interacts with its mirror image and, by the OPE (14.25), produces a series of other primaries which then describe the boundary critical behaviour.

Likewise, a field $\phi_{(r,s)}$ with a given interpretation in the bulk will typically have a different interpretation when situated on the boundary. Examples pertinent to loop models will be given in section 14.6.

The finite-size formulae (14.15) and (14.17) can be adapted to the case of a strip of width $L$. For this, one uses the method of images and the mapping (14.52). The end results read:

$$f_0(L) = f_0(\infty) + \frac{f_0^S}{L} - \frac{\pi c}{24L^2} + o(L^{-2}),$$

$$f_\phi(L) - f_0(L) = \frac{f_\phi^S - f_0^S}{L} + \frac{\pi \Delta}{L^2} + o(L^{-2}). \tag{14.53}$$

where there is now a non-universal $1/L$ dependence due to the presence of surface free energies $f^S$. For some (but not all) choices of excited levels $f_\phi(L)$ it can be argued that $f_\phi^S = f_0^S$, thus simplifying the second of these formulae.

Note that (14.7) applied to a boundary operator is the reason why we have not discussed *finite* Dirichlet boundary conditions at the beginning of this section. More generally, any uniform boundary condition is expected to flow under the renormalisation group towards a *conformally invariant boundary condition*. It is one of the goals of BCFT to classify such boundary conditions. One of the main results obtained is the following [30]: For diagonal models (i.e., $n_{h,\bar{h}} = \delta_{h,\bar{h}}$ in (14.26)) there is a bijection between the primary fields in the bulk CFT and the conformally invariant boundary conditions in the BCFT. For example, for the Ising model ($m = 3$ and $m' = 4$ in (14.35)) the three different bulk primary operators (the identity $I = \phi_{(1,1)}$, the spin $\sigma = \phi_{(1,2)}$, and the energy $\varepsilon = \phi_{(2,1)}$) correspond to three types of uniform boundary conditions in the lattice model of spins (fixed $s = +1$ and $s = -1$, and free boundary conditions).

To this point we have discussed only uniform boundary conditions. It is important to realise that the radial quantisation picture with a boundary operator $\phi_j(0)$ situated at the origin is compatible also with mixed boundary conditions, i.e., one boundary condition on the negative real half-axis and another on the positive real half-axis. In this case, $\phi_j(0)$ is called a *boundary condition changing operator*. One then needs a second operator $\phi_j(\infty)$ situated at infinity to change back the boundary condition. A more symmetric picture is obtained by mapping the upper half plane to the strip, through (14.52). There are then different boundary conditions on the two sides of the strip, and a boundary condition changing operators is located at either end of the strip. More generally, one may study a BCFT on any simply connected domain with a variety of different boundary conditions along the boundary, each separated by a boundary condition changing operator.

For bulk CFT, crucial insight was gained by considering the theory on a torus. The analogous tool for BCFT is to consider the theory on an annulus.[4] In analogy

---

[4] It makes sense to think of this in the radial quantisation, or transfer matrix, picture. The theories are initially considered on a semi-infinite cylinder (resp. a strip) with specified transverse boundary conditions (periodic, resp. non-periodic) and unspecified longitudinal boundary conditions. This gives access to the transfer matrix eigenvalues. To access the fine structure, such as amplitudes of

with the torus case, we denote by $\omega_1 \in \mathbb{R}$ the width of the annulus and by $\omega_2 \in i\mathbb{R}$ its length (in the periodic direction), defining $\tau = \omega_2/\omega_1 \in i\mathbb{R}$. The boundary conditions on the two rims are denoted, symbolically, $a$ and $b$. Then

$$Z_{ab}(\tau) = \mathrm{Tr}\left(q^{L_0 - c/24}\right) \tag{14.54}$$

with $q = \exp(\pi i \tau)$. This should be compared with (14.40). The analogue of (14.41),

$$Z_{ab}(\tau) = \sum_h n_h^{(ab)} \chi_{(c,h)}(\tau), \tag{14.55}$$

then becomes linear in the characters. Equivalently, one might exchange the space and time direction and view the annulus as a cylinder of circumference $\omega_2$ and finite length $\omega_1$, with boundary conditions $a$ (resp. $b$) in the initial (resp. final) state. This leads to

$$Z_{ab}(\tau) = \left\langle b \left| e^{\tau^{-1}\mathscr{H}_{\mathrm{bulk}}} \right| a \right\rangle, \tag{14.56}$$

where now $\mathscr{H}_{\mathrm{bulk}}$ is the Hamiltonian of the *bulk* CFT propagating between boundary states $|a\rangle$ and $\langle b|$. The links between bulk and boundary CFT result from a detailed study of the equivalence between (14.54) and (14.56).


## 14.3 Coulomb Gas Construction

It has been known since the 1970's [34] that the critical point of many two-dimensional models of statistical physics can be identified with a Gaussian free-field theory. A general framework for the computation of critical exponents was first given in 1977 by José et al. in the so-called spin wave picture [9]. This was further elaborated in the early 1980's by den Nijs [10] and Nienhuis [11] into what has become known as the Coulomb gas (CG) construction. These developments have been reviewed by Nienhuis [35].

The CG approach is particularly suited to deal with the continuum limit of lattice models of closed loops, in which each loop carries a Boltzmann weight $n$. Such *loop models* arise as the diagrammatic expansion of spin systems in which the spins take values in $\mathbb{R}^n$ and the interactions possess an $O(n)$ symmetry. Depending on the normalisation constraint imposed on the spins, and on the underlying lattice structure, the loops may or may not admit self-intersections. The former case can be treated by supersymmetric techniques [36], within the framework of the non-linear sigma model, but does not admit a CG representation. In the present review we are however only concerned with cases without self-intersections, for which the CG approach does apply. A particularly elegant and useful example was given by Nienhuis [37]. Another important model, the $Q$-state Potts model, can be formulated

---

the eigenvalues, one must impose periodic longitudinal boundary conditions and take the length of the cylinder (resp. strip) to be *finite*.

as a model of self-avoiding loops with $n = \sqrt{Q}$, as first shown by Baxter, Kelland and Wu [38]. We shall review the relevant mappings in section 14.3.1.

The marriage between the CG and conformal field theory (CFT) happened in 1986–87, when Di Francesco, Saleur and Zuber [39, 40] made the loop model $\leftrightarrow$ CG correspondence more precise and showed how the ideas of modular invariance [23, 29] can be put to good use in the study of loop models (see section 14.7 below). At the same time, Duplantier and Saleur developed a range of applications to SAW's and SAP's (see in particular [14]).

In section 14.3.2 we show how the loop models can be transformed into height models with local (albeit complex) Boltzmann weights. It is the continuum limit of this height which acts as the conformally invariant free field. The underlying lattice model implies that this height field is compactified, thus making contact with the modular invariance results of section 14.2.7.

The naive free field action however needs to be modified with extra terms, traditionally known as background and screening electric charges [35]. The resulting CFT, known as a Liouville field theory, is written down in section 14.3.3.

The requirement that the Liouville potential be RG marginal determines the coupling constant of the free field as a function of $n$, as first pointed out by Kondev [17]. This is an important ingredient, since otherwise one would have to rely on an independent exact solution to fix the coupling. The analogous marginality requirement for the case of surface critical behaviour has been established recently by Cardy [18]. We discuss these developments in sections 14.3.4–14.3.5.

### 14.3.1 From Potts and O(n) Models to Loops

In this section we show how to transform the $Q$-state Potts model and the O($n$) model into loop models. There are several mappings of this type, depending on the lattice structure, the types of (local) interactions, and so on, but for simplicity we shall concentrate here on the simplest cases in which the Potts model is defined on a square lattice [38] and the O($n$) model on a hexagonal lattice [37].

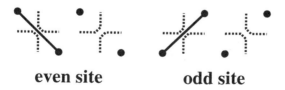

**even site**          **odd site**

**Fig. 14.1** Relation between the clusters $E'$ (solid lines) on $\mathscr{L}$ (filled circles) and the transition system $T(E')$ (broken lines) on $S_{\mathscr{L}}$. Note that the rules differ for the two sublattices of $S_{\mathscr{L}}$.

Consider first the Potts model, defined initially by assigning spins $\sigma_i = 1, 2, \ldots, Q$ to each of the vertices of the square lattice $\mathscr{L}$. A pair of nearest-neighbour spins has the interaction energy $-K\delta_{\sigma_i,\sigma_j}$, and we set $J = K/k_B T$. The partition function then becomes that of the so-called random cluster model [41]:

$$Z = \sum_{\{\sigma\}} \prod_{(ij)\in E} e^{J\delta_{\sigma_i,\sigma_j}} = \sum_{E'\subseteq E} (e^J - 1)^{|E'|} Q^{c(E')}, \qquad (14.57)$$

where $E$ is the set of lattice edges, $E'$ runs through all $2^{|E|}$ subsets thereof, and $c(E')$ is the number of connected components in the graph induced by $E'$. Define now the surrounding lattice $S_{\mathscr{L}}$ with vertices which are the midpoints of edges in $\mathscr{L}$, here a rotated square lattice. On $S_{\mathscr{L}}$ we define for each $E' \subseteq E$ a transition system $T(E')$ system according to the rule in Fig. 14.1; then $T(E')$ constitutes a set of cycles (loops[5]) that separate the connected components in the edge set $E'$ from those dual to the complement $E \setminus E'$. (In other words, the loops form the boundaries of connected components in either set.) Using the Euler relation for a *planar* lattice with $N$ vertices this results in

$$Z = Q^{N/2} \sum_{E'\subseteq E} \left(\frac{e^J - 1}{\sqrt{Q}}\right)^{|E'|} Q^{l(T(E'))/2}, \qquad (14.58)$$

where $l(T(E'))$ is the number of loops in the transition system $T(E')$. On a non-planar graph, (14.58) would be slightly modified (see section 14.7 below).

The local weights of the transition system in (14.58) are in general inhomogeneous, due to the first factor inside the sum, since vertices on the even (resp. odd) sublattice of $S_{\mathscr{L}}$ stand on horizontal (resp. vertical) edges of $\mathscr{L}$. This inhomogeneity can be directly read off from Fig. 14.1. Even though critical (and even integrable) points of the inhomogeneous model do exist [42, 43] we are here interested in homogeneous solutions only. Indeed, for $e^J - 1 = \pm\sqrt{Q}$ and $0 \le Q \le 4$ the Potts model (14.57) is at its self-dual critical point [44, 45]. With the plus sign, (14.58) then becomes simply

$$Z = Q^{N/2} \sum_{E'\subseteq E} n^{l(T(E'))} \qquad (14.59)$$

with $n \equiv \sqrt{Q}$. With the minus sign, (14.59) still holds true provided we take the other determination of the square root, $n = -\sqrt{Q}$, since $N + |E'| + l(T(E'))$ is even for any $E'$. In conclusion, (14.59) describes a selfdual critical $Q = n^2$ state Potts model for $-2 \le n \le 2$, and it takes the form of a simple *loop model* in which each loop carries the weight $n$.

The limit $n \to 0$ is of special interest here. The dominant contribution to (14.59) is such that each $E'$ in the sum represents an (unrooted) spanning tree, and its contour is a so-called *osculating SAP*.

We now turn to the O($n$) model, which is defined initially by assigning vector spins $\mathbf{S}_i \in \mathbb{R}^n$ to each of the vertices of the hexagonal lattice $\mathscr{L}$. A pair of nearest-neighbour spins has the interaction energy $-J\mathbf{S}_i \cdot \mathbf{S}_j$. The integration measure is

---

[5] The use of the word *loop* as a synonym of cycle is common in the physics literature, and should not be confused with its different meaning in graph theory.

defined such that $\int d\mathbf{S}_i d\mathbf{S}_j\, S_i^\alpha S_j^\beta = \delta_{\alpha,\beta}$ and odd moments of $\mathbf{S}_i$ vanish by the symmetry $\mathbf{S}_i \to -\mathbf{S}_i$. Expanding out the Boltzmann weights $\tilde{w}_{ij} = \exp(J\mathbf{S}_i \cdot \mathbf{S}_j / k_B T)$ and forming the partition function, the contributing configurations are in bijection with systems of loops for which each loop carries a weight $n$. These loops are in general rather complicated. Namely, on a general lattice containing vertices of degree $\geq 4$ the loops may cross; and for any lattice they may cover each edge more than once. The choice of the hexagonal lattice overcomes the first complication. To overcome the second we follow Nienhuis [37] and redefine the weights as $w_{ij} \equiv 1 + K\mathbf{S}_i \cdot \mathbf{S}_j$, i.e, by truncating the formal high-temperature expansion of the original weights. The effect of these simplifications on the critical behaviour may be judged a posteriori, in section 14.4.

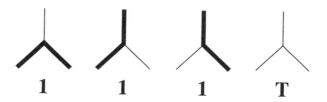

**Fig. 14.2** Vertices in the O($n$) loop model at temperature $T = 1/K$.

The partition function then reads

$$Z = \sum_{\mathscr{G}} K^{|\mathscr{G}|} n^{l(\mathscr{G})}, \qquad (14.60)$$

where $\mathscr{G}$ are edge subsets of $\mathscr{L}$ with the property that every vertex is adjacent to an even number (zero or two) of edges in $\mathscr{G}$, as shown in Fig. 14.2. So $\mathscr{G}$ forms a system of $l(\mathscr{G})$ self-avoiding, mutually avoiding loops drawn on $\mathscr{L}$. Nienhuis [37] has argued that the loop model (14.60) is critical for $-2 \leq n \leq 2$ and

$$1/K^2 = 2 \pm \sqrt{2-n}. \qquad (14.61)$$

The high-temperature solution [plus sign in (14.61)] is supposed to correctly describe the critical point of the original O($n$) model. The loops described by it are commonly referred to as *dilute*, as they fill a vanishing fraction of the lattice in the thermodynamic limit. The low-temperature solution [minus sign in (14.61)] describes *dense loops* which cover a finite fraction of the lattice. One would expect these to be intimately related to the osculating SAP's of the Potts model, and this is indeed the case. Their critical behaviour is however not coincident with that of the original, unmodified O($n$) model. These remarks will be clarified further below, and in section 14.4.

## 14.3.2 Transformation to a Height Model

In the definition of the $Q$-state Potts and the O($n$) models, the parameters $Q$ and $n$ were originally positive integers. However, in the corresponding loop models, (14.59) and (14.60), they appear as formal parameters and may thus take arbitrary complex values. The price to pay for this generalisation is the appearance of a non-locally defined quantity, the number of loops $l$. The locality of the models may be recovered (though not completely, see section 14.3.3) by transforming them to height models with complex Boltzmann weights [38], as we now show.

In a first step, each loop is independently decorated by a global orientation $s = \pm 1$, which by planarity and self-avoidance can be described as either counter-clockwise ($s = 1$) or clockwise ($s = -1$). Each oriented loop must be given a weight $w(s)$, so that $n = \sum_s w(s)$. An obvious possibility, sometimes referred to as the *real loop ensemble*, is $w(1) = w(-1) = n/2$. This can be interpreted as an O($n/2$) model of complex spins.

We are however more interested in the *complex loop ensemble* with $w(s) = e^{is\gamma}$. Note that in the expected critical regime,

$$n = 2\cos\gamma \in [-2, 2],\qquad\qquad(14.62)$$

the parameter $\gamma \in [0, \pi]$ is real. Locality is retrieved by remarking that the weights $w(\pm 1)$ are equivalent to assigning a local weight $w(\alpha/2\pi)$ to each vertex where a loop turns an angle $\alpha$ (counted positive for left turns). If a vertex is traversed by more than one loop, it gets weighted by the product of $w(\alpha/2\pi)$ over all traversals.

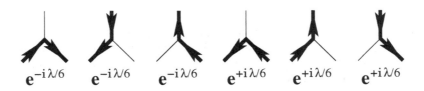

**Fig. 14.3** Local redistribution of the loop weight $n$ in the O($n$) loop model on the hexagonal lattice.

The models (14.59) and (14.60) are now transformed into *local vertex models* by assigning to each edge traversed by a loop the orientation of that loop. An edge not traversed by any loop is assigned no orientation. The total vertex weight is determined from the configuration of its incident oriented edges, as the above local loop weights summed over the oriented transition systems compatible with edge orientations; this is illustrated for the hexagonal-lattice O($n$) model in Fig. 14.3. In addition, one must multiply this by any loop-independent local weights, such as $K$ in (14.60).

As a result, (14.59) is transformed into a six-vertex model on the square lattice, each vertex being incident on two outgoing and two ingoing edges [44], as shown in Fig. 14.4. The weights $\omega_i$ (resp. $\omega_i'$) on the even (resp. odd) sublattice read explicitly

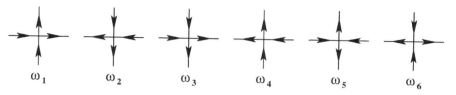

**Fig. 14.4** Weights in the six-vertex model.

$$\omega_1, \ldots, \omega_6 = 1, 1, x, x, e^{i\gamma/2} + xe^{-i\gamma/2}, e^{-i\gamma/2} + xe^{i\gamma/2} \qquad (14.63)$$

$$\omega_1', \ldots, \omega_6' = x, x, 1, 1, e^{-i\gamma/2} + xe^{i\gamma/2}, e^{i\gamma/2} + xe^{-i\gamma/2} \qquad (14.64)$$

where we have defined $x = (e^J - 1)/\sqrt{Q}$. Note that the anisotropy parameter $\Delta = -\cos\gamma$ of the equivalent XXZ spin chain is independent of $x$ and of the sublattice.

Similarly, (14.60) becomes a seven-vertex model on the hexagonal lattice, each vertex being either empty, or incident on one outgoing, one ingoing, and one empty edge. The six non-empty vertices are shown in Fig. 14.3.

Finally, the vertex models are turned into *height models*. For this, assign a scalar variable $h(\mathbf{x})$ to each lattice face $\mathbf{x}$ (i.e., to each vertex of the lattice *dual* to the one on which the loop model has been defined), so that $h$ increases (resp. decreases) by $a$ each time one traverses a left-going (resp. right-going) edge. This definition of the height $h$ is consistent, since each vertex is incident on as many ingoing as outgoing edges. Since this defines only height differences, one may imagine fixing $h$ completely by arbitrarily fixing $h(\mathbf{0}) = 0$.

In the continuum limit, we expect the local height field $h$ to converge to a free bosonic field $\phi(\mathbf{x})$, whose entropic fluctuations are described by an action of the form (14.42), with coupling $g = g(n)$ which is a monotonically increasing function of $n$. In particular, for $n \to \infty$ the lattice model is dominated by the configuration where loops of the minimal possible length cover the lattice densely; the height field is then flat, $\phi(\mathbf{x}) = $ constant, and the correlation length $\xi$ is of the order of the lattice spacing. For finite but large $n$, $\phi$ will start fluctuating, loop lengths will be exponentially distributed, and $\xi$ will be of the order of the linear size of the largest loop. When $n \to n_c^+$, for some critical $n_c$ (we shall see that $n_c = 2$), this size will diverge, and for $n \le n_c$ the loop model will be conformally invariant with critical exponents that depend on $g(n)$. The interface described by $\phi(\mathbf{x})$ is then in a *rough* phase. The remainder of this section is devoted to making this intuitive picture more precise, and to refine the free bosonic description of the critical phase.

As a first step towards greater precision, we now argue that $\phi(\mathbf{x})$ is in fact a *compactified boson*, cf. section 14.2.7. To see this, it is convenient to consider the *oriented* loop configurations that give rise to a maximally flat microscopic height $h$; following Henley and Kondev [61] we shall refer to them as *ideal states*. For the Potts model (14.59), an ideal state is a dense packing of length-four loops, all having the same orientation. There are four such states, corresponding to two choices of orientation and two choices of the sublattice of lattice faces surrounded by the loops. An ideal state can be gradually changed into another by means of $\sim N$ local changes of the transition system and/or the edge orientations. As a result, the mean height will change, $\phi \to \phi \pm a$. Iterating this, one sees that one may return to the initial ideal state whilst having $\phi \to \phi \pm 2a$. For consistency, we must therefore require $\phi(\mathbf{x}) \in \mathbb{R}/(2a\mathbb{Z})$, i.e., the field is compactified with radius $R = 1/\pi$, cf. (14.44).

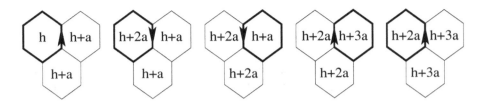

**Fig. 14.5** Ideal states of the O($n$) model on the hexagonal lattice. For each of the five panels, the state of the complete infinite lattice is obtained by tiling the plane with three faces shown, while respecting the three-sublattice structure. The different panels are related, from left to right, by the construction explained in the main text, under which one ideal state is gradually changed into another. The leftmost and rightmost panels represent the same ideal state, but with a global height change $\phi \to \phi + 2a$ that determines the compactification radius.

The same construction applied to the O($n$) model (14.61) yields six ideal states of oriented length-six loops (resulting from a choice of three sublattices and two orientations). Changing the ideal state in four steps, as shown in Fig. 14.5, produces the initial state but with a height change of $\pm 2a$. So one has the same compactification radius, $\phi(\mathbf{x}) \in \mathbb{R}/(2a\mathbb{Z})$, as in the case of the Potts model.

Before we go on, a few remarks are in order:

1. In section 14.2.7 we have seen in detail that the normalisation constant $a$ drops out from the final physical results. We shall therefore follow standard conventions and set $a = \pi$ in what follows.
2. While the complex loop ensemble is geometrically appealing, it is difficult to make quite rigorous a setup which is based on complex Boltzmann weights.
3. We may already suspect—and we shall see below in more detail—that the O($n$) model in the dense phase and the $Q$-state Potts model give identical critical theories in the continuum limit, for $n = \sqrt{Q}$. However, the correspondence between operators in the microscopic model and the continuum limit is not necessarily identical, leading to subtle differences. For instance, the energy operators of the two models become different objects in the continuum limit (see section 14.4).

### 14.3.3 Liouville Field Theory

The essence of the above discussion is that the critical properties of the loop models under consideration can be described by a continuum-limit partition function that takes the form of a functional integral

$$Z = \int \mathscr{D}\phi(\mathbf{x}) \exp\left(-S[\phi(\mathbf{x})]\right). \tag{14.65}$$

Here $S[\phi(\mathbf{x})]$ is the Euclidean action of the compactified scalar field $\phi(\mathbf{x}) \in \mathbb{R}/(2\pi\mathbb{Z})$. The hypothesis that the critical phase is described by bounded elastic fluctuations around the ideal states means that $S$ must contain a term

$$S_{\mathrm{E}} = \frac{g}{4\pi} \int d^2\mathbf{x} \, (\nabla\phi)^2 \tag{14.66}$$

with coupling constant $g > 0$. Higher derivative terms that one may think of adding to (14.66) can be ruled out by the $\phi \to -\phi$ symmetry, or by arguing a posteriori that they are RG irrelevant in the full field theory that we are about to construct.

Note that the partition function (14.65) does not purport to coincide with (14.59) or (14.60) on the scale of the lattice constant. (A similar remark holds true for the correlation functions that one may similarly write down.) We do however claim that their long-distance properties are the same. In that sense, the CG approach is an exact, albeit by no means rigorous, method for computing critical exponents and related quantities. A more precise equivalence between discrete and continuum-limit partition functions can however be achieved on a torus; see section 14.7.

The action (14.66) coincides with (14.42) for the compactified boson. To obtain the full physics of the loop model one however needs to add two more terms to the action, as we now shall see.

To proceed, we consider the underlying lattice model as being defined on a cylinder, $\mathbf{x} = (x, t)$. This has the advantage of making direct contact with the radial quantisation formalism of section 14.2.5 used in both numerical (transfer matrix) and analytical (Bethe Ansatz) studies. The boundary conditions are thus periodic in the space direction, $x = x + L$, and free in the time ($t$) direction. Ultimately, the results obtained on the cylinder can always be transformed into other geometries by means of a conformal mapping.

With this geometry, the equivalence between the loop model and a local height model with complex weights, established in section 14.3.2, must be revisited. While loops homotopic to a point still acquire their correct global weight $n$ from the local angle-dependent weights $w(\alpha/2\pi)$, this is no longer true for loops that wind around the cylinder. Summing over loop orientations, their weight would be $\bar{n} = 1 + 1 = 2$. Consider now adding a term

$$S_{\mathrm{B}} = \frac{ie_0}{4\pi} \int d^2\mathbf{x} \, \phi(\mathbf{x})\mathscr{R}(\mathbf{x}) \tag{14.67}$$

to the effective action $S$, where $\mathscr{R}$ is the scalar curvature[6] of the space $\mathbf{x}$. The parameter $e_0$ is known in CG language as the *background electric charge*. On the cylinder, one has simply $S_{\mathrm{B}} = ie_0\left(\phi(x, \infty) - \phi(x, -\infty)\right)$, meaning that in the partition function (14.65) an oriented loop with winding number $q = 0, \pm 1$ (all other winding numbers are forbidden by the self-avoidance of the loops) can equivalently be assigned an ex-

---

[6] We consider the scalar curvature in a generalised sense, so that delta function contributions may be located at the boundaries. Implicitly, we are just applying the Gauss-Bonnet theorem.

tra weight of $\exp(i\pi q e_0)$. For non-winding loops ($q = 0$) this does not change the reasoning of section 14.3.2, whilst summing over the two orientations ($q = \pm 1$) of a winding loop produces the weight $\bar{n} \equiv 2\cos(\pi e_0)$. The choice $e_0 = \gamma/\pi$ will thus assign to a winding loop the same weight $\bar{n} = n$ [see (14.62)] as to a non-winding one (but note that other choices leading to $\bar{n} \neq n$ may be useful in some applications of the CG technique).

The object $e^{ie\phi}$ (or more precisely, its normal ordered product : $e^{ie\phi}$ :) is known in field theory as a *vertex operator* of (electric) charge $e$. The boundary term (14.67) thus corresponds to the insertion of two oppositely charged vertex operators at either end of the cylinder.

At this stage two problems remain: the field theory does not yet take account of the weight $n$ of contractible loops, and the coupling constant $g$ has not yet been determined. These two problems are closely linked, and allow [17] us to fix exactly $g = g(n)$. The idea is to add a further *Liouville term*

$$S_{\mathrm{L}} = \int \mathrm{d}^2\mathbf{x}\, w[\phi(\mathbf{x})] \tag{14.68}$$

to the action, which then reads in full

$$S[\phi(\mathbf{x})] = S_{\mathrm{E}} + S_{\mathrm{B}} + S_{\mathrm{L}}. \tag{14.69}$$

In (14.68), $e^{-w[\phi(\mathbf{x})]}$ is the scaling limit of the microscopic vertex weights $w_i$. To identify it we show the argument for the O($n$) model, the Potts case being similar.

Due to the compactification, $S_L[\phi]$ is a periodic functional of the field, and as such it can be developed as a Fourier sum over vertex operators

$$w[\phi] = \sum_{e \in \mathscr{L}_w} \tilde{w}_e\, e^{ie\phi}, \tag{14.70}$$

where $\mathscr{L}_w$ is some sublattice of $\mathscr{L}_0 \equiv \mathbb{Z}$. Note that $\mathscr{L}_w$ may be a proper sublattice of $\mathscr{L}_0$ if $w[\phi]$ has a higher periodicity than that trivially conferred by the compactification of $\phi$. By inspecting Fig. 14.5 we see that this is indeed the case here: the (geometric) averages of the microscopic weights coincide on the first, third, and fifth panels, indicating that the correct choice is $\mathscr{L}_w = 2\mathscr{L}_0$. This intuitive derivation of $\mathscr{L}_w$ (which can easily be corroborated by considering more complicated microscopic configurations) demonstrates the utility of the ideal state construction.

Some important properties of the compactified boson with action $S_{\mathrm{E}}$ have already been derived in section 14.2.7. In particular, its central charge is $c = 1$ and the dimension $\Delta_{e,m}$ of an operator with electromagnetic charge $(e,m)$ is given by (14.50). Having now identified the electric charge $e$ with that of the vertex operator $e^{ie\phi}$, one could alternatively rederive (14.50) by computing the two-point function $\left\langle e^{ie\phi(\mathbf{x})} e^{-ie\phi(\mathbf{y})} \right\rangle$ by standard Gaussian integration.

The physical interpretation of the magnetic charge $m$ is already obvious from (14.44): it corresponds to dislocations in the height field $\phi$ due to the presence of defect lines. In section 14.4 we shall see how to identify these defect lines with SAW's and compute the related critical exponents.

It remains to assess how the properties of the compactified boson are modified by the inclusion of the term $S_B$. Physical reasoning consists in arguing that the vertex operators $e^{\pm i e_0 \phi}$ will create a "floating" electric charge of magnitude $2e_0$ that "screens" that of the other fields in any given correlation function. We infer that (14.50) must be changed into

$$\Delta_{e,m} = \frac{1}{2} \left[ \frac{e(e - 2e_0)}{g} + gm^2 \right] . \tag{14.71}$$

Note that to obtain (14.71) we have changed our normalisation so that both $e$ and $m$ are integers. This is consistent with the normalisation (14.66) of the coupling constant, rather than (14.42), which is the standard choice in the CG literature.

### 14.3.4 Marginality Requirement

Following Kondev [17] we now claim that the Liouville potential $S_L$ must be exactly marginal. This follows from the fact that all loops carry the same weight $n$, independently of their size, and so the term $S_L$ in the action that enforces the loop weight must not renormalise under a scale transformation. The most relevant vertex operator appearing in (14.70) has charge $e_w = 2\pi/a = 2$, and so $\Delta_{e_w,0} = 2$. Using (14.71), this fixes the coupling constant as $g = 1 - e_0$. In other words, the loop weight has been related to the CG coupling as

$$n = \pm\sqrt{Q} = -2\cos(\pi g) \tag{14.72}$$

with $0 < g \leq 1$ for the Potts model or the dense $O(n)$ model.

The term $S_B$ shifts the ground state energy with respect to the $c = 1$ theory described by $S_E$ alone. The corrected central charge is then $c = 1 + 12\Delta_{e_0,0}$, where the factor of 12 comes from comparing (14.15) and (14.17). This gives

$$c = 1 - \frac{6(1-g)^2}{g} . \tag{14.73}$$

It should be noted that the choice $e_w = 2$ is not the only one possible. Namely, the coefficient $\tilde{w}_{e_w}$ of the corresponding vertex operator in (14.70) may be made to vanish, either by tuning the temperature $T$ in the $O(n)$ model, or by introducing non-magnetic vacancies in the Potts model. The former case corresponds to taking the high-temperature solution [plus sign] in (14.61), while the latter amounts to being at the tricritical point of the Potts model. The next-most relevant choice is then $\tilde{e}_w = -2$, and going through the same steps as above we see that one can simply maintain (14.72), but take the coupling in the interval $1 \leq g \leq 2$ for the dilute $O(n)$ model or the tricritical Potts model.

The electric charge $e_w$ whose vertex operator is required to be exactly marginal is known as the *screening charge* in standard CG terminology.

The central charge (14.73) can now be formally identified with that of the Kac table (14.35), with $m' = m + 1$. The result is a formal relation between the minimal model index $m$ and the CG coupling $g$, valid for integer $m$. We have

$$m = \begin{cases} \frac{g}{1-g} & \text{for the dense O($n$) model, or the critical Potts model} \\ \frac{1}{g-1} & \text{for the dilute O($n$) model} \end{cases} \tag{14.74}$$

The special cases $n \to 0$ are related to self-avoiding walks and polygons. This gives $g = 1/2$ for dense polymers (with $c = -2$ and $m = 1$), and $g = 3/2$ for dilute ones (with $c = 0$ and $m = 2$).

### 14.3.5 Annular Geometry

Consider now instead the loop model defined on an annulus which we shall take as an $L \times M$ rectangle with coordinates $x \in [0,L]$ and $y \in [0,M]$. The boundary conditions are free (f) in the $x$-direction and periodic in the $y$-direction. Very recently, Cardy [18] has shown how to impose the correct marginality requirement for this geometry.

Consider first the continuum-limit partition function $Z = Z_{\text{ff}}(\tau)$ from (14.54) in the limit $M/L \gg 1$ of a very long and narrow annulus. The modular parameters $\tau = iM/L$ and $q = \exp(i\pi\tau) = \exp(-\pi M/L)$. We expect in this limit that only the identity operator contributes to $Z$, and so

$$Z \sim q^{-c/24} \sim \exp\left(\frac{\pi c M}{24L}\right). \tag{14.75}$$

The central charge $c$ is (14.73) from the bulk theory, and in particular is known to vary with the coupling constant $g$.

The question then arises how (14.75) is compatible with the continuum-limit action (14.66). According to Cardy [18] the answer is that there is a background magnetic flux $m_0$, a sort of electromagnetic dual of the background electric charge $e_0$ present in the cylinder geometry. Thus, in the continuum limit there is effectively a number (in general fractional) $m_0$ of oriented loops running along the rims of the annulus, giving rise to a height difference between the left and the right rim. Accepting this hypothesis, we can write

$$\phi(x,y) = \tilde{\phi}(x,y) + \frac{\pi m_0 x}{L} \tag{14.76}$$

where $\tilde{\phi}$ is a "gauged" height field that still contains the elastic fluctuations but obeys identical Dirichlet boundary conditions on both rims, say $\tilde{\phi}(0,y) = \tilde{\phi}(L,y) = 0$.

According to the functional integrations in section 14.2.7, the field $\tilde{\phi}$ contributes $q^{-1/24}$ to $Z$, corresponding to $c = 1$. The last term in (14.76) modifies the action (14.66) by $\Delta S = \frac{g}{4\pi}(\pi m_0)^2 \frac{M}{L}$ and thus multiplies $Z$ by a factor $e^{-\Delta S} = q^{gm_0^2/4}$, which correctly reproduces the contribution of the last term in (14.73) to (14.75) provided that we set

$$m_0 = \pm \frac{(1-g)}{\pi g}. \tag{14.77}$$

This value of $m_0$ can be retrieved from a marginality requirement which has the double advantage of being more physically appealing and of not invoking the formula (14.73) for $c$. Indeed, if $m_0$ is too large a pair of oriented loop strands will shed from the rims, corresponding to a vortex pair of strength $m = \pm 2$ situated at the top and the bottom of the annulus. This vortex pair can then annihilate in order to reduce the free energy. And if $m_0$ is too small the opposite will occur. The equilibrium requirement is then that inserting such a vortex pair must be an exactly marginal perturbation in the RG sense, i.e., the corresponding *boundary* scaling dimension is $\Delta_v = 1$.

The free energy increase for creating the vortex pair is, by the same gauge argument as before,

$$\Delta S = \frac{g}{4\pi}\left((m_0 + 2)^2 - m_0^2\right)\left(\frac{\pi}{L}\right)^2 ML \tag{14.78}$$

and noting the factor of 24 between $c$ and the scaling dimension $\Delta_v$ in (14.53), we now have $e^{-\Delta S} = q^{-\Delta_v}$ from (14.75), so that

$$\Delta_v = \frac{g}{4}\left((m_0 + 2)^2 - m_0^2\right) = 1 \tag{14.79}$$

and we recover (14.77).

## 14.4 Bulk Critical Exponents

We shall now see how to use the Coulomb gas (CG) technology of section 14.3 to compute a variety of critical exponents in loop models.

The watermelon exponents were derived by Nienhuis [35] and by Duplantier and Saleur (see [14] and references therein). The issues of their relation to the standard exponents of polymer physics [46], and to the Kac table (14.31), were discussed in [14].

Although the watermelon exponents are essentially magnetic-type exponents in the CG, they do not produce the standard magnetic exponent of the Potts model. The latter was derived by den Nijs [10], but we present here a somewhat different argument.

We have seen above that the dense $O(n)$ model and the critical Potts model coincide on the level of the central charge, but their thermal exponents are different.

These are electric-type exponents in the CG, and were first computed by Nienhuis [37] and den Nijs [10].

Duplantier and Saleur have developed a range of geometrical applications of the exponents mentioned above. In [47, 14] they have generalised the configurational exponent $\gamma$ to arbitrary polymer network conformations. From these a family of physically relevant contact exponents can be derived. They have also obtained the probability distribution of the winding angle of a SAW around one of its end points [48]. Finally, they have derived the exponents for a polymer at the collapse transition (theta point) from a specific model [19].

### 14.4.1 Watermelon Exponents

An important object in loop models is the operator $\mathscr{O}_\ell(\mathbf{x}_1)$ that inserts $\ell$ oriented lines at a given point $\mathbf{x}_1$. Microscopically, this can be achieved by violating the arrow conservation constraint at $\mathbf{x}_1$. For instance, in the $O(n)$ model one can allow a vertex which is adjacent to one outgoing and two empty edges. Doing so at $\ell$ vertices in a small region around $\mathbf{x}_1$ yields a microscopic realisation of the composite operator $\mathscr{O}_\ell(\mathbf{x}_1)$.

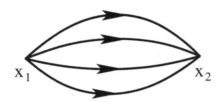

**Fig. 14.6** Watermelon configuration with $\ell = 4$ legs.

If one had strict arrow conservation at all other vertices, the insertion of $\mathscr{O}_\ell(\mathbf{x}_1)$ would not lead to a consistent configuration. However, also inserting $\mathscr{O}_{-\ell}(\mathbf{x}_2)$, the operator that absorbs $\ell$ oriented lines in a small region around $\mathbf{y}$, will lead to consistent configurations (see Fig. 14.6) in which $\ell$ defect lines propagate from $\mathbf{x}_1$ to $\mathbf{x}_2$. Let $Z_\ell(\mathbf{x}_1, \mathbf{x}_2)$ be the corresponding constrained partition function. One then expects

$$\langle \mathscr{O}_\ell(\mathbf{x}_1)\mathscr{O}_{-\ell}(\mathbf{x}_2)\rangle \equiv \frac{Z_\ell(\mathbf{x}_1, \mathbf{x}_2)}{Z} \sim \frac{1}{|\mathbf{x}_1 - \mathbf{x}_2|^{2\Delta_\ell}} \text{ for } |\mathbf{x}_1 - \mathbf{x}_2| \gg 1. \qquad (14.80)$$

The corresponding critical exponents $\Delta_\ell$ are known as watermelon (or *fuseau*, or $\ell$-leg) exponents. To compute them we first notice that the sum of the height differences around a closed contour encircling $\mathbf{x}_1$ but not $\mathbf{x}_2$ will be $a\ell$. Equivalently, one could place the two defects at the extremities of a cylinder [i.e., taking $\mathbf{x}_1 = (x, -\infty)$ and $\mathbf{x}_2 = (x, \infty)$], and the height difference would be picked up by any non-contractible loop separating $\mathbf{x}_1$ and $\mathbf{x}_2$. This latter formulation makes contact with the defect lines (14.44) introduced when studying the compactified boson, the equivalent magnetic charge being $m_\ell = \frac{\ell a}{2\pi} = \frac{\ell}{2}$.

A little care is needed to interpret the configurations of $Z_\ell(\mathbf{x}_1, \mathbf{x}_2)$ in the model of un-oriented loops. The fact that all $\ell$ lines are oriented away from $\mathbf{x}_1$ prevents them from annihilating at any other vertex than $\mathbf{x}_2$. One should therefore like to think about them as $\ell$ marked lines linking $\mathbf{x}_1$ and $\mathbf{x}_2$, where each line carries the Boltzmann weight 1. This is consistent with not summing over the orientations of the defect lines in the oriented loop model. However, each oriented line can also pick up spurious phase factors $w(\alpha/2\pi)$, due to the local redistribution of loop weights, whenever it turns around the end points $\mathbf{x}_1$ and $\mathbf{x}_2$. These factors are however exactly cancelled if we insert in addition a vertex operator $e^{ie_0\phi}$ (resp. $e^{-ie_0\phi}$) at $\mathbf{x}_1$ (resp. $\mathbf{x}_2$) [11]. Note that these vertex operators do not modify the weighting of closed loops, since these must encircle either none of both of $\mathbf{x}_1$, $\mathbf{x}_2$. We conclude that $\Delta_\ell = \Delta_{e_0,m_\ell}$, and using (14.71) this gives

$$\Delta_\ell = \frac{1}{8}g\ell^2 - \frac{(1-g)^2}{2g}. \tag{14.81}$$

Interestingly, these exponents can be attributed to the Kac table under the identification (14.74). One has

$$\Delta_\ell = \begin{cases} 2h_{0,\ell/2} & \text{for the dense } O(n) \text{ model} \\ 2h_{\ell/2,0} & \text{for the dilute } O(n) \text{ model} \end{cases} \tag{14.82}$$

The appearance of half-integer indices [12, 14] is somewhat puzzling, whereas the fact that these exponents are located outside the fundamental domain of the Kac table reflects the non-local nature of the watermelon operators.

It should be noticed [49] that $\Delta_4$ is irrelevant (resp. relevant) in the dilute (resp. dense) phase of the $O(n)$ model, i.e., for $1 < g < 2$ (resp. $0 < g < 1$). This means that on lattices with vertices of degree $\geq 4$, loop self-intersections are irrelevant in the dilute phase. On the other hand, for the dense phase such self-intersections are relevant and will induce a flow to a supersymmetric Goldstone phase [49] that is not described by the CG approach. In other words, Nienhuis' original approximation of the true $O(n)$ model that led to (14.60) is exact in the continuum limit, but only in the dilute phase.

### 14.4.2 Standard Exponents of Polymer Physics

The relation of the standard exponents of critical phenomena (usually denoted $\alpha$, $\beta$, $\gamma$, $\delta$, $\nu$, and $\eta$) to polymer physics has been discussed in details by de Gennes [46]. The end-to-end distance $R$ of a SAW (and also the radius of gyration of a SAP) of chain length $l \gg 1$ behaves as

$$R^2 \sim l^{2\nu} \tag{14.83}$$

whereas the number of such objects in $d$ dimensions (here both are supposed to have one monomer attached to a fixed point) scales like

$$\mathcal{N}_{\text{SAW}} \sim \mu^l \, l^{\gamma-1}$$
$$\mathcal{N}_{\text{SAP}} \sim \mu^l \, l^{-vd} \tag{14.84}$$

where the connectivity constant $\mu$ can be related to the inverse of the critical temperature. There is a more detailed relation where the end-to-end distance $R$ of the SAW has been fixed, valid for $R \gg 1$:

$$\frac{\mathcal{N}_{\text{SAW}}(R)}{\mathcal{N}_{\text{SAP}}} = R^{(\gamma-1)/\nu} H\left(\frac{R}{l^\nu}\right) \tag{14.85}$$

defining the scaling function $H(u)$, which obeys $H(u) \to$ const as $u \to 0$.

Standard scaling theory applied to the $O(n)$ model then leads to the exponent relations

$$\Delta_1 = 1 - \frac{\gamma}{2\nu}$$
$$\Delta_2 = 2 - \frac{1}{\nu} \tag{14.86}$$

In view of (14.81) this gives $\nu = \frac{1}{2}$, $\gamma = \frac{19}{16}$ for dense polymers ($g = \frac{1}{2}$) and $\nu = \frac{3}{4}$, $\gamma = \frac{43}{32}$ for dilute ones ($g = \frac{3}{2}$). The remaining critical exponents follow from standard scaling relations.

### 14.4.3 Magnetic Exponent in the Potts Model

The watermelon exponents can be said to be of the "magnetic" type, since they induce a magnetic type defect charge $m_\ell$ in the CG. The standard magnetic exponent, describing the decay of the spin-spin correlation function in the Potts model, is however not of the watermelon type. It can nevertheless be inferred from (14.71) as follows:

The probability that two spins situated at $\mathbf{x}_1$ and $\mathbf{x}_2$ are in the same Potts state is proportional, in the random cluster picture, to the probability that they belong to the same cluster. In the cylinder geometry this means that no winding loop separates $\mathbf{x}_1$ from $\mathbf{x}_2$. This can be attained in the CG by giving a weight $\bar{n} = 0$ to such loops. We have seen that inserting a pair of vertex operators with charge $\pm e$ at $\mathbf{x}_1$ and $\mathbf{x}_2$ leads exactly to this situation with $\bar{n} = 2\cos(\pi e)$, and so we need $e = \frac{1}{2}$. The scaling dimension of this excitation, with respect to the ground state which has $e = e_0$, is then

$$\Delta_{\text{m}} = \Delta_{\frac{1}{2},0} - \Delta_{e_0,0} = \frac{1 - 4(1-g)^2}{8g}. \tag{14.87}$$

In particular we verify that for the Ising model, $g = \frac{3}{4}$, this yields $\Delta_m = \frac{1}{8}$ as it should.

The location in the Kac table (14.31), using (14.74), is

$$\Delta_m = 2h_{1/2,0}. \tag{14.88}$$

Note that this differs from the lowest possible watermelon excitation $\Delta_1 = 2h_{0,1/2}$, corresponding to one loop strand propagating along the length direction of the cylinder.[7] Indeed, the dominant configurations participating in the magnetic correlation function have *no* propagating strands, since the cluster containing $x_1$ and $x_2$ will typically wrap around the cylinder.

### 14.4.4 Thermal Exponents

As discussed in section 14.3.1, the Potts model at the critical temperature can be identified with the loop model (14.59) with homogeneous weights, $(e^J - 1)/\sqrt{Q} = 1$. A deviation from the critical temperature will make the weights inhomogeneous, i.e., give different weights to the two possible states of the transition system in Fig. 14.1. Comparing this observation with the discussion of the marginality requirement in section 14.3.4, we see that the result is the appearance of electric charges $e = \pm 1$. Determining the correct sign of the charge requires a more careful microscopic analysis, which was first carried out by den Nijs [10] (see also [11, 35]). The end result for the thermal exponent is then

$$\Delta_t^{\text{Potts}} = \Delta_{-1,0} = \frac{3}{2g} - 1 \tag{14.89}$$

where we have used (14.71).

Under the identification (14.74), the location of this operator in terms of the Kac table (14.31) becomes [50]

$$\Delta_t^{\text{Potts}} = 2h_{2,1} = \frac{m+3}{2m}. \tag{14.90}$$

Note that this is an RG relevant operator for $m > 1$ [i.e., for coupling $\frac{1}{2} < g \le 1$, or loop weight $0 < \sqrt{Q} \le 2$], meaning that the critical point is unstable to a deviation from the critical temperature. On the other hand, for $0 < m < 1$ [i.e., for coupling $0 < g < \frac{1}{2}$, or loop weight $-2 < \sqrt{Q} < 0$] the thermal operator is irrelevant, implying the existence of a Berker-Kadanoff phase [51].

For the $O(n)$ model, the microscopic derivation of the CG is different (see section 14.3.1), and by construction the sublattice symmetry can no longer be broken. Ac-

---

[7] Naively, the relation to the six-vertex model shows that in the Potts model these strands necessarily come in *pairs*, but this can be arranged by a suitable generalisation of the periodic boundary conditions, such that the six-vertex model is defined on an *odd* number of strands.

cordingly, a deviation in $K$ from the critical values (14.61) must now couple to the next most relevant electrical charges $e = \pm 2$. Of these two, $e = 2$ has already been used for the marginality requirement $\Delta_{2,0} = 2$, and we expect in contrast a thermal exponent that depends on $g$. We are therefore led to

$$\Delta_t = \Delta_{-2,0} = \frac{4}{g} - 2. \tag{14.91}$$

A detailed derivation was first given by Nienhuis [37]. Note that the thermal operator is relevant for the dilute case ($1 < g < 2$), and irrelevant for the dense case ($0 < g < 1$). The entire low-temperature phase of the $O(n)$ model will therefore renormalise towards the dense case. Exactly at zero temperature, a new critical theory emerges (see section 14.5).

The identification of $\Delta_t$ with the Kac table (14.31), via (14.74), is now

$$\Delta_t = \begin{cases} 2h_{3,1} \text{ for the dense } O(n) \text{ model} \\ 2h_{1,3} \text{ for the dilute } O(n) \text{ model} \end{cases} \tag{14.92}$$

As a check, note that the critical Ising model is a special case of both the Potts model ($Q = 2$, or $g = \frac{3}{4}$) and of the $O(n)$ model (dilute $n = 1$, or $g = \frac{4}{3}$). In both cases, the above formulae give $\Delta_t = 1$ as they should.

### 14.4.5 Network Exponents

Duplantier and Saleur [47, 14] have shown how to generalise the exponent $\gamma$ of (14.84) to more complicated network geometries. In the notation of section 14.4.1, consider a multi-point correlation function

$$\mathscr{C}_{\mathscr{G}} = \langle \mathscr{O}_{\ell_1}(\mathbf{x}_1) \mathscr{O}_{\ell_2}(\mathbf{x}_2) \cdots \mathscr{O}_{\ell_\mathscr{V}}(\mathbf{x}_\mathscr{V}) \rangle \tag{14.93}$$

involving $\mathscr{V}$ watermelon operators, where the $k$'th operator inserts an $\ell_k$ leg vertex at position $\mathbf{x}_k$. The orientations of the loop segments inserted do not matter in the following discussion, but must be chosen so that $\mathscr{C}_{\mathscr{G}}$ describes a well-defined network $\mathscr{G}$ (see Fig. 14.7). Accordingly we can assume that the indices $\ell_k$ are all positive.

Let $n_\ell$ be the number of $\ell$-leg operators in $\mathscr{G}$, and let $\mathscr{E}$ be the total number of edges in $\mathscr{G}$. We then have the topological relations $\sum_\ell n_\ell \ell = 2\mathscr{E}$ and $\sum_\ell n_\ell = \mathscr{V}$.

Consider the case of a monodisperse network, where each of the $\mathscr{E}$ edges is constrained to have the same length $l$ (with $l \gg 1$). Laplace transform and some scaling analysis then generalises (14.84) into

$$\frac{\mathscr{N}_{\mathscr{G}}}{\mathscr{N}_{\text{SAP}}} \sim l^{\gamma_{\mathscr{G}} - 1 + vd} \tag{14.94}$$

with the network exponent

**Fig. 14.7** A network $\mathscr{G}$ made of $\mathscr{E} = 10$ chains and $\mathscr{V} = 9$ vertices, with $n_1 = 5$, $n_3 = 3$, and $n_4 = 2$.

$$\gamma_{\mathscr{G}} = v\left[2(\mathscr{V} - 1) - \sum_\ell n_\ell \Delta_L\right] - (\mathscr{E} - 1) \tag{14.95}$$

Note that for the ordinary SAW topology ($n_\ell = 2\delta_{\ell,1}$), we retrieve $\gamma_{\mathscr{G}} = 2v(1 - \Delta_1)$, in agreement with (14.86).

For a polydisperse network, the total length $l$ is freely distributed among the $\mathscr{E}$ edges in the network, and so one has simply to omit the last term $(\mathscr{E} - 1)$ in (14.95).

Special cases of (14.95) yield contact exponents, describing e.g. the probability that one of the end points of a SAW comes close to the midpoint of the walk.

### 14.4.6 Winding Angle Distribution

In section 14.4.1 we have seen that when computing the conformal weights of the watermelon operators, it was necessary to insert vertex operators $e^{\pm ie_0\phi}$ at the chain ends in order to cancel the spurious phase factors that occur in the oriented loop model due to the winding of the SAW around its end points. By a generalisation of this argument, Duplantier and Saleur [48] have shown how to actually compute the winding angle distribution of a SAW.

Consider an $O(n)$ model with an arbitrary number of closed loops, each of fugacity

$$n = 2\cos(\pi e_0) = -2\cos(\pi g), \tag{14.96}$$

and a single open walk (the SAW, for $n \to 0$) with end points $\mathbf{x}_1$ and $\mathbf{x}_2$. In the oriented loop picture, the walk is taken to be oriented from $\mathbf{x}_1$ to $\mathbf{x}_2$. To count precisely its number of windings around each $\mathbf{x}_i$, these two points are connected to infinity through parallel half lines $\mathscr{L}_i$, as shown in Fig. 14.8. Then let the winding number $n_i$ be the signed number of times the walk crosses $\mathscr{L}_i$, the sign being positive for an anti-clockwise crossing. In the scaling limit we expect $n_i \gg 1$, and so even though $n_i$ has been defined as an integer it can be used to deduce the winding angle.

One now compares on one hand the correlation function $\langle \exp(i\pi e_1 n_1 + i\pi e_2 n_2) \rangle$ in the $O(n)$ model, and on the other hand $\langle \exp(ie'_1 \phi(\mathbf{x}_1) + ie'_2 \phi(\mathbf{x}_2)) \rangle$ in the equiva-

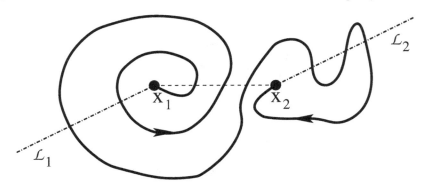

**Fig. 14.8** SAW oriented from $\mathbf{x}_1$ to $\mathbf{x}_2$ with winding numbers $n_1 = 2$ and $n_2 = -1$.

lent height model, where we recall that $\phi$ is the height function. These will in general give incorrect weights (i.e., $\neq n$) to loops that surround both $\mathbf{x}_1$ and $\mathbf{x}_2$. A careful study of the complex phase factors arising in these correlators, and in the oriented loop/height model shows that the two correlators are identical, and the surrounding loops are weighted correctly, provided that we satisfy the conditions

$$e_1' = e_1 - e_0, \qquad e_2' = e_2 - e_0, \qquad e_1 + e_2 = 0 \text{ or } 2e_0. \qquad (14.97)$$

We shall take $e_1 = -e_2 \equiv e$ in the last condition.

This deals with the electric charges in the CG picture. The magnetic charges needed to insert the walk are $m_1 = -m_2 \equiv m$ with $m = 1$ as usual. The joint value of the correlators is then, from (14.50),

$$\sim |\mathbf{x}_1 - \mathbf{x}_2|^{-2\Delta(e)} \qquad \text{with } \Delta(e) = \frac{1}{2}\left(\frac{e^2}{2g} + g\right) \qquad (14.98)$$

valid in the scaling limit $|\mathbf{x}_1 - \mathbf{x}_2| \gg 1$. This result can be written

$$\left\langle e^{i\pi e(n_1 - n_2)} \right\rangle = e^{-\frac{e^2}{g} \log |\mathbf{x}_1 - \mathbf{x}_2|} \qquad (14.99)$$

where we have normalised with respect to the same correlator with $e = 0$ (but still $m = 1$). Noting that this is Gaussian in $e$, the distribution of the winding angle $\theta = \theta_1 - \theta_2 = 2\pi(n_1 - n_2)$ itself is also Gaussian, and after a Fourier transformation we obtain finally the normalised distribution

$$P(\theta) = \left(\frac{16\pi \log |\mathbf{x}_1 - \mathbf{x}_2|}{g}\right)^{-1/2} \exp\left(-\frac{g\theta^2}{16 \log |\mathbf{x}_1 - \mathbf{x}_2|}\right) \qquad (14.100)$$

Usually we are interested in the fixed-length rather than the fixed-extremities ensemble, in which case it suffices to replace $|\mathbf{x}_1 - \mathbf{x}_2|$ by $l^\nu$ in (14.100), for a walk of length $l$. Another remark is that the winding numbers $n_1$ and $n_2$ can be argued to be independent in the scaling limit, and so if one is interested in just one of them it suffices to replace (14.100) by a distribution of half the width.

The above argument was shown for a SAW (as in [48]), but can easily be adapted to the windings of a SAP constrained to go through two fixed points $\mathbf{x}_1$ and $\mathbf{x}_2$. The relevant magnetic charge is then $m = 2$, but the remainder of the argument is essentially unchanged.

The issue of winding angle distribution for a SAW, and its relation with that of Brownian walks, was studied further by Saleur [52].

### 14.4.7 Polymer Collapse: the Theta Point

The dilute SAW is a model of a polymer in a good solvent. When this assumption fails, e.g., upon lowering the temperature of the solvent, the effective attraction between the monomers increases, and eventually the polymer undergoes a collapse transition, first described by Flory [53]. The corresponding critical temperature is traditionally called $\Theta$, or the theta point. It was argued by de Gennes [54, 46] that this is a tricritical point: intuitively this means that with respect to the critical, or dilute, SAW—obtained in our framework by tuning the monomer fugacity to a particular value—one additional parameter, viz. the effective monomer-monomer interaction, has to be adjusted to its critical value.

Duplantier and Saleur [19] have proposed a particularly simple model of the monomer-monomer interaction that is capable of capturing the physics of the theta point. Their argument shows that the corresponding universality class is that of the dense $O(n = 1)$ model, i.e., of critical percolation, and the exponents follow readily. The argument runs as follows.

Consider a SAW on the usual hexagonal lattice, but in the presence of annealed dilution. To be specific, each lattice face contains a "defective solvent" with probability $p$, and an "ideal solvent" with probability $1 - p$, independently for each face. The SAW is constrained to touch only lattice faces containing an ideal solvent, as shown in Fig. 14.9.

In the partition function, one may first sum over the configurations of the solvent consistent with a fixed configuration of the SAW, and then over those of the SAW itself. In the first sum, any face not touching the SAW contributes a trivial factor of one, whereas each of the remaining faces yields a weight $(1 - p)$. Let $N_2$ (resp. $N_3$) be the number of faces which are adjacent to two (resp. three) successions of monomers which are non-subsequent along the SAW. [In other words, let $N_k$ be the number of hexagons having $2k$ occupied external legs.] The total face weight is then simply

$$(1 - p)^{l+1-N_2-2N_3} \tag{14.101}$$

where $l$ is the length (number of edges) of the SAW. Clearly this is a kind of short-range attraction between the monomers in the SAW, albeit a somewhat peculiar one.

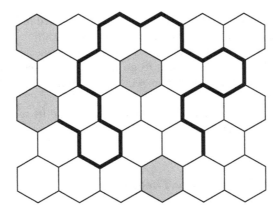

**Fig. 14.9** Model of a polymer at the theta point. The SAW lives on a hexagonal lattice, but is not allowed to share an edge with any of the faces containing a defective solvent, here shown shaded in grey.

One now argues that the parameter $p$ has a critical value $p_c$. If $p < p_c$ the system should renormalise towards the standard dilute SAW, and if $p > p_c$ the solvent deficiencies will percolate and the SAW will be in the dense phase. The threshold $p_c$ must be that of site percolation on the triangular lattice, and so $p_c = \frac{1}{2}$ exactly. But exactly at $p_c$, defective and ideal faces are equiprobable and so form clusters which may equally well be described in terms of their contours, as an O($n$) model of loops (14.60) with trivial parameters, $n = 1$ and $K = 1$. The latter value is precisely[8] the dense loop solution of (14.61).

We conclude that polymers at the theta point are described by the CG of the dense O($n = 1$) model, with coupling constant $g = \frac{2}{3}$. In particular, the watermelon exponents (14.81) read

$$\Delta_\ell = \frac{\ell^2 - 1}{12}. \tag{14.102}$$

The standard polymer exponents (14.84) describing the theta point are related to these through the scaling relations (14.86):

$$\eta = 2\Delta_1 = 0, \qquad \nu = \frac{1}{2 - \Delta_2} = \frac{4}{7}, \qquad \gamma = (2 - \eta)\nu = \frac{8}{7}. \tag{14.103}$$

Finally, the thermal exponent $\nu'$ describing how the size of defective solvent clusters diverges, $\xi \sim |p - p_c|^{-\nu'}$ as $p \uparrow p_c$ is given by $\nu' = 1/(2 - \Delta_4) = \frac{4}{3}$, and so the so-called crossover exponent is

$$\phi = \frac{\nu}{\nu'} = \frac{3}{7}. \tag{14.104}$$

---

[8] Even if this had not precisely been the case, the outcome of the argument would be the same, since the dense loop solution of (14.61) is RG attractive in $K$ by (14.91). This observation is certainly part of the explanation that the variant models explored in [55] yield unchanged critical exponents.

Although the theta point exponents (14.103)–(14.104) are in very good agreement with those of numerical simulations, and even experiments, the original paper [19] was subsequently challenged by a number of authors. Indeed, the interaction between monomers employed is quite peculiar—it corresponds to attractions between nearest neighbour vertices and a *subset* of the next-nearest neighbour vertices—and one may fear that the universality class described is not the required tricritical point, but an even higher multicritical point.

To meet this criticism, Duplantier and Saleur produced a second paper [55] in which they examined numerically a certain number of variant models, involving different local monomer interactions, anisotropy, and using different lattices. In all cases the exponents (14.103)–(14.104) were shown to be unchanged, and far from the values of any variant proposal. We can therefore conclude that in spite of its simplicity the original model captures the correct theta point physics and produces the exact values of the corresponding critical exponents.

## 14.5 Fully Packed Loop Models

In section 14.3 we have seen how to solve loop models by writing a Liouville field theory for their associated height model. We have discussed in detail two models, the $Q$-state Potts and the $O(n)$ model. In both cases, we exploited a bijection between the configurations of oriented loops and those of a scalar (one-component) height variable $h(\mathbf{x})$ defined on the lattice faces $\mathbf{x}$. There does however exist geometric models whose microscopic formulation allows for the definition of a vector height $\mathbf{h}(\mathbf{x}) \in \mathbb{R}^D$ with $D > 1$. A necessary (but not sufficient) condition for this to happen seems to be that the objects (loops, tiles, colours, ...) be maximally packed on the lattice.

Tiling models are nice examples of statistical models possessing vector height mappings, provided that the shapes of the tiles and the lattice are chosen carefully. For example, one obtains $D = 2$ dimensional heights by tiling the square lattice with two different types of dimers [56], or by linear trimers [57], or by tiling the triangular lattice with triangular trimers [58]. In general, such models are either non-critical, or described in the continuum by $D$ free bosons. For example, all of the tiling problems just mentioned have been shown (analytically and/or numerically) to have central charge $c = 2$.

In this section we are rather interested in models of fully packed loops (FPL) possessing a vector height. When critical, such models can be described in terms of a vectorial Coulomb gas (CG), generalising the working of section 14.3. In particular, the non-local nature of the loops allows for the existence of a background electric charge, and so one or more control parameters (typically the loop weights) permit one to change $c$ and the critical exponents continuously. This amounts to a rather more interesting continuum limit than that of the tiling problems.

In section 14.5.1 we give some examples of FPL models and establish their height mappings. These models turn out to be highly non-universal, in the sense that $D$—

and hence ultimately the values of the critical exponents—depends on the underlying lattice. A necessary condition for the existence of a non-trivial continuum limit seems to be that the underlying lattice is bipartite [59].

The CG construction for these models is presented in some detail (see section 14.5.2), as it sheds further light both on the ideal state construction and on the marginality requirement. Bulk critical exponents are derived in section 14.5.3. We conclude by a few further remarks on the underlying quantum group symmetries of the FPL models, and the possibility of coupling them to two-dimensional quantum gravity.

## 14.5.1 Three Loop Models with a Vector Height Mapping

The loop model based on the $Q$-state Potts model is an example of an FPL model, since its loops jointly visit every vertex of the lattice (twice). But since its height mapping is one-dimensional (see section 14.3.2) this is not what were are interested in here.

Let us instead revisit the $O(n)$ model (14.60) on the hexagonal lattice. The allowed vertices are shown in Fig. 14.2. It is easy to see that on a lattice with periodic boundary conditions respecting the three-sublattice structure of the lattice faces, the number of voids (of weight $T \equiv 1/K$) must be even, whence the model is symmetric under $T \to -T$. The fact (14.91) that the low-temperature branch of (14.61) is RG attractive in $T$ then implies that the $T = 0$ manifold is a line of repulsive fixed points [60]. A numerical study [60] further reveals that these fixed points are critical for $|n| \leq 2$. For brevity we shall refer to this $O(n)$ model at $T = 0$ simply as the FPL model in the remainder of this section. In the polymer limit $n \to 0$ it describes Hamiltonian circuits (SAP's) and paths (SAW's) on the hexagonal lattice.

In the oriented loop model corresponding to the FPL model there are three types of edges: A) empty edges, B) occupied edges pointing from an even to an odd vertex, and C) occupied edges pointing from odd to even. We shall often refer to these edge labels as colours. Denoting the height change when traversing any one of these edges by $\mathbf{A}$, $\mathbf{B}$, or $\mathbf{C}$, occupied vertices lead to the consistency requirement

$$\mathbf{A} + \mathbf{B} + \mathbf{C} = \mathbf{0}, \tag{14.105}$$

whereas empty vertices require $\mathbf{A} = \mathbf{0}$. By symmetry, one must then choose $\mathbf{B} = +a$ and $\mathbf{C} = -a$ for some scalar $a$, as was indeed done in section 14.3.2. But if $T = 0$, empty vertices are forbidden, and the only requirement is (14.105). The height differences can then be taken as two-dimensional vectors pointing from the centre to the vertices of an equilateral triangle. Since we have shown carefully in section 14.2.7 that the normalisation of the heights drops out from the final results for the critical exponents, we henceforth adopt the choice

$$\mathbf{A} = \left( \frac{1}{\sqrt{3}}, 0 \right), \qquad \mathbf{B} = \left( -\frac{1}{2\sqrt{3}}, \frac{1}{2} \right), \qquad \mathbf{C} = \left( -\frac{1}{2\sqrt{3}}, -\frac{1}{2} \right). \tag{14.106}$$

Kondev, de Gier and Nienhuis [20] used the height mapping (14.106) as the starting point for solving the FPL model exactly.

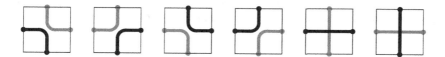

**Fig. 14.10** The six vertices defining the FPL$^2$ model.

Another model that we shall consider is the so-called FPL$^2$ model on the square lattice, which was defined by Kondev and Henley [61] and solved by Jacobsen and Kondev [21]. This is a model of *two* different types of fully packed loops (whence the superscript in the denomination FPL$^2$), henceforth referred to as black and grey. The allowed vertices are shown in Fig. 14.10. The partition function is defined by assigning independent fugacities, $n_b$ and $n_g$, to the two types of loops,

$$Z = \sum_{\mathscr{G}} n_b^{N_b} n_g^{N_g}, \qquad (14.107)$$

where $N_b$ (resp. $N_g$) is the number of black (resp. grey) loops, and $\mathscr{G}$ is the set of all allowed loop configurations.

A number of special cases of the FPL$^2$ model are of special interest: suffice it here to say that the limit $n_g = 1$, $n_b \to 0$ describes Hamiltonian circuits and paths on the square lattice.

The two types of loops can be oriented independently, giving rise to four types of edges: A) black edges oriented from the even to the odd sublattice, B) black edges oriented from the odd to the even sublattice, C) grey even-to-odd edges, and D) grey odd-to-even edges. These define the height differences for the height model on the dual lattice, with the consistency requirement

$$\mathbf{A} + \mathbf{B} + \mathbf{C} + \mathbf{D} = \mathbf{0} \qquad (14.108)$$

that the total height change when encircling any vertex be zero. We have then a $D = 3$ dimensional height model, and by symmetry we can take the height differences to point from the centre to the vertices of a regular tetrahedron:

$$\mathbf{A} = (-1,1,1), \quad \mathbf{B} = (1,1,-1), \quad \mathbf{C} = (-1,-1,-1), \quad \mathbf{D} = (1,-1,1).$$
$$(14.109)$$

The last model to be discussed in this section is obtained from the FPL$^2$ model by attributing local vertex weights in addition to the loop weights. By rotational symmetry, it suffices to give a special weight $w_X$ to the last two vertices in Fig. 14.10,

$$Z = \sum_{\mathscr{G}} n_b^{N_b} n_g^{N_g} w_X^V, \qquad (14.110)$$

where $V$ is the number of vertices where the two types of loops cross, and $\mathscr{G}$ has the same meaning as in (14.107). We shall refer to this as the semi-flexible loop (SFL) model. It was first solved by Jacobsen and Kondev [62].

The polymer limit ($n_g = 1$, $n_b \to 0$) of the SFL model has been proposed as a model of protein melting by Flory half a century ago [63].

### 14.5.2 Coulomb Gas Construction

We now discuss how to dress the Coulomb gas (CG) for the three models of oriented loops (FPL, FPL², and SFL) just introduced. As in section 14.3 the CG will eventually take the form of a Liouville field theory, but with electromagnetic charges which are $D$-dimensional vectors. The construction for the FPL² and SFL models will only start differing when imposing the marginality requirement.

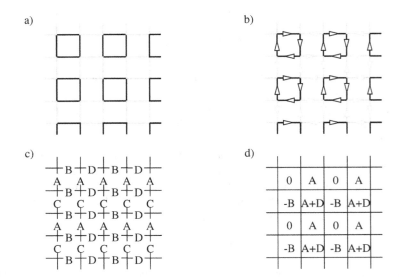

**Fig. 14.11** An ideal state in the FPL² model shown in terms of a) loops, b) oriented loops, c) edge colourings, and d) the height mapping.

The first issue is to determine the analogue of the compactification radius, which now takes the form of a $D$-dimensional lattice. To this end, the ideal state construction [61] is a very convenient tool.

We define first the *ideal states* as periodic arrangements of the colours, such that the microscopic height is maximally flat. An example of an FPL² ideal state is shown in Fig. 14.11. In general, an ideal state is obtained by selecting a permutation of the edge labels around a fixed vertex; the arrangement is then extended to the whole lattice in such a way that alternations of any pair of colours form as many short cycles as possible. For the FPL model there are 6 ideal states; in these all 3 colour pairs form short cycles (of length 6); and the colour pair BC defines the loops. In the FPL$^2$ model there are 24 ideal states; in these 4 out of 6 colour pairs form short cycles (of length 4); and the colour pairs AB and CD define black and grey loops respectively.

We construct next the *ideal state graph* $\mathscr{I}$. For this, we define a *transition* between two ideal states as a transposition of a colour pair forming a short cycle. To each transition we associate a vector in $\mathbb{R}^D$ equal to the difference between the average height in the two concerned ideal states. By definition, a transition changes only the heights on the faces surrounded by transposed short cycles. In $\mathscr{I}$, each vertex represents an ideal state, and each edge represents a transition. The graph $\mathscr{I}$ is embedded into $\mathbb{R}^D$ by letting each edge correspond to the vector associated with the transition between ideal states.

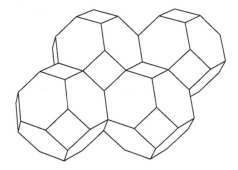

**Fig. 14.12** The ideal state graph of the FPL$^2$ model.

For the FPL model, $\mathscr{I}$ turns out to be a hexagonal lattice [20], and for the FPL$^2$ model it is a tiling of $\mathbb{R}^3$ with truncated octahedra [61] (also known as the Wigner-Seitz cell of a body-centered cubic lattice). The latter $\mathscr{I}$ is shown in Fig. 14.12. Crucially, any *fixed* ideal state is represented infinitely many times in $\mathscr{I}$ and forms a lattice which we shall call the *repeat lattice* $\mathscr{R}$. It turns out that $\mathscr{R}$ is spanned by the vectors $\mathbf{A} - \mathbf{B}$, $\mathbf{B} - \mathbf{C}$ (and $\mathbf{C} - \mathbf{D}$ for the FPL$^2$ model). In other words, for the FPL model $\mathscr{R}$ is a triangular lattice of edge length 1 in the normalisation (14.105), and for the FPL$^2$ model $\mathscr{R}$ is a face-centered cubic lattice with a conventional cubic cell of edge length 4 in the normalisation (14.108).

In the CG construction, the height is then compactified with respect to $\mathscr{R}$, i.e.,

$$\mathbf{h}(\mathbf{x}) \in \mathbb{R}^D / \mathscr{R}. \tag{14.111}$$

In particular, the magnetic charges $\mathbf{m} \in \mathscr{R}$, and the electric charges $\mathbf{e} \in \mathscr{R}^*$, where $\mathscr{R}^*$ denotes the reciprocal lattice of $\mathscr{R}$.

The Liouville field theory is again described by the action (14.69), consisting of an elastic term $S_E$, a boundary term $S_B$, and the Liouville potential $S_L$. We now describe these three terms in turn.

The *elastic term* is constrained by rotational invariance in real $d = 2$ dimensional space to take the form

$$S_E = \frac{1}{2} \int d^2\mathbf{x}\, K_{\alpha\beta}\, \partial h^\alpha \cdot \partial h^\beta \,, \tag{14.112}$$

where $\partial = (\partial_1, \partial_2)$ is the usual gradient. The $D$-dimensional tensor $K_{\alpha\beta}$ is further constrained by the loop reversal symmetries: $\mathbf{B} \leftrightarrow \mathbf{C}$ for the FPL model, while for the FPL$^2$ model one has both $\mathbf{A} \leftrightarrow \mathbf{B}$ and $\mathbf{C} \leftrightarrow \mathbf{D}$. The result for the FPL model is simply

$$S_E = \frac{1}{2} \int d^2\mathbf{x}\, g_\alpha (\partial h^\alpha)^2 \,. \tag{14.113}$$

with coupling constants $g_1 \equiv K_{11}$, $g_2 \equiv K_{22}$, and $g_1 = g_2$. For the FPL$^2$ model one obtains first a more complicated result

$$S_E = \frac{1}{2} \int d^2\mathbf{x}\, \left\{ K_{11}[(\partial h^1)^2 + (\partial h^3)^2] + 2K_{13}(\partial h^1 \cdot \partial h^3) + K_{22}(\partial h^2)^2 \right\}, \tag{14.114}$$

which can however be diagonalised by a change of coordinates in height space,

$$\tilde{h}^1 = \frac{h^1 - h^3}{2}, \qquad \tilde{h}^2 = h^2, \qquad \tilde{h}^3 = \frac{h^1 + h^3}{2} \tag{14.115}$$

yielding once again (14.113) for $\tilde{h}^\alpha$, but now with three *unrelated* coupling constants $g_1 \equiv 2(K_{11} - K_{13})$, $g_2 \equiv K_{22}$, and $g_3 \equiv 2(K_{11} + K_{13})$. Note that magnetic charges $\mathbf{m}$ transform as (14.115), whereas electric charges $\mathbf{e}$ transform according to the reciprocal transformation (swap tilded and untilded quantities). Henceforth we shall write everything in terms of the new heights $\tilde{h}^\alpha$ and electromagnetic charges $(\tilde{\mathbf{e}}, \tilde{\mathbf{m}})$, but drop the tildes to lighten the notation.[9]

Let us now parametrise the loop weights as

$$n = 2\cos(\pi e_0) \tag{14.116}$$

for the FPL model, and

$$n_b = 2\cos(\pi e_b), \qquad n_g = 2\cos(\pi e_g) \tag{14.117}$$

for the FPL$^2$ model. The *boundary term* $S_B$ in (14.69) reads in vector notation

---

[9] It would seem tempting in a review like this to impose the transformation (14.115) right from the beginning and choose $\mathbf{A} = (-1, 1, 0)$, $\mathbf{B} = (1, 1, 0)$, $\mathbf{C} = (0, -1, -1)$, $\mathbf{D} = (0, -1, 1)$ instead of (14.108). But note that (14.108) respects the full permutation symmetry of the four colours: the scalar product of any two vectors in (14.108) is $-1$. With (14.115) this symmetry is broken in a very particular way that could not have been guessed from the outset, and which reflects the choice of defining the loops from the colour pairs AB and CD.

$$S_B = \frac{i}{4\pi} \int d^2\mathbf{x} \, (\mathbf{e}_0 \cdot \mathbf{h}) \mathscr{R}(\mathbf{x}) \tag{14.118}$$

and $\mathbf{e}_0$ is determined as in section 14.3 by requiring that wrapping loops of each type be weighted correctly. For instance, for the FPL$^2$ model we must have

$$\mathbf{e}_0 \cdot \mathbf{A} = \pi e_b, \quad \mathbf{e}_0 \cdot \mathbf{B} = -\pi e_b, \quad \mathbf{e}_0 \cdot \mathbf{C} = \pi e_g, \quad \mathbf{e}_0 \cdot \mathbf{D} = -\pi e_g, \tag{14.119}$$

which fixes the background electric charge as

$$\mathbf{e}_0 = -\pi(e_b, 0, e_g) \tag{14.120}$$

The analogous result for the FPL model is

$$\mathbf{e}_0 = 2\pi(0, e_0). \tag{14.121}$$

The sceptical reader may object that the ideal state graph $\mathscr{I}$ is not really necessary to determine the repeat lattice $\mathscr{R}$. Indeed, it suffices to notice that the fundamental height dislocations (magnetic charges) allowed by the microscopic model is the difference between two of the colour vectors (14.105) or (14.108). However, the real use of $\mathscr{I}$ is for determining the *Liouville term*

$$S_L = \int d^2\mathbf{x} \, w[\mathbf{h}(\mathbf{x})] \,, \tag{14.122}$$

where $\exp(-w[\mathbf{h}(\mathbf{x})])$ is the scaling limit of the (complex) microscopic vertex weights in the oriented loop representation. To this end, we focus for a while on the FPL$^2$ model.

Denoting by $(\sigma_1, \sigma_2, \sigma_3, \sigma_4)$ the configuration of colours when going anticlockwise around an even vertex, the microscopic weights may be written compactly as

$$w(\mathbf{h}) = \frac{i}{16} \mathbf{e}_0 \cdot [(\sigma_1 - \sigma_3) \times (\sigma_2 - \sigma_4)] \,. \tag{14.123}$$

Fourier analysing this formula, i.e., writing it as a sum of vertex operators, one arrives at

$$w(\mathbf{h}) = \sum_{\mathbf{e} \in \mathscr{R}_w^*} \tilde{w}_\mathbf{e} \exp(i\mathbf{e} \cdot \mathbf{h}) \,. \tag{14.124}$$

where $\mathscr{R}_w^*$ is a *proper* sublattice of $\mathscr{R}^*$ whose shortest vectors are the twelve next shortest vectors in the bcc-lattice $\mathscr{R}^*$.

Coming back to the FPL model one finds similarly that $\mathscr{R}_w^*$ is spanned by the next shortest vectors in $\mathscr{R}^*$.

To impose the *marginality requirement* and extract critical exponents we need to know the dimension of electromagnetic operators. In the basis where the coupling constant tensor $g^\alpha$ is diagonal, as in (14.113), this is a straightforward generalisation of the usual CG formula (14.71), which reads in the normalisation of the present section

$$\Delta_{\mathbf{e},\mathbf{m}} = \frac{1}{4\pi} \left[ \frac{1}{g_\alpha} e_\alpha (e_\alpha - 2e_{0\alpha}) + g_\alpha (m^\alpha)^2 \right]. \tag{14.125}$$

The FPL model is now solved by noting that $\mathbf{e}^{(0)} = (0, 4\pi)$ is the most relevant electric charge in $\mathscr{R}_w^*$. Solving the marginality requirement $\Delta_{\mathbf{e}^{(0)},0} = 2$ then fixes the CG coupling in terms of the loop weight:

$$g = 2\pi (1 - e_0). \tag{14.126}$$

The case of the FPL$^2$ model is considerably more intricate. Moreover, determining the correct marginality requirement is the one and only point where the analysis differs from that of the SFL model. Referring to [21, 62] for a detailed discussion, the outcome is that the vectors

$$\mathbf{e}^{(1)} = (-2\pi, 0, 0), \qquad \mathbf{e}^{(2)} = (0, 0, -2\pi) \tag{14.127}$$

act as screening charges in both the FPL$^2$ and the SFL models, whereas an additional symmetry of the FPL$^2$ model implies that it possesses two extra screening charges

$$\mathbf{e}^{(3)} = (-\pi, \pi, -\pi), \qquad \mathbf{e}^{(4)} = (-\pi, \pi, -\pi). \tag{14.128}$$

The marginality requirement then fixes

$$g_1 = \frac{\pi}{2}(1 - e_{\mathrm{b}}), \qquad g_3 = \frac{\pi}{2}(1 - e_{\mathrm{g}}) \tag{14.129}$$

for both models. The extra screening charges in the FPL$^2$ model imply a third relation

$$\frac{1}{g_2} = \frac{1}{g_1} + \frac{1}{g_3}, \tag{14.130}$$

whereas in the SFL model $g_2$ remains a non-universal function of the parameter $w_{\mathrm{X}}$ appearing in (14.110).

In the critical regime $|n_{\mathrm{b}}|, |n_{\mathrm{g}}| \leq 2$ we have $g_1, g_3 \leq \frac{\pi}{2}$ from (14.129), and so $g_2 \leq \frac{\pi}{4}$ when $w_{\mathrm{X}} = 1$ (i.e., in the FPL$^2$ model). It is easy to see that increasing $w_{\mathrm{X}}$ will make the height interface stiffer in the 2-direction, and so will increase $g_2$. When $g_2 > \frac{\pi}{2}$ the operator that discretises the height in the 2-direction becomes relevant, and so the height profile becomes flat in the continuum limit, meaning that the model is no longer critical. We deduce that precisely at $g_2 = \frac{\pi}{2}$ the model stands at a Kosterlitz-Thouless transition; in the polymer limit this is the protein melting transition [62] that Flory [63] originally aimed at describing. Exact exponents can therefore be computed at this KT transition.

However, the CG method does not permit us to solve the SFL model for generic values of $w_{\mathrm{X}}$, the relation to $g_2$ being unknown. However, note that all critical exponents can be expressed in terms of just one unknown parameter $g_2$. Furthermore, for any given values of $n_{\mathrm{b}}$, $n_{\mathrm{g}}$ and $w_{\mathrm{X}}$, numerical transfer matrix methods allow us to determine $g_2$ to very high precision [62].

## 14.5.3 Bulk Critical Exponents

In a $D$-dimensional CG the central charge is $c = D + 12\Delta_{e_0,0}$, giving

$$
c = \begin{cases}
2 - \dfrac{6e_0^2}{1-e_0} & \text{for the FPL model} \\[2mm]
3 - 6\left(\dfrac{e_b^2}{1-e_b} + \dfrac{e_g^2}{1-e_g}\right) & \text{for the FPL}^2 \text{ and the SFL models.}
\end{cases}
\tag{14.131}
$$

Watermelon configurations are obtained by violating the colouring constraint at two vertices $x_1$ and $x_2$. For example, in the FPL model a vertex whose adjacent edges are coloured $(\mathbf{A},\mathbf{A},\mathbf{B})$ will insert a 1-leg operator of magnetic charge $\mathbf{m}_1 = \mathbf{A} - \mathbf{C}$, whereas the vertex $(\mathbf{A},\mathbf{B},\mathbf{B})$ gives a 2-leg operator of charge $\mathbf{m}_2 = \mathbf{B} - \mathbf{C}$. Higher-leg operators are obtained by taking multiples of these basic charges. In all cases, a further electric charge $e_0$ is needed to correct the spurious phase factors due to the polymer strands winding around their insertion point. The $\ell$-leg watermelon exponents are then found from (14.125):

$$
\Delta_\ell = \begin{cases}
\dfrac{1}{8}g\ell^2 - \dfrac{(1-g)^2}{2g} & \text{for } \ell \text{ even} \\[2mm]
\dfrac{1}{8}g\ell^2 - \dfrac{(1-g)^2}{2g} + \dfrac{3g}{8} & \text{for } \ell \text{ odd}
\end{cases}
\tag{14.132}
$$

where we have set $g = 1 - e_0$ to facilitate the comparison with (14.81).

Similar results can be obtained for the FPL$^2$ [21] and SFL [62] models. Note that the watermelon strands can now be either black or grey, and the parity of the number of black strands must equal the parity of the number of grey strands. We here state the results only for the simplest case of the FPL$^2$ model where grey strands are ignored and grey loops are assigned a trivial fugacity, $n_g = 1$. This gives for the black $\ell$-leg watermelon exponents

$$
\Delta_\ell = \begin{cases}
\dfrac{1}{8}g\ell^2 - \dfrac{(1-g)^2}{2g} & \text{for } \ell \text{ even} \\[2mm]
\dfrac{1}{8}g\ell^2 - \dfrac{(1-g)^2}{2g} + \dfrac{g}{3g+2} & \text{for } \ell \text{ odd}
\end{cases}
\tag{14.133}
$$

where again we have set $g = 1 - e_b$ for easy comparison.

With the watermelon exponents (14.132)–(14.133) many of the exponents discussed in section 14.4 (e.g., network and contact exponents) follow as before. We focus here on the *conformational exponents* for Hamiltonian circuits and paths on the hexagonal and square lattices, obtained from taking the polymer limit in the FPL and FPL$^2$ models and using the scaling relations (14.86). On both lattices $\nu = \frac{1}{2}$, a trivial result which however serves as a check of the above CG construction. More interestingly, we have

$$
\gamma_{\text{hex}} = 1, \qquad \gamma_{\text{sq}} = \frac{117}{112}.
\tag{14.134}
$$

This means that the end points of a Hamiltonian SAW on the hexagonal lattice do not interact in the continuum limit, whereas they repel each other weakly on the square lattice.

Finally, one can compute the *thermal exponent* associated with breaking the $T = 0$ constraint that all lattice vertices be visited by a loop. For the FPL model the corresponding defect vertex is $(\mathbf{A}, \mathbf{A}, \mathbf{A})$ of magnetic charge $\mathbf{m}_t = 3\mathbf{A}$, yielding

$$\Delta_t = \Delta_{0,\mathbf{m}_t} = \frac{3g}{2}. \tag{14.135}$$

Note that this is always RG relevant, confirming the result of section 14.5.1 that the $T = 0$ manifold is a line of repulsive fixed points.

For the FPL$^2$ model the defect $(\mathbf{C}, \mathbf{D}, \mathbf{C}, \mathbf{D})$ of magnetic charge $\mathbf{m}_t = 2(\mathbf{C} + \mathbf{D})$ excludes black loops from visiting the defect vertex. Again we specialise the general result [21] to the case where grey loops are weighted trivially ($n_g = 1$):

$$\Delta_t = \Delta_{0,\mathbf{m}_t} = \frac{4g}{3g+2}. \tag{14.136}$$

Once again this is always RG relevant.

## 14.5.4 Further Remarks

We conclude this section with a few further remarks about the fully packed loop models.

The FPL model [64] and the FPL$^2$ model with equal fugacities $n_b = n_g$ [65] are also solvable by the Bethe Ansatz (BA) technique. The FPL$^2$ model with $n_b \neq n_g$, or the SFL model with $w_X \neq 1$, do however not appear to be BA solvable. The critical exponents computed from the BA [64, 65, 66] confirm those of the CG, giving the above results a more rigorous status. Defining $n = q + q^{-1}$, there are even underlying quantum group symmetries, viz., $SU(3)_q$ for the FPL model [67] and $SU(4)_q$ for the equally weighted FPL$^2$ model [66]. The corresponding quantum group symmetry for the Potts and usual O($n$) model discussed in section 14.3 is $SU(2)_q$ [68].

The FPL model has also been solved on random lattices using matrix integration techniques [69]. To be more precise, the loops in [69] were required to live on planar random graphs where each vertex is adjacent to three edges. However, while the loops on the regular hexagonal lattice automatically have even length (due to the bipartiteness), this restriction was not imposed in [69]. Since this is crucial for constructing the two-dimensional height model, the critical exponents on such unrestricted random lattices are not directly related to those on a regular lattice [20] by means of the KPZ equation.

A slightly modified version of the FPL$^2$ model [70] coupled to two-dimensional quantum gravity provides the exact asymptotic behaviour of meanders [71] and their multi-component generalisation [72].

Research on the surface critical behaviour of the fully packed loop models discussed in this section appears to have begun only very recently. This issue will be further discussed in section 14.6.5.

## 14.6 Surface Critical Behaviour

### 14.6.1 Ordinary, Special and Extraordinary Surface Transitions

The O($n$) model with suitably modified surface couplings permits one to realise the ordinary, special, and extraordinary surface transitions described qualitatively in section 14.2.8. To this end, one studies the model defined in the annular geometry of section 14.3.5.

To be precise, the special transition requires the loops to be in the dilute phase, and so we shall assume this to be the case throughout section 14.6.1. The results for the ordinary and extraordinary transitions hold true in the dense phase as well.

**Fig. 14.13** Hexagonal lattice in an annular geometry. The top and the bottom of the figure are identified. Boundary edges on the left are shown in grey.

A well-studied case is the hexagonal-lattice loop model (14.60). The lattice is oriented such that one third of the lattice bonds are parallel to the $x$-axis, as shown in Fig. 14.13. The fugacity of a monomer is still denoted $K$ in the bulk, but we now take a different weight $K_s$ for a monomer touching the *left* rim of the annulus, $x = 0$. In contrast, the right rim of the annulus, $x = L$, enjoys free boundary conditions, meaning that its surface monomers still carry the usual weight $K$.

In this section we wish to limit the discussion to the case where only the left boundary sustains particular ($\neq$ free) boundary conditions; this is sometimes referred to as *mixed* boundary conditions. The case where both boundaries are distinguished is also of interest and will be discussed in section 14.6.8.

The loop model described above has been thoroughly studied by Batchelor and coworkers [73, 74, 75, 76], in particular using Bethe Ansatz analysis. They find in particular that when $K_s = K$ the model is integrable and belongs to the universality class of the *ordinary* transition, while for

$$K_s = K_s^S \equiv (2 - n)^{-1/4} \qquad (14.137)$$

it is also integrable and describes the special transition. [10] This is consistent with a boundary RG scenario, where $K_s^S$ is a repulsive fixed point that flows towards either of the attractive fixed points $K_s^O < K$ and $K_s^E = \infty$, the former (resp. latter) point describing the ordinary (resp. the extraordinary) transition.

This scenario is corroborated by a detailed analysis [76] showing that a perturbation to the fixed point $K_s^E$ is RG irrelevant. Moreover, the operator conjugate to $K_s$ is obviously the energy density on the boundary. At the special transition, this operator can be identified [77] with $\phi_{(1,3)}$ of weight $h_{1,3} = \frac{2}{g} - 1$, and so this is a relevant perturbation (i.e., $h_{1,3} < 1$) only for $g > 1$ (i.e., in the dilute phase). On the other hand, the surface energy density has weight $h = 2$ at the ordinary transition [78], and so is always irrelevant.

The flow between the ordinary and special transitions has been further studied by Fendley and Saleur [79], from the point of view of boundary S matrices.

### 14.6.2 Watermelon Exponents

Surface watermelon exponents can be defined as in section 14.4.1, the only difference being that the $\ell$ legs are inserted at the boundary. We shall denote these exponents by $\Delta_\ell^O$, $\Delta_\ell^S$, $\Delta_\ell^E$ at the ordinary, special, extraordinary surface transition respectively. Whenever a result applies to any of these transitions, we use the generic notation $\Delta_\ell'$, where the prime indicates a surface rather than a bulk exponent.

For the ordinary transition, $\Delta_\ell^O$ can be derived by a slight refinement of the marginality argument given in section 14.3.5. First recall that in the continuum limit there is a background flux $m_0$ given by (14.77), corresponding to a (fractional) number of oriented loop strands running along the rims of the annulus. Suppose now that we wish to evaluate the scaling dimension $\Delta_\ell^O$ corresponding to having $\ell > 0$ non-contractible oriented loop strands running around the periodic direction of the annulus. This can be done by evaluating the free energy increase $\Delta S = S_\ell - S_0$ due to these strands, as in (14.78)

$$\Delta S = \frac{g}{4\pi} \left( (\ell + m_0)^2 - m_0^2 \right) \left( \frac{\pi}{L} \right)^2 ML \qquad (14.138)$$

and using $e^{-\Delta S} = q^{-\Delta_\ell^O}$ from (14.75).

The question now arises which sign for $m_0$ to pick in (14.77). With the plus sign we would have $\Delta_2 = 1$ independently of $g$, in clear contradiction with numerical results [47]. Taking therefore the minus sign leads to the result

$$\Delta_\ell^O = \frac{1}{4}g\ell^2 - \frac{1}{2}(1-g)\ell. \qquad (14.139)$$

---

[10] Technically speaking this is the mixed ordinary-special transition, but we have simplified the terminology according to the above remarks.

The derivation just presented follows the argument of Cardy [18], but in fact (14.139) was found a long time before by other means. Duplantier and Saleur [47] were the first to propose (14.139) for any $\ell$, by noting that their numerical transfer matrix results were in excellent agreement with the following locations in the Kac table (14.31)

$$\Delta_\ell^O = \begin{cases} h_{1,1+\ell} & \text{for the dense O}(n) \text{ model} \\ h_{1+\ell,1} & \text{for the dilute O}(n) \text{ model} \end{cases} \tag{14.140}$$

from which (14.139) follows by the identification (14.74). On a more rigorous level, (14.139) has been established by Bethe Ansatz (BA) techniques [80, 73, 74].

For the special transition, $\Delta_\ell^S$ does not seem to permit a CG derivation. It is however known from the BA analysis [75, 74] that one has

$$\Delta_\ell^S = \frac{1}{4}g(1+\ell)^2 - (1+\ell) + \frac{4-(1-g)^2}{4g}$$
$$= h_{1+\ell,2} \text{ for the dilute O}(n) \text{ model} \tag{14.141}$$

in this case.

Alternatively, one may imagine producing the special $\ell$-leg operator $\mathscr{O}_\ell^S$ by fusion of the ordinary $\ell$-leg operator $\mathscr{O}_\ell^O$ and an ordinary-to-special boundary condition changing operator $\phi_{OS}$. The scaling dimension (14.141) pertains to the insertion of this composite operator at either strip end. Comparing the Kac indices in (14.140) and (14.141), and using the CFT fusion rules (14.36), immediately leads to the identification $\phi_{OS} = \phi_{1,2}$. If one wants special boundary conditions on both the left and the right rim, two insertions of $\phi_{OS}$ are needed (to change from special to ordinary and back again). One would then expect $h_{1+\ell,3}$, as is indeed confirmed by the BA analysis [75, 74].

Finally, the extraordinary transition is rather trivially related to the ordinary transition. Indeed, for $K_s = \infty$ the entire left rim of the annulus will be coated by a straight polymer strand, so that the remaining system (of width $L - 1$) effectively sees free boundary conditions—this is dubbed the *teflon effect* in [76]. Thus, for $\ell = 0$ the coating strand will be the left half of a long stretched-out loop, whose right half will act as a one-leg operator, and one effectively observes the exponent $\Delta_1^O$. For $\ell > 0$, one of the legs will act as the coating strand, and one observes $\Delta_{\ell-1}^O$.

### 14.6.3 Network Exponents

The network exponents discussed in section 14.4.5 can be generalised [81] to the case where at least one vertex of the network $\mathscr{G}$ is constrained to stay close to the surface. Let $\mathscr{G}$ consist of $n_\ell$ (resp. $n_\ell'$) bulk (resp. surface) $\ell$-leg vertices, $\mathscr{E}$ edges, and let $\mathscr{V} = \sum_\ell n_\ell$ (resp. $\mathscr{V}' = \sum_\ell n_\ell'$) be the total number of bulk (resp. surface) vertices. The derivation then goes through with straightforward modifications.

For the case of a monodisperse network, where each of the $\mathscr{E}$ edges is constrained to have the same length $l$ (with $l \gg 1$), the end result for the network exponent is

$$\gamma_{\mathscr{G}} = v \left[ 2\mathscr{V} + \mathscr{V}' - 1 - \sum_{\ell} (n_\ell \Delta_L + n'_\ell \Delta'_\ell) \right] - (\mathscr{E} - 1). \tag{14.142}$$

For a polydisperse network of total length $l$, the last term $(\mathscr{E} - 1)$ has to be omitted as before.

Note that (14.142) does not reduce to (14.95) upon setting all $n'_\ell = 0$ (there is one excess $v$). This is because of the initial hypothesis that at least one vertex of $\mathscr{G}$ is attached to the surface.

Instead of having $\mathscr{G}$ grafted to a linear surface, one may consider tying the network in a wedge of opening angle $\alpha \neq \pi$ by means of an extra $\hat{\ell}$-leg vertex. Since the wedge can be transformed back on the half plane through the conformal mapping $w(z) = z^{\pi/\alpha}$ [22], this geometry leads only to a minor modification [81] of the previous result (14.142):

$$\gamma_{\mathscr{G}}(\alpha) = \gamma_{\mathscr{G}}(\pi) - v \left( \frac{\pi}{\alpha} - 1 \right) \Delta'_\ell \tag{14.143}$$

Special cases of these formulae were obtained prior to [81] by Cardy [22], and yet others were conjectured by Guttmann and Torrie [82].

### 14.6.4 Standard Exponents of Polymer Physics

Standard exponents describing surface critical behaviour can be defined in analogy with those valid in the bulk. However, these can all be derived from the watermelon exponents (14.139) by using the network relation (14.142) and standard scaling relations. We focus here on the ordinary transition.

Consider as an example the exponent $\eta_\parallel$ describing the decay of the spin-spin correlation function along the surface. This is related to the conformational exponent $\gamma_1$ of a chain with one extremity tied to the surface and the other belonging to the bulk, through the scaling relation [31]

$$2\gamma_1 = \gamma + v(2 - \eta_\parallel) \tag{14.144}$$

Now, $\gamma_1$ is a special case of (14.142) with $n_\ell = n'_\ell = \delta_{\ell,1}$, giving $\gamma_1 = v(2 - \Delta_1 - \Delta'_1)$. Isolating $\eta_\parallel$ reveals that it belongs to the Kac table:

$$\eta_\parallel = \begin{cases} 2h_{1,2} \text{ for the dense } O(n) \text{ model} \\ 2h_{2,1} \text{ for the dilute } O(n) \text{ model} \\ 2h_{1,3} \text{ for the Potts model} \end{cases} \tag{14.145}$$

as first conjectured by Cardy [22].

## 14.6.5 Ordinary Transition in Fully Packed Loop Models

As mentioned in section 14.6.1, the special surface transition is absent in the dense $O(n)$ model. This agrees with physical intuition: since each edge has a finite probability of being covered by a monomer, it is redundant to try to attract the loops to the surface by enhancing the fugacity of surface monomers. In analogy, one would expect that fully packed loops are unable to sustain a special transition.

On the other hand, the ordinary transition for fully packed loop models does exist. It can be investigated [83] by adapting the Coulomb gas analysis of sections 14.3.5 and 14.6.2 to the vectorial setup of section 14.5. To this end, we focus on the FPL model on the hexagonal lattice and the FPL$^2$ model on the square lattice.

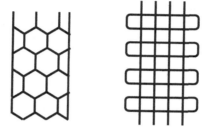

**Fig. 14.14** Appropriate modifications of the hexagonal (left panel) and square (right panel) lattices for the definition of the surface versions of the FPL and FPL$^2$ models. The top and bottom of the figures are identified to make an annular geometry.

The vectorial Coulomb gas treatment of section 14.5 depended crucially on the lattice being bipartite and of constant coordination number. These features must therefore be maintained when defining the surface geometry of the FPL and FPL$^2$ models. In particular, the FPL model cannot be defined on the lattice shown in Fig. 14.13. The appropriate choices of lattices are shown in Fig. 14.14. Note that in both cases, the corresponding transfer matrix adds two rows of vertices at a time.

For simplicity, we henceforth limit the discussion to the boundary FPL$^2$ model, defined on an annulus of *even* width $L$. The case $L = 4$ is shown in Fig. 14.14.

Note that if the vertical edges of a given time-slice are labelled alternatingly as even and odd, the following quantity is strictly conserved by the transfer matrix:

$$Q = \frac{1}{2} \left[ (\text{even}_b - \text{odd}_b) - (\text{even}_g - \text{odd}_g) \right] \tag{14.146}$$

where, e.g., $\text{even}_b$ means the number of even-labelled vertical edges covered by black loops. Accordingly, critical exponents are labelled by three indices, viz., $Q$ and the number of watermelon strands of each colour ($\ell_b$ and $\ell_g$). For $L$ even, all three indices must have the same parity (even or odd).

Using again the argument of section 14.3.5, the spontaneous vector magnetic flux $\mathbf{m}_0$ is obtained by matching the known central charge (14.131) to (14.75). After the change of basis (14.115) which makes the action diagonal, the result reads

$$\widetilde{\mathbf{m}}_0 = \left( \frac{e_b}{1 - e_b}, 0, \frac{e_g}{1 - e_g} \right) \tag{14.147}$$

where we have used the usual parameterisation (14.117).

Consider first the case $Q = 0$ and $\ell_b, \ell_g$ even. Then the flux is increased by the magnetic charges of the watermelon strands

$$\mathbf{m}_{\ell_b,\ell_g} = \frac{\ell_b}{2}(\mathbf{A} - \mathbf{B}) + \frac{\ell_g}{2}(\mathbf{C} - \mathbf{D}) \tag{14.148}$$

which reads in coordinates $\widetilde{\mathbf{m}}_{\ell_b,\ell_g} = (-\ell_b, 0, -\ell_g)$. This excitation multiplies the partition function by $q^{-\Delta c/24}$ and we can identify the critical exponent $\Delta_{\ell_b,\ell_g;0} = -\frac{\Delta c}{24}$. Recalling the one-colour boundary exponents $\Delta_\ell^O$ of (14.139), this result can be expressed as

$$\Delta_{\ell_b,\ell_g;0}(e_b, e_g) = \Delta_{\ell_b}^O(e_b) + \Delta_{\ell_g}^O(e_g). \tag{14.149}$$

Thus, in this sector the two loop species do not interact in the continuum limit.

Consider next the case $Q = 1$ and $\ell_b, \ell_g$ odd. The defect magnetic charge is now

$$\mathbf{m}_{\ell_b,\ell_g} = \frac{\ell_b - 1}{2}(\mathbf{A} - \mathbf{B}) + \frac{\ell_g - 1}{2}(\mathbf{C} - \mathbf{D}) + (\mathbf{C} - \mathbf{B}) \tag{14.150}$$

which reads in coordinates $\widetilde{\mathbf{m}}_{\ell_b,\ell_g} = (-\ell_b, -2, -\ell_g)$. The critical exponent is then

$$\Delta_{\ell_b,\ell_g;1}(e_b, e_g) = \Delta_{\ell_b}^O(e_b) + \Delta_{\ell_g}^O(e_g) + \delta(e_b, e_g) \tag{14.151}$$

where the additional contribution

$$\delta(e_b, e_g) = \frac{2g_2}{\pi} = \frac{(1 - e_b)(1 - e_g)}{(1 - e_b) + (1 - e_g)} \tag{14.152}$$

comes from the second height component.

To settle the general case, we note that to obtain sectors with higher charge, $Q$ can be increased by two units by a succession of four consecutive vertical edges with alternating loop colors. This corresponds to a height defect $\mathbf{m} = -2(\mathbf{A} + \mathbf{B}) = (0, -4, 0) = \widetilde{\mathbf{m}}$, and in general for even $Q$ $\mathbf{m}_Q = \widetilde{\mathbf{m}}_Q = (0, -2Q, 0)$. The final result for any values of $(s_1, s_2; Q)$ can thus be written succinctly as

$$\Delta_{\ell_b,\ell_g;Q}(e_b, e_g) = \Delta_{\ell_b}^O(e_b) + \Delta_{\ell_g}^O(e_g) + Q^2 \delta(e_b, e_g). \tag{14.153}$$

This expression has been checked numerically to a very good precision [83].

Finally, let us consider the special case where grey strands are ignored and grey loops are assigned a trivial fugacity, $n_g = 1$. This gives for the black $\ell$-leg boundary watermelon exponents (with $Q = \ell \bmod 2$)

$$\Delta_\ell = \begin{cases} \frac{1}{4}g\ell^2 - \frac{1}{2}(1-g)\ell & \text{for } \ell \text{ even} \\ \frac{1}{4}g\ell^2 - \frac{1}{2}(1-g)\ell + \frac{2g}{3g+2} & \text{for } \ell \text{ odd} \end{cases} \qquad (14.154)$$

where $g = 1 - e_b$. This should be compared with (14.133) and (14.139).

## 14.6.6 Conformal Boundary Loop Model

Very recently, Jacobsen and Saleur [84] have studied a so-called *conformal bound-ary loop* (CBL) model in which a continuous parameter $n_1$ permits one to vary the boundary condition. In sharp contrast with the model of section 14.6.1 this bound-ary condition remains conformal for any real value of $n_1$, i.e., any $n_1$ constitutes a boundary RG fixed point and gives rise to a distinct critical exponent of the associ-ated boundary condition changing operator.

For definiteness, consider the loop model (14.59) based on the critical Potts model, with each loop having the fugacity $n$, and defined on the annulus. Now as-sign to each loop touching *at least once* the left rim a different fugacity $n_1$. Alge-braically this situation is closely related to the so-called blob algebra—subsequently often known as the one-boundary Temperley-Lieb algebra—originally introduced by Martin and Saleur [85]. Obviously it is possible to apply this boundary condition also to other types of loop models.

Physically one can consider the CBL model as an O($n$)-type model in which the bulk spins belong to $\mathbb{R}^n$, while the boundary spins have been constrained to live in a smaller space $\mathbb{R}^{n_1}$ (this makes sense also for $n_1 > n$, by analytic continuation). Alternatively, the same developments which led to (14.59) establish the equivalence with a Potts model in which bulk spins can take $Q = n^2$ states, and boundary spins $Q_1 = nn_1$ states.

One central claim of [84] is that the operator that changes the boundary condi-tions from free ($n = n_1$) to the CBL boundary conditions just described ($n \neq n_1$) has conformal weight $h_{r_1,r_1}$, where we have parameterised

$$n = 2\cos\gamma$$
$$n_1 = \frac{\sin[(r_1+1)\gamma]}{\sin(r_1\gamma)} \qquad (14.155)$$

and $\gamma = \frac{\pi}{m+1}$ defines the central charge and the conformal weights through (14.31). The parameter $r_1 \in (0, m+1)$ is in general a real number. When $r_1$ and $m$ are inte-gers, the above statement can be rigorously derived from the representation theory of the corresponding XXZ spin chain with boundary terms. Another check is when $Q_1 = 1$ (i.e., $n = 1/n_1$, or $r_1 = m - 1$); indeed it is a well-known result by Cardy [33] that the operator that changes the Potts model boundary conditions from free to fixed is $\phi_{m-1,m-1} = \phi_{1,2}$. Finally, the statement for arbitrary $r_1$ and $m$ has been subjected to extensive numerical tests in [84].

The above result can be generalised to the watermelon topology where $\ell$ non-contractible loops wrap around the periodic direction of the annulus. A careful study of the transfer matrix structure reveals that in this case one needs to distinguish two possible situations, or sectors: either the leftmost non-contractible loop is constrained to touch the left rim at least one (blobbed sector), or it is constrained to never touching it (unblobbed sector). The corresponding conformal weight is then

$$\Delta_\ell^O(n, n_1) = h_{r_1, r_1 \pm \ell} \tag{14.156}$$

where the upper (resp. lower) sign is for the blobbed (resp. unblobbed) sector.

The formula (14.156) has subsequently been derived for the O($n$) model on random lattices by Kostov [86].

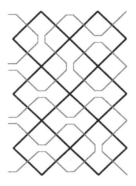

**Fig. 14.15** Configuration of loops (in green) on the annulus, with Neumann (resp. Dirichlet) boundary conditions on the left (resp. right) rim. The original Potts spins and their duals live on the black lattice.

Obviously, the result (14.156) has many applications. One of these is to identify the operator that changes the O($n$) model boundary condition from Dirichlet to Neumann. By definition, these names refer to the equivalent height model. Therefore, Dirichlet boundary conditions mean that loops are reflected off the boundary, while Neumann conditions mean that loop strands terminate perpendicularly on the boundary. A configuration with Neumann boundary conditions on the left rim of the annulus is shown in Fig. 14.15. Note that it involves half loops beginning and ending on the left rim, and by definition these must have unit weight. By connecting the termination points of these half loops two by two, it is seen that this situation is actually equivalent to the CBL model with parameter $n_1 = 1$. Thus, the required operator is $\phi_{\text{DN}} = \phi_{m/2, m/2}$ of weight

$$h_{\text{DN}} = h_{m/2, m/2} = \frac{m^2 - 4}{16m(m+1)}. \tag{14.157}$$

The result (14.156) holds true also for the hexagonal-lattice model (14.60), provided that one performs the usual swap of the indices when going from the dense to the dilute phase [87].

Note that when $n_1 = n$, we have $r_1 = 1$ from (14.155), and so (14.156) reproduces (14.140) for the ordinary transition, as it should.

## 14.6.7 Generalised Special Transition

The hexagonal-lattice loop model (14.60) that admitted us to access the special transition in section 14.6.1 can be modified in a natural way so as to accommodate the CBL type boundary conditions of section 14.6.6. To this end, we consider again the annular geometry of Fig. 14.13 with the left boundary being distinguished. A pair of consecutive boundary edges can be either empty, or carry a surface monomer, or carry a *marked* surface monomer.

Loops containing only bulk and surface monomers are called bulk loops, while loops containing at least one marked surface monomer are referred to as boundary loops. Note that this terminology differs slightly from that used in section 1.6.6, since it is now possible for a bulk loop to touch the surface, provided that it contains only unmarked surface monomers. While this may appear physically slightly unnatural, it is necessary in order to make contact with the relevant algebraic framework, viz., the *dilute* one-boundary Temperley-Lieb algebra.

The weight of a bulk (resp. boundary) loop is $n$ (resp. $n_1$). An unmarked (resp. marked) surface monomer comes with a weight $K_s$ (resp. $K_{s,1}$). The physically natural situation, in which boundary loops are simply those that touch the surface, is recovered upon setting $K_s = 0$. Depending on the parameters, this may renormalise towards any of the fixed points to be discussed below.

The following parameterisation turns out useful:

$$n = -2\cos(4\Phi)$$
$$n_1 = -\frac{\sin[4(\kappa - 1)\Phi]}{\sin(4\kappa\Phi)}. \tag{14.158}$$

The dilute critical point (14.61) is then obtained by setting $1/K = 2\cos(\Phi)$. One can show [88] that the model defined above admits two new integrable points, that we shall refer to as the *generalised ordinary* transition and the *generalised special* transition. For the former $(K_s, K_{s,1}) = (K_s^O, K_{s,1}^O)$ with

$$(K_s^O)^2 = \frac{\sin[(2\kappa - 1)\Phi]}{2\cos(\Phi)\sin(2\kappa\Phi)} \qquad (K_{s,1}^O)^2 = \frac{\cos(2\kappa\Phi)\tan(\Phi)}{\sin(2\kappa\Phi)}, \tag{14.159}$$

and for the latter $(K_s, K_{s,1}) = (K_s^S, K_{s,1}^S)$ with

$$(K_s^S)^2 = \frac{\cos[(2\kappa - 1)\Phi]}{2\cos(\Phi)\cos(2\kappa\Phi)} \qquad (K_{s,1}^S)^2 = -\frac{\sin(2\kappa\Phi)\tan(\Phi)}{\cos(2\kappa\Phi)}. \qquad (14.160)$$

It is interesting to examine a few special cases of (14.159)–(14.160). For $n_1 = n$, we have $\kappa = -1$, and boundary loops are indistinguishable from bulk loops. The weight of a pair of boundary monomers is therefore $y^2 \equiv (K_s)^2 + (K_{s,1})^2$. Using (14.159) this gives $y^2 = K^2$, which is the usual ordinary transition. Meanwhile, (14.160) gives $y^2 = (2 - n)^{-1/2}$ as in (14.137), which is the usual special transition.

The case $n_1 = 1$, or $\kappa = \frac{1}{2}$, corresponds to Neumann boundary conditions. Indeed, when boundary loops are weighted trivially, they might as well be transformed into half loops, as in Fig. 14.15. The generalised ordinary transition (14.159) then corresponds to $K_s^O = 0$ and $K_{s,1}^O = 1$, i.e., only boundary loops are allowed at the surface, and are weighted trivially.

Finally, for $n_1 = 0$, or $\kappa = 1$, boundary loops are forbidden, and the weight of marked surface monomers is therefore immaterial. As expected one has then $K_s^O = K$, and $K_s^S = (2 - n)^{-1/4}$ in agreement with (14.137).

The critical exponents corresponding to these generalised surface transitions can be identified from numerical diagonalisation of the transfer matrix [88]. To state the results, it is convenient to go back to the parameterisation (14.155). For the ordinary case (14.159) one finds

$$\Delta_\ell^O(n, n_1) = h_{r_1 \pm \ell, r_1} \qquad (14.161)$$

where we recall that the upper (resp. lower) sign refers to the blobbed (resp. unblobbed) sector. This agrees as expected with (14.156) after swapping the indices (since we are here in the dilute phase of the O($n$) model). For the special case (14.160) one finds instead

$$\Delta_\ell^S(n, n_1) = h_{r_1 \pm \ell, 1 + r_1} \qquad (14.162)$$

which is a nice generalisation of (14.141).

### 14.6.8 Two-Boundary CBL Model

The conformal boundary loop (CBL) model of section 14.6.6 can be generalised to the case where both boundaries of the annulus are distinguished. In this two-boundary CBL model, bulk loops have a weight $n$, while boundary loops touching only the left (resp. right) boundary have weight $n_1$ (resp. $n_2$), and loops touching both boundaries have weight $n_{12}$.

This model—which is related to the so-called two-boundary Temperley-Lieb algebra—has been the subject of several recent studies (see [89, 90, 91, 92] and references therein). It is equivalent to a Potts model in which bulk spins have $Q = n^2$ states, while spins on the left (resp. right) boundary are constrained to a smaller number $Q_1 = nn_1$ (resp. $Q_2 = nn_2$) of states, of which there are $Q_{12} = nn_{12}$ common states.

The following parameterisation turns out to be instrumental for further study:

$$n = 2\cos\gamma$$

$$n_1 = \frac{\sin[(r_1+1)\gamma]}{\sin(r_1\gamma)}$$

$$n_2 = \frac{\sin[(r_2+1)\gamma]}{\sin(r_2\gamma)}$$

$$n_{12} = \frac{\sin[(r_1+r_2+1-r_{12})\frac{\gamma}{2}]\sin[(r_1+r_2+1+r_{12})\frac{\gamma}{2}]}{\sin(r_1\gamma)\sin(r_2\gamma)} \qquad (14.163)$$

The full meaning of this parameterisation only becomes clear within the representation theory of the underlying algebra.

From a CFT point of view, distinguishing both rims of the annulus is expected to be described by the fusion (OPE) of the "free to one-boundary CBL" boundary condition changing operators which are responsible for the weights $n_1$ and $n_2$ on either rim. However, a detailed study [92] unravels a number of technical subtleties, mainly having to do with the possibility of a loop of weight $n_{12}$ touching both rims.

Critical exponents $\Delta_\ell^{\alpha_1\alpha_2}$ depend on all the weights (14.163), on the number of non-contractible loops $\ell$, and on the sector labels $\alpha_1$ and $\alpha_2$. The label $\alpha_1 = b$ (resp. $\alpha_1 = u$) if the leftmost non-contractible loop is constrained to touching (resp. to never touching) the left rim; these two possibilities are referred to as the blobbed (resp. unblobbed) sector. The label $\alpha_2$ similarly constrains the behaviour of the rightmost non-contractible loop. Note that we have supposed that the system size $L$ is even (and so the number of non-contractible loops is always even).

We give here only the final results for the leading critical exponents in each sector [92]:

$$\Delta_0 = h_{r_{12},r_{12}}$$

$$\Delta_\ell^{bb} = h_{r_1+r_2-1,r_1+r_2-1+\ell}$$

$$\Delta_\ell^{bu} = h_{r_1-r_2-1,r_1-r_2-1+\ell}$$

$$\Delta_\ell^{ub} = h_{-r_1+r_2-1,-r_1+r_2-1+\ell}$$

$$\Delta_\ell^{uu} = h_{-r_1-r_2-1,-r_1-r_2-1+\ell} \qquad (14.164)$$

Note that changing the sector label $\alpha_k$ simply results in changing the sign of the parameter $r_k$ (for $k = 1, 2$).

We refer the reader to [92] for details on how these expressions are derived.

## 14.7 Exact Partition Functions

Writing down exact partition functions $Z$ in the continuum limit is a very strong tool for revealing the complete operator content of the underlying theory. This was first pointed out by Cardy [23] who worked on the torus, where the possible forms of $Z$ are very strongly constrained by modular invariance. Thus, for the three-state Potts model, Cardy was able to determine $Z$ and the complete operator content from a prior knowledge of just the central charge and a few scaling dimensions.

Most of the early efforts [23, 29] concentrated on extending this approach to all the unitary minimal models, and an extensive set of modular invariant $Z$ on the torus were unravelled. The question however soon arose how to adapt this approach to non-minimal models [39], and in particular how to make the connection [40] with the Potts and O($n$) model in their loop model formulation, reviewed in section 14.3.

In section 14.2.7 we have already seen how a modular invariant $Z$ of the free boson (14.42) is constructed by summing over all possible frustrations (magnetic charges) $m$. This led to the form (14.48) that revealed the electromagnetic operator content (14.50). However, this is valid only for a truly free field, with Gaussian action (14.42) and central charge $c = 1$. It does not apply to the Potts and O($n$) models whose CG action (14.69) also contains the boundary term (14.67), linked to the background electric charge $e_0$ and the modification (14.73) of $c$.

In section 14.7.1 we show how to remedy this shortcoming on a torus, following the original work of Di Francesco, Saleur and Zuber [40], with a few subsequent improvements [93, 94]. We also present some applications to polymers, following Duplantier and Saleur [14].

Exact continuum limit partition functions on the annulus are given in section 14.7.2. Due to the two possible ways of orienting the annulus, these give access to both the bulk and the boundary operator content. The multiplicities with which the various terms appear in $Z$ are derived from a combinatorial argument by Richard and Jacobsen [95], which has the advantage of being readily generalisable to more complicated geometries [94, 84, 91]. The applications to polymers are due to Cardy [18].

Finally, we treat the CBL model in section 14.7.3, its two-boundary extension in section 14.7.4, and the fully packed loop model FPL$^2$ in section 14.7.5.

### 14.7.1 Toroidal Geometry

Recall first the expression (14.46) for the free boson partition function $Z_{m,m'}(g)$ at coupling $g$ and fixed frustrations $m, m'$. Summing this over frustrations which are multiples of $2\pi f$ defines the coulombic partition function

$$Z_c[g, f] = f \sum_{m,m' \in f\mathbb{Z}} Z_{m,m'}(g) = \frac{1}{\eta\bar{\eta}} \sum_{e \in \mathbb{Z}/f, \, m \in f\mathbb{Z}} q^{\Delta_{e,m}} \bar{q}^{\bar{\Delta}_{e,m}}. \qquad (14.165)$$

where we denote by $\Delta_{e,m}$ the conformal weights with respect to the $c = 1$ theory, i.e., without the correction coming from the background electric charge $e_0$:

$$\Delta_{e,m} + \bar{\Delta}_{e,m} = \frac{e^2}{2g} + \frac{g}{2}m^2$$

$$\Delta_{e,m} - \bar{\Delta}_{e,m} = em. \tag{14.166}$$

The free field admits a duality transformation that exchanges electric and magnetic charges

$$g \to \frac{4}{g}, \qquad e \to 2m, \qquad m \to \frac{e}{2} \tag{14.167}$$

leading to the following symmetries

$$Z_c[g,f] = Z_c[g^{-1}, f^{-1}] = Z_c[gf^2, 1]. \tag{14.168}$$

Consider now the O(n) model on a torus in its formulation as an oriented loop model, or a height model. When $n$ is distributed locally as complex Boltzmann weights in the usual way, the partition function is simply $Z_c[g, 1/2] = Z_c[g/4, 1]$. This however assigns a wrong weight $\bar{n} = 2$ to any loop which is non-homotopic to a point, since by self-avoidance each of its oriented versions makes as many left as right turns.

We therefore consider more carefully oriented loops of non-trivial homotopy. Let there be $\mathcal{N}$ such loops. Clearly, they all belong to the same homotopy class, up to the choice of their global orientation which can be described by a sign $\varepsilon_i = \pm 1$. The homotopy class can be defined by giving the (signed) winding numbers $n_1$ and $n_2$ with respect to the two principal cycles of the torus. One then has $|n_1| \wedge |n_2| = 1$. Using this and a trigonometric identity yields

$$\left(2\cos(\pi e_0)\right)^{\mathcal{N}} = \prod_i \sum_{\varepsilon_i = \pm 1} e^{i\pi e_0 \varepsilon_i} = \sum_{\{\varepsilon_i = \pm 1\}} \cos\left(\pi e_0 \sum_i \varepsilon_i\right) \tag{14.169}$$

from which one deduces that the correctly weighted modular invariant partition function is

$$\hat{Z}[g, e_0] = \sum_{m,m' \in \mathbb{Z}} Z_{m,m'}\left(\frac{g}{4}\right) \cos(\pi e_0 m \wedge m'). \tag{14.170}$$

Computing this term by term in the $m$ summation leads to the central result

$$\hat{Z}[g, e_0] = \frac{1}{\eta\bar{\eta}} \left\{ \sum_{p \in \mathbb{Z}} (q\bar{q})^{\Delta_{e_0 + 2p,0}} + \sum_{p \in \mathbb{Z}} \sum_{m>0} \sum_{k>0}' \Lambda(m,k) q^{\Delta_{2p/k,m/2}} \bar{q}^{\bar{\Delta}_{2p/k,m/2}} \right\} \tag{14.171}$$

where we have singled out the $m = 0$ term. The prime on the sum over $k$ indicates the constraints $k|m$ and $p \wedge k = 1$. Note that (14.171) has the correct form to enable a physical interpretation. Clearly, $m$ is the magnetic charge corresponding to the number of non-contractible polymer strands that propagate along the time direction. The indices $k$ and $p$ control how fast the strands wind around the space direction, and give access to subdominant operators. Finally, the coefficients $\Lambda(m,k)$ count the multiplicities of each operator with $m > 0$.

The original paper [40] provides an operational way of computing the $\Lambda(m,k)$ from the prime decomposition of $m$ and $k$. Unfortunately, this is quite cumbersome to apply, even for moderately small values of $m, k$. An elegant closed-form expression which brings out the number theoretical content of $\Lambda(m,k)$ was derived much later by Read and Saleur [93]:

$$\Lambda(m,k) = 2 \sum_{d>0:d|m} \frac{\mu\left(\frac{k}{k\wedge d}\right)\phi\left(\frac{m}{d}\right)}{m\,\phi\left(\frac{k}{k\wedge d}\right)} \cos(2\pi d e_0). \tag{14.172}$$

Here, $k \wedge d$ denotes the greatest common divisor of $k$ and $d$, and $\mu$ and $\phi$ are respectively the Möbius and Euler's totient function. The Möbius function $\mu$ is defined by $\mu(x) = (-1)^r$, if $x$ is an integer that is a product $x = \prod_{i=1}^{r} p_i$ of $r$ distinct primes, $\mu(1) = 1$, and $\mu(x) = 0$ otherwise or if $x$ is not an integer. Similarly, Euler's totient function $\phi(x)$ is defined for positive integers $x$ as the number of integers $x'$ such that $1 \le x' \le x$ and $x \wedge x' = 1$. Note that in (14.172) we may also write $\cos(2\pi d e_0) = T_{2d}(\bar{n})$, where $T_\ell$ is the $\ell$'th order Chebyshev polynomial of the first kind, and $\bar{n}$ is the weight of a non-contractible loop as usual.

The expression (14.172) has been rederived by Richard and Jacobsen [94] following a completely different route. Indeed, these authors view $\Lambda(m,k)$ as eigenvalue amplitudes with respect to a suitably defined transfer matrix for the $O(n)$ model on a torus of width and length which are a *finite* number of lattice spacings (but wide enough to accommodate $m$ non-contractible strands), on an arbitrary regular lattice, and at an arbitrary temperature. By an intricate, but completely rigorous, combinatorial argument they arrive at the same formula (14.172). We shall illustrate their method for a much simpler case in section 14.7.2.

Very recently, a more concise and equally rigorous derivation of (14.172) was provided by Dubail et al. [92] through the construction of the Jones-Wenzl projectors of the periodic Temperley-Lieb algebra.

Note that the $\Delta(m,k)$ are not integers for general values of the loop fugacity $n$. This was to be expected in view of the non-minimality of the underlying CFT.

The operator content of (14.171) is readily extracted [40]. We state the results in terms of the true ($c < 1$) conformal weights

$$h = \Delta - \frac{e_0^2}{4g} = \Delta + \frac{c-1}{24} \tag{14.173}$$

and their Kac table values

$$h_{r,s} = \frac{(gr-s)^2 - (1-g)^2}{4g} \tag{14.174}$$

where we have taken the notation appropriate for the dilute phase of the $O(n)$ model. The terms with $m = 0$ contain the thermal series

$$\Delta_t(\ell) = 2h_{1,1+2\ell} = \frac{2\ell(\ell+1)}{g} - 2\ell \tag{14.175}$$

of which the principal member $\ell = 1$ is (14.91). Similarly, the terms with $m > 0$ contain precisely the $\ell$-leg watermelon exponents (14.81).

Note also that (14.171) can be rewritten in the form (14.41). Once again, the coupling constants $n_{h,\bar{h}}$ need not be integers in general.

When $g$ and/or $e_0$ is rational, (14.171) simplifies due to multiple cancellations, and one can in some cases derive simpler expressions. These can in turn be compared to those derived for the minimal models [23, 29].

Turning now to the Potts model, the derivation is almost identical, with one important modification. As we have already remarked in section 14.4.3 it may happen that Potts clusters wrap around (at least) *two* independent non-contractible cycles on the torus; this is closely linked to the magnetic exponent $\Delta_{\mathrm{m}}$. Care must be taken to give such clusters their correct weight $Q$, rather than 1. The result is simply that (14.171) must be replaced by

$$\hat{Z}[g, e_0] + \frac{1}{2}(Q-1)\hat{Z}\left[g, \frac{1}{2}\right]. \tag{14.176}$$

We end this subsection by giving some applications to polymers, following Duplantier and Saleur [14]. First note that (14.170) still permits one to distinguish the weights of contractible and non-contractible loops, which are respectively $n = -2\cos(\pi g)$ and $\bar{n} = 2\cos(\pi e_0)$. In other words, we need not take $e_0 = 1 - g$.

There are obviously several interesting ways of taking the polymer (SAP) limit. To obtain polymers of indeterminate homotopy, one first sets $\bar{n} = n$ and then lets $n \to 0$. To have contractible polymers only, one first sets $\bar{n} = 0$ and then lets $n \to 0$. Finally, to have non-contractible polymers only, one first sets $n = 0$ and then lets $\bar{n} \to 0$. In all cases, a derivative with respect to the fugacity is needed before taking the last limit, in order to single out configurations having a single loop—otherwise, the surviving configuration will have zero loops and give rise to a trivial partition function, just as in the discrete model (14.60).

To illustrate this, consider the case of contractible polymers. Setting $\bar{n} = 0$ in (14.170) gives

$$\hat{Z}\left[g, \frac{1}{2}\right] = 2\sum_{m,m'\in 4\mathbb{Z}} Z_{m,m'}\left(\frac{g}{4}\right) - \sum_{m,m'\in 2\mathbb{Z}} Z_{m,m'}\left(\frac{g}{4}\right) = \frac{1}{2}\left(Z_{\mathrm{c}}[4g, 1] - Z_{\mathrm{c}}[g, 1]\right),$$
$$\tag{14.177}$$

where we have used (14.165) and (14.168). If we now set simply $n = 0$, one recovers in the dilute (resp. dense) case $g = \frac{3}{2}$ (resp. $g = \frac{1}{2}$), by using Euler's pentagonal identity (resp. (14.168))

$$\hat{Z}\left[\frac{3}{2}, \frac{1}{2}\right] = 1, \qquad \hat{Z}\left[\frac{1}{2}, \frac{1}{2}\right] = 0. \tag{14.178}$$

Both these results are trivial, as expected. By contrast, if one takes the derivative $\partial/\partial n$ before setting $n = 0$, a non-trivial result is obtained. For dense polymers this reads explicitly, after some algebra,

$$\frac{\partial}{\partial n}\hat{Z}\left[g,\frac{1}{2}\right]\bigg|_{g=\frac{1}{2}} = -\frac{1}{4\pi}\eta^2(q)\eta^2(\bar{q})\log(q\bar{q}). \tag{14.179}$$

## 14.7.2 Annular Geometry

We now consider instead the geometry of an $L \times M$ annulus with free $\times$ periodic boundary conditions, as defined in section 14.3.5. Recall that in the preceding subsection we have constructed the continuum limit partition function $Z$ starting from the explicit weights of the microscopic model—and invoking modular invariance—and extracted the operator content as a corollary at the end of the calculation. Let us instead now work the other way around, starting from the known operator content, viz., the watermelon exponents (14.139) at the ordinary surface transition.

According to (14.55) we have

$$Z \equiv Z_{\text{ff}}(q) = \sum_h n_h \chi_{(c,h)}(q) \tag{14.180}$$

where the sum is over the boundary scaling dimensions $h$, $\chi_{(c,h)}(q)$ is the generic character (14.28), and the modular parameter $q = \exp(i\pi\tau) = \exp(-\pi M/L)$. The degeneracy factor $n_h$ states how many times a given character appears in the partition function, and as usual for non-minimal theories it needs not in general be an integer. We omit in the following the subscript ff which reminds us that the boundary conditions on both rims of the annulus are free.

As the watermelon operators are indexed by their number of legs $\ell$, we may replace the sum over $h$ by one over $\ell$. Below we shall give a combinatorial argument that the correct degeneracy factor is

$$n_\ell = \frac{\sin\left((1+\ell)\pi e_0\right)}{\sin(\pi e_0)} = U_\ell\left(\frac{\bar{n}}{2}\right), \tag{14.181}$$

where $U_\ell(x)$ is the $\ell$'th order Chebyshev polynomial of the second kind. Note that $n_\ell$ depends only on the weight $\bar{n} = 2\cos(\pi e_0)$ of a non-contractible loop, which may in general be different from that of a contractible loop, $n = -2\cos(\pi g)$.

Accepting for the moment (14.181), we then have the central result

$$Z[g,e_0] = \frac{q^{-c/24}}{P(q)}\sum_{\ell\in\mathbb{Z}}\frac{\sin\left((1+\ell)\pi e_0\right)}{\sin(\pi e_0)}q^{\frac{g\ell^2}{4}-\frac{(1-g)\ell}{2}} \tag{14.182}$$

which is the analogue of (14.171) on the torus. The attentive reader may object that 1) the expansion (14.180) should not be over generic characters, but the degenerate ones

$$K_{r,s} = \frac{q^{h_{r,s}} - q^{h_{r,-s}}}{q^{c/24}P(q)}, \tag{14.183}$$

and 2) the sum in (14.182) should be over $\ell \geq 0$ and not $\ell \in \mathbb{Z}$. While these observations are certainly correct, a little analysis shows that taking into account 1) and 2) leads to exactly the same result (14.182).

The expression (14.182) was first obtained by Saleur and Bauer [80], using techniques of integrability and quantum groups. It has later been rederived and discussed by Cardy from a Coulomb gas point of view [18].

We now turn to the derivation of (14.181). One line of reasoning is to invoke the correspondence between the oriented loop model and an $SU(2)$ spin chain Hamiltonian, as in [80]. The number of non-contractible loop strands $\ell$ is then the conserved spin $S$ of the chain. For each value $S = \ell$ there is a degeneracy corresponding to the $(2\ell + 1)$ corresponding values of $S^z$. To be more precise, the symmetry of the spin chain is not classical $SU(2)$ but the quantum algebra $SU(2)_q$, with deformation parameter[11] given by $\bar{n} = q + q^{-1}$. The degeneracy factor is therefore the $q$-deformed number $(2\ell + 1)_q$, by definition equal to $n_\ell$ in (14.181).

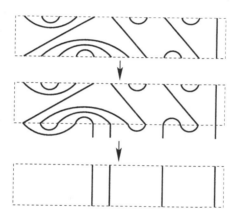

**Fig. 14.16** Construction of compatible states (see text).

A very different argument was given by Richard and Jacobsen [95] who used elementary combinatorics. The lattice partition function can be constructed from a transfer matrix $T$ that acts on connectivity states between two time slices at $t = 0$ and $t = t_0$. An example of a *state* is shown in Fig. 14.16.a; it consists of *arcs* that connect pairs of points within one time slice, and $\ell$ *strings* that connect one point on either time slice. Upon cutting all the strings, a state is transformed into a pair of *reduced states*. $T$ transforms a state from time $t_0$ to $t_0 + 1$ by acting on the upper time slice. One may write

$$Z = \langle v | T^M | u \rangle = \sum_i n_i \lambda_i^M \tag{14.184}$$

---

[11] This $q$ should not be confused with the modular parameter $q$ used elsewhere in the text.

where $|u\rangle$ is an initial state that identifies the two time slices, and $\langle v|$ is a final state that reglues the two time slices and imposes the correct powers of $n$ and $\bar{n}$ according to the loops thus formed. As $Z$ is not a trace, each eigenvalue $\lambda_i$ of $T$ has a corresponding amplitude $n_i$.

Note that $\ell$ cannot increase under the action by $T$, whence $T$ is upper block-triangular with respect to $\ell$. Therefore, each eigenvalue of $T$ is an eigenvalue of one of the blocks on the diagonal denoted $T_\ell$. Furthermore, $T$ cannot change the reduced state corresponding to the lower time slice, so each $T_\ell$ is block-diagonal with a number of identical blocks given by the number of reduced states with $\ell$ strings. In particular, $n_i = n_{\ell(i)}$ and the eigenvalues can be labeled by $\ell$, so we may write

$$Z = \sum_{\ell=0}^{L} n_\ell \sum_k \lambda_{\ell,k}^M. \qquad (14.185)$$

Define now $K_\ell = \text{Tr}(T_\ell^M)$ as a trace over reduced states, and $Z_j$ as the partition function constrained to having exactly $j$ non-contractible loops. To determine $n_\ell$ we must determine how many times each $K_\ell$ contributes to $Z$. Consider instead the inverse problem

$$K_\ell = \sum_{j=\ell}^{L} n(j,\ell) Z_j \bar{n}^{-j}. \qquad (14.186)$$

where $n(j,\ell)$ is the number of times a configuration with $j$ non-contractible loops occurs in the trace $K_\ell$. To determine it we depict a configuration contributing to $Z_j$ as a state $\mathscr{S}_j$, i.e., we suppress all internal loops and empty sites in the time slices; see Fig. 14.16.a. Then $n(j,\ell)$ is the number of $\ell$-string reduced states $\mathscr{R}_\ell$ that are *compatible* with $\mathscr{S}_j$, i.e., that are invariant when propagated through $\mathscr{S}_j$. A necessary condition is that $\mathscr{R}_\ell$ contains the same arcs as the upper time slice of $\mathscr{S}_j$; see Fig. 14.16.b. What remains is topologically equivalent to just $j$ strings; see Fig. 14.16.c. To ensure compatibility, these must be linked up by arcs so as to leave exactly $\ell$ non-enclosed strings. This is an easy counting problem with solution

$$n(L,\ell) = \binom{L}{(L-\ell)/2} - \binom{L}{(L-\ell)/2 - 1}. \qquad (14.187)$$

Finally inverting (14.186) gives the number of times $K_\ell$ appears in each $Z_j$, and since by definition each eigenvalue appears with unit amplitude in $K_\ell$, we can sum this over $j$ to obtain $n_\ell = U_\ell(\bar{n}/2)$ proving (14.181).

Once again, an alternative to this combinatorial method is furnished by the study of Jonez-Wenzl projectors of the Temperley-Lieb algebra. The degeneracy factors (or eigenvalue amplitudes) $n_\ell$ then appear as the Markov traces of the $\ell$-strand projectors. The reader is referred to [92] for further details.

We now return to the result (14.182). It may be rewritten [18] in terms of the conjugate modulus $\tilde{q} = \exp(-2\pi L/M)$, giving

$$Z[g,e_0] = \sqrt{\frac{2}{g}} \frac{\tilde{q}^{-c/12}}{P(\tilde{q}^2)} \sum_{m\in\mathbb{Z}} \frac{\sin\left(\frac{\pi(e_0+2m)}{g}\right)}{\sin(\pi e_0)} \tilde{q}^{\frac{(e_0+2m)^2}{2g} - \frac{(1-g)^2}{2g}}. \tag{14.188}$$

Note that when going from $q$ to $\tilde{q}$, the time and space directions have effectively been swapped, and so (14.188) pertains to a *cylinder* geometry (with free boundary conditions on the rims), meaning that the expansion is in terms of the *bulk* theory. More precisely, in view of the decomposition (14.41) for a *diagonal* theory, (14.180) is now replaced by

$$Z = \sum_h |b_h|^2 \chi_{(c,h)}(\tilde{q}^2), \tag{14.189}$$

where the sum is over the bulk conformal weights $h$, $\chi_{(c,h)}(q)$ is a bulk character, and $b_h$ is a matrix element with the boundary state corresponding to free boundary conditions at the rim. Note that there is no multiplicity $\Lambda(m,k)$, as in the first term of (14.171), since the free boundary conditions do not allow loops to wrap around the system in the *time*-like direction.

In particular setting $h = 0$ in (14.188) we can read off [18]

$$b_0^2 = -\sqrt{\frac{2}{g}} \frac{\sin(\pi/g)}{\sin(\pi g)} \tag{14.190}$$

for $\bar{n} = n$, from which the boundary entropy [96] can be determined as $\log b_0$.

As in the toroidal case, we end by giving some applications to polymers, following Cardy [18]. For $\bar{n} = n$, simply setting $n = 0$, one recovers in the dilute (resp. dense) case $g = \frac{3}{2}$ (resp. $g = \frac{1}{2}$), by using Euler's pentagonal identity (resp. simple algebra), that

$$Z\left[\frac{3}{2}, \frac{1}{2}\right] = 1, \qquad Z\left[\frac{1}{2}, \frac{1}{2}\right] = 0. \tag{14.191}$$

This should be compared with (14.178). Non-trivial results are obtained by singling out the $O(\bar{n})$ term, i.e., by taking a derivative before going to the limit. This gives for the dilute case

$$Z_1 = \prod_{r=1}^{\infty}(1-q^r)^{-1} \sum_{k\in\mathbb{Z}} k(-1)^{k-1} q^{\frac{3}{2}k^2-k+\frac{1}{8}} \sim q^{5/8} \tag{14.192}$$

and since $Z = 1$ this can be interpreted as the probability of having a single non-contractible loop. For the dense case the $O(\bar{n})$ term in $Z$ is similarly

$$Z_1 = q^{-1/24} \prod_{m=1}^{\infty} \left(1 - q^{m-\frac{1}{2}}\right)^2 \sim q^{-1/24}. \tag{14.193}$$

**Fig. 14.17** Continuum-limit view of the four different types of loops in the CBL model. In this figure the annulus has been conformally mapped to the plane, and the "left rim" referred to in the text has become the outer rim.

### 14.7.3 Conformal Boundary Loop Model

Continuum limit partition functions for the CBL model defined in section 14.6.6 have been written down by Jacobsen and Saleur [84]. In this context it is convenient to define a more general model in which bulk loops have fugacity $n$ or $\bar{n}$, and loops touching the left boundary have weight $n_1$ or $\overline{n_1}$, where in all cases the overline refers to non-contractible loops (i.e., loops that are not homotopic to a point). This is illustrated in Fig. 14.17.

We have seen above how the transfer matrix $T$ of any loop model on the annulus can be decomposed into blocks $T_\ell$ labeled by the number of non-contractible loops $\ell$. For the CBL model one may further decompose $T_\ell$ into the blobbed (resp. unblobbed) sector $T_\ell^{\rm b}$ (resp. $T_\ell^{\rm u}$) in which the leftmost non-contractible loop is required (resp. forbidden) to touch the left rim of the annulus. Indeed, since a non-contractible loop is conserved by definition, once it has been blobbed (i.e., touched the boundary) it cannot subsequently be unblobbed. Therefore, $T_\ell$ is upper block-triangular in the basis $\{|{\rm b}\rangle, |{\rm u}\rangle\}$ and the previous argument applies mutatis mutandis.

The CBL model contains the ordinary $O(n)$ loop model as the special case $n_1 = n$, but it is clear that its transfer matrix must contain many more states in order to produce the correct weights for $n_1 \neq n$. Therefore, the conformal towers must be more densely filled, and the spectrum generating functions must contain fewer degeneracies. Since the loop model characters (14.183) contain just one subtraction, it seems reasonable that the CBL characters for generic $n_1 \neq n$ will not involve any subtractions, i.e., they must be the generic characters (14.28). This is indeed confirmed by numerical diagonalisation of the transfer matrix [84]. Combining this with the result for the conformal weights (14.156), we conclude that the spectrum generating functions for the blobbed and unblobbed sectors read

$$Z_\ell^{\rm b} = \frac{q^{h_{r,r+\ell}-c/24}}{P(q)}, \qquad Z_\ell^{\rm u} = \frac{q^{h_{r,r-\ell}-c/24}}{P(q)}. \tag{14.194}$$

To find out how to combine these sectors to obtain the complete partition function $Z$, one needs to know the multiplicities (eigenvalue amplitudes) of each sector. These can be derived combinatorially [84], by using the line of reasoning [95, 94] that we illustrated in section 14.7.2 for a simpler case. Parametrising the weights of non-contractible loops as

$$\bar{n} = 2\cosh\alpha, \qquad \overline{n_1} = \frac{\sinh(\alpha+\beta)}{\sinh\beta} \tag{14.195}$$

the result reads

$$n_\ell^b = \frac{\sinh(\ell\alpha+\beta)}{\sinh\beta}, \qquad n_\ell^u = \frac{\sinh(\ell\alpha-\beta)}{\sinh(-\beta)}. \tag{14.196}$$

Supposing $L$ is even, and setting $\ell = 2j$, the results (14.194) and (14.196) lead to

$$Z = q^{-c/24}\left[\sum_{j=0}^\infty \frac{\sinh(2j\alpha+\beta)}{\sinh\beta} \frac{q^{h_{r,r+2j}}}{P(q)} - \sum_{j=1}^\infty \frac{\sinh(2j\alpha-\beta)}{\sinh\beta} \frac{q^{h_{r,r-2j}}}{P(q)}\right] \tag{14.197}$$

If one further supposes $\beta = r\alpha$ with $r$ integer this can be rewritten as

$$Z = \sum_{j=-\lfloor r/2\rfloor}^\infty \frac{\sinh(2j+r)\alpha}{\sinh r\alpha} K_{r,r+2j} \tag{14.198}$$

where $\lfloor\ldots\rfloor$ denotes the integer part, and $K_{r,s}$ is given by (14.183).

As an example of an application to dense polymers ($c = -2$), we consider the case $n = \bar{n} = 0$ and $n_1 = \overline{n_1} = 1$ where only loops touching the boundary are allowed. Then (14.197) can be cast in the form

$$Z = \frac{q^{-1/24}}{P(q)} \sum_{j=-\infty}^\infty (-1)^j q^{(4j-1)^2/32}. \tag{14.199}$$

### 14.7.4 Two-Boundary CBL Model

The two-boundary extension of the CBL model was defined in section 14.6.8. In particular, we recall the four different weights (14.163) of contractible loops. Since we have constrained the width of the annulus $L$ to be even, a non-contractible loop cannot touch both rims of the annulus. We thus need only the following additional three weights for non-contractible loops:

$$\bar{n} = 2\cos\chi$$
$$\bar{n_1} = \frac{\sin[(u_1+1)\chi]}{\sin(u_1\chi)}$$
$$\bar{n_2} = \frac{\sin[(u_2+1)\chi]}{\sin(u_2\chi)} \tag{14.200}$$

The exact continuum limit partition function, expressed in terms of all these seven weights, has been derived by Dubail et al. [92]:

$$Z = \frac{q^{-c/24}}{P(q)} \sum_{n\in Z} q^{h_{r_{12}-2n,r_{12}}} \tag{14.201}$$

$$+ \frac{q^{-c/24}}{P(q)} \sum_{j\geq 1} \sum_{n\geq 0} \frac{\sin[(u_1+u_2-1+2j)\chi]\sin\chi}{\sin(u_1\chi)\sin(u_2\chi)} q^{h_{r_1+r_2-1-2n,r_1+r_2-1+2j}}$$

$$+ \frac{q^{-c/24}}{P(q)} \sum_{j\geq 1} \sum_{n\geq 0} \frac{\sin[(-u_1+u_2-1+2j)\chi]\sin\chi}{\sin(-u_1\chi)\sin(u_2\chi)} q^{h_{-r_1+r_2-1-2n,-r_1+r_2-1+2j}}$$

$$+ \frac{q^{-c/24}}{P(q)} \sum_{j\geq 1} \sum_{n\geq 0} \frac{\sin[(u_1-u_2-1+2j)\chi]\sin\chi}{\sin(u_1\chi)\sin(-u_2\chi)} q^{h_{r_1-r_2-1-2n,r_1-r_2-1+2j}}$$

$$+ \frac{q^{-c/24}}{P(q)} \sum_{j\geq 1} \sum_{n\geq 0} \frac{\sin[(-u_1-u_2-1+2j)\chi]\sin\chi}{\sin(-u_1\chi)\sin(-u_2\chi)} q^{h_{-r_1-r_2-1-2n,-r_1-r_2-1+2j}}$$

The five-term structure of this expression mirrors that of the principal critical exponents (14.164). The trigonometric factors inside the four last terms are the eigenvalue amplitudes, which can be derived using combinatorial [91] or algebraic [92] means.

Obviously, an expression like (14.202) contains a wealth of exact probabilistic information, which can be extracted explicitly for any special case of interest (such as percolation). Moreover, it determines the complete operator content of the two-boundary model, and the precise fusion rules of two one-boundary CBL type boundary condition changing operators.

### 14.7.5 Fully Packed Loop Models

To write the exact continuum limit partition function of the FPL$^2$ model of section 14.5, we need the sector labels $(\ell_b, \ell_g; Q)$ identified in section 14.6.5 as well as the corresponding critical exponents $\Delta_{\ell_b,\ell_g;Q}(e_b, e_g)$ of (14.153). The remaining ingredients are the corresponding eigenvalue amplitudes $D_{\ell_b,\ell_g;Q}$ and the structure of descendent states within each sector.

These can be obtained by noting [83] that the two Temperley-Lieb like structures associated with each of the loop flavours (black and grey), with corresponding semi-conserved quantum numbers $\ell_b$ and $\ell_g$, decouple on the algebraic level. The last field, related to the conserved quantum number $Q$, essentially behaves as a free boson. One has therefore a simple product form of (14.181)

$$D_{\ell_b,\ell_g;Q} = \frac{\sin[(1+\ell_b)\pi f_b]}{\sin(\pi f_b)} \frac{\sin[(1+\ell_g)\pi f_g]}{\sin(\pi f_g)} \tag{14.202}$$

independently of $Q$. We have here as usual given different weights to non-contractible loops:

$$\begin{aligned} \overline{n_b} &= 2\cos(\pi f_b) \\ \overline{n_g} &= 2\cos(\pi f_g). \end{aligned} \tag{14.203}$$

A further consequence of this algebraic decoupling is that the characters describing the structure of descendent states factorise. The factors corresponding to quantum numbers $\ell_b$ and $\ell_g$ are degenerate characters of the $K$ type, while the last factor corresponding to quantum number $Q$ is just that of a free boson.

Assembling all this information we thus arrive at

$$Z = \frac{q^{-c/24}}{P(q)^3} \sum_{Q=-\infty}^{\infty} \sum_{\ell_b=0}^{\infty} \sum_{\ell_g=0}^{\infty} D_{\ell_b,\ell_g;Q}(1-q^{\ell_b+1})(1-q^{\ell_g+1})q^{\Delta_{\ell_b,\ell_g;Q}(e_b,e_g)} \tag{14.204}$$

where the sums over $\ell_b$ and $\ell_g$ are constrained so that all three labels $(\ell_b, \ell_g; Q)$ have the same parity.

It should be possible to endow one or both loop flavours of the FPL$^2$ model with (one- or two-boundary) CBL type boundary conditions, and work out the corresponding partition function using the methods of sections 14.7.3–14.7.4.

## 14.8 Epilogue

We hope to have convinced the reader that loop models are a useful tool for deriving exact results about two-dimensional self-avoiding polygons and walks, and that these models offer a fruitful testing ground for many, if not most, of the concepts developed in two-dimensional conformal field theory.

There are many relevant issues about loop models that we have omitted in order to keep this review to a reasonable length. Most importantly, we have focused here mainly on the application of the Coulomb gas approach, and only mentioned very briefly the results obtainable from integrability, combinatorics, quantum groups, etc. Also, several exact results—often due to Cardy—are known about universal amplitude ratios in loop models, such as the ratio between the mean area of a loop and its squared ratio of gyration [97]. Other issues have been omitted because we feel that they have not yet been sufficiently elucidated. This is the case for loop models in the presence of quenched disorder, and for certain aspects of surface critical behaviour in which the two sides of the annulus both sustain non-trivial boundary conditions.

Another promising field of future research is that of several coupled loop models (see [98] for an example).

Another limitation of this review resides of course in the number of loop models that we have treated. Roughly speaking, we have included here only models of *self-avoiding* loops whose bulk critical properties are more-or-less fully understood, and amenable to Coulomb gas analysis. Some interesting examples of loop models which fall outside this criterion have been discussed by Fendley [99]. One important model that we could actually have chosen to include is the dilute $O(n)$ model on the square lattice [100], which is related to the integrable Izergin-Korepin model [101], and its two-loop generalisation, which is referred to as the $DPL^2$ model in [70].

In any case, despite the effort dedicated to understanding two-dimensional loop models, they remain a very active area of research to this date.

## Acknowledgments

The author warmly thanks the Bannier family for hospitality at Les Loges where the present review was written. He is very much indebted to J.L. Cardy, J. Dubail, Y. Ikhlef, J. Kondev, B. Nienhuis, J.-F. Richard, H. Saleur, and P. Zinn-Justin who have helped shaping his understanding of the subject over the years. This work was supported through the European Community Network ENRAGE (grant MRTN-CT-2004-005616) and by the Agence Nationale de la Recherche (grant ANR-06-BLAN-0124-03).

## References

1. L. Onsager, *Crystal statistics. I. A two-dimensional model with an order-disorder transition*, Phys. Rev. **65**, 117 (1944).
2. K.G. Wilson, *Non Lagrangian models of current algebra*, Phys. Rev. **179**, 1499 (1969).
3. K.G. Wilson and J. Kogut, *The renormalization group and the $\varepsilon$ expansion*, Phys. Rep. C **12**, 75 (1974).
4. J. Zinn-Justin, *Quantum field theory and critical phenomena* (Oxford Science Publications, Oxford, 1989).
5. E.H. Lieb, *Residual entropy of square ice*, Phys. Rev. **162**, 162 (1967).
6. R.J. Baxter, *Eight-vertex model in lattice statistics*, Phys. Rev. Lett. **26**, 832 (1971).
7. A.A. Belavin, A.M. Polyakov and A.B. Zamolodchikov, *Infinite conformal symmetry in two-dimensional quantum field theory*, Nucl. Phys. B **241**, 333 (1984).
8. A.M. Polyakov, *Conformal symmetry of critical fluctuations*, JETP Lett. **12**, 381 (1970).
9. J.V. José, L.P. Kadanoff, S. Kirkpatrick and D.R. Nelson, *Renormalization, vortices, and symmetry-breaking perturbations in the two-dimensional planar model*, Phys. Rev. B **16**, 1217 (1977).
10. M. den Nijs, *Extended scaling relations for the magnetic critical exponents of the Potts model*, Phys. Rev. B **27**, 1674 (1983); *Extended scaling relations for the chiral and cubic crossover exponents*, J. Phys. A **17**, L295 (1984).
11. B. Nienhuis, *Critical behavior of two-dimensional spin models and charge asymmetry in the Coulomb gas*, J. Stat. Phys. **34**, 731 (1984).

12. Vl. S. Dotsenko and V. Fateev, *Conformal algebra and multipoint correlation functions in 2D statistical models*, Nucl. Phys. B **240**, 312 (1984).
13. Vl. S. Dotsenko and V. Fateev, *Four-point correlation functions and the operator algebra in 2D conformal invariant theories*, Nucl. Phys. B **251**, 691 (1985).
14. B. Duplantier and H. Saleur, *Exact critical properties of two-dimensioanl dense self-avoiding walks*, Nucl. Phys. B **290**, 291 (1987).
15. P. Di Francesco, P. Mathieu and D. Sénéchal, *Conformal field theory* (Springer Verlag, New York, 1987).
16. J.L. Cardy, *Conformal invariance and statistical mechanics*, and P. Ginsparg, *Applied conformal field theory*, both in *Fields, strings and critical phenomena* (Les Houches, session XLIX), eds. E. Brézin and J. Zinn-Justin (Elsevier, New York, 1989).
17. J. Kondev, *Liouville field theory of fluctuating loops*, Phys. Rev. Lett. **78**, 4320 (1997).
18. J. Cardy, *The O(n) model on the annulus*, J. Stat. Phys. **125**, 1 (2006).
19. B. Duplantier and H. Saleur, *Exact tricritical exponents for polymers at the Θ point in two dimensions*, Phys. Rev. Lett. **59**, 539 (1987).
20. J. Kondev, J. de Gier and B. Nienhuis, *Operator spectrum and exact exponents of the fully packed loop model*, J. Phys. A **29**, 6489 (1996).
21. J.L. Jacobsen and J. Kondev, *Field theory of compact polymers on the square lattice*, Nucl. Phys. B **532**, 635 (1998); *Conformational entropy of compact polymers*, Phys. Rev. Lett. **81**, 2922 (1998).
22. J.L. Cardy, *Conformal invariance and surface critical behavior*, Nucl. Phys. B **240**, 514 (1984).
23. J. Cardy, *Operator content of two-dimensional conformally invariant theories*, Nucl. Phys. B **270**, 186 (1986).
24. V.A. Fateev and A.B. Zamolodchikov, *Conformal quantum field theory models in two dimensions having $Z_3$ symmetry*, Nucl. Phys. B **280**, 644 (1987).
25. H.W.J. Blöte, J.L. Cardy and M.P. Nightingale, *Conformal invariance, the central charge, and universal finite-size amplitudes at criticality*, Phys. Rev. Lett. **56**, 742 (1986); I. Affleck, *Universal term in the free energy at a critical point and the conformal anomaly*, Phys. Rev. Lett. **56**, 746 (1986).
26. J.L. Cardy, *Conformal invariance and universality in finite-size scaling*, J. Phys. A **17**, L385 (1984).
27. D. Friedan, Z. Qiu and S. Shenker, *Conformal invariance, unitarity and critical exponents in two dimensions*, Phys. Rev. Lett. **52**, 1575 (1984).
28. A.B. Zamolodchikov, *Conformal symmetry and multicritical points in two-dimensional quantum field theory*, Sov. J. Nucl. Phys. **44**, 530 (1986).
29. C. Itzykson and J.-B. Zuber, *Two-dimensional conformal invariant theories on a torus*, Nucl. Phys. B **275**, 580 (1986).
30. J. Cardy, *Boundary conformal field theory*, in J.-P. Françoise, G. Naber and T.S. Tsun (eds.), *Encyclopedia of mathematical physics* (Elsevier, 2005).
31. K. Binder, *Critical behaviour at surfaces*, in C. Domb and J.L. Lebowitz (eds.), *Phase transitions and critical phenomena*, vol. 8, p. 1 (Academic Press, London, 1983).
32. T.W. Burkhardt and E. Eisenriegler, *Conformal theory of the two-dimensional O(N) model with ordinary, extraordinary, and special boundary conditions*, Nucl. Phys. B **424**, 487 (1994).
33. J.L. Cardy, *Critical percolation in finite geometries*, J. Phys. A **25**, L201 (1992).
34. A. Luther and I. Peschel, *Calculation of critical exponents in two dimensions from quantum field theory in one dimension*, Phys. Rev. B **12**, 3908 (1975); L.P. Kadanoff, *Lattice Coulomb gas representations of two-dimensional problems*, J. Phys. A **11**, 1399 (1978); L.P. Kadanoff and A.C. Brown, *Correlation functions on the critical lines of the Baxter and Ashkin-Teller models*, Ann. Phys. **121**, 318 (1979); H.J.F. Knops, *Renormalization connection between the eight-vertex model and the Gaussian model*, Ann. Phys. **128**, 448 (1981).
35. B. Nienhuis, *Coulomb gas formulations of two-dimensional phase transitions*, in C. Domb and J.L. Lebowitz (eds.), *Phase transitions and critical phenomena*, vol. 11, p. 1–53 (Academic Press, London, 1987).

36. G. Parisi and N. Sourlas, *Self avoiding walk and supersymmetry*, J. Physique Lett. **41**, L403 (1980).
37. B. Nienhuis, *Exact critical point and critical exponents of O(n) models in two dimensions*, Phys. Rev. Lett. **49**, 1062 (1982).
38. R.J. Baxter, S.B. Kelland and F.Y. Wu, *Equivalence of the Potts model or Whitney polynomial with an ice-type model*, J. Phys. A **9**, 397 (1975).
39. P. Di Francesco, H. Saleur and J.B. Zuber, *Modular invariance in non-minimal two-dimensional conformal theories*, Nucl. Phys. B **285**, 454 (1987).
40. P. Di Francesco, H. Saleur and J.B. Zuber, *Relations between the Coulomb gas picture and conformal invariance of two-dimensional critical models*, J. Stat. Phys. **49**, 57 (1987).
41. P. W. Kasteleyn et C. M. Fortuin, *Phase transitions in lattice systems with random local properties*, J. Phys. Soc. Jpn. **26** (suppl.), 11 (1969); C.M. Fortuin and P.W. Kasteleyn, *On the random-cluster model. I. Introduction and relation to other models*, Physica **57**, 536 (1972).
42. R.J. Baxter, *Critical antiferromagnetic square-lattice Potts model*, Proc. Roy. Soc. London Ser. A **383**, 43 (1982).
43. J.L. Jacobsen and H. Saleur, *The antiferromagnetic transition for the square-lattice Potts model*, Nucl. Phys. B **743**, 207 (2006); Y. Ikhlef, J.L. Jacobsen and H. Saleur, *A staggered six-vertex model with non-compact continuum limit*, Nucl. Phys. B **789**, 483–524 (2008).
44. R.J. Baxter, *Potts model at the critical temperature*, J. Phys. C **6** L445 (1973).
45. H. Saleur, *The antiferromagnetic Potts model in two dimensions: Berker-Kadanoff phase, antiferromagnetic transition, and the role of Beraha numbers*, Nucl. Phys. B **360**, 219 (1991).
46. P.-G. de Gennes, *Scaling concepts in polymer physics* (Cornell University Press, New York, 1979).
47. B. Duplantier and H. Saleur, *Exact surface and wedge exponents for polymers in two dimensions*, Phys. Rev. Lett. **57**, 3179 (1986).
48. B. Duplantier and H. Saleur, *Winding-angle distributions of two-dimensional self-avoiding walks from conformal invariance*, Phys. Rev. Lett. **60**, 2343 (1988).
49. J.L. Jacobsen, N. Read and H. Saleur, *Dense loops, supersymmetry, and Goldstone phases in two dimensions*, Phys. Rev. Lett. **90**, 090601 (2003).
50. Vl. S. Dotsenko, *Critical behaviour and associated conformal algebra of the $Z_3$ Potts model*, Nucl. Phys. B **235**, 54 (1984).
51. H. Saleur, *The antiferromagnetic Potts model in two dimensions: Berker-Kadanoff phase, antiferromagnetic transition, and the role of Beraha numbers*, Nucl. Phys. B **360**, 219 (1991).
52. H. Saleur, *Winding-angle distribution for Brownian and self-avoiding walks*, Phys. Rev. E **50**, 1123 (1994).
53. P.J. Flory, *The configuration of real polymer chains*, J. Chem. Phys. **17**, 303 (1949).
54. P.G. de Gennes, *Collapse of a polymer chain in poor solvents*, J. Physique Lett. **36**, 55 (1975).
55. B. Duplantier and H. Saleur, *Stability of the polymer $\Theta$ point in two dimensions*, Phys. Rev. Lett. **62**, 1368 (1989).
56. R. Raghavan, C.L. Henley and S.L. Arouh, *New two-color dimer models with critical ground states*, J. Stat. Phys. **86**, 517 (1997).
57. A. Ghosh, D. Dhar and J.L. Jacobsen, *Random trimer tilings*, Phys. Rev. E **75**, 011115 (2007).
58. A. Verberkmoes and B. Nienhuis, *Triangular trimers on the triangular lattice: An exact solution*, Phys. Rev. Lett. **83**, 3986 (1999).
59. J.L. Jacobsen, *On the universality of fully packed loop models*, J. Phys. A **32**, 5445 (1999).
60. H.W.J. Blöte and B. Nienhuis, *Fully packed loop model on the honeycomb lattice*, Phys. Rev. Lett. **72**, 1372 (1994).
61. J. Kondev and C.L. Henley, *Four-coloring model on the square lattice: A critical ground state*, Phys. Rev. B **52**, 6628 (1995).
62. J.L. Jacobsen and J. Kondev, *Conformal field theory of the Flory model of protein melting*, Phys. Rev. E **69**, 066108 (2004); *Continuous melting of compact polymers*, Phys. Rev. Lett. **92**, 210601 (2004).
63. P.J. Flory, *Statistical thermodynamics of semi-flexible chain molecules*, Proc. Roy. Soc. London A **234**, 60 (1956).

64. M.T. Batchelor, J. Suzuki and C.M. Yung, *Exact results for hamilton walks from the solution of the fully packed loop model on the honeycomb lattice*, Phys. Rev. Lett. **73**, 2646 (1994).
65. D. Dei Cont and B. Nienhuis, *The packing of two species of polygons on the square lattice*, J. Phys. A **37**, 3085 (2004); *Critical exponents for the FPL$^2$ model*, cond-mat/0412018.
66. J.L. Jacobsen and P. Zinn-Justin, *Algebraic Bethe Ansatz for the FPL$^2$ model*, J. Phys. A **37**, 7213 (2004).
67. N.Y. Reshetikhin, *A new exactly solvable case of an O(n) model on a hexagonal lattice*, J. Phys. A **24**, 2387 (1991).
68. V. Pasquier and H. Saleur, *Common structures between finite systems and conformal field theories through quantum groups*, Nucl. Phys. B **330**, 523 (1990).
69. B. Eynard, E. Guitter and C. Kristjansen, *Hamiltonian cycles on a random three-coordinate lattice*, Nucl. Phys. B **528**, 523 (1998).
70. J.L. Jacobsen and J. Kondev, *Transition from the compact to the dense phase of two-dimensional polymers*, J. Stat. Phys. **96**, 21 (1999).
71. P. Di Francesco, O. Golinelli and E. Guitter, *Meanders: Exact asymptotics*, Nucl. Phys. B **570**, 699 (2000).
72. P. Di Francesco, E. Guitter and J.L. Jacobsen, *Exact meander asymptotics: A numerical check*, Nucl. Phys. B **580**, 757 (2000).
73. M.T. Batchelor and J. Suzuki, *Exact solution and surface critical behaviour of an O(n) model on the honeycomb lattice*, J. Phys. A **26**, L729 (1993).
74. C.M. Yung and M.T. Batchelor, *O(n) model on the honeycomb lattice via reflection matrices: Surface critical behaviour*, Nucl. Phys. B **453**, 552 (1995).
75. M.T. Batchelor and C.M. Yung, *Exact results for the adsorption of a flexible self-avoiding polymer chain in two dimensions*, Phys. Rev. Lett. **74**, 2026 (1995).
76. M.T. Batchelor and J. Cardy, *Extraordinary transition in the two-dimensional O(n) model*, Nucl. Phys. B **506**, 553 (1997).
77. T.W. Burkhardt, E. Eisenriegler and I. Guim, *Conformal theory of energy correlations in the semi-infinite two-dimensional O(N) model*, Nucl. Phys. B **316**, 559 (1989).
78. T.W. Burkhardt and J.L. Cardy, *Surface critical behaviour and local operators with boundary-induced critical profiles*, J. Phys. A **20**, L233 (1987).
79. P. Fendley and H. Saleur, *Exact theory of polymer adsorption in analogy with the Kondo problem*, J. Phys. A **27**, L789 (1994).
80. H. Saleur and M. Bauer, *On some relations between local height probabilities and conformal invariance*, Nucl. Phys. B **320**, 591 (1989).
81. B. Duplantier and H. Saleur, *Exact surface and wedge exponents for polymers in two dimensions*, Phys. Rev. Lett. **57**, 3179 (1986).
82. A.J. Guttmann and G.M. Torrie, *Critical behaviour at an edge for the SAW and Ising model*, J. Phys. A **17**, 3539 (1984).
83. J.L. Jacobsen, *Surface critical behaviour of fully packed loop models*, in preparation (2008).
84. J.L. Jacobsen and H. Saleur, *Conformal boundary loop models*, Nucl. Phys. B **788**, 137–166 (2008).
85. P.P. Martin and H. Saleur, *The blob algebra and the periodic Temperley-Lieb algebra*, Lett. Math. Phys. **30**, 189 (1994).
86. I. Kostov, *Boundary loop models and 2D quantum gravity*, J. Stat. Mech. P08023 (2007).
87. J.L. Jacobsen and H. Saleur, unpublished (2007).
88. J. Dubail, J.L. Jacobsen and H. Saleur, *Generalised special surface transition in the two-dimensional O(n) model*, in preparation (2008).
89. A. Nichols, *The Temperley-Lieb algebra and its generalizations in the Potts and XXZ models*, J. Stat. Mech. P01003 (2006); *Structure of the two-boundary XXZ model with non-diagonal boundary terms*, J. Stat. Mech. L02004 (2006).
90. J. de Gier and A. Nichols, *The two-boundary Temperley-Lieb algebra*, math.RT/0703338.
91. J.L. Jacobsen and H. Saleur, *Combinatorial aspects of conformal boundary loop models*, J. Stat. Mech. P01021 (2008).

92. J. Dubail, J.L. Jacobsen and H. Saleur, *Conformal two-boundary loop model on the annulus*, Nucl. Phys. B **813**, 430 (2009); *Boundary extensions of the Temperley-Lieb algebra: representations, lattice models and BCFT*, in preparation (2008).

93. N. Read and H. Saleur, *Exact spectra of conformal supersymmetric nonlinear sigma models in two dimensions*, Nucl. Phys. B **613**, 409 (2001).

94. J.-F. Richard and J.L. Jacobsen, *Eigenvalue amplitudes of the Potts model on a torus*, Nucl. Phys. B **769**, 256 (2007).

95. J.-F. Richard and J.L. Jacobsen, *Character decomposition of Potts model partition functions, I: Cyclic geometry*, Nucl. Phys. B **750**, 250 (2006).

96. I. Affleck and A.W.W. Ludwig, *Universal noninteger "ground-state degeneracy" in critical quantum systems*, Phys. Rev. Lett. **67**, 161 (1991).

97. J.L. Cardy, *Mean area of self-avoiding loops*, Phys. Rev. Lett. **72**, 1580 (1994).

98. P. Fendley and J.L. Jacobsen, *Critical points in coupled Potts models and critical phases in coupled loop models*, J. Phys. A **41**, 215001 (2008).

99. P. Fendley, *Loop models and their critical points*, J. Phys. A **39**, 15445 (2006).

100. B. Nienhuis, *Critical spin-1 vertex models and O(n) models*, Int. J. Mod. Phys. B **4**, 929 (1990).

101. A.G. Izergin and V.E. Korepin, *The inverse scattering method approach to the quantum Shabat-Mikhailov model*, Comm. Math. Phys. **79**, 303 (1981).

# Chapter 15
# Stochastic Löwner Evolution and the Scaling Limit of Critical Models

Bernard Nienhuis and Wouter Kager

## 15.1 Introduction

Great progress in the understanding of conformally invariant scaling limits of stochastic models, has been given by the Stochastic Löwner Evolutions (SLE). This approach has been pioneered by Schramm [46] and by Lawler, Schramm and Werner [31]. It describes a one-parameter family of conformally invariant measures of curves in the plane or a two-dimensional domain. This family is commonly referred to as SLE$_\kappa$, where $\kappa$ parametrizes the family. It has been shown to be the scaling limit of many well-known and less well-known statistical lattice models. These models are typically members of the families of critical and tricritical [40] $q$-state Potts models [61] and of O($n$) models [17], or believed to be in the corresponding universality class.

SLE describes the scaling limit of various open, non-crossing, stochastic paths on the lattice, which are, at least on one side, attached to the boundary. Therefore its application to polygons is restricted in various ways. In the first place it describes only the scaling limit. In many studies of lattice polygons, of course, the scaling limit is considered the most interesting aspect. The restriction to open paths attached to the boundary is more severe. This restriction has been lifted to some extent by recursively considering domains bounded by closed paths resulting from a previous SLE process. This approach applies only to paths that have a tendency to touch themselves (without, of course crossing), and this generalization is not the subject of this chapter. In most cases the paths under consideration by their nature occur in extensive numbers. However, one may concentrate on one of them, and treat the

Bernard Nienhuis
Institute for Theoretical Physics, University of Amsterdam, 1018 XE Amsterdam, The Netherlands, e-mail: b.nienhuis@uva.nl

Wouter Kager
VU University Amsterdam, Department of Mathematics, De Boelelaan, 1081 HV Amsterdam, The Netherlands, e-mail: wkager@few.vu.nl

interaction with the others only as an ingredient that defines the stochastic measure of the path under consideration. This, in fact is precisely what SLE does.

So far the essential progress made by the SLE approach, does not consist of the derivation of explicit unknown properties of the scaling limit. The properties it proves have been known for decades in the physics community, without, however, a proof being available. They were obtained by means of the Coulomb Gas (CG) [42] approach, and by Conformal Field Theory (CFT) [14], or Bethe Ansatz and similar techniques for integrable models [7]. Other properties were not known but can be obtained by the same methods without more difficulty.

In this chapter we give a brief description of the meaning of SLE, proofs of its basic properties, and a selection of its results (mostly without proof). For more extensive treatments and proofs, we refer the reader to several existing reviews which target different communities: Werner [54] from the mathematical perspective, Cardy [16] for physicists, and Kager e.a. [25] for both mathematicians and physicists.

A conformally invariant stochastic measure of curves naturally brings together the theory of conformal maps, stochastic calculus, and a description of curves. Already in 1923 Löwner[1] combined curves and conformal maps [39]. He considered a singly connected domain $D$ of the complex plane (for example the upper half-plane or the unit disk) and a path $\gamma_s \in D$ starting from the boundary, and parametrized by $s > 0$. He then considered the domains $D_s$ from which the initial part of the path is excluded: $D_s = D \setminus \gamma_{[0,s]}$. For the conformal maps that map $D_s$ back to $D$ he found a surprisingly simple differential equation in terms of $s$, provided a suitable parametrization. Schramm [46] later used this equation to define a stochastic measure of paths. It turned out that he and others could prove many properties of this measure by means of Löwner's equation.

In the following sections we will discuss some of these properties and their applications to problems in statistical physics. Though mathematical rigor is an essential ingredient of the progress made by this approach we will largely omit proofs, and rather refer the reader to the above mentioned reviews and the original literature. Only in section 3, treating the basics of SLE, do we include proofs of the various theorems.

## 15.2 Conformal Maps

Conformal maps play an essential role in this chapter. This is not the place to treat this subject extensively, but the notation and those few properties that are used frequently are introduced in this section. A comprehensive treatment of the subject is found in Ahlfors [1], and for a discussion more specific to SLE we refer to Kager e.a. [25] and its appendix A.

We shall write $\mathbb{C}$ for the complex plane, and $\mathbb{R}$ for the set of real numbers. The open upper half-plane $\{z : \operatorname{Im} z > 0\}$ is denoted by $\mathbb{H}$, and the open unit disk

---

[1] The spelling Loewner, while adopted by himself, is of later date.

$\{z : |z| < 1\}$ by $\mathbb{D}$. The defining property of conformal maps is that they preserve angles. Denoted in complex numbers this makes them holomorphic (there is no need to include the antiholomorphic variety). The basis of conformal mapping theory is the Riemann mapping theorem, which tells us that any simply connected domain $D$ can be mapped conformally onto the open unit disk $\mathbb{D}$.

**Theorem 1 (Riemann mapping theorem).** *Let $D \neq \mathbb{C}$ be a simply connected domain in $\mathbb{C}$. Then there is a conformal map of $D$ onto the open unit disk $\mathbb{D}$.*

This map is not unique, which follows from the fact that the unit disk can be conformally mapped onto itself in multiple ways, as follows from the following:

**Theorem 2.** *The conformal self-maps of the open unit disk $\mathbb{D}$ are precisely the transformations of the form*

$$f(z) = e^{i\varphi} \frac{z - a}{1 - \bar{a}z}, \qquad |z| < 1, \tag{15.1}$$

*where $a$ is complex, $|a| < 1$, and $0 \leq \varphi \leq 2\pi$.*

Thus it follows that the conformal map from $D$ to $\mathbb{D}$ can be determined uniquely by three real parameters. It immediately follows that between any two singly connected domains there exists a three parameter family of conformal maps. We shall consider only domains whose boundary is a continuous curve, and this implies that the conformal maps we work with have well-defined limit values on the boundary.

Now suppose that $D$ is a simply connected domain with a continuous boundary, and that $z_1$, $z_2$, $z_3$ and $z_4$ are distinct points on $\partial D$, ordered in the counter-clockwise direction. Then we can map $D$ onto a rectangle $(0, L) \times (0, \pi)$ in such a way that the arc $[z_1, z_2]$ of $\partial D$ maps onto $[0, i\pi]$, and $[z_3, z_4]$ maps onto $[L, L + i\pi]$. The length $L > 0$ of this rectangle is determined uniquely, and is called the $\pi$-**extremal distance** between $[z_1, z_2]$ and $[z_3, z_4]$ in $D$.

A compact subset $K$ of $\overline{\mathbb{H}}$ such that $\mathbb{H} \setminus K$ is simply connected and $K = \overline{K \cap \mathbb{H}}$ is called a **hull** (it is basically a compact set bordering on the real line). For any hull $K$ there exists a unique conformal map, denoted by $g_K$, which sends $\mathbb{H} \setminus K$ onto $\mathbb{H}$ and satisfies the normalization

$$\lim_{z \to \infty} \left( g_K(z) - z \right) = 0. \tag{15.2}$$

This map has an expansion for $z \to \infty$ of the form

$$g_K(z) = z + \frac{a_1}{z} + \ldots + \frac{a_n}{z^n} + \ldots \tag{15.3}$$

where all expansion coefficients are real. The coefficient $a_1 = a_1(K)$ is called the **capacity** of the hull $K$. Conformal maps that have this limit at $\infty$ are said to satisfy the hydrodynamic normalization.

The capacity of a nonempty hull $K$ is a positive real number, and satisfies a **scaling rule** and a **summation rule**. The scaling rule says that if $r > 0$ then $a_1(rK) = r^2 a_1(K)$. The summation rule says that if $J \subset K$ are two hulls and $L$ is the closure of $g_J(K \setminus J)$, then $g_K = g_L \circ g_J$ and $a_1(K) = a_1(J) + a_1(L)$. The capacity of a hull is bounded from above by the square of the radius of the smallest half-disk that contains the hull and has its centre on the real line.

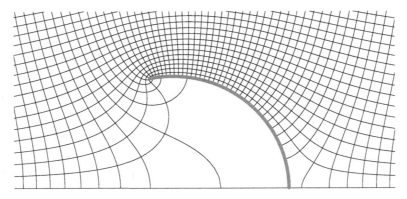

**Fig. 15.1** The conformal map from $\mathbb{H}$ slit by a circular arc, to the whole of $\mathbb{H}$. The map is shown by the inverse image of the coordinate grid. The excluded arc is shown in gray.

## 15.3 Löwner Evolutions

In this section, we will first discuss the Löwner equation in a deterministic setting, as it was conceived by Löwner himself. Before we enter into a mathematical discussion, we will first give explicit examples of the ingredients and meaning of this celebrated equation. We will then show how one can describe a given continuous path by a family of conformal maps, and we will prove that these maps satisfy Löwner's differential equation. Then we will prove that conversely, the Löwner equation generates a family of conformal maps, that may or may not describe a continuous curve. Finally, we move on to the definition of the stochastic Löwner evolution. This section is based on ideas from Lawler, Schramm and Werner [31], Lawler [29], and Rohde and Schramm [44].

### 15.3.1 Describing a Path by the Löwner Equation

Suppose the upper half of the complex plane, $\mathbb{H}$, is slit by a circular arc starting at the real axis. When we exclude the points on this arc, what remains of $\mathbb{H}$ is still simply connected, provided we do not extend the arc so far that it meets the real axis again. Thus it follows that the slit upper half plane can be mapped onto the full upper half-plane by a conformal map. This map can be written in closed form, and is shown in Fig. 15.1, by the coordinate grid of the target half plane, on the original slit half plane. The figure shows that the interior of the arc shrinks strikingly under this map. When the arc is extended so that it approaches the real axis again, the image of the interior shrinks more and more, until after closure of the arc, its interior has no image anymore.

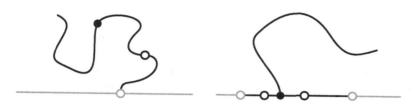

**Fig. 15.2** A simple path $\gamma_t$ from the origin (left). The path up to the full dot is sent to the real axis by $g_t$ (right). The original real axis is indicated in gray, and the image of the path and its image is in black. The open point on $\gamma_{[0,t]}$ has two images under $g_t$.

Consider now a general simple (i.e. non-intersecting) path $\gamma_t$, in $\mathbb{H}$, parametrized by $t$, with $\gamma_0 = 0$, see Fig. 15.2. We choose a point $w = \gamma_t$ on the curve, indicated by a full dot, and consider the subset of the upper half-plane that excludes the part of the curve between the origin and $w$ (or $\gamma_{[0,t]}$). Now we construct a conformal map $g_t$, from $\mathbb{H} \setminus \gamma_{[0,t]}$ to the entire upper half plane. Clearly since $\gamma_{[0,t]}$ are now boundary points, they are sent by $g_t$ to the real axis. Because the curve can be approached from two sides, every point of $\gamma_{[0,t)}$ has two images on the real axis, but the image of $\gamma_t$ itself is unique. Löwner discovered, that by specifying the map such that it approaches the identity at $\infty$, by choosing a suitable parametrization of $\gamma_t$, these conformal maps $g_t$ satisfy the simple differential equation:

$$\frac{\partial}{\partial t} g_t(z) = \frac{2}{g_t(z) - g_t(\gamma_t)} \,, \tag{15.4}$$

which will be derived below. We note that from the definition $g_t(\gamma_t)$ is real. Thus it follows that the path $\gamma_t$ can be alternately defined by the family of confomal maps $g_t$ or by the real function $U_t \equiv g_t(\gamma_t)$. This surprising observation led Schramm to apply it to a conformally invariant measure for curves in the plane. If the probability measure of $\gamma_t$ is conformally invariant, the law for $\gamma_{[0,\tau]}$ is the same as for $g_t(\gamma_{[t,t+\tau]})$. Since $U_t$ fully determines the curve $\gamma_t$, it is pertinent to investigate the law of $U_t$, see subsection 15.3.3. The condition that the path is simple is unnecessary, as we shall see now in a more formal treatment.

Suppose that $\gamma_t$ (where $t \geq 0$) is a continuous path in $\overline{\mathbb{H}}$ which starts from $\gamma_0 \in \mathbb{R}$. The parameter $t$, to be defined later, will be referred to as time. We allow the path to hit itself or the real line, but if it does, we require the path to reflect off into open space immediately. In other words, the path is not allowed to enter a region which has been disconnected from infinity by $\gamma_{[0,t]} \cup \mathbb{R}$. To be specific, let us denote by $H_t$

for $t \geq 0$ the unbounded connected component of $\mathbb{H} \setminus \gamma_{[0,t]}$, and let $K_t$ be the closure of $\mathbb{H} \setminus H_t$. Then we require that for all $0 \leq s < t$, $K_s$ is a proper subset of $K_t$. See Fig. 15.3 for a picture of a path satisfying these conditions.

We further impose the conditions that for all $t \geq 0$ the set $K_t$ is bounded, so that $\{K_t : t \geq 0\}$ is a family of growing hulls, and that the capacity of these hulls eventually goes to infinity, i.e. $\lim_{t \to \infty} a_1(K_t) = \infty$. The latter condition implies that the path eventually has to escape to infinity (this is a necessary but not sufficient condition for the hull to diverge, see e.g. section A.4 of [25]). Now let us state the purpose of this subsection.

For every $t \geq 0$ we set $g_t := g_{K_t}$, and we further define the real-valued function $U_t := g_t(\gamma_t)$ (this is the point to which the tip of the path is mapped). The purpose of this subsection is to prove that the maps $g_t$ satisfy a simple differential equation, which is 'driven' by $U_t$. Ideas for the proof were taken from [31]. For a different, probabilistic approach, see [29]. The first thing that we show, is that we can choose the time parameterization of $\gamma$ such that the capacity grows linearly in time. Clearly, this fact is a direct consequence of the following theorem.

**Theorem 3.** *Both $a_1(K_t)$ and $U_t$ are continuous in $t$.*

*Proof.* The proof relies heavily on properties of $\pi$-extremal distance, and we refer to the chapter on extremal length, sections 4.1–4.5 and 4.11–4.13, in Ahlfors [1] for the details. We shall prove left-continuity first.

Without loss of generality we may assume that $\gamma_0 = 0$. Fix $t > 0$, let $R$ be a large number, say at least several times the radius of $K_t$, and let $C_R$ be the upper half of the circle with radius $2R$ centred at the origin. Fix $\varepsilon > 0$. Then by continuity of $\gamma_t$, there exists a $\delta > 0$ such that $|\gamma_t - \gamma_u| < \varepsilon/2$ for all $u \in (t - \delta, t)$. Now let $C_\varepsilon$ be the circle with radius $\varepsilon$ and centre $\gamma_t$, and let $S$ be the arc of this circle in the domain $H_t$. Then this set $S$ disconnects $K_t \setminus K_{t-\delta}$ from infinity in $H_{t-\delta}$, see Fig. 15.3. Observe that the set $K_t \setminus K_{t-\delta}$ may be just a piece of $\gamma$, but that it can also be much larger, as in the figure.

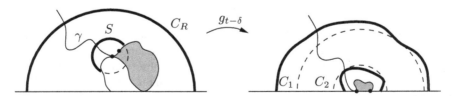

**Fig. 15.3** A path $\gamma$. The two points represent $\gamma_t$ and $\gamma_{t-\delta}$, and the shaded area is the set $K_t \setminus K_{t-\delta}$. For clarity, the arc $C_R$ is drawn much smaller than it is in the proof.

For convenience let us denote by $\Omega$ the part of the domain $H_{t-\delta}$ that lies below $C_R$. Let $\mathscr{L}$ be the $\pi$-extremal distance between $S$ and $C_R$ in $\Omega$. By the properties of $\pi$-extremal distance, because the circle with radius $R$ and centre at $\gamma_t$ lies below $C_R$, $\mathscr{L}$ must be at least $\log(R/\varepsilon)/2$. Note that since $\pi$-extremal distance is invariant under conformal maps, $\mathscr{L}$ is also the $\pi$-extremal distance between $g_{t-\delta}(C_R)$ and $g_{t-\delta}(S)$ in $g_{t-\delta}(\Omega)$. This allows us to find an upper bound on $\mathscr{L}$.

To get this upper bound, we draw two concentric semi-circles $C_1$ and $C_2$, the first hitting $g_{t-\delta}(C_R)$ on the inside, and the second hitting $g_{t-\delta}(S)$ on the outside as in Fig. 15.3 (this is always possible if $R$ was chosen large enough). Note that by the hydrodynamic normalization of the map $g_{t-\delta}$, we have an upper bound on the radius of $C_1$, which depends only on $R$. As is explained in Ahlfors, this means that the $\pi$-extremal distance $\mathscr{L}$ satisfies an inequality of the form $\mathscr{L} \le \log(C(R)/r)$, where $C(R)$ depends only on our choice of $R$, and $r$ is the radius of the inner half-circle $C_2$. But $\mathscr{L}$ was at least $\log(R/\varepsilon)/2$, implying that $r$ can be made arbitrarily small by choosing $\delta$ small enough. It follows that for every $\varepsilon > 0$ there exists a $\delta > 0$ such that the set $K_{t,\delta} := g_{t-\delta}(K_t \setminus K_{t-\delta})$ is contained in a half-disk of radius $\varepsilon$. But then by the summation rule of capacity $a_1(K_t) - a_1(K_{t-\delta}) = a_1(K_{t,\delta}) \le \varepsilon^2$, proving left-continuity of $a_1(K_t)$.

To prove left-continuity of $U_t$, let $\delta$ and $\varepsilon$ be as above, and denote by $g_{t,\delta}$ the normalized map $g_{K_{t,\delta}}$ associated with the hull $K_{t,\delta}$. It is clearly sufficient to show that $g_{t,\delta}$ converges uniformly to the identity as $\delta \downarrow 0$ (remember that $U_t$ is defined as $g_t(\gamma_t)$ and refer to Fig. 15.3). To prove this, we may assume without loss of generality that the set $K_{t,\delta}$ is contained within the disk of radius $\varepsilon$ centred at the origin, since the claim remains valid under translations over the real line. But for a hull bounded within a radius $\varepsilon$ the $n$-th order coefficient of the asymptotic expansion of $g_t$ is bounded by $\varepsilon^{n+1}$ (see equation (A.11) in [25]). Therefore, for $|z| > 2\varepsilon$,

$$|g_{t,\delta}(z) - z| \le \sum_{n=1}^{\infty} \frac{a_n(K_{t,\delta})}{|z|^n} \le \varepsilon \sum_{n=1}^{\infty} \frac{\varepsilon^n}{(2\varepsilon)^n} = \varepsilon. \tag{15.5}$$

This shows that the map $g_{t,\delta}$ converges uniformly to the identity. Left-continuity of $U_t$ follows. In the same way we can prove right-continuity of $a_1(K_t)$ and $U_t$.

**Theorem 4.** *Let $\gamma_t$ be parameterized such that $a_1(K_t) = 2t$. Then for all $z \in \mathbb{H}$, as long as $z$ is not an element of the growing hull, $g_t(z)$ satisfies the Löwner differential equation*

$$\frac{\partial}{\partial t} g_t(z) = \frac{2}{g_t(z) - U_t}, \quad g_0(z) = z. \tag{15.6}$$

*Proof.* Our proof is based on the proof of theorem 3 and the Poisson integral formula, which states that the map $g_{t,\delta}$ satisfies

$$g_{t,\delta}(z) - z = \frac{1}{\pi} \int_{-\infty}^{\infty} \frac{\operatorname{Im} g_{t,\delta}^{-1}(\xi)}{g_{t,\delta}(z) - \xi} d\xi, \quad z \in \mathbb{H} \setminus K_{t,\delta} \tag{15.7}$$

while the capacity $a_1(K_{t,\delta})$ is given by the integral

$$a_1(K_{t,\delta}) = \frac{1}{\pi} \int_{-\infty}^{\infty} \operatorname{Im} g_{t,\delta}^{-1}(\xi) d\xi. \tag{15.8}$$

First consider the left-derivative of $g_t(z)$. Using the same notation as in the proof of theorem 3 we can write $g_t = g_{t,\delta} \circ g_{t-\delta}$. We know that $g_{t,\delta}$ converges to the identity as $\delta \downarrow 0$, and that the support of $\operatorname{Im} g_{t,\delta}^{-1}$ on the real line shrinks to the point $U_t$. Moreover, using the summation rule of capacity and our choice of time parameterization, equation (15.8) gives $\int \operatorname{Im} g_{t,\delta}^{-1}(\xi) \, d\xi = 2\pi\delta$. Hence from equation (15.7) we get

$$\lim_{\delta \downarrow 0} \frac{g_t(z) - g_{t-\delta}(z)}{\delta} = \lim_{\delta \downarrow 0} \frac{1}{\pi\delta} \int \frac{\operatorname{Im} g_{t,\delta}^{-1}(\xi)}{g_{t,\delta}(g_{t-\delta}(z)) - \xi} \, d\xi = \frac{2}{g_t(z) - U_t}. \tag{15.9}$$

In the same way one obtains the right-derivative.

### 15.3.2 The Solution of the Löwner Equation

In the previous subsection, we started from a continuous path $\gamma$ in the upper half-plane. We proved that the corresponding conformal maps satisfy the Löwner equation, driven by a suitably defined continuous function $U_t$. In this subsection, we will try to go the other way around. Starting from a driving function $U_t$, we will prove that the Löwner equation generates a (continuous) family of conformal maps $g_t$ onto $\mathbb{H}$. The proof follows Lawler [29].

So suppose that we have a continuous real-valued function $U_t$. Consider for some point $z \in \overline{\mathbb{H}} \setminus \{0\}$ the Löwner differential equation

$$\frac{\partial}{\partial t} g_t(z) = \frac{2}{g_t(z) - U_t}, \qquad g_0(z) = z. \tag{15.10}$$

This equation gives us some immediate information on the behaviour of $g_t(z)$. For instance, taking the imaginary part we obtain

$$\frac{\partial}{\partial t} \operatorname{Im} g_t(z) = \frac{-2 \operatorname{Im} g_t(z)}{(\operatorname{Re} g_t(z) - U_t)^2 + (\operatorname{Im} g_t(z))^2}. \tag{15.11}$$

This shows that for fixed $z \in \mathbb{H}$, $\partial_t \operatorname{Im}[g_t(z)] < 0$, and hence that $g_t(z)$ moves towards the real axis. Further, points on the real axis will stay on the real axis.

For a given point $z \in \overline{\mathbb{H}} \setminus \{0\}$, the solution of the Löwner equation is well-defined as long as $g_t(z) - U_t$ stays away from zero. This suggests that we define a time $\tau(z)$ as the first time $\tau$ such that $\lim_{t \uparrow \tau}(g_t(z) - U_t) = 0$, setting $\tau(z) = \infty$ if this never happens. Note that as long as $g_t(z) - U_t$ is bounded away from zero, equation (15.11) shows that the time derivative of $\operatorname{Im}[g_t(z)]$ is bounded in absolute value by some constant times $\operatorname{Im}[g_t(z)]$. For points $z \in \mathbb{H}$ this shows that in fact, $\tau(z)$ must be the first time when $g_t(z)$ hits the real axis. We set

$$H_t := \{z \in \mathbb{H} : \tau(z) > t\}, \qquad K_t := \{z \in \overline{\mathbb{H}} : \tau(z) \leq t\}. \tag{15.12}$$

Then $H_t$ is the set of points in the upper half-plane for which $g_t(z)$ is still well-defined, and $K_t$ is the closure of its complement, i.e. it is the hull which is excluded from $H_t$. Our goal is now to prove the following theorem.

**Theorem 5.** *Let $U_t$ be a continuous real-valued function, and for every $t \geq 0$ let $g_t(z)$ be the solution of the Löwner equation (15.10). Define the set $H_t$ as in (15.12). Then $g_t(z)$ is a conformal map of the domain $H_t$ onto $\mathbb{H}$ which satisfies*

$$g_t(z) = z + \frac{2t}{z} + O\left(z^{-2}\right), \qquad z \to \infty. \tag{15.13}$$

*Proof.* It is easy to see from (15.10) that $g_t$ is analytic on $H_t$. We will prove (i) that the map $g_t$ is conformal on the domain $H_t$, (ii) that this map is of the form (15.13), and (iii) that $g_t(H_t) = \mathbb{H}$.

To prove (i), we have to verify that $g_t$ has non-zero derivative on $H_t$, and that it is injective. So consider equation (15.10) for times $t < \tau(z)$. Then the differential equation behaves nicely, and we can differentiate with respect to $z$ to obtain

$$\frac{\partial}{\partial t} \log g_t'(z) = -\frac{2}{(g_t(z) - U_t)^2}. \tag{15.14}$$

This gives $|\partial_t \log g_t'(z)| \leq 2/[\operatorname{Im} g_t(z)]^2$. But we know that $\operatorname{Im}[g_t(z)]$ is decreasing. Hence, if we fix $t_0 < \tau(z)$, then the change in $\log g_t'(z)$ is uniformly bounded for all times $t < t_0$. It follows that $\log g_{t_0}'(z)$ is well-defined and bounded and hence, that $g_t'(z)$ is well-defined and non-zero for all $t < \tau(z)$.

Next, choose two different points $z, w \in \mathbb{H}$ and let $t < \min\{\tau(z), \tau(w)\}$. Then

$$\frac{\partial}{\partial t} \log[g_t(z) - g_t(w)] = -\frac{2}{(g_t(z) - U_t)(g_t(w) - U_t)}. \tag{15.15}$$

It follows that $g_t(z) \neq g_t(w)$ for all $t < \min\{\tau(z), \tau(w)\}$, using a similar argument as above. We conclude that $g_t(z)$ is conformal on the domain $H_t$.

For the proof of (ii), we note that (i) implies that the map $g_t(z)$ can be expanded around infinity. We can determine the form of the expansion by integrating the Löwner differential equation from 0 to $t$. This yields

$$g_t(z) - z = \int_0^t \frac{2 \, ds}{g_s(z) - U_s}. \tag{15.16}$$

Consider this equation in the limit $z \to \infty$. Then it is easy to see that the expansion of $g_t(z)$ has no terms of quadratic or higher power in $z$, and no constant term. The form (15.13) follows immediately.

Finally, we prove (iii), i.e. we will show that $g_t(H_t) = \mathbb{H}$. To see this, let $w$ be any point in $\mathbb{H}$, and let $t_0$ be a fixed time. Define $h_t(w)$ for $0 \leq t \leq t_0$ as the solution of the problem

$$\frac{\partial}{\partial t} h_t(w) = -\frac{2}{h_t(w) - U_{t_0 - t}}, \qquad h_0(w) = w. \tag{15.17}$$

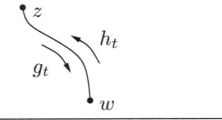

**Fig. 15.4** If the flow of a point $z$ up to a time $t_0$ is described by $g_t(z)$, then $h_t(w)$ as defined in the text describes the inverse flow.

The imaginary part of this equation says that $\partial_t \operatorname{Im}[h_t(w)] > 0$ and hence, that $\operatorname{Im}[h_t(w)]$ is increasing in time. Since $|\partial_t h_t(w)| \le 2/\operatorname{Im}[h_t(w)]$, it follows that $h_t(w)$ is well-defined for all $0 \le t \le t_0$.

We defined $h_t(w)$ such that it describes the inverse of the flow of some point $z \in H_{t_0}$ under the Löwner evolution (15.10) (see Fig. 15.4). To see that this is indeed the case, suppose that for some $t$ between 0 and $t_0$, $h_{t_0-t}(w) = g_t(z)$ for some $z$. Then it follows from the differential equation for $h_t(w)$, that $g_t(z)$ satisfies equation (15.10). This observation holds for all times $t$ between 0 and $t_0$. It follows that such a point $z$ exists, and that it is in fact determined by $z = g_0(z) = h_{t_0}(w)$. In other words, for all $w \in \mathbb{H}$ we have $g_{t_0}(z) = w$ for some $z \in H_{t_0}$. This completes the proof.

We have just proved that a continuous function $U_t$ leads, via the Löwner evolution equation (15.10), to a collection of conformal maps $\{g_t : t \ge 0\}$. These conformal maps are defined on subsets of the upper half-plane, namely the sets $H_t = \mathbb{H} \setminus K_t$, with $K_t$ a growing hull. At this point we still don't know if the maps $g_t(z)$ also correspond to a path $\gamma_t$. But in the next subsection we shall take $U_t$ to be a scaled Brownian motion, and it is known [44] that in this case the Löwner evolution does correspond to a path.

In a few cases the solution of the Löwner equation for $\gamma_t$ and $U_t$, the pair of a path and its corresponding driving term, is known [24]. To find such solutions one may choose a form of $U_t$, so that Löwner's equation can be solved. An alternative is to find paths $\gamma_t$ for which the conformal map $g_t$ is known, and simply calculate $U_t$ from the Löwner equation. Kager e.a. [24] calculated the traces $\gamma_t$ for the cases $U_t$ constant, linearly dependent on $t$ and proportional to $t^{1/2}$ and to $(1-t)^{1/2}$. In the last case the behavior depends critically on the prefactor of the driving term: for small coefficient the path spirals in to a point in $\mathbb{H}$, and for large coefficient the path ends on the real axis.

## 15.3.3 *Chordal SLE in the Half-Plane*

In the previous subsection we showed that the Löwner equation (15.10) driven by a continuous real-valued function generates a set of conformal maps. Furthermore, these conformal maps may correspond to a path in the upper half-plane, as is suggested by the conclusions of section 15.3.1. Chordal SLE$_\kappa$ in the half-plane is obtained by taking scaled Brownian motion as the driving process. We give a precise definition in this subsection.

Let $B_t, t \in [0, \infty)$, be a standard Brownian motion on $\mathbb{R}$, starting from $B_0 = 0$, and let $\kappa > 0$ be a real parameter. For each $z \in \overline{\mathbb{H}} \setminus \{0\}$, consider the Löwner differential equation

$$\frac{\partial}{\partial t} g_t(z) = \frac{2}{g_t(z) - \sqrt{\kappa} B_t}, \quad g_0(z) = z. \tag{15.18}$$

This has a solution as long as the denominator $g_t(z) - \sqrt{\kappa} B_t$ stays away from zero.

For all $z \in \overline{\mathbb{H}}$, just as in the previous subsection, we define $\tau(z)$ to be the first time $\tau$ such that $\lim_{t \uparrow \tau} (g_t(z) - \sqrt{\kappa} B_t) = 0$, and $\tau(z) = \infty$ if this never happens, and we set

$$H_t := \{z \in \mathbb{H} : \tau(z) > t\}, \qquad K_t := \{z \in \overline{\mathbb{H}} : \tau(z) \leq t\}. \tag{15.19}$$

That is, $H_t$ is the set of points in the upper half-plane for which $g_t(z)$ is well-defined, and $H_t = \mathbb{H} \setminus K_t$. The definition is such that $K_t$ is a hull, while $H_t$ is a simply-connected domain. We showed in the previous subsection that for every $t \geq 0$, $g_t$ defines a conformal map of $H_t$ onto the upper half-plane $\mathbb{H}$, that satisfies the normalization $\lim_{z \to \infty} (g_t(z) - z) = 0$.

**Definition 1 (Stochastic Löwner Evolution).** The family of conformal maps $\{g_t : t \geq 0\}$ defined through the stochastic Löwner equation (15.18) is called **chordal SLE$_\kappa$**. The sets $K_t$ (15.19) are the **hulls** of the process.

The SLE$_\kappa$ process defined through equation (15.18) is called chordal, because its hulls are growing from a point on the boundary (the origin) to another point on the boundary (infinity). We will keep using the term chordal for processes going between two boundary points (and not only for SLE processes). Other kinds of processes might for instance grow from a point on the boundary to a point in the interior of a domain. An example of such a process is radial SLE, see section 15.3.5.

It turns out that the hulls of chordal SLE in fact are the hulls of a continuous path $\gamma_t$, that is called the **trace** of the SLE process. It is through this trace that the connection with discrete models can be made. We shall discuss properties of the trace in section 15.4, and we will look at the connection with discrete models in section 15.5. The precise definition of the trace is as follows.

**Definition 2 (Trace).** The **trace** $\gamma$ of SLE$_\kappa$ is defined by

$$\gamma_t := \lim_{z \to 0} g_t^{-1}(z + \sqrt{\kappa} B_t), \tag{15.20}$$

where the limit is taken from within the upper half-plane.

At this point we would like to make some remarks about the choice of time parameterization. Chordal SLE is defined such that the capacity of the hull $K_t$ satisfies $a_1(K_t) = 2t$, and this may seem somewhat arbitrary. But in practice, the choice of time parameterization does not matter for our calculations. The point is, that in SLE calculations we are usually interested in expected values of random variables at the first time when some event happens, that is, at a stopping time. These values are clearly independent from the chosen time parameterization (even if we make a random change of time). For examples of such calculations, see sections 15.4.2 and 15.6.1.

Still, it is interesting to examine how a time-change affects the Löwner equation. So, let $c(t)$ be an increasing and differentiable function defining a change of time. Then $\hat{g}_t := g_{c(t)/2}$ is a collection of conformal transformations parameterized such that $a_1(\hat{K}_t) := a_1(K_{c(t)/2}) = c(t)$. This family of transformations satisfies the equation

$$\frac{\partial}{\partial t}\hat{g}_t(z) = \frac{\frac{d}{dt}c(t)}{\hat{g}_t(z) - \sqrt{k}B_{c(t)/2}}, \qquad \hat{g}_0(z) = z. \tag{15.21}$$

In particular, if we choose $c(t) = 2\alpha t$ for some constant $\alpha > 0$, then the conformal maps $\hat{g}_t$ satisfy

$$\frac{\partial}{\partial t}\frac{1}{\sqrt{\alpha}}\hat{g}_t(\sqrt{\alpha}z) = \frac{2}{\frac{1}{\sqrt{\alpha}}\hat{g}_t(\sqrt{\alpha}z) - \sqrt{\frac{\kappa}{\alpha}}B_{\alpha t}}, \qquad \frac{1}{\sqrt{\alpha}}\hat{g}_0(\sqrt{\alpha}z) = z. \tag{15.22}$$

But the scaling property of Brownian motion shows that the driving term of this Löwner equation is again a standard Brownian motion multiplied by $\sqrt{\kappa}$. This proves the following lemma.

**Lemma 1 (Scaling property of $SLE_\kappa$).** *If $g_t$ are the transformations of $SLE_\kappa$ and $\alpha$ is a positive constant, then the process $(t,z) \mapsto \hat{g}_t(z) := \alpha^{-1/2}g_{\alpha t}(\sqrt{\alpha}z)$ has the same distribution as the process $(t,z) \mapsto g_t(z)$. Furthermore, the process $t \mapsto \alpha^{-1/2}K_{\alpha t}$ has the same distribution as the process $t \mapsto K_t$.*

This lemma is used frequently in SLE calculations. Its significance will be shown already in the following subsection, where we define the $SLE_\kappa$ process in an arbitrary simply connected domain. Meanwhile, the strong Markov property of Brownian motion implies that chordal $SLE_\kappa$ has another basic property, which is referred to as stationarity. Indeed, for any stopping time $\tau$ the process $\sqrt{\kappa}(B_{t+\tau} - B_\tau)$ is itself a standard Brownian motion multiplied by $\sqrt{\kappa}$. So if we use this process as a driving term in the Löwner equation, we will obtain a collection of conformal maps $\hat{g}_t$ which is equal in distribution to the normal $SLE_\kappa$ process.

It is not difficult to see that the process $\hat{g}_t(z)$ in question is in fact the process defined by

$$\hat{g}_t(z) := g_{t+\tau}\left(g_\tau^{-1}(z + \sqrt{\kappa}B_\tau)\right) - \sqrt{\kappa}B_\tau. \tag{15.23}$$

Indeed, taking the derivative of $\hat{g}_t(z)$ with respect to $t$, we find that this process satisfies the Löwner equation

$$\frac{\partial}{\partial t}\hat{g}_t(z) = \frac{2}{\hat{g}_t(z) - \sqrt{\kappa}\left(B_{t+\tau} - B_\tau\right)}, \qquad \hat{g}_0(z) = z. \qquad (15.24)$$

This result establishes the following lemma.

**Lemma 2 (Stationarity of SLE$_\kappa$).** *Let $g_t(z)$ be an SLE$_\kappa$ process in $\mathbb{H}$, and let $\tau$ be a stopping time. Define $\hat{g}_t(z)$ by (15.23). Then $\hat{g}_t$ has the same distribution as $g_t$, and it is independent from $\{g_t : t \in [0, \tau]\}$.*

Observe that the process $\hat{g}_t$ of this lemma is just the original SLE$_\kappa$ process from the time $\tau$ onwards, but shifted in such a way that the new process starts again in the origin. The content of the lemma is that this new process is the same in distribution as the standard SLE$_\kappa$ process, and independent from the history up to time $\tau$. So it is in this sense that the SLE$_\kappa$ process is stationary.

### 15.3.4 Chordal SLE in an Arbitrary Domain

Suppose that $D \subsetneq \mathbb{C}$ is a simply connected domain. Then the Riemann mapping theorem says that there is a conformal map $f : D \to \mathbb{H}$. Now, let $f_t$ be the solution of the Löwner equation (15.18) with initial condition $f_0(z) = f(z)$ for $z \in D$. Then we will call the process $\{f_t : t \geq 0\}$ the SLE$_\kappa$ in $D$ under the map $f$. The connection with the solution $g_t$ of (15.18), with initial condition $g_0(z) = z$, is easily established. Obviously we have $f_t = g_t \circ f$, and if $K_t$ are the hulls associated with $g_t$, then the hulls associated with $f_t$ are $f^{-1}(K_t)$.

Now suppose that we want to consider an SLE$_\kappa$ trace that crosses some domain $D$ from a specified point $a \in \partial D$ to another specified point $b \in \partial D$, $a \neq b$. Then we can find a conformal map $f : D \to \mathbb{H}$ such that $f(a) = 0$ and $f(b) = \infty$. The SLE$_\kappa$ process from $a$ to $b$ in $D$ under the map $f$ is then defined as we discussed above, with starting point $f(a) = 0$.

The map $f$, however, is not determined uniquely. But the maps $\tilde{f}$ of $D$ onto $\mathbb{H}$ that sends $a$ to $0$ and $b$ to $\infty$, have only one free parameter (see section 15.2), scaling the whole map: $\tilde{f}(z) = \alpha f(z)$.

Lemma 1 then tells us that the trace of the SLE$_\kappa$ process in $D$ under $\tilde{f}$ is given simply by a linear time-change of the SLE$_\kappa$ process under $f$. But we explained in the previous subsection that a time-change does not affect our calculations, and may therefore be ignored. Hence, in the sequel, we can simply speak of SLE processes in an arbitrary domain, without mentioning the conformal maps that take these processes to the upper half-plane.

## 15.3.5 Radial SLE

So far we have looked only at chordal Löwner evolution processes, which grow from one point on the boundary of a domain to another point on the boundary. One can also study Löwner evolution processes which grow from a boundary point to a point in the interior of the domain. These are known as **radial** Löwner evolutions. Radial SLE$_\kappa$ in the unit disk, for example, is defined as follows.

Let $B_t$ again be Brownian motion, and $\kappa > 0$. Set $W_t := \exp(i\sqrt{\kappa}B_t)$, so that $W_t$ is Brownian motion on the unit circle starting from 1. Then radial SLE$_\kappa$ is defined to be the solution of the Löwner equation

$$\frac{\partial}{\partial t}g_t(z) = g_t(z)\frac{W_t + g_t(z)}{W_t - g_t(z)}, \quad g_0(z) = z, \quad z \in \overline{\mathbb{D}}. \tag{15.25}$$

The solution again exists up to a time $\tau(z)$ which is defined to be the first time $\tau$ such that $\lim_{t\uparrow\tau}(g_t(z) - W_t) = 0$.

If we set

$$H_t := \{z \in \mathbb{D} : \tau(z) > t\}, \qquad K_t := \{z \in \overline{\mathbb{D}} : \tau(z) \leq t\}, \tag{15.26}$$

then $g_t$ is a conformal map of $\mathbb{D} \setminus K_t = H_t$ onto $\mathbb{D}$. The maps are in this case normalized by $g_t(0) = 0$ and $g_t'(0) > 0$. In fact it is easy to see from the Löwner equation that $g_t'(0) = \exp(t)$, and this specifies the time parameterization.

The trace of radial SLE$_\kappa$ is defined by $\gamma_t := \lim_{z \to W_t} g_t^{-1}(z)$, where now the limit is to be taken from within the unit disk. The trace goes from the starting point 1 on the boundary to the origin. By conformal mappings, one can likewise define radial SLE in an arbitrary simply connected domain, growing from a given point on the boundary to a given point in the interior.

## 15.3.6 Dipolar SLE

A third version of the SLE process is one that can terminate anywhere on a singly connected segment of the boundary. This process is called dipolar SLE [6]. Consider a domain $D$ with three boundary points, $x_-$, $x_0$, and $x_+$, with $x_0 \in (x_-, x_+)$. We consider paths $\gamma_t$ with $\gamma_0 = x_0$ terminating in the interval $(x_+, x_-)$. The conformal map $g_t$ sends $\gamma_{[0,t]}$ to the interval $(x_-, x_+)$. It leaves $x_-$ and $x_+$ invariant and satisfies $g_t'(x_-) = g_t'(x_+)$. The domain in which the defining equation is simplest is the strip $\mathbb{S} = \{z \in \mathbb{C} : 0 < \text{Im } z < 2\pi\}$. The two boundary fixed points are $x_\pm = \pm\infty$ and the starting point $x_0 = 0$:

$$\partial_t g_t(z) = \frac{2}{\tanh(g_t(z) - \sqrt{\kappa}B_t)}, \tag{15.27}$$

with $g_0(z) = z$ and $z \in \mathbb{S}$. Generalized versions of SLE called SLE$_{\kappa,\rho}$ [38, 56] have been constructed, in which the driving term contains a drift with respect to special points on the boundary. This review will not be concerned with this generalization.

## 15.4 Properties of SLE

So far we have not said very much about the parameter $\kappa$. At first sight it looks very innocent, as it scales only the parameter $t$. It can not, however, be eliminated from Löwner's equation, which indicates that it may not be as ineffectual as it seems. In the deterministic context the behavior depends in a qualitative way on the prefactor of the driving term (end of subsection 15.3.2 and [24]), when the driving term has a square-root singularity. This convincingly contradicts the naive intuition that the prefactor $\kappa$ in (15.18) is irrelevant. In this section we show how SLE$_\kappa$ depends qualitatively on its index.

First we shall see that the family of conformal maps $\{g_t : t \geq 0\}$ that is the solution of the stochastic Löwner equation (15.18) does describe a continuous path. We will look at the properties of this path, and we shall describe the connection with the hulls $\{K_t : t \geq 0\}$ of the process. All of this work was done originally by Rohde and Schramm [44]. We shall also see that SLE has some special properties in the cases $\kappa = 6$ (locality) and $\kappa = 8/3$ (restriction), as was shown in [31] and [38]. We end the section by giving the Hausdorff dimensions of the SLE paths, calculated by Beffara [11, 10].

### 15.4.1 Continuity and Transience

In section 15.3.2 we proved that the solution of the Löwner equation is a family of conformal maps onto the half-plane. We then raised the question whether these conformal maps describe a continuous path. Rohde and Schramm [44] proved that for chordal SLE$_\kappa$ this is indeed the case, at least for all $\kappa \neq 8$. The proof by Rohde and Schramm does not work for $\kappa = 8$. But later, Lawler, Schramm and Werner [36] proved that SLE$_8$ is the scaling limit of the Peano curve winding around a uniform spanning tree (more details follow in section 15.5). Thereby, they showed indirectly that the trace is a continuous curve in the case $\kappa = 8$ as well. More precisely, the following theorem holds.

**Theorem 6 (Continuity).** *For all $\kappa \geq 0$ almost surely the limit*

$$\gamma_t := \lim_{z \to 0} g_t^{-1}(z + \sqrt{\kappa}B_t) \qquad (15.28)$$

*exists for every $t \geq 0$, where the limit is taken from within the upper half-plane. Moreover, almost surely $\gamma : [0, \infty) \to \overline{\mathbb{H}}$ is a continuous path and $H_t$ is the unbounded connected component of $\mathbb{H} \setminus \gamma_{[0,t]}$ for all $t \geq 0$.*

In the same paper, Rohde and Schramm also showed that the trace of $SLE_\kappa$ is transient for all $\kappa \geq 0$, that is, $\lim_{t\to\infty} |\gamma_t| = \infty$ almost surely. This proves that the SLE process in the half-plane is indeed a chordal process growing from 0 to infinity.

### 15.4.2 Phases of SLE

The behaviour of the trace of $SLE_\kappa$ depends naturally on the value of the parameter $\kappa$. It is the purpose of this subsection to point out that we can discern three different phases in the behaviour of this trace. The two phase transitions take place at the values $\kappa = 4$ and $\kappa = 8$. A sketch of what the three different phases look like is given in Fig. 15.5.

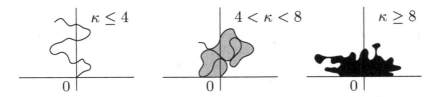

**Fig. 15.5** Simplified impression of SLE in the three different phases. The trace of the SLE process is shown in black. The union of the black path and the grey areas represents the hull.

For $\kappa \in [0,4]$ the $SLE_\kappa$ trace $\gamma$ is almost surely a simple path, i.e. $\gamma_s \neq \gamma_t$ for all $0 \leq t < s$. Moreover, the trace a.s. does not hit the real line but stays in the upper half-plane after time 0. Clearly then, the hulls $K_t$ of the process coincide with the trace $\gamma_{[0,t]}$.

When $\kappa$ is larger than 4, the trace is no longer simple. In fact, for all $\kappa > 4$ every point $z \in \overline{\mathbb{H}} \setminus \{0\}$ a.s. becomes part of the hull in finite time. This means that every point is either on the trace, or is disconnected from infinity by the trace. But as long as $\kappa < 8$, it can be shown that the former happens with probability zero. Therefore, for $\kappa \in (4,8)$ we have a phase where the trace is not dense but does eventually disconnect all points from infinity. In other words, the trace now intersects both itself and the real line, and the hulls $K_t$ now consist of the union of the trace $\gamma_{[0,t]}$ and all bounded components of $\overline{\mathbb{H}} \setminus \gamma_{[0,t]}$.

Finally, when $\kappa \geq 8$ the trace becomes dense in $\mathbb{H}$. In fact, we are then in a phase where $\gamma_{[0,\infty)} = \overline{\mathbb{H}}$ with probability 1, and the hulls $K_t$ coincide with the trace $\gamma_{[0,t]}$ again.

## *15.4.3 Locality and Restriction*

We discussed above the two special values of $\kappa$, 4 and 8, where SLE undergoes a phase transition. Two other special values of $\kappa$, $\kappa = 6$ and $\kappa = 8/3$, have received much attention from the beginning. At these values, $\text{SLE}_\kappa$ has some very specific properties, that will be discussed in detail below.

### 15.4.3.1 The Locality Property of $\text{SLE}_6$

Let us start by giving a precise definition of the locality property. Assume for now that $\kappa > 0$ is fixed. Suppose that $L$ is a hull in $\mathbb{H}$ which is bounded away from the origin. Let $K_t$ be the hulls of a chordal $\text{SLE}_\kappa$ process in $\mathbb{H}$, and let $K_t^*$ be the hulls of a chordal $\text{SLE}_\kappa$ process in $\mathbb{H} \setminus L$, both processes going from 0 to $\infty$. Denote by $T_L$ the first time at which $K_t$ intersects the set $L$. Likewise, let $T_L^*$ be the first time when $K_t^*$ intersects $L$ (note that in this case, $T_L^*$ is the hitting time of an arc on the boundary of the domain). See Fig. 15.6 for an illustration comparing the traces of the two processes in their respective domains.

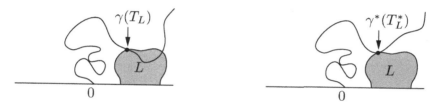

**Fig. 15.6** Comparison of two $\text{SLE}_\kappa$ processes from 0 to $\infty$, in the domain $\mathbb{H}$ (left) and in the domain $\mathbb{H} \setminus L$ (right). If these processes have the same distribution up to the hitting time of the set $L$, then we say that $\text{SLE}_\kappa$ has the locality property.

Chordal $\text{SLE}_\kappa$ is said to satisfy the locality property if for all hulls $L$ bounded away from the origin, the distribution of the hulls $\{K_t : t < T_L\}$ is the same as the distribution of the hulls $\{K_t^* : t < T_L^*\}$, modulo a time re-parameterization. Loosely speaking, suppose that $\text{SLE}_\kappa$ has the locality property, and that we are only interested in the process up to the first time that it hits $L$. Then it doesn't matter whether we consider chordal $\text{SLE}_\kappa$ from 0 to $\infty$ in the domain $\mathbb{H}$, or in the smaller domain $\mathbb{H} \setminus L$. Because the equivalence between these processes may involve a time reparametrization, the hitting times $T_L$ and $T_L^*$ need not be the same, but all the hulls in $\{K_t : t < T_L\}$ appear in $\{K_t^* : t < T_L^*\}$ in the same order.

It was first proved in [31] that chordal $\text{SLE}_\kappa$ has the locality property for $\kappa = 6$, and for no other values of $\kappa$. Later, a much simpler proof appeared in [38]. A sketch of the proof with a discussion of some consequences appears in [29].

So far, we defined the locality property for a chordal process in $\mathbb{H}$, but it is clear that by conformal invariance (invariance under conformal maps) we can translate the property to an arbitrary simply connected domain. It is also true that radial and dipolar $SLE_6$ have the same property. We shall not go into this further, but we would like to point out one particular consequence of the locality property of $SLE_6$.

Suppose that $D$ is a simply connected domain with continuous boundary, and let $a$, $b$ and $b'$ be three distinct points on the boundary of $D$. Denote by $I$ the arc of $\partial D$ between $b$ and $b'$ which does not contain $a$ (see Fig. 15.7 for an illustration). Let $K_t$ (respectively $K_t'$) be the hulls of a chordal $SLE_6$ process from $a$ to $b$ (respectively $b'$) in $D$, and let $T$ (respectively $T'$) be the first time when the process hits $I$. Then modulo a time-change, $\{K_t : t < T\}$ and $\{K_t' : t < T'\}$ have the same distribution. As a result, the hulls of dipolar $SLE_6$ are the same as those of chordal $SLE_6$ up to the time the exit arc of the dipolar process is hit.

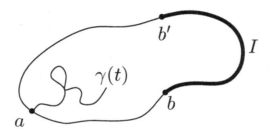

**Fig. 15.7** An $SLE_\kappa$ process aimed towards an arc $I$ on the boundary of a domain.

### 15.4.3.2 The Restriction Property of $SLE_{8/3}$

To define the restriction property, assume that $\kappa \leq 4$ is fixed. Then the trace $\gamma$ of $SLE_\kappa$ is a simple path. Now suppose, as in our discussion of the locality property above, that $L$ is a hull in the half-plane which is bounded away from the origin. Let $\Psi$ be the map defined by $\Psi(z) := g_L(z) - g_L(0)$. Then $\Psi$ is the unique conformal map of $\mathbb{H} \setminus L$ onto $\mathbb{H}$ such that $\Psi(0) = 0$, $\Psi(\infty) = \infty$ and $\Psi'(\infty) = 1$. Now suppose that $\gamma$ never hits $L$. Then we let $\gamma^*$ be the image of $\gamma$ under the map $\Psi$, that is $\gamma_t^* := \Psi(\gamma_t)$.

We say that $SLE_\kappa$ has the restriction property if for all hulls $L$ that are bounded away from the origin, conditional on the event $\{\gamma_{[0,\infty)} \cap L = \emptyset\}$, the distribution of $\gamma_{[0,\infty)}^*$ is the same as the distribution of the trace of a chordal $SLE_\kappa$ process in $\mathbb{H}$, modulo a time re-parameterization. In words, suppose that $SLE_\kappa$ has the restriction property. Then the distribution of all paths that are restricted not to hit $L$, and which are generated by $SLE_\kappa$ in the half-plane, is the same as the distribution of all paths generated by $SLE_\kappa$ in the domain $\mathbb{H} \setminus L$.

SLE has the restriction property for $\kappa = 8/3$ and for no other values of $\kappa$. A proof is given in [38] (a sketch of a proof appears in [29]), and in the same article it was also shown that

$$\mathbf{P}\big[\gamma_{[0,\infty)} \cap L = \emptyset\big] = |\Psi'(0)|^{5/8}. \tag{15.29}$$

Again, the restriction property can be translated into a similar property for arbitrary domains, and radial $SLE_{8/3}$ also satisfies the restriction property. We refer to Lawler, Schramm and Werner [38] and Lawler [29] for more information.

### 15.4.4 Hausdorff Dimensions

Consider an $SLE_\kappa$ process in the upper half-plane. If $\kappa \geq 8$ the trace of the process is space-filling, and therefore the Hausdorff dimension of the set $\gamma_{[0,\infty)}$ is 2. But for $\kappa \in (0,8)$ the Hausdorff dimension of $\gamma_{[0,\infty)}$ is a non-trivial number. Rohde and Schramm [44] showed that its value is bounded from above by $1 + \kappa/8$, and the proof that for $\kappa \neq 4$ the Hausdorff dimension is in fact $1 + \kappa/8$ was completed by Beffara [11, 10]. In the physics literature the Hausdorff dimensions of the curves that are believed to converge to SLE were predicted by Duplantier and Saleur [19, 45].

In the case $\kappa > 4$ the hull of $SLE_\kappa$ is not a simple path, and it is natural to consider also the Hausdorff dimension of the boundary of $K_t$ for some fixed value of $t > 0$. Its value is conjectured to be $1 + 2/\kappa$, because (based on a duality relation derived by Duplantier [19]) it is believed that the boundary of the hull for $\kappa > 4$ is described by $SLE_{16/\kappa}$. The dimension of the hull boundary is known rigorously only for $\kappa = 6$ (where it is $4/3$) and for $\kappa = 8$ (where it is $5/4$). For $\kappa = 6$ this follows from the study of the "conformal restriction measures" in [38], for $\kappa = 8$ this is a consequence of the strong relation between loop-erased random walks and uniform spanning trees [36] (section 15.5.3).

## 15.5 SLE and Discrete Models

For a number of discrete lattice models, the scaling limit (of some of its observables) has been proven to be SLE. For many more models such a connection is only conjectured. Typically the stochastic measure of these models induces a measure on paths which in the scaling limit converges to the trace of an SLE process.

### 15.5.1 Critical Percolation

We define site percolation on the triangular lattice as follows. All vertices of the lattice are independently coloured blue with probability $p$ or yellow with probability $1 - p$. An equivalent, visually more attractive, viewpoint is to say that we colour all hexagons of the dual lattice blue or yellow with probabilities $p$ and $1 - p$, respectively. It is well-known that for $p \leq 1/2$, there is almost surely no infinite cluster of connected blue hexagons, while for $p > 1/2$ there a.s. exists a unique infinite blue cluster. This makes $p = 1/2$ the critical point for site percolation on the triangular lattice. This critical percolation model is discussed here.

Let us for now restrict ourselves to the half-plane. Suppose that as our boundary conditions, we colour all hexagons intersecting the negative real line yellow, and all hexagons intersecting the positive real line blue. All other hexagons in the half-plane are independently coloured blue or yellow with equal probabilities. Then there exists a unique path over the edges of the hexagons, starting in the origin, which separates the cluster of blue hexagons attached to the positive real half-line from the cluster of yellow hexagons attached to the negative real half-line. This path is called the chordal exploration process from 0 to $\infty$ in the half-plane. It is the unique path from the origin such that at each step there is a blue hexagon on the right, and a yellow hexagon on the left. See Fig. 15.8 for an illustration.

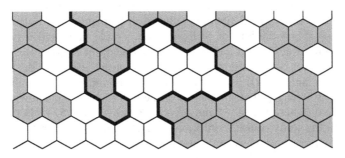

**Fig. 15.8** Part of the percolation exploration process in the half-plane.

The exploration process depends only on the hexagons it passes. This makes it possible to generate it dynamically as follows. Initially, only the hexagons on the boundary receive a colour. Then after each step, the exploration process meets a hexagon, which may or may not have been coloured. If it has not yet been coloured, its colour is decided with equal probability for yellow or blue. Then the exploration process proceeds always keeping the yellow hexagon on its left. Note that the tip of the process cannot become trapped, because it is forced to reflect off into the open if it meets an already coloured hexagon. This property it has in common with the trace of a chordal SLE process. The process, here described for the half-plane, can be generalized without difficulty to other domains. In addition, it is easy to envision closed polygons to be generated by the percolation process. Each percolation configuration in a domain of which the entire boundary has one colour, induces closed polygons separating the blue from the yellow hexagons. The partition function of such polygons is a sum over all possible closed polygons on the infinite hexagonal lattice, with a weight 1/2 for each hexagon touching the polygon. If this polygon is interpreted as one of the domain walls in the percolation process, each hexagon touching the polygon has to be coloured with probability 1/2, all others being free. Besides the stochastics of the local exploration behaviour of such paths, which is the same as that of the open path far from the boundary, this partition function also induces the distribution of the length of such paths.

Smirnov [51] proved that in the continuum limit, the exploration process is conformally invariant. Together with the results on $SLE_6$ developed by Lawler, Schramm and Werner, this should prove that the exploration process converges to the trace of $SLE_6$ in the half-plane. Thus, $SLE_6$ may be used to calculate properties of critical percolation. Some examples are described in section 15.6.

## 15.5.2 The Harmonic Explorer

The harmonic explorer is a random path similar to the exploration process of critical percolation. It was defined recently by Schramm and Sheffield as a discrete process that converges to $SLE_4$ [48]. To define the harmonic explorer, consider an approximation of a bounded domain with hexagons, as in Fig. 15.9. As we did for critical percolation, we partition the set of hexagons on the boundary of our domain into two components, and colour the one component yellow and the other blue. The hexagons in the interior are uncoloured initially.

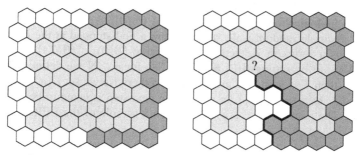

**Fig. 15.9** Left: the initial configuration for the harmonic explorer, with blue hexagons (dark faces), yellow hexagons (white faces) and uncoloured hexagons (light faces). Right: a part of the harmonic explorer process. The colour of the marked hexagon is determined as described in the text.

The harmonic explorer is a path over the edges of the hexagons that starts out on the boundary with a blue hexagon on its right and a yellow hexagon on its left. It turns left when it meets a blue hexagon, and it turns right when it meets a yellow hexagon. The only difference with the exploration process of critical percolation is in the way the colour of an as yet uncoloured hexagon is determined. For the harmonic explorer this is done as follows.

Suppose that the harmonic explorer meets an uncoloured hexagon (see Fig. 15.9). Let $f$ be the function, defined on the faces of the hexagons, that takes the value 1 on the blue hexagons, the value 0 on the yellow hexagons, and is discrete harmonic on the uncoloured hexagons (for each hexagon the value of $f$ is the average of that of its neighbors). Then the probability that the hexagon whose colour we want to determine is made blue, is given by the value of $f$ on this hexagon. Proceeding in this way, we obtain a path crossing the domain between the two points on the boundary where the blue and yellow hexagons meet. In the scaling limit this path converges to the trace of chordal $SLE_4$.

## 15.5.3 Loop-Erased Random Walks and Uniform Spanning Trees

In this subsection we consider loop-erased random walks (LERW's) and uniform spanning trees (UST's). We shall define both models first, and we will point out the close relation between the two. Schramm [46] already proved that the LERW converges to $SLE_2$ under the assumption that the scaling limit exists and is conformally invariant. In the same work, he also conjectured the relation between UST's and $SLE_8$. The final proofs of these connections were given by Lawler, Schramm and Werner in [36]. Their proofs hold for general lattices, but for simplicity, we shall restrict our description here to finite subgraphs of the square grid $\delta \mathbb{Z}^2$ with mesh $\delta > 0$.

Suppose that $G$ is a finite connected subgraph of $\delta \mathbb{Z}^2$. Let $u$ be a vertex of $G$ and let $V$ be a collection of vertices of $G$ not containing $u$. Then the LERW from $u$ to $V$ in $G$ is defined by taking a simple random walk in $G$ from $u$ to $V$ and erasing all its loops in chronological order. More precisely, if $\big(\omega(0), \ldots, \omega(T_V)\big)$ are the vertices visited by a simple random walk starting from $u$ and stopped at the first time $T_V$ when it visits a vertex in $V$, then its loop-erasure $\big(\beta(0), \ldots, \beta(T)\big)$ is defined as follows. We start by setting $\beta(0) = \omega(0)$. Then for $n \in \mathbb{N}$ we define inductively: if $\beta(n) \in V$ then $T = n$ and we are done, and otherwise we set $\beta(n+1) = \omega(1 + \max\{m \le T_V : \omega(m) = \beta(n)\})$. In words: the next step is taken to be the last exit from $\beta(n)$. The path $\big(\beta(0), \ldots, \beta(T)\big)$ is then a sample of the LERW in $G$ from $u$ to $V$.

A spanning tree $T$ in $G$ is a subgraph of $G$ such that every two vertices of $G$ are connected via a unique simple path in $T$. A *uniform* spanning tree (UST) in $G$ is a spanning tree chosen with the uniform distribution from all spanning trees in $G$. There is an interesting connection between UST and LERW:

**Theorem 7.** *The unique path between two distinct vertices $u$ and $v$ on a UST, is distributed as the LERW from $u$ to $v$.*

*Proof.* There are several ways to see this connection. One proceeds via a Monte Carlo (MC) process to generate UST. A MC process is a stochastic process of which the stationary distribution is the measure one wants to sample. The MC process used here is driven by a random walk on the graph $G$. Every time an edge is traversed by the random walk it is included in the tree, at the expense of the edge of the newly occupied node, by which the random walk left the node on its previous visit. Since that edge is the first step from that node to the current position in the random walk, the loop is removed. The edge added and the one removed need not be different. That the uniform measure is stationary under this process, follows from the fact that for every move in the process there is precisely one counter move, and it is equally probable. Moreover, when the walk starts in a node $u$, the first time it arrives in a preselected subset $V$, the unique path from $u$ to the arrival point in $V$ is precisely a LERW.

An interesting collorary of this result is that the LERW is symmetric between beginning and end. This is not at all obvious from its definition. An algorithm to generate UST's, somewhat related to the MC process above is known as Wilson's algorithm [60].

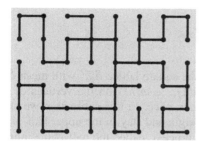 

**Fig. 15.10** Examples of spanning trees on a rectangular graph. In the right-hand panel, the nodes in the left and bottom boundary are fully connected. The thin curve from the lower right to the upper left corner closely surrounds the tree, and is space filling. Such paths are called Peano curves.

The connection between LERW and $SLE_2$ can be described as follows. Choose a subgraph of $\mathbb{Z}^2$, with a single, connected boundary, and a UST on it. Figure 15.10 in the left panel shows a rectangular example with the tree indicated by solid bonds. The unique path between a node $u$ on the boundary and another node $v$ has $SLE_2$ as its scaling limit. When $v$ (or its limiting point in the scaling procedure) is also on the boundary in the chordal process, and if $v$ is in the interior, it is radial SLE.

To see in what way UST converges to $SLE_8$, consider the right panel of Fig. 15.10. It also shows a rectangular subgraph of $\mathbb{Z}^2$ with a spanning tree. Now the left and bottom boundary nodes are fully connected to each other. One can trace a path which closely surrounds the spanning tree. For spanning trees on the square lattice these paths are called Peano curves. It visits all the sites of a square lattice with half the mesh size of the original lattice, while it never intersects itself. In general, non-intersecting paths visiting every site of a graph once are called Hamiltonian walks.

For this specific boundary condition on the spanning tree, such that all vertices on the left and bottom of the rectangle are connected, the tree induces a Peano curve which runs from the lower right corner to the upper left, and visits all sites in between. It is this path which has chordal $SLE_8$ as its scaling limit.

While we need open paths from the boundary to make the connection to SLE, the LERW and UST also define closed polygons of well-defined distributions as follows. Take a UST, and add one bond uniformly chosen from all edges not included in the tree. This creates a closed polygon, with the properties of the LERW.

The same distribution would be obtained by starting a LERW at some point in the lattice, and terminating it the first time it visits the starting point after it has been away from the starting point at least two steps. This last restriction is only to suppress trivial walks of two steps. A closed Peano curve can be defined by removing one bond from the spanning tree, which necessarily cuts it in two. The closed Peano curve surrounding each of the two parts is a closed polygon. These two polygons together fill the original domain. The same distribution of polygons can be obtained by adding an arbitrary bond to the tree, and taking the Peano curve which traces the inside of the loop uniquely created by the extra bond.

## 15.5.4 Self-Avoiding Walks

A self-avoiding walk (SAW) of length $n$ on the square lattice $\delta \mathbb{Z}^2$ with mesh $\delta > 0$ is a nearest-neighbour path $\omega = \big(\omega(0), \omega(1), \ldots, \omega(n)\big)$ on the vertices of the lattice, such that no vertex is visited more than once. In this subsection we shall restrict ourselves to SAW's that start in the origin and stay in the upper half-plane afterwards. The idea is to define a stochastic process, called the half-plane infinite SAW, that in the scaling limit $\delta \downarrow 0$ is believed to converge to chordal $\mathrm{SLE}_{8/3}$.

Following [37] we write $\Lambda_n^+$ for the set of all SAW's $\omega$ of length $n$ that start at the origin, and stay above the real line afterwards. For a given $\omega$ in $\Lambda_n^+$, let $Q_k^+(\omega)$ be the fraction of walks $\omega'$ in $\Lambda_{n+k}^+$ whose beginning is $\omega$, i.e. such that $\omega'(i) = \omega(i)$ for $0 \le i \le n$. Define $Q^+(\omega)$ as the limit of $Q_k^+(\omega)$ as $k \to \infty$. Then $Q^+(\omega)$ is roughly the fraction of very long SAW's in the upper half-plane whose beginning is $\omega$. It was shown by Lawler, Schramm and Werner that the limit $Q^+(\omega)$ exists [37].

Now we can define the *half-plane infinite self-avoiding walk* as the stochastic process $X_i$ such that for all $\omega = \big(0, \omega(1), \ldots, \omega(n)\big) \in \Lambda_n^+$,

$$\mathbf{P}[X_0 = 0, X_1 = \omega(1), \ldots, X_n = \omega(n)] = Q^+(\omega). \tag{15.30}$$

We believe that the scaling limit of this process as the mesh $\delta$ tends to 0 exists and is conformally invariant. By the restriction property the scaling limit has to be $\mathrm{SLE}_{8/3}$, as pointed out in [37]. At this moment it is unknown how the existence, let alone the conformal invariance, of the scaling limit can be proved. However, the knowledge of its properties is still growing [57].

Lawler, Schramm and Werner [37] also explain how one can define a natural measure on SAW's with arbitrary starting points, leading to conjectures relating SAW's to chordal and radial $\mathrm{SLE}_{8/3}$ in bounded simply-connected domains. The article further discusses similar conjectures for self-avoiding polygons, and predictions for the critical exponents of SAW's that can be obtained from SLE. We shall not go into these topics here.

## 15.5.5 The Critical Ising Model

The Ising model is the prototypical model for a phase transition. It is solvable not only at the critical point, but also at other temperatures. In the physics literature the conformal invariance of its scaling is almost always taken for granted. As a consequence there were not many attempts by physicists to formally prove this property. However, the model is so extensively studied and so much is known about it in detail that it may well be that some assertions in the physics literature of the conformal invariance of certain correlation functions, are open to complete proof.

Only recently Smirnov [53] presented a formal proof that some observables in the scaling limit of the Ising model are conformally invariant, and as a consequence are $SLE_{16/3}$. This proof is based on an exploration process of the Fortuin-Kasteleyn random cluster [23] representation of the model. This connection will be discussed in the next section, but the percolation exploration process, and that exploring the UST, discussed in the previous subsections 15.5.1 and 15.5.3 are examples of it. What Smirnov was able to prove is that the probability that a particular point $z$ is on the trace of the exploration process, is the absolute value of a discrete analytic function of $z$. The phase of this function is proportional to the winding angle at which the trace passes $z$.

## 15.5.6 The Potts Model

So far in this section we discussed relations between SLE at specific values of $\kappa$ to certain statistical lattice models. The results of SLE however suggest a further connection to continuous families of models. This subsection deals with the $q$-state Potts model, a natural generalization of the Ising model, which has $q = 2$. Below we will give a standard treatment [9], which relates the partition sum of the Potts model to an ensemble of multiple paths on the lattice. In the scaling limit these paths will be the candidates for the SLE processes.

The Potts model has on each site of a lattice a variable $s_j$ which can take values in $\{1, 2, \ldots, q\}$. Of these variables only nearest neighbours interact such that the energy is $-1$ if both variables are in the same state and 0 otherwise. The canonical partition sum is

$$Z = \sum_{\{s\}} \exp\left( \beta \sum_{\langle j,k \rangle} \delta_{s_j, s_k} \right). \tag{15.31}$$

The summation in the exponent is over all nearest-neighbour pairs of sites, and the external summation over all configurations of the $s_j$. The model is known to be disordered at high temperatures, and ordered at low temperatures. Here we are interested in the behaviour at the transition.

In order to make the connection with a path on the lattice, we express this partition sum in a high-temperature expansion, i.e. in powers of a parameter which is small when $\beta$ is small. The first step is to write the summand as a product:

**Fig. 15.11** Graph in the cluster expansion of the Potts model in a rectangular domain. The nodes of the left and bottom boundary are fully connected, so that there is one open path, from the top left to the bottom right corner.

$$Z = \sum_{\{s\}} \prod_{\langle j,k \rangle} \left[ 1 + (e^\beta - 1) \delta_{s_j,s_k} \right].$$                                    (15.32)

The product can be expanded in terms in which at every edge of the lattice a choice is made between the two terms 1 and $(e^\beta - 1)\delta_{s_j,s_k}$. In a graphical notation we place a bond on every edge of the lattice where the second term is chosen, see Fig. 15.11. For each term in the expansion of the product the summation over the $s$-variables is trivial: if two sites are connected by bonds, their respective $s$-variables take the same value, and are independent otherwise. As a result the summation over $\{s\}$ results in a factor $q$ for each connected component of the graph. Hence

$$Z = \sum_{\text{graphs}} (e^\beta - 1)^b q^c,$$                                    (15.33)

where $c$ is the number of connected components of the graph and $b$ the number of bonds. This expansion is known by the name of Fortuin-Kasteleyn [23] cluster model. Note that, while $q$ has been introduced as the (integer) number of states, in this expansion it can take any value.

It is convenient to rewrite the cluster expansion as an expansion of paths on a new lattice, called the surrounding lattice. The edges of the original lattice correspond to the vertices of the surrounding lattice. The clusters on the original lattice are rewritten into polygon decompositions of the new lattice. Every vertex of the surrounding lattice is separated into two non-intersecting path segments. These path segments intersect the corresponding edge of the original lattice if and only if this edge does not carry a bond of the graph, as follows:

As a result of these transformations the new lattice is decomposed into a collection of non-intersecting paths, as indicated in Fig. 15.11. Notice that every component of the original graph is surrounded by one of these closed paths, but also the closed circuits of the graph are inscribed by these paths. By Euler's relation the number of components $c$ of the original graph can be expressed in the number of bonds $b$, the total number of sites $N$ and the number of polygons $p$: $c = (N - b + p)/2$. An alternative expression for the partition sum is then

$$Z = \sum_{\text{graphs}} \left( \frac{e^\beta - 1}{\sqrt{q}} \right)^b q^{(N+p)/2}. \tag{15.34}$$

At the critical point $\beta_c$ the relation $\exp(\beta_c) = 1 + \sqrt{q}$ holds, so that the partition sum simplifies.

We will now consider this model at the critical point on a rectangular domain. The lattice approximation of this domain is chosen such that the lower-left corner of the rectangle coincides with a site of the lattice, while the upper-right corner coincides with a site of the dual lattice. The sides of the rectangle are parallel to the edges of the lattice, as in Fig. 15.11. We choose as boundary condition that all edges that are contained in the left and lower sides of the rectangle carry bonds, and all edges that intersect the right and upper sides perpendicularly carry no bonds. For the spin variables this means that all the spins on the left and lower sides are in the same state, while all other spins are unconstrained.

In such an arrangement the diagrams in (15.34) include one path from the lower-right to the upper-left corner. All further paths are closed polygons, see Fig. 15.11. We take the scaling limit by covering the same domain with a finer and finer mesh. It is believed [44] that in the scaling limit the measure on the paths approaches that of chordal $SLE_\kappa$ traces. From e.g. the Hausdorff dimension [10, 45] the relation between $\kappa$ and $q$ is

$$q = 2 + 2\cos(8\pi/\kappa) \tag{15.35}$$

where $4 \leq \kappa \leq 8$. Only in a few cases has this relationship between $SLE_\kappa$ and the Potts partition sum been made rigorous. For instance, in the limit $q \to 0$, the graph expansion reduces to the uniform spanning tree, which has $SLE_8$ as its scaling limit.

### 15.5.7 The O(n) Model

We now turn to the O($n$) model, which is another well-known model already discussed in Chapter 1, where a high-temperature expansion results in a sum over paths. Here the dynamic variables are $n$-component vectors of a fixed length, and the Hamiltonian is invariant under rotations in the $n$-dimensional space. The simplest high-temperature expansion is obtained when the Boltzmann weight is chosen as

$$\prod_{\langle j,k \rangle} (1 + x \, s_i \cdot s_j), \tag{15.36}$$

where the product is over nearest neighbours on a hexagonal lattice. The partition sum is obtained by integrating this expression over the directions of the spin vectors. As for the Potts model, one can expand the product and do the bookkeeping of the terms by means of graphs. In each factor in (15.36) the choice of the second term is indicated by a bond. Then the graphs that survive the integration over the spin variables have only even vertices, i.e. on the hexagonal lattice vertices with zero or two bonds. As a result the graphs consist of paths on the lattice. In a well-chosen normalization of the measure and the length of the spins, the partition sum is a sum over even graphs

$$Z = \sum_{\text{graphs}} x^L n^M, \tag{15.37}$$

where $M$ is the number of closed loops, and $L$ their combined length. Note that this expression for the partition sum is well-defined also when the number of spin components $n$ is not integer. It is known [8, 41] that the critical point is at $x_c = [2 + (2 - n)^{1/2}]^{-1/2}$ for $0 \leq n \leq 2$. When $x$ is larger than this critical value, the model also shows critical behaviour.

Consider now this model on a bounded domain, and take a correlation function between two spins on the boundary. The diagrams that contribute to this function contain one path between the two specified boundary points and any number of closed polygons in the interior. We conjecture that at the critical value of $x$ in the scaling limit, the measure on the paths between the two boundary spins approaches that of chordal $SLE_\kappa$ for $n = -2\cos(4\pi/\kappa)$ and $8/3 \leq \kappa \leq 4$. For larger values of $x$, the scaling limit would again be $SLE_\kappa$, with the same relation between $\kappa$ and $n$, but now with $4 \leq \kappa \leq 8$.

To conclude this section, we remark that the same partition sum (15.37) can also be viewed as the partition sum of a dilute Potts model on the triangular lattice, described in [43]. In this variant of the Potts model the spins take values in $\{0, 1, 2, \ldots, q\}$. The model is symmetric under permutations of the $q$ positive values. The name dilute comes from the interpretation of the neutral value 0 as a vacant site. If neighbouring sites take different values, then one of them takes the value 0. The Boltzmann weight is a product over the elementary triangles of weights that depend on the three sites at the corners of the triangle. We take this weight to be 1 when all three sites are in the same state, vacant or otherwise. Triangles with one or two vacant sites have weights $xy$ and $x/y$, respectively. The partition sum can be expanded in terms of domain walls between sites of different values. This expansion takes the form of (15.37) for $y^{12} = q = n^2$, which is the locus of the phase transition between an ordered phase and a disordered phase. Within this locus, the region with $x > x_c$ is a second-order transition. In the regime $x < x_c$ the transition is discontinuous, and the position $x = x_c$ separates the two regimes and is called the tricritical point. When $q = x = 1$ the site percolation problem on the triangular lattice is recovered, which is known to converge to $SLE_6$ in the scaling limit.

## 15.6 SLE Computations and Results

In this section we discuss some of the results that have been obtained from calculations involving SLE processes. Our aim in this section is not only to provide an overview of these results, but also to give an impression of the typical SLE computations involved, using techniques from stochastic calculus and conformal mapping theory.

This section is organized as follows: In the first subsection we discuss several SLE calculations independently from their connection with other models. The results we obtain will be key ingredients for further calculations. The second subsection gives a brief overview of how SLE can be applied to calculate the intersection exponents of Brownian motion. Finally, we will discuss results on critical percolation that have been obtained from its connection with $SLE_6$.

### 15.6.1 Several SLE Calculations

The purpose of this subsection is to show what kind of probabilities and corresponding exponents of events involving chordal SLE processes can be calculated. The results we find in this subsection are for whole ranges of $\kappa$, and might therefore have applications in various statistical models. In this overview we do not include proofs and extensive calculations. These can be found in [25] with references to the original literature.

#### 15.6.1.1 Crossing and Passage Probabilities

Consider a chordal $SLE_\kappa$ process inside the rectangle $\mathscr{R}_L := (0,L) \times (0,\pi)$, which goes from $i\pi$ to $L$. If $\kappa > 4$ this process will at some random time $\tau$ hit the right edge $[L, L+i\pi]$ of the rectangle, as in Fig. 15.12. Suppose that $E$ denotes the event that up to this time $\tau$, the SLE process has not hit the lower edge of the rectangle. Then the following holds:

**Theorem 8.** *The $SLE_\kappa$ process as described above satisfies, for $\kappa > 4$,*

$$P[E] \asymp \exp\left[-\left(1 - \frac{4}{\kappa}\right)L\right] \quad as\ L \to \infty, \tag{15.38}$$

*where $\asymp$ indicates that each side is bounded by some constant times the other side.*

The theorem in this form is proved in [25]. Since the definition of chordal SLE in any domain is based on that in the upper half-plane, the first step in the proof is a mapping of the rectangle to $\mathbb{H}$.

The original form of the theorem is more general: Consider again an $SLE_\kappa$ process crossing the rectangle $\mathscr{R}_L$ from $i\pi$ to $L$. On the event $E$ the trace $\gamma$ has crossed

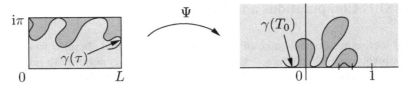

**Fig. 15.12** An SLE process crossing a rectangle, and its translation to the upper half-plane. The darker grey areas represent the hulls of the processes.

the rectangle without hitting the bottom edge. So conditional on this event, the $\pi$-extremal distance between $[0, i\pi]$ and $[L, L + i\pi]$ in $\mathscr{R}_L \setminus K_\tau$ is well-defined. Let us call this $\pi$-extremal distance $\mathscr{L}$. Then one can prove the following generalization of theorem 8 [31].

**Theorem 9.** *For any $\lambda \geq 0$ and $\kappa > 4$,*

$$E[1_E e^{-\lambda \mathscr{L}}] \asymp \exp\left[-u(\kappa, \lambda)L\right] \qquad as\ L \to \infty, \tag{15.39}$$

*where*

$$u(\kappa, \lambda) = \lambda + \frac{\kappa - 4 + \sqrt{(\kappa - 4)^2 + 16\kappa\lambda}}{2\kappa}. \tag{15.40}$$

The exponent $u(\kappa, \lambda)$ is called the one-sided crossing exponent, because it measures the extremal distance on one side of an SLE process crossing a rectangle. Observe that $u(\kappa, \lambda)$ reduces to the exponent $1 - 4/\kappa$ for $\lambda = 0$ as it should, because in this case theorem 9 is completely analogous to theorem 8.

There is an analogue of the one-sided crossing exponent for radial SLE, which we shall discuss only briefly here. The setup is as follows. We consider radial SLE$_\kappa$ for any $\kappa > 0$, and set $A_t := \partial \mathbb{D} \setminus K_t$. Then the set $A_t$ is either a piece of arc of the unit circle, or $A_t = \emptyset$. Let $r > 0$ and let $T(r)$ be the first time when the SLE process hits the circle $\{z : |z| = r\}$. Denote by $E$ the event that $A_{T(r)}$ is non-empty. On the event $E$, let $\mathscr{L}$ be the $\pi$-extremal distance between the circles $\{z : |z| = 1\}$ and $\{z : |z| = r\}$ in $\mathbb{D} \setminus K_{T(r)}$, see Fig. 15.13.

**Theorem 10.** *For all $\lambda > 0$ and $\kappa > 0$,*

$$E[1_E e^{-\lambda \mathscr{L}}] \asymp r^{-v(\kappa, \lambda)} \qquad as\ r \downarrow 0, \tag{15.41}$$

*where*

$$v(\kappa, \lambda) = \frac{8\lambda + \kappa - 4 + \sqrt{(\kappa - 4)^2 + 16\kappa\lambda}}{16}. \tag{15.42}$$

We call $v(\kappa, \lambda)$ the annulus crossing exponent of SLE$_\kappa$. A detailed proof of the theorem can be found in [32].

So far, we have considered several crossing events of SLE processes. A different kind of event, namely the event that the trace of SLE passes to the left of a given point $z_0$, was studied by Schramm in [47].

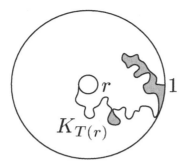

**Fig. 15.13** An SLE process crossing an annulus.

**Theorem 11.** *Let $\kappa \in [0,8)$ and $z_0 = x_0 + iy_0 \in \mathbb{H}$. Suppose that $E$ is the event that the trace $\gamma$ of chordal $SLE_\kappa$ passes to the left of $z_0$. Then*

$$P[E] = \frac{1}{2} + \frac{\Gamma\left(\frac{4}{\kappa}\right)}{\sqrt{\pi}\,\Gamma\left(\frac{8-\kappa}{2\kappa}\right)}\, {}_2F_1\left(\frac{1}{2}, \frac{4}{\kappa}; \frac{3}{2}; -\frac{x_0^2}{y_0^2}\right)\frac{x_0}{y_0}. \tag{15.43}$$

This expression played an interesting role in the proof by Smirnov [51] that the percolation exploration process has $SLE_6$ as its scaling limit. When the half plane is mapped onto a unilateral triangle with the origin, $\infty$ and another real point mapped to corners of the triangle, the hypergeometric function appearing in the theorem for $\kappa = 6$ simplifies to a simple linear function. This suggested that things might become simpler for this value of $\kappa$ and for a lattice model with hexagonal symmetry, in the case of site percolation on the triangular lattice.

## 15.6.2 Intersection Exponents of Planar Brownian Motion

One of the first successes of SLE was the determination of the intersection exponents of planar Brownian motion. One way of defining these exponents is as follows (see reference [30], which also presents alternative definitions). Let $k \geq 2$ and $p_1, \ldots, p_k$ be positive integers. At time $t = 0$, from each of the points $(0, j)$ (for $j \in \{1, \ldots, k\}$) we start $p_j$ planar Brownian motions. Then we can define an exponent $\xi(p_1, \ldots, p_k)$ from the probability that at time $t$ none of the Brownian walkers has hit the path of another one that started at a different initial position. Formally, denote by $\mathscr{B}_t^j$ the union of the traces of the $p_j$ Brownian motions started from $(O, j)$ up to time $t$. Then

$$\mathbf{P}\left[\forall i \neq j \in \{1, \ldots, k\}, \mathscr{B}_t^i \cap \mathscr{B}_t^j = \emptyset\right] \asymp \left(\sqrt{t}\right)^{-\xi(p_1, \ldots, p_k)} \tag{15.44}$$

when $t \to \infty$. The exponent $\xi(p_1, \ldots, p_k)$ is called the intersection exponent between $k$ packets of $p_1, \ldots, p_k$ Brownian motions.

If we further require that the Brownian motions stay in the upper half-plane, we get different exponents $\tilde{\xi}(p_1, \ldots, p_k)$ defined by

$$\mathbf{P}\left[\forall i \neq j \in \{1, \ldots, k\}, \mathscr{B}_t^i \cap \mathscr{B}_t^j = \emptyset \text{ and } \mathscr{B}_t^i \subset \mathbb{H}\right] \asymp \left(\sqrt{t}\right)^{-\tilde{\xi}(p_1, \ldots, p_k)} \qquad (15.45)$$

when $t \to \infty$. We could also *condition* on the event that the Brownian motions stay in the upper half-plane. The corresponding exponents are $\hat{\xi}(p_1, \ldots, p_k)$. They are related to the previous half-plane exponents by

$$\hat{\xi}(p_1, \ldots, p_k) = \tilde{\xi}(p_1, \ldots, p_k) - (p_1 + \ldots + p_k), \qquad (15.46)$$

since the probability that a Brownian motion started in the half-plane stays in the half-plane up to time $t$ decays like $t^{-1/2}$.

Duplantier and Kwon [20] predicted the values of the intersection exponents $\xi(p_1, \ldots, p_k)$ and $\hat{\xi}(p_1, \ldots, p_k)$ in the case where all $p_i$ are equal to 1. In the series of papers [31, 32, 33, 34], Lawler, Schramm and Werner confirmed these predictions rigorously, and generalized them. Here, we will give an impression of the arguments used in the first paper [31], and then we will summarize the main conclusions of the series.

### 15.6.2.1 Intersection Exponents Generalized

In [30] Lawler and Werner show how the definition of the Brownian intersection exponents can be extended in a natural way. This leads to the definition of the exponents $\tilde{\xi}(\lambda_1, \ldots, \lambda_k)$ for all $k \geq 1$ and all non-negative real numbers $\lambda_1, \ldots, \lambda_k$, and of the exponents $\xi(\lambda_1, \ldots, \lambda_k)$ for all $k \geq 2$ and nonnegative real numbers $\lambda_1, \ldots, \lambda_k$, at least two of which must be at least 1.

It is then convenient to define the exponents not in terms of ordinary Brownian motions, but of Brownian excursions [30, 31]. A Brownian excursion in a domain is a measure of Brownian motions up to the time they hit the boundary, in the limit that the starting point is taken to the boundary, with an appropriate normalization for the measure to remain bounded and non-zero. Let $\mathscr{R}_L$ be the rectangle $(0, L) \times (0, \pi)$, and denote by $\omega$ the path of a Brownian excursion in $\mathscr{R}_L$ started from the left side. Let $A$ be the event that the Brownian excursion crosses the rectangle from the left to the right. In other words that the right boundary is hit before the bottom or top boundary is hit. On this event, let $D_+$ and $D_-$ be the domains remaining above and below $\omega$ in $\mathscr{R}_L \setminus \omega$, respectively, and let $\mathscr{L}_+$ and $\mathscr{L}_-$ be the $\pi$-extremal distances between the left and right edges of the rectangle in these domains. We refer to Fig. 15.14 for an illustration.

By symmetry, the distributions of $\mathscr{L}_+$ and $\mathscr{L}_-$ are the same. The exponent $\tilde{\xi}(1, \lambda)$ is characterized by

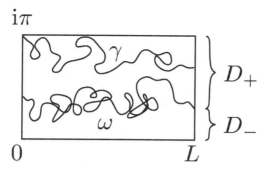

$i\pi$

$0$                    $L$

**Fig. 15.14** An SLE$_6$ trace $\gamma$ and a Brownian excursion $\omega$ crossing a rectangle.

$$\mathbf{E}_B[1_A\,e^{-\lambda\mathscr{L}_+}] = \mathbf{E}_B[1_A\,e^{-\lambda\mathscr{L}_-}] \asymp e^{-\tilde{\xi}(1,\lambda)L} \qquad \text{as } L \to \infty \qquad (15.47)$$

where $\mathbf{E}_B$ is used to indicate expectation with respect to the Brownian excursion measure. Likewise, $\tilde{\xi}(\lambda_+,1,\lambda_-)$ is characterized by

$$\mathbf{E}_B[1_A\,e^{-\lambda_+\mathscr{L}_+}e^{-\lambda_-\mathscr{L}_-}] \asymp e^{-\tilde{\xi}(\lambda_+,1,\lambda_-)L} \qquad \text{as } L \to \infty. \qquad (15.48)$$

It is quite remarkable that the weighting of the crossing event with the $\pi$-extremal distance has the same effect as the conditioning on not hitting the path of another Brownian excursion.

Another major result from [30] is the theorem below, which gives the so-called cascade relations between the Brownian intersection exponents. Together with an analysis of the asymptotic behaviour of the exponents (theorems 11 and 12 in [30]), these relations show that it is sufficient to determine the exponents $\xi(1,1,\lambda)$, $\tilde{\xi}(1,\lambda)$ and $\tilde{\xi}(\lambda,1,\lambda)$ for $\lambda \geq 0$ to know all the intersection exponents.

**Theorem 12.** *The exponents* $\tilde{\xi}(\lambda_1,\dots,\lambda_k)$ *and* $\xi(\lambda_1,\dots,\lambda_k)$ *are invariant under permutations of their arguments. Moreover, they satisfy the following cascade relations:*

$$\tilde{\xi}(\lambda_1,\dots,\lambda_k) = \tilde{\xi}(\lambda_1,\dots,\lambda_{j-1},\tilde{\xi}(\lambda_j,\dots,\lambda_k)); \qquad (15.49)$$

$$\xi(\lambda_1,\dots,\lambda_k) = \xi(\lambda_1,\dots,\lambda_{j-1},\tilde{\xi}(\lambda_j,\dots,\lambda_k)). \qquad (15.50)$$

These relations not only determine all of the $\tilde{\xi}$ in terms of a few, they also give a restriction on what form these can take. The final expressions can be computed by making use of SLE. Since the argument is relatively short, we give it here.

### 15.6.2.2 An Example Calculation

Suppose that we start an SLE$_6$ process from $i\pi$ to $L$ to the same rectangle $\mathscr{R}_L$ in which the Brownian excursion $\omega$ is already defined. In what follows, it is crucial that this process, as well as the Brownian excursion, have the locality property. In our present setup, this implies that as long as the SLE$_6$ trace does not hit $\omega$, it doesn't matter whether we regard it as an SLE$_6$ in the domain $\mathscr{R}_L$ or in the domain $D_+$. Since SLE$_\kappa$ has this property only for $\kappa = 6$, the following argument works only for this special value of $\kappa$.

Let us denote by $\gamma$ the trace of the SLE$_6$ process up to the first time that it hits $[L, L + i\pi]$, and let $E$ be the event that $\gamma$ is disjoint from $\omega$ and that $\omega$ crosses the rectangle from left to right. See Fig. 15.14. On the event $E$, the $\pi$-extremal distance between $[0, i\pi]$ and $[L, L + i\pi]$ in the domain between $\gamma$ and $\omega$ is well-defined. We call this $\pi$-extremal distance $\mathscr{L}$. To obtain the value of $\tilde{\xi}(1, \lambda)$, our strategy is to express the asymptotic behaviour of $f(L) = \mathbf{E}[1_E \exp(-\lambda \mathscr{L})]$ in two different ways.

On the one hand, when $\omega$ is given, $1_E \exp(-\lambda \mathscr{L})$ is comparable to $\exp[-u(6, \lambda)\mathscr{L}_+]$ by theorem 9. We therefore get

$$f(L) \asymp \mathbf{E}_B[1_A e^{-u(6,\lambda)\mathscr{L}_+}] \asymp e^{-\tilde{\xi}(1, u(6,\lambda))L}. \tag{15.51}$$

On the other hand, when $\gamma$ is given, the distributions of $\mathscr{L}$ and $\mathscr{L}_-$ are the same by the conformal invariance of the Brownian excursion. But also, given $\mathscr{L}_+$, the probability of the event $E$ is comparable to $\exp(-\mathscr{L}_+/3)$ by theorem 8. Therefore

$$f(L) \asymp \mathbf{E}_B[1_A e^{-\mathscr{L}_+/3} e^{-\lambda \mathscr{L}_-}] \asymp e^{-\tilde{\xi}(1/3, 1, \lambda)L}. \tag{15.52}$$

By the cascade relations, $\tilde{\xi}(1/3, 1, \lambda) = \tilde{\xi}(1, \tilde{\xi}(1/3, \lambda))$. Hence, comparing the two results we obtain

$$\tilde{\xi}(1/3, \lambda) = u(6, \lambda) = \frac{6\lambda + 1 + \sqrt{1 + 24\lambda}}{6} \tag{15.53}$$

since $\tilde{\xi}(1, \lambda)$ is strictly increasing in $\lambda$. Finally, this result gives us for example $\tilde{\xi}(1, \lambda)$, because $\tilde{\xi}(1/3, 1/3) = 1$, and then the cascade relations give

$$\tilde{\xi}(1, \lambda) = \tilde{\xi}(\tilde{\xi}(1/3, 1/3), \lambda) = \tilde{\xi}(1/3, \tilde{\xi}(1/3, \lambda)). \tag{15.54}$$

### 15.6.2.3 Summary of Results

As we mentioned before, the series of papers by Lawler, Schramm and Werner [31, 32, 33, 34] led to the determination of all Brownian intersection exponents we defined above. We state their conclusions in a few equations.

For all integers $k \geq 2$ and all $\lambda_1, \ldots, \lambda_k \geq 0$,

$$\tilde{\xi}(\lambda_1,\dots,\lambda_k) = \frac{1}{24}\left[1 + \sum_{j=1}^{k}\left(\sqrt{1+24\lambda_j} - 1\right)\right]^2 - \frac{1}{24} \tag{15.55}$$

For all integers $k \geq 2$ and all $\lambda_1,\dots,\lambda_k \geq 0$, at least two of which are at least 1,

$$\xi(\lambda_1,\dots,\lambda_k) = \frac{1}{48}\left[\sum_{j=1}^{k}\left(\sqrt{1+24\lambda_j} - 1\right)\right]^2 - \frac{1}{12} \tag{15.56}$$

For $k = 2$ the requirement that at least two of the $\lambda$ be $\geq 1$, may be replaced by $\lambda_1 > 1$ and $\lambda_1 \in \mathbb{Z}$ and $\lambda_2 \geq 0$.

To the physicist these exponents, based on $\pi$-extremal distance, may seem artificial, but for integer $\lambda$ they are still the intersection exponents of packets of Brownian motions. It is worth noting also that those exponents need not be rational.

From earlier work of Lawler [26, 27, 28], it is known that some of these exponents are related to the Hausdorff dimensions of special subsets of the Brownian paths. Indeed, suppose that we denote by $B[0,1]$ the trace of a planar Brownian motion up to time 1. Then the Hausdorff dimension of its frontier (the boundary of the unbounded connected component of $\mathbb{C} \setminus B[0,1]$), is $2 - \xi(2,0) = 4/3$. The Hausdorff dimension of the set of cut points (those points $z$ such that $B[0,1] \setminus \{z\}$ is disconnected) is $2 - \xi(1,1) = 3/4$. Finally, the set of pioneer points of $B[0,1]$ (those points $z$ such that for some $t \in [0,1]$, $z = B_t$ is in the frontier of $B[0,t]$) has Hausdorff dimension $2 - \xi(1,0) = 7/4$. Not only do these exponents figure in the percolation problem, they are precisely the Hausdorff dimension of the same subsets of the percolation exploration process, i.e. the frontier, the cut points and the pioneer points. This shows how closely the random walk and critical percolation are related.

## 15.6.3 Results on Critical Percolation

The connection between SLE$_6$ and critical site percolation on the triangular lattice can be used to verify rigorously the values of certain percolation exponents. In this subsection we review how for example the multi-arm exponents for percolation can be calculated from the one-sided crossing exponent and the annulus crossing exponent of SLE$_6$. Predictions of the values of these exponents have appeared in several places in the physics literature, see e.g. [18] and references therein.

### 15.6.3.1 Half-Plane Exponents

Consider critical site percolation on the triangular lattice with fixed mesh. Let $A^+(r,R)$ be the semi-annulus $\{z : r < |z| < R, \operatorname{Im} z > 0\}$, and denote by $f_k^+(r,R)$ the probability that there exist $k$ disjoint crossings of arbitrary colours from the inner circle to the outer circle in $A^+(r,R)$. By a crossing we mean a sequence of distinct connected hexagons, all in the same colour, whose first and last hexagons are adjacent to a hexagon intersecting the inner and outer circle, respectively. Obviously, $r$ has to be large enough if the definition of $f_k^+(r,R)$ is to make sense, i.e. $r > \text{const}(k)$.

The probability $f_k^+(r,R)$ does not depend on the choice of colours of the different crossings. The reason for this is that one can always flip the colours of crossings without changing probabilities. This needs to be done with some care, because flippings conditioned on certain events may change the probability of the configuration. One can start by considering the right-most crossing. If desired, its colour can be decided by flipping the colours of all hexagons. Then one proceeds by each time considering the right-most crossing to the left of the previous one. If desired, its colour can be changed by flipping the colours of all hexagons to the left of this previous crossing. In the end one obtains a configuration with all crossings in the desired colours, without having changed probabilities. In particular, we can take $f_k^+(r,R)$ to be the probability of $k$ crossings of alternating colours.

To make the connection with SLE, suppose that we colour all hexagons that intersect the boundary of the semi-annulus blue if they are on the counter-clockwise part of the boundary from $-r$ to $R$, and yellow if they are on the clockwise boundary from $-r$ to $R$. Then the probability $f_k^+(r,R)$ is exactly the probability that the exploration process from $-r$ to $R$ makes $k$ crossings before it hits the interval $[r,R]$. By Smirnov's result, this translates in the scaling limit into the probability that a chordal $\text{SLE}_6$ process from $-r$ to $R$ in the semi-annulus makes $k$ crossings before it hits the interval $[r,R]$, see Fig. 15.15.

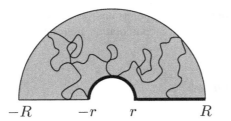 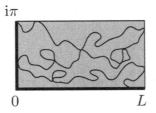

**Fig. 15.15** An $\text{SLE}_6$ process which crosses a semi-annulus three times, and the equivalent process in a rectangle. The thick part of the boundary is the part coloured blue.

It is more convenient now to map the problem to a rectangle using the logarithmic map. Suppose that $g_k^+(L)$ denotes the probability that an $\text{SLE}_6$ trace from $i\pi$ to $L$ in the rectangle $\mathcal{R}_L := (0,L) \times (0,\pi)$ makes $k$ horizontal crossings before it hits the bottom. Then, by conformal invariance, we want to determine $g_k^+(L)$ for $L = \log(R/r)$. For $k = 1$ theorem 8 immediately gives $g_1^+(L) \asymp \exp(-L/3)$. Exponents for larger $k$ can be determined using theorem 9. This is a good example of the use of that theorem.

Let $T$ be the time at which the $\text{SLE}_6$ process has crossed the rectangle for the first time, and let $E$ be the event that up to time $T$ the process has not hit the bottom. Then the process still has to make $k - 1$ crossings in the domain below this first crossing. Hence, if $\mathscr{L}$ denotes the $\pi$-extremal distance between the left and right edges in this remaining domain, we have

$$g_k^+(L) = \mathbf{E}[1_E\, g_{k-1}^+(\mathscr{L})]. \tag{15.57}$$

When we define $g_k^+(L) \asymp \exp(-v_k^+ L)$ we find $v_1^+ = 1/3$ from Theorem 8 and $v_k^+ = k(k+1)/6$ by applying recursively Theorem 9. For discrete percolation in the semi-annulus, this implies that

$$f_k^+(r,R) \asymp R^{-k(k+1)/6} \qquad \text{when } R \to \infty. \tag{15.58}$$

To make this transition to discrete percolation completely rigorous some more work is required. We refer to [52] for more details.

### 15.6.3.2 Plane Exponents

In order to obtain exponents for a point interior in a domain rather than at its boundary, we must use a radial process. Suppose that $A(r,R)$ is an approximation of the full annulus $\{z : r < |z| < R\}$ by hexagons, where $r$ is again assumed to be large enough. We can define an exploration process in this annulus as follows. We colour all hexagons intersecting the inner circle blue. The exploration process starts at $R$ with a blue hexagon on its right (above it), and a yellow hexagon on its left. Each time the exploration process hits a hexagon on the outer circle that was not visited before, we look at the phase angle of the position of the tip of the trajectory at that time (defining this phase to be continuous in time). If the phase is positive, the hexagon on the boundary is coloured blue, and otherwise it is coloured yellow.

When the exploration process described above first hits the inner circle, it defines unambiguously a clockwise-most blue crossing of the annulus and a counter-clockwise-most yellow crossing, such that the point $R$ lies between them. Moreover, it can be easily seen that afterwards, the exploration process continues like a chordal process in the remaining domain between these two crossings, where the outer circle may now be assumed to be coloured yellow. This remaining domain is equivalent to a semi-annulus. Therefore, the probability that the process crosses this remaining domain $k-2$ times before it disconnects the inner circle from the outer circle is equal to the probability that there are $k-2$ crossings of arbitrary colours of this domain, as we discussed in the previous subsection.

Let $f_k(r,R)$ be the probability that the exploration process crosses the annulus a total number of $k-1$ times. Then for even $k$, $f_k(r,R)$ is just the probability that there exist $k$ crossing paths of the annulus, which are not all of the same colour. (To avoid confusion one must carefully distinguish the crossings by the exploration process, and the existence of crossing paths on the percolation clusters.) In this case we have the freedom of choosing alternating colours for the crossing paths, and then the point $R$ is always between a right-most blue and a left-most yellow crossing, which proves the point. For odd $k$, the situation is different, and $f_k(r,R)$ is not equal to the probability that there exist $k$ crossings of the annulus which are not all of the same colour. However, it can be shown that the two probabilities differ only by a multiplicative constant, see [52].

We now make the connection with $SLE_6$. In the continuum limit, the discrete exploration process converges to the following SLE process. First, we do radial $SLE_6$ in the annulus from $R$ to 0, up to the first time $T$ that the process hits the inner circle. Afterwards, the process continues like a chordal $SLE_6$ process in the remaining domain. We further define $E$ to be the event that up to time $T$, the process has not disconnected the inner circle from the outer circle. On this event, we let $\mathcal{L}$ denote the $\pi$-extremal distance between the two circles in the remaining domain.

Denote by $g_k(r,R)$ the probability that this $SLE_6$ process crosses the annulus $k-1$ times before it disconnects the inner circle from the outer circle. Then

$$g_k(r,R) = \mathbf{E}[1_E\, g_{k-2}^+(\mathcal{L})] \asymp \mathbf{E}[1_E\, e^{-v_{k-2}^+ \mathcal{L}}] \qquad (15.59)$$

where $g_k^+(L)$ is the probability of $k$ crossings of the rectangle $(0,L) \times (0,\pi)$, as before. Theorem 10 now tells us that $g_k(r,R) \asymp (R/r)^{-v_k}$, where

$$v_k = v(6, v_{k-2}^+) = \frac{k^2 - 1}{12}. \qquad (15.60)$$

Returning to discrete percolation, it follows from this result that the probability of $k$ crossings of the annulus $A(r,R)$ which are not all of the same colour behaves like

$$f_k(r,R) \asymp R^{-(k^2-1)/12} \qquad \text{when } R \to \infty. \qquad (15.61)$$

Again, all of this has been made rigorous [52]. Observe also that we can again interpret the result in terms of crossings of clusters. In this case we have that for $k$ even, $f_k(r,R)$ is comparable to the probability that there exist $j = k/2$ disjoint blue clusters crossing the annulus.

So far we have considered only the dichromatic exponents associated with the probability of $k$ percolation crossings of an annulus that are *not* all of the same colour. The corresponding monochromatic exponents for $k$ crossings that *are* of the same colour are known to have different values. They are not so easily accessible through SLE as the dichromatic exponents. However, SLE computations [35] have confirmed that the one-arm exponent ($k = 1$) has the value $5/48$, and in the same article, a description of the backbone exponent ($k = 2$) as the leading eigenvalue of a differential operator was given.

## 15.7 Discussion

While Schramm originally introduced SLE as the only possible candidate for the scaling limit of the loop-erased random [46] walk, the definition and properties of SLE were sufficiently general to allow Schramm to conjecture that SLE also describes the scaling limits of uniform spanning trees and critical percolation. Subsequently it was proved to be the scaling limit of several other models. In fact, it is believed that conformal invariance and stationarity is sufficient for a whole range of critical models to converge to SLE (section 15.5).

For this approach to be applicable to a lattice statistical model with local interaction (on a domain with a boundary), it must at least have the following properties.

(i) It must have a conformally invariant scaling limit.
(ii) SLE being a measure of curves, the model must induce a measure of non-crossing paths in the lattice.
(iii) Since the law for $\gamma_{t'} - g_t(\gamma_t)$ is the same as that for $g_t(\gamma_{t+t'})$ the image under $g_t$ of $\gamma_{[0,t]}$ may have no other properties than the simple original boundary.
(iv) And because there is no drift in the driving term, there should be symmetry between the two sides of the curve.

The first condition is believed to be true for almost all models at an isotropic phase transition. But only in very rare cases has this been proved. Note in the second condition the difference between crossing and intersecting. This condition is easily satisfied, as one can take as paths the domain walls, between different values, of the variable. In a continuous model, level lines may figure as paths. In many cases where a diagrammatic expansion of the partition sum is possible, it may induce paths on the lattice. The third requirement, however, is rather restrictive. Together with (iv) it is known as the stationarity property. Consider as an example a model with a discrete variable on the lattice, taking four values. Without loss of generality, it can be described by two coupled Ising variables, $\sigma_1$ and $\sigma_2$. The curves we consider first are the domain walls of $\sigma_1$, which will be defined in such a way that they cannot cross. We must now choose boundary conditions compatible with this definition of the path, i.e. *in accordance with the condition on either side of these domain walls.* Clearly $\sigma_1$ must be fixed at $\pm 1$ on the positive and negative real axis respectively. Now there are several choices to make. We may choose $\sigma_2$ to be fixed as well, on the boundary and on either side of the domain wall of $\sigma_1$. If $\sigma_2$ is the same on the positive and negative real axis, then this implies that the domain walls of $\sigma_1$ and those of $\sigma_2$ are not permitted to cross each other. If we choose $\sigma_2$ to be opposite on the positive and negative real axes, and consequently on either side of the path, the model must have domain walls for both Ising components, which cannot split up in separate domain walls. A third possibility is to choose $\sigma_2$ to be free on the boundary. This implies that it is also free on the domain wall of $\sigma_1$. However, after the path is mapped onto the real axis, the same point of the path has two images, far apart. It does not seem acceptable that $\sigma_2$ has strong correlations between these two points, as a result of their common history. The only way to avoid such correlations is to forbid all interactions between $\sigma_2$ variables accross a domain wall of $\sigma_1$. Finally, instead of a domain wall for $\sigma_1$, we may consider the curves separating one of the four states from the three others. Such domain walls have an intrinsic asymmetry, because on one side three states are allowed and on the other only one. Nonetheless, by special properties of the model this asymmetry may be lifted.

In summary we can say about models with multiple degrees of freedom, that it may be possible for SLE to describe their scaling limit, but the model must have a number of very restrictive properties. It may be noted that the four-state Potts model, which is expected to be $SLE_4$, does indeed satisfy these restrictions. If it is written in terms of two Ising spins, the one does not interact accross the domain wall of the other. Furthermore the domain wall separating one state from the three others is probably symmetric as a result of the dual symmetry between the ordered and disordered phases.

Apart from being the candidate for the scaling limit of critical models, SLE also gives us an idea of how the convergence can be proved. One could try to describe the discrete path of the critical model by a Löwner evolution, and then prove that the driving function converges to Brownian motion. Indeed, this is the way in which the convergence of loop-erased random walks to $SLE_2$, and of the Peano curve winding around the uniform spanning tree to $SLE_8$ were proved. Recently, the harmonic explorer was added to the list, and it seems reasonable to believe that in the future more connections between discrete models and SLE will be established.

However, a limitation of SLE appears to be that it is only capable of describing a very specific aspect of the discrete models. In the Fortuin-Kasteleyn cluster formulation of the Potts model, for example, SLE describes the boundary of one special cluster connected to the boundary, as explained in section 15.5.6. An interesting question is then what can SLE tell us about the full configuration of clusters, rather than only about the boundary of one? Methods to describe this have been developed recently, starting with $SLE_6$ [12], and later for general $SLE_\kappa$ [12, 58, 50, 13, 49, 59].

Interesting developments have taken place regarding the connection between SLE and conformal field theory, a subject not considered in this article. Various aspects of this connection have been studied in a series of papers by Michel Bauer and Denis Bernard [2, 3, 4, 5], showing for example how results from SLE can be computed in the CFT language. Another connection was proposed by John Cardy [15] who introduced a multiple SLE process. This he could connect with Dyson's Brownian process, and through it to the distribution of eigenvalues of ensembles of random matrices. Using the conformal restriction properties studied in [38], the work of Roland Friedrich and Wendelin Werner [21, 22, 55] further clarifies the link between the discrete systems and conformal field theory. Thus SLE may prove to be very useful in putting the ideas of conformal field theory on a mathematically more rigorous footing. Clearly an obvious open question is the proof that the $q$-state Potts model for all $q$, and the $O(n)$ model for all $n$ has SLE as scaling limit. It is conceivable that the approach by Smirnov [53], successful for the Ising model, is open to generalization. Unlike the case of percolation it is not based on a property in which the Ising model is qualitatively different from the other Potts models.

SLE is a promising field of research, and the literature on SLE is already quite vast and still growing. In this discussion we only touched upon some of the developments that have taken place, without the intention of providing a complete list. In conclusion, SLE seems invaluable for adding mathematical rigour to our understanding of the scaling limits of critical two-dimensional systems and their conformal invariance. This same fact makes SLE a mathematically and technically challenging object of study.

# References

1. L. V. Ahlfors. *Conformal invariants: topics in geometric function theory*. McGraw-Hill, New York, 1973.
2. M. Bauer and D. Bernard. $SLE_\kappa$ growth processes and conformal field theories. *Phys. Lett. B* 543:135–138, 2002. arXiv: math-ph/0206028.
3. M. Bauer and D. Bernard. Conformal Field Theories of Stochastic Loewner Evolutions. *Comm. Math. Phys.* 239:493–521, 2003. arXiv: hep-th/0210015.
4. M. Bauer and D. Bernard. SLE martingales and the Virasoro algebra. *Phys. Lett. B* 557:309–316, 2003. arXiv: hep-th/0301064.
5. M. Bauer and D. Bernard. Conformal transformations and the SLE partition function martingale. *Ann. Henri Poincaré* 5:289–326. arXiv: math-ph/0305061.
6. M. Bauer, D. Bernard, J. Houdayer. Dipolar SLE's. *J. Stat. Mech.* 0503:P001, 2005. arXiv:math-ph/0411038.
7. R. J. Baxter. *Exactly solved models in statistical mechanics*. Academic Press, London, 1982.
8. R. J. Baxter. $q$ colourings of the triangular lattice. *J. Phys. A* 19:2821–2839, 1986.
9. R. J. Baxter, S. B. Kelland, and F. Y. Wu. Equivalence of the Potts model or Whitney polynomial with an ice-type model. *J. Phys. A* 9:397–406, 1976.
10. V. Beffara. The dimension of the SLE curves. *Ann. Prob.* 36:1421–1452, 2002. arXiv: math.PR/0211322.
11. V. Beffara. Hausdorff dimensions for $SLE_6$. *Ann. Prob.* 32:2606–2629, 2002. arXiv: math.PR/0204208.
12. F. Camia and C. M. Newman. Continuum nonsimple loops and 2D critical percolation. *J. Stat. Phys.* 37:157–173, 2004. arXiv: math.PR/0308122.
13. F. Camia, C. M. Newman. SLE(6) and CLE(6) from Critical Percolation. arXiv:math/0611116.
14. J. Cardy. Conformal invariance. In C. Domb and J. L. Lebowitz, editors, *Phase transitions and critical phenomena*, volume 11, pages 55–126. Academic Press, London, 1987.
15. J. Cardy. Stochastic Loewner Evolution and Dyson's Circular Ensembles. *J. Phys. A* 36:L379–L408, 2003. arXiv: math-ph/0301039.
16. J. Cardy. SLE for theoretical physicists. *Ann. Phys.* 318:81–118, 2005.
17. E. Domany, D. Mukamel, B. Nienhuis and A. Schwimmer. Duality relations and equivalences for models with O($n$) and cubic symmetry. *Nucl. Phys.* B190:279, 1981.
18. B. Duplantier. Harmonic measure exponents for two-dimensional percolation. *Phys. Rev. Lett.* 82:3940–3943, 1999.
19. B. Duplantier. Conformally invariant fractals and potential theory. *Phys. Rev. Lett.* 84:1363–1367, 2000.
20. B. Duplantier and K-H. Kwon. Conformal invariance and intersections of random walks. *Phys. Rev. Lett.* 61:2514–2517, 1988.
21. R. Friedrich and W. Werner. Conformal fields, restriction properties, degenerate representations and SLE. *C. R. Acad. Sci. Paris Ser. I* 335:947–952, 2002. arXiv: math.PR/0209382.
22. R. Friedrich and W. Werner. Conformal restriction, highest-weight representations and SLE. *Comm. Math. Phys.* 243:105–122, 2003. arXiv: math-ph/0301018.
23. C. M. Fortuin and P. W. Kasteleyn. On the random cluster model 1: Introduction and relation to other models. *Physica* 57:536–564, 1972.
24. W. Kager, B. Nienhuis, L.P. Kadanoff. Exact solutions for Loewner evolutions. *J. Stat. Phys.* 115:805–822, 2004. arXiv: math-ph/0309006.
25. W. Kager, B. Nienhuis. A guide to stochastic Löwner evolution and its application. *J. Stat. Phys.* 115:1149–1229, 2004. arXiv: math-ph/0312056.
26. G. F. Lawler. Hausdorff dimension of cut points for Brownian motion. *Electron. J. Probab.* 1:1–20, 1996.
27. G. F. Lawler. The dimension of the frontier of planar Brownian motion. *Elect. Comm. Probab.* 1:29–47, 1996.
28. G. F. Lawler. Geometric and fractal properties of Brownian motion and random walk paths in two and three dimensions. In *Random Walks (Budapest, 1998), Bolyai Society Mathematical Studies*, volume 9, pages 219–258, 1999.

29. G. F. Lawler. An introduction to the Stochastic Loewner Evolution. Available online at URL http://www.math.duke.edu/%7Ejose/esi.html, 2001.

30. G. F. Lawler and W. Werner. Intersection exponents for planar Brownian motion. *Ann. Prob.* 27(4):1601–1642, 1999.

31. G. F. Lawler, O. Schramm, and W. Werner. Values of Brownian intersection exponents I: Half-plane exponents. *Acta Math.* 187(2):237–273, 2001. arXiv: math.PR/9911084.

32. G. F. Lawler, O. Schramm, and W. Werner. Values of Brownian intersection exponents II: Plane exponents. *Acta Math.* 187(2):275–308, 2001. arXiv: math.PR/0003156.

33. G. F. Lawler, O. Schramm, and W. Werner. Values of Brownian intersection exponents III: Two-sided exponents. *Ann. Inst. H. Poincaré Statist.* 38(1):109–123, 2002. arXiv: math.PR/0005294.

34. G. F. Lawler, O. Schramm, and W. Werner. Analyticity of intersection exponents for planar Brownian motion. *Acta Math.* 189:179–201, 2002. arXiv: math.PR/0005295.

35. G. F. Lawler, O. Schramm, and W. Werner. One-arm exponent for critical 2D percolation. *Electron. J. Probab.* 7(2):13 pages, 2001. arXiv: math.PR/0108211.

36. G. F. Lawler, O. Schramm, and W. Werner. Conformal invariance of planar loop-erased random walks and uniform spanning trees. *Ann. Prob.* 32:939–995, 2001. arXiv: math.PR/0112234.

37. G. F. Lawler, O. Schramm, and W. Werner. On the scaling limit of planar self-avoiding walk. In *Fractal geometry and application, A jubilee of Benoît Mandelbrot, Amer. Math. Soc.* Proc. Symp. Pure Math. 72, Amer. Math. Soc., Providence RI, 2004. arXiv: math.PR/0204277.

38. G. F. Lawler, O. Schramm, and W. Werner. Conformal restriction: the chordal case. *J. Amer. Math. Soc.* 16(4):917–955, 2003. arXiv: math.PR/0209343.

39. K. Löwner. Untersuchungen über schlichte konforme Abbildungen des Einheitskreises. I. *Math. Ann.* 89:103–121, 1923.

40. B. Nienhuis, A.N. Berker, E.K. Riedel and M. Schick. First- and second-order phase transitions in Potts models; a renormalization-group solution. *Phys. Rev. Lett.* 43:737 1979.

41. B. Nienhuis. Exact critical point and exponents of the $O(n)$ model in two dimensions. *Phys. Rev. Lett.* 49:1062–1065, 1982.

42. B. Nienhuis. Critical behavior of two-dimensional spin models and charge asymmetry in the Coulomb Gas. *Journal of Statistical Physics* 34:731–761, 1984. Coulomb Gas formulation of two-dimensional phase transitions. In C. Domb and J. L. Lebowitz, editors, *Phase transitions and critical phenomena*, volume 11, pages 1–53. Academic Press, London, 1987.

43. B. Nienhuis. Locus of the tricritical transition in a two-dimensional $q$-state Potts model. *Physica A* 177:109–113, 1991.

44. S. Rohde and O. Schramm. Basic properties of SLE. *Ann. Math.* 161:879–920, 2005. arXiv: math.PR/0106036.

45. H. Saleur and B. Duplantier. Exact determination of the percolation hull exponent in two dimentions. *Phys. Rev. Lett.* 58:2325–2328, 1987.

46. O. Schramm. Scaling limits of loop-erased random walks and uniform spanning trees. *Israel J. Math.* 118:221–288, 2000. arXiv: math.PR/9904022.

47. O. Schramm. A percolation formula. *Elect. Comm. Probab.* 6:115–120, 2001. arXiv: math.PR/0107096.

48. O. Schramm and S. Sheffield. The harmonic explorer and its convergence to $SLE_4$. *Ann. Prob.* 33:2127–2148, 2003. arXiv: math.PR/0310210.

49. O. Schramm, S. Sheffield, D. B. Wilson. Conformal radii for conformal loop ensembles. arXiv:math/0611687.

50. S. Sheffield. Exploration trees and conformal loop ensembles. arXiv:math/0609167.

51. S. Smirnov. Critical percolation in the plane: conformal invariance, Cardy's formula, scaling limits. *C. R. Acad. Sci. Paris Sér. I Math.* 333(3):239–244, 2001. A longer version is available at URL http://www.math.kth.se/~stas/papers/.

52. S. Smirnov and W. Werner. Critical exponents for two-dimensional percolation. *Math. Res. Lett.* 8:729–744, 2001. arXiv: math.PR/0109120.

53. S. Smirnov. Conformal Invariance in random cluster models. I Holomorphic fermions in the Ising model, 2007. arXiv:0708.0039.

54. W. Werner. Random planar curves and Schramm-Löwner Evolutions. *Lecture notes from the 2002 Saint-Flour summer school* Springer, 2003. arXiv: math.PR/0303354.
55. W. Werner. Conformal restriction and related questions. 2003. arXiv: math.PR/0307353.
56. W. Werner, Girsanov's transformation for $SLE_{\kappa,\rho}$ processes, intersection exponents and hiding exponents. 2003. arXiv:math/0302115.
57. W. Werner. The conformally invariant measure on self-avoiding loops. *J. Amer. Math. Soc.* 21:137–169, 2005. arXiv:math/0511605.
58. W. Werner. Some recent aspects of random conformally invariant systems. arXiv:math/0511268.
59. W. Werner. SLEs as boundaries of clusters of Brownian loops. *C. R. Acad. Sci. Paris* to appear. arXiv:math/0308164
60. D. B. Wilson. Generating random spanning trees more quickly than the cover time. In *Proceedings of the Twenty-eighth Annual ACM Symposium on the Theory of Computing (Philadelphia, PA, 1996)*, pages 296–303, New York, 1996. ACM.
61. F.Y. Wu. The Potts model. *Rev. Mod. Phys.* 54:235, 1982.

# Chapter 16
# Appendix: Series Data and Growth Constant, Amplitude and Exponent Estimates

Anthony J Guttmann and Iwan Jensen

In this appendix we have gathered together the series expansions for self-avoiding polygons on square, honeycomb and triangular lattices enumerated by either perimeter or area and the counts for the number of polyominoes on the same lattices. In addition we provide data for the number of SAP on three-dimensional lattices and the number of three-dimensional polyominoes (or polycubes).

Below we also provide a listing for the estimated growth constants[1], critical amplitudes and critical exponents for these problems. For any lattice, the growth constant for SAP and SAW is the same. The amplitude $B$ for polygons is defined through $p_n \sim B\mu^n n^{\alpha-3}$. For polyominoes if the growth constant is $\tau$, the amplitude $B$ is defined by assuming the number of $n$-celled polyominoes grows as $B\tau^n/n$, while for polycubes the corresponding expression is $B\tau^n/n^{1.5}$. In estimating the amplitudes of polygons, we used the value of the growth constant $\mu$ in the table below and assumed $\alpha = 0.5$ for two-dimensional lattices, and $\alpha = 0.23721$ for three-dimensional lattices. The analysis of the amplitudes assumed only analytic correction-to-scaling terms. For the two-dimensional problems we believe this to be appropriate, while for the three-dimensional problems it is generally believed that there are non-analytic corrections to scaling, but the data we have is so limited that incorporating this refinement into the analysis is probably not justified.

Anthony Guttmann and Iwan Jensen

Department of Mathematics and Statistics, The University of Melbourne, Victoria, Australia, e-mail: tonyg@ms.unimelb.edu.au, e-mail: iwan@ms.unimelb.edu.au

[1] The estimate for cubic SAP exponents has been obtained by Nathan Clisby using as yet unpublished Monte Carlo data.

**Table 16.1** Growth constants and amplitudes for various problems and lattices. For any lattice, the growth constant for SAP and SAW is the same.

| Problem | Growth constant | Amplitude |
|---|---|---|
| Honeycomb SAP by perimeter | $\sqrt{2+\sqrt{2}} = 1.84775\ldots$ | 1.2719299(1) |
| Square SAP by perimeter | 2.63815853031(3) | 0.56230130(2) |
| Triangular SAP by perimeter | 4.150797226(26) | 0.2639393(1) |
| Diamond SAP by perimeter | 2.87905(12) | 0.3057(5) |
| Simple cubic SAP by perimeter | 4.684043(12) | 0.2625(5) |
| Body-centred cubic SAP by perimeter | 6.5304(4) | 0.2403(5) |
| Face-centred cubic SAP by perimeter | 10.0363(6) | 0.1173(5) |
| Honeycomb SAP by area | 5.161930154(8) | 0.2808499(1) |
| Square SAP by area | 3.97094397(9) | 0.408105(2) |
| Triangular SAP by area | 2.9446600(8) | 1.33652(6) |
| Polycubes | 8.344(10) | 0.184(3) |
| Honeycomb polyominoes | 5.1831453(4) | 0.273525(5) |
| Square polyominoes | 4.0625696(5) | 0.316915(5) |
| Triangular polyominoes | 3.0359688(3) | 0.81243(3) |

**Table 16.2** Critical exponents. Note the hyperscaling relation $d\nu = 2 - \alpha$, which has been used in estimating $\alpha$ from estimates of $\mu$. For polygons, the critical exponent is $\alpha$, while for SAW it is $\gamma$. The size exponent, for both SAW and SAP, is $\nu$.

| Lattice dimension and model | Exponent $\alpha$ | Exponent $\gamma$ | Exponent $\nu$ |
|---|---|---|---|
| 2-dimensional SAP, SAW | 1/2 | 43/32 | 3/4 |
| 2-dimensional polyominoes | 0 | n/a | 3/4 |
| 3-dimensional SAP, SAW | $\alpha = 0.237209(21)$ | $\gamma = 1.156957(9)$ | $\nu = 0.587597(7)$ |

**Table 16.3** Honeycomb lattice SAP by perimeter [8].

| $n$ | $p_n$ | $n$ | $p_n$ |
|---|---|---|---|
| 6 | 1 | 84 | 4911780837106379222 |
| 8 | 0 | 86 | 15812530326396066033 |
| 10 | 3 | 88 | 50973962313111365766 |
| 12 | 2 | 90 | 164534436391372800477 |
| 14 | 12 | 92 | 531740594220327078666 |
| 16 | 18 | 94 | 1720505108356961603588 |
| 18 | 65 | 96 | 5573173800726924560322 |
| 20 | 138 | 98 | 18072583319913099824466 |
| 22 | 432 | 100 | 58666387862953110375900 |
| 24 | 1074 | 102 | 190630526823665553726955 |
| 26 | 3231 | 104 | 620029587600178306565222 |
| 28 | 8718 | 106 | 2018522402354141268474600 |
| 30 | 25999 | 108 | 6577196223792547336083488 |
| 32 | 73650 | 110 | 21449641090067624222337488 |
| 34 | 220215 | 112 | 70009632627911763793237744 |
| 36 | 643546 | 114 | 228686997090423718112494633 |
| 38 | 1937877 | 116 | 747582263059542414616356666 |
| 40 | 5783700 | 118 | 2445676261807412384681165522 |
| 42 | 17564727 | 120 | 8006644737200277548189079522 |
| 44 | 53222094 | 122 | 26230313295781319616970977000 |
| 46 | 163009086 | 124 | 85989994675198615346289924300 |
| 48 | 499634508 | 126 | 282081480223653050509926004200 |
| 50 | 1542392088 | 128 | 925922302454406552630977316300 |
| 52 | 4770925446 | 130 | 3041160686167983143811114836733 |
| 54 | 14832934031 | 132 | 9994497333435118671953405572000 |
| 56 | 46227584010 | 134 | 32864845464240407376587266375022 |
| 58 | 144632622552 | 136 | 108129427805992845057691655808000 |
| 60 | 453628244950 | 138 | 355951506507521095422824630083188 |
| 62 | 1427228330481 | 140 | 1172372538211441476233897902329544 |
| 64 | 4500947210772 | 142 | 3863330372190418094024222116617575 |
| 66 | 14231512500103 | 144 | 12737179782347216193128055892401145 |
| 68 | 45095972401236 | 146 | 42013996406258692673354273584799586 |
| 70 | 143219294049399 | 148 | 138649497908032099199726964733325107 |
| 72 | 455745199043542 | 150 | 457763053454199193712858503296767566 |
| 74 | 1453111646955645 | 152 | 1512014848485103906172142106303815422 |
| 76 | 4641449091849300 | 154 | 4996423205806358965647443486174726583 |
| 78 | 14851454597198009 | 156 | 16517540247783422472385676551331621988 |
| 80 | 47598148798881660 | 158 | 54627316451219720780399777403081456722 |
| 82 | 152789607567089925 | | |

**Table 16.4** Square lattice SAP by perimeter [6].

| $n$ | $p_n$ | $n$ | $p_n$ |
|---|---|---|---|
| 4 | 1 | 58 | 59270905595010696944 |
| 6 | 2 | 60 | 379108737793289505364 |
| 8 | 7 | 62 | 2431560774079622817356 |
| 10 | 28 | 64 | 15636142410456687798584 |
| 12 | 124 | 66 | 100792521026456246096640 |
| 14 | 588 | 68 | 651206027727607425003232 |
| 16 | 2938 | 70 | 4216407618470423070733556 |
| 18 | 15268 | 72 | 27355731801639756123505014 |
| 20 | 81826 | 74 | 177822806050324126648352460 |
| 22 | 449572 | 76 | 1158018792676190545425711414 |
| 24 | 2521270 | 78 | 7554259214694896127239818088 |
| 26 | 14385376 | 80 | 49360379260931646965916677280 |
| 28 | 83290424 | 82 | 323028185951187646733521902740 |
| 30 | 488384528 | 84 | 2117118644744425875029583096670 |
| 32 | 2895432660 | 86 | 13895130612692826326409919713700 |
| 34 | 17332874364 | 88 | 91319729650588816198004801698400 |
| 36 | 104653427012 | 90 | 600931442757555468862970353941700 |
| 38 | 636737003384 | 92 | 3959306049439766117380237943449096 |
| 40 | 3900770002646 | 94 | 26117050944268596220897591868398452 |
| 42 | 24045500114388 | 96 | 172472018113289556124895798382016316 |
| 44 | 149059814328236 | 98 | 1140203722938033441542255979068861816 |
| 46 | 928782423033008 | 100 | 7545649677448506970646886033356862162 |
| 48 | 5814401613289290 | 102 | 49985425311771305735407129290605556804 |
| 50 | 36556766640745936 | 104 | 331440783010043009106782321492277936522 |
| 52 | 230757492737449636 | 106 | 2199725502650970871182263620080571090156 |
| 54 | 1461972662850874916 | 108 | 14612216410979678692651320184958285074180 |
| 56 | 9293993428791901042 | 110 | 97148177367657853074723038687712338567772 |

**Table 16.5** Triangular lattice SAP by perimeter [7].

| $n$ | $p_n$ | $n$ | $p_n$ |
|---|---|---|---|
| 3 | 2 | 32 | 2692047018699717 |
| 4 | 3 | 33 | 10352576717684506 |
| 5 | 6 | 34 | 39902392511347329 |
| 6 | 15 | 35 | 154126451419554156 |
| 7 | 42 | 36 | 596528356905096920 |
| 8 | 123 | 37 | 2313198287784319026 |
| 9 | 380 | 38 | 8986249863419780682 |
| 10 | 1212 | 39 | 34969337454759091232 |
| 11 | 3966 | 40 | 136301962040079085257 |
| 12 | 13265 | 41 | 532093404471021533628 |
| 13 | 45144 | 42 | 2080235431107538787148 |
| 14 | 155955 | 43 | 8144154378525048003270 |
| 15 | 545690 | 44 | 31927176350778729318192 |
| 16 | 1930635 | 45 | 125322778845662829008494 |
| 17 | 6897210 | 46 | 492527188641409773340797 |
| 18 | 24852576 | 47 | 1937931188484341585677962 |
| 19 | 90237582 | 48 | 7633665703654150673637363 |
| 20 | 329896569 | 49 | 30101946001283232799847562 |
| 21 | 1213528736 | 50 | 118823919397444557546535851 |
| 22 | 4489041219 | 51 | 469508402822449711313115200 |
| 23 | 16690581534 | 52 | 1856933773092076293566747007 |
| 24 | 62346895571 | 53 | 7351015093472721439659392448 |
| 25 | 233893503330 | 54 | 29126027071450640626653986531 |
| 26 | 880918093866 | 55 | 115500592701344029351721102550 |
| 27 | 3329949535934 | 56 | 458398255374927436357237021173 |
| 28 | 12630175810968 | 57 | 1820727406941365079260306390484 |
| 29 | 48056019569718 | 58 | 7237327695683743010999188700157 |
| 30 | 183383553173255 | 59 | 28789332223533619621001538109842 |
| 31 | 701719913717994 | 60 | 114602547490254934327469368968190 |

**Table 16.6** Honeycomb lattice SAP by area [13, 9].

| $n$ | $p_n$ | $n$ | $p_n$ |
|---|---|---|---|
| 1 | 1 | 26 | 36138633393334038 |
| 2 | 3 | 27 | 179768675964165939 |
| 3 | 11 | 28 | 895425672624735867 |
| 4 | 44 | 29 | 4465589678921947602 |
| 5 | 186 | 30 | 22295966620155816954 |
| 6 | 813 | 31 | 111439693993112940196 |
| 7 | 3640 | 32 | 557558620919353655115 |
| 8 | 16590 | 33 | 2792233438943251452902 |
| 9 | 76663 | 34 | 13995852369729891369431 |
| 10 | 358195 | 35 | 70212003186716473817832 |
| 11 | 1688784 | 36 | 352506828543839738006802 |
| 12 | 8022273 | 37 | 1771125269041561567830953 |
| 13 | 38351973 | 38 | 8905113919188230264955009 |
| 14 | 184353219 | 39 | 44804571829235959198699855 |
| 15 | 890371070 | 40 | 225570974088699920561748746 |
| 16 | 4318095442 | 41 | 1136340745302289809680018862 |
| 17 | 21018564402 | 42 | 5727773558054438208070950886 |
| 18 | 102642526470 | 43 | 28887056504374868913302241736 |
| 19 | 502709028125 | 44 | 145763914212751560334802981991 |
| 20 | 2468566918644 | 45 | 735894997233174457602406978869 |
| 21 | 12150769362815 | 46 | 3716988842355112053567240722854 |
| 22 | 59937663454017 | 47 | 18783102592560998779533576292617 |
| 23 | 296245438278258 | 48 | 94958908613774943408509332060260 |
| 24 | 1466858366128911 | 49 | 480273434248924455452231252618009 |
| 25 | 7275229222292218 | 50 | 2430068453031180290203185942420933 |

**Table 16.7**  Square lattice SAP by area [11].

| n | $p_n$ | n | $p_n$ |
|---|---|---|---|
| 1 | 1 | 22 | 261803388854 |
| 2 | 2 | 23 | 996971935098 |
| 3 | 6 | 24 | 3802944302442 |
| 4 | 19 | 25 | 14528816598358 |
| 5 | 63 | 26 | 55585800967658 |
| 6 | 216 | 27 | 212949334034600 |
| 7 | 756 | 28 | 816822217132804 |
| 8 | 2684 | 29 | 3136762752545213 |
| 9 | 9638 | 30 | 12058858335360206 |
| 10 | 34930 | 31 | 46405735929935474 |
| 11 | 127560 | 32 | 178752169549746269 |
| 12 | 468837 | 33 | 689161111033801080 |
| 13 | 1732702 | 34 | 2659240868309971570 |
| 14 | 6434322 | 35 | 10269318260428629674 |
| 15 | 23993874 | 36 | 39687503569859369443 |
| 16 | 89805691 | 37 | 153488864908550236363 |
| 17 | 337237337 | 38 | 594011587420226879158 |
| 18 | 1270123530 | 39 | 2300345838908310537296 |
| 19 | 4796310672 | 40 | 8913696266990663512620 |
| 20 | 18155586993 | 41 | 34560203892113934050327 |
| 21 | 68874803609 | 42 | 134071571821918373415776 |

**Table 16.8** Triangular lattice SAP by area [10].

| $n$ | $p_n$ | $n$ | $p_n$ |
|---|---|---|---|
| 1 | 2 | 31 | 13768900283696 |
| 2 | 3 | 32 | 39381761647878 |
| 3 | 6 | 33 | 112731209513148 |
| 4 | 14 | 34 | 322944141486223 |
| 5 | 36 | 35 | 925821793030182 |
| 6 | 94 | 36 | 2655999419889775 |
| 7 | 250 | 37 | 7624549165478464 |
| 8 | 675 | 38 | 21901410528537396 |
| 9 | 1832 | 39 | 62948996221716186 |
| 10 | 5005 | 40 | 181030734048561330 |
| 11 | 13746 | 41 | 520896427277498160 |
| 12 | 37901 | 42 | 1499600346000360661 |
| 13 | 104902 | 43 | 4319315951924817740 |
| 14 | 291312 | 44 | 12446880627889433646 |
| 15 | 811346 | 45 | 35884225522293806438 |
| 16 | 2265905 | 46 | 103498974852276615147 |
| 17 | 6343854 | 47 | 298641621862752294144 |
| 18 | 17801383 | 48 | 862063552257379673111 |
| 19 | 50057400 | 49 | 2489408387765856393710 |
| 20 | 141034248 | 50 | 7191418561913160942210 |
| 21 | 398070362 | 51 | 20782056815997725229126 |
| 22 | 1125426581 | 52 | 60077568702764010825658 |
| 23 | 3186725646 | 53 | 173732332629110214974028 |
| 24 | 9036406687 | 54 | 502560484888013883133120 |
| 25 | 25658313188 | 55 | 1454221557880565649765344 |
| 26 | 72946289247 | 56 | 4209231246688674394442949 |
| 27 | 207628101578 | 57 | 12187106400969184313465204 |
| 28 | 591622990214 | 58 | 35295544624608480713053597 |
| 29 | 1687527542874 | 59 | 102248441850332810905160592 |
| 30 | 4818113792640 | 60 | 296283374352751571959397999 |

**Table 16.9** Honeycomb lattice polyominoes [14, 10].

| $n$ | $p_n$ | $n$ | $p_n$ |
|---|---|---|---|
| 1 | 1 | 24 | 1570540515980274 |
| 2 | 3 | 25 | 7821755377244303 |
| 3 | 11 | 26 | 39014584984477092 |
| 4 | 44 | 27 | 194880246951838595 |
| 5 | 186 | 28 | 974725768600891269 |
| 6 | 814 | 29 | 4881251640514912341 |
| 7 | 3652 | 30 | 24472502362094874818 |
| 8 | 16689 | 31 | 122826412768568196148 |
| 9 | 77359 | 32 | 617080993446201431307 |
| 10 | 362671 | 33 | 3103152024451536273288 |
| 11 | 1716033 | 34 | 15618892303340118758816 |
| 12 | 8182213 | 35 | 78679501136505611375745 |
| 13 | 39267086 | 36 | 396658618080234793950206 |
| 14 | 189492795 | 37 | 2001232317628022658203349 |
| 15 | 918837374 | 38 | 10103836183314489605735070 |
| 16 | 4474080844 | 39 | 51046672861235124190631667 |
| 17 | 21866153748 | 40 | 258063337786459258279344114 |
| 18 | 107217298977 | 41 | 1305417245856690662912152269 |
| 19 | 527266673134 | 42 | 6607298985024639624903163419 |
| 20 | 2599804551168 | 43 | 33460963467529713458350419245 |
| 21 | 12849503756579 | 44 | 169543788582768431534598929547 |
| 22 | 63646233127758 | 45 | 859496482176765849253640160036 |
| 23 | 315876691291677 | 46 | 4359288232974777294574313228655 |

**Table 16.10**  Square lattice polyominoes [5].

| $n$ | $p_n$ | $n$ | $p_n$ |
|---|---|---|---|
| 1 | 1 | 29 | 4820975409710116 |
| 2 | 2 | 30 | 18946775782611174 |
| 3 | 6 | 31 | 74541651404935148 |
| 4 | 19 | 32 | 293560133910477776 |
| 5 | 63 | 33 | 1157186142148293638 |
| 6 | 216 | 34 | 4565553929115769162 |
| 7 | 760 | 35 | 18027932215016128134 |
| 8 | 2725 | 36 | 71242712815411950635 |
| 9 | 9910 | 37 | 281746550485032531911 |
| 10 | 36446 | 38 | 1115021869572604692100 |
| 11 | 135268 | 39 | 4415695134978868448596 |
| 12 | 505861 | 40 | 17498111172838312982542 |
| 13 | 1903890 | 41 | 69381900728932743048483 |
| 14 | 7204874 | 42 | 275265412856343074274146 |
| 15 | 27394666 | 43 | 1092687308874612006972082 |
| 16 | 104592937 | 44 | 4339784013643393384603906 |
| 17 | 400795844 | 45 | 17244800728846724289191074 |
| 18 | 1540820542 | 46 | 68557762666345165410168738 |
| 19 | 5940738676 | 47 | 272680844424943840614538634 |
| 20 | 22964779660 | 48 | 1085035285182087705685323738 |
| 21 | 88983512783 | 49 | 4319331509344565487555270660 |
| 22 | 345532572678 | 50 | 17201460881287871798942420736 |
| 23 | 1344372335524 | 51 | 68530413174845561618160604928 |
| 24 | 5239988770268 | 52 | 273126660016519143293320026256 |
| 25 | 20457802016011 | 53 | 1088933685559350300820095990030 |
| 26 | 79992676367108 | 54 | 4342997469623933155942753899000 |
| 27 | 313224032098244 | 55 | 17326987021737904384935434351490 |
| 28 | 1228088671826973 | 56 | 69150714562532896936574425480218 |

**Table 16.11** Triangular lattice polyominoes [10].

| $n$ | $p_n$ | $n$ | $p_n$ |
|---|---|---|---|
| 1 | 2 | 39 | 131764274746623618 |
| 2 | 3 | 40 | 390209282091660817 |
| 3 | 6 | 41 | 1156271319511222890 |
| 4 | 14 | 42 | 3428243851059071792 |
| 5 | 36 | 43 | 10170021606617062092 |
| 6 | 94 | 44 | 30185576357912854854 |
| 7 | 250 | 45 | 89638467588131276054 |
| 8 | 675 | 46 | 266316031025897652002 |
| 9 | 1838 | 47 | 791588201780520478260 |
| 10 | 5053 | 48 | 2353922513181100648048 |
| 11 | 14016 | 49 | 7002741498223502133792 |
| 12 | 39169 | 50 | 20841060277596144244446 |
| 13 | 110194 | 51 | 62049806988299870226456 |
| 14 | 311751 | 52 | 184809160446574540356778 |
| 15 | 886160 | 53 | 550633812416956110450696 |
| 16 | 2529260 | 54 | 1641167126237394780804458 |
| 17 | 7244862 | 55 | 4893142168882883602047972 |
| 18 | 20818498 | 56 | 14593611643638701475828219 |
| 19 | 59994514 | 57 | 43538430128312213641221102 |
| 20 | 173338962 | 58 | 129931105423136465757345880 |
| 21 | 501994070 | 59 | 387864007832776437943416162 |
| 22 | 1456891547 | 60 | 1158157489920023082651029625 |
| 23 | 4236446214 | 61 | 3459183249840776065090197424 |
| 24 | 12341035217 | 62 | 10334596819468361754858559890 |
| 25 | 36009329450 | 63 | 30883315424482772009364074195 |
| 26 | 105229462401 | 64 | 92312659826727115613777214819 |
| 27 | 307942754342 | 65 | 275995688697147821120388899585 |
| 28 | 902338712971 | 66 | 825360885842983560010493834969 |
| 29 | 2647263986022 | 67 | 2468783621745427137367974848117 |
| 30 | 7775314024683 | 68 | 7386128647683127584480035530328 |
| 31 | 22861250676074 | 69 | 22102564476279407636273464326490 |
| 32 | 67284446545605 | 70 | 66154257908909010874896059091279 |
| 33 | 198214729430994 | 71 | 198043122493458453529791815751245 |
| 34 | 584439943107748 | 72 | 592988333797578245954808147666097 |
| 35 | 1724665203979836 | 73 | 1775884216384559876692792048399568 |
| 36 | 5093434042872294 | 74 | 5319404853729116558903334777867316 |
| 37 | 15053558945238166 | 75 | 15936363137225733301433441827683823 |
| 38 | 44521869233046747 | | |

**Table 16.12** Diamond lattice SAP by perimeter [4].

| $n$ | $p_n$ | $n$ | $p_n$ |
|---|---|---|---|
| 6 | 2 | 22 | 759846 |
| 8 | 3 | 24 | 4930656 |
| 10 | 24 | 26 | 32852424 |
| 12 | 94 | 28 | 221672022 |
| 14 | 582 | 30 | 1519813822 |
| 16 | 3126 | 32 | 10538532360 |
| 18 | 19402 | 34 | 73902970188 |
| 20 | 118110 | | |

**Table 16.13** Simple cubic lattice SAP by perimeter [3].

| $n$ | $p_n$ | $n$ | $p_n$ |
|---|---|---|---|
| 4 | 3 | 20 | 1768560270 |
| 6 | 22 | 22 | 29764630632 |
| 8 | 207 | 24 | 512705615350 |
| 10 | 2412 | 26 | 9005206632672 |
| 12 | 31754 | 28 | 160810554015408 |
| 14 | 452640 | 30 | 2912940755956084 |
| 16 | 6840774 | 32 | 534245552150523386 |
| 18 | 108088232 | | |

**Table 16.14** Body-centred cubic lattice SAP by perimeter [2].

| $n$ | $p_n$ | $n$ | $p_n$ |
|---|---|---|---|
| 4 | 12 | 14 | 43702920 |
| 6 | 148 | 16 | 1282524918 |
| 8 | 2736 | 18 | 39354507576 |
| 10 | 61896 | 20 | 1250685059616 |
| 12 | 1759324 | 22 | 40887160690224 |

**Table 16.15** Face-centred cubic lattice SAP by perimeter [12].

| $n$ | $p_n$ | $n$ | $p_n$ |
|---|---|---|---|
| 3 | 8 | 9 | 301376 |
| 4 | 33 | 10 | 2241420 |
| 5 | 168 | 11 | 17173224 |
| 6 | 970 | 12 | 134806948 |
| 7 | 6168 | 13 | 1079802216 |
| 8 | 42069 | 14 | 8798329080 |

**Table 16.16** Number of polycubes [1].

| $n$ | $p_n$ | $n$ | $p_n$ |
|---|---|---|---|
| 1 | 1 | 10 | 8294738 |
| 2 | 3 | 11 | 60494549 |
| 3 | 15 | 12 | 446205905 |
| 4 | 86 | 13 | 3322769321 |
| 5 | 534 | 14 | 24946773111 |
| 6 | 3481 | 15 | 188625900446 |
| 7 | 23502 | 16 | 1435074454755 |
| 8 | 162913 | 17 | 10977812452428 |
| 9 | 1152870 | 18 | 84384157287999 |

# References

1. Aleksandrowicz G and Barequet G 2006 Counting $d$-Dimensional Polycubes and Nonrectangular Planar Polyominoes *Lect. Notes Comp. Science* **4112** 418–427
2. P. Butera and M. Comi, 1999 Enumeration of the self-avoiding polygons on a lattice by the Schwinger-Dyson equations *Ann. Comb.* **3** 277
3. Clisby N, Liang R and Slade G 2007 Self-avoiding walk enumeration via the lace expansion *J. Phys. A: Math, Theor.* **40** 10973–11017
4. Clisby N 2008 Private communication. A calculation for this appendix performed in 29.5 hours on the Victorian Partnership for Advanced Computing (VPAC) machine 'tango', which consists of AMD Opteron processors.
5. Jensen I 2003 Counting Polyominoes: A Parallel Implementation for Cluster Computing *Lect. Notes Comp. Science* **2659** 203–212
6. Jensen I 2003 A parallel algorithm for the enumeration of self-avoiding polygons on the square lattice *J. Phys. A: Math. Gen.* **36** 5731–5745
7. Jensen I 2004 Self-avoiding walks and polygons on the triangular lattice *J. Stat. Mech: Theor. Exp.* P10008
8. Jensen I 2006 Honeycomb lattice polygons and walks as a test of series analysis techniques *J. Phys.: Conf. Ser.* **42** 163–178
9. Jensen I 2008 A parallel algorithm for the enumeration of benzenoid hydrocarbons Preprint arXiv:0808.0963
10. Jensen I Unpublished
11. Jensen I and Guttmann A J 2000 Self-avoiding polygons on the square lattice *J. Phys. A: Math. Gen.* **33** L25–L263
12. Sykes M F, McKenzie D S, Watts M G and Martin J L 1972 *J. Phys. A: Gen. Phys.* **5** 661–666
13. Vöge M, Guttmann A J and Jensen I 2002 On the Number of Benzenoid Hydrocarbons *J. Chem. Inf. Comput. Science* **42** 456–466
14. Vöge M and Guttmann A J 2003 On the Number of Hexagonal Polyominoes *Theor. Comp. Science* **307** 433–453

# Index

Printed in the United States
By Bookmasters